MW01008411

PIPE FLOW

PIPE FLOW

A Practical and Comprehensive Guide

DONALD C. RENNELS
General Electric Company (ret.)

HOBART M. HUDSON
Aerojet General Corporation (ret.)

A JOHN WILEY & SONS, INC., PUBLICATION

Published by John Wiley & Sons, Inc., Hoboken, New Jersey
Published simultaneously in Canada

Library of Congress Cataloging-in-Publication Data:

Rennels, Donald C., 1937–
Pipe flow : a practical and comprehensive guide / Donald C Rennels, Hobart M Hudson.
p. cm.
Includes bibliographical references and index.
ISBN 978-0-470-90102-1 (cloth)
1. Pipe–Fluid dynamics. 2. Water-pipes–Hydrodynamics. 3. Fluid mechanics. I. Hudson, Hobart M., 1931– II. Title.
TJ935.R46 2012
620.1'064–dc23

2011043325

Printed in the United States of America

9780470901021

10 9 8 7 6 5 4 3 2 1

CONTENTS

Knowledge shared is everything.

Knowledge kept is nothing.

—Richard Beere,
Abbot of Glastonbury
(1493–1524)

PREFACE

This book provides practical and comprehensive information on the subject of pressure drop and other phenomena in fluid flow in pipes. The importance of piping systems in distribution systems, in industrial operations, and in modern power plants justifies a book devoted exclusively to this subject. The emphasis is on flow in piping components and piping systems where greatest benefit will derive from accurate prediction of pressure loss.

A great deal of experimental and theoretical research on fluid flow in pipes and their components has been reported over the years. However, the basic methodology in fluid flow textbooks is usually fragmented, scattered throughout several chapters and paragraphs; and useful, practical information is difficult to sort out. Moreover, textbooks present very little loss coefficient data, and those that are given are desperately out of date. Elsewhere, experimental data and published formulas for loss coefficients have provided results that are in considerable disagreement. Into the bargain, researchers have not accounted for all possible flow configurations and their results are not always presented in a readily useful form. This book addresses and fixes these deficiencies.

Instead of having to search and read through various sources, this book provides the user with virtually all the information required to design and analyze piping systems. Example problems, their setups, and solutions are provided throughout the book. Most parts of the book will be easily understood by those who are not experts in the field.

Part I (Chapters 1 through 7) contains the essential methodology required to solve accurately pipe flow problems. Chapter 1 provides knowledge of the physical properties of fluids and the nature of fluid flow. Chapter 2 presents the basic principles of conservation of mass, momentum, and energy, and introduces the concepts of head loss and energy grade line. Chapter 3 presents the conventional head loss equation and characterizes the two sources of head loss—surface friction and induced turbulence. Several compressible flow calculation methods are presented in Chapter 4. The straightforward setup of series, parallel, and branching flow networks, including sample problems, is presented in Chapter 5. Chapter 6 introduces the basic methodology for solving transient flow problems, with specific examples. A method to assess the uncertainty associated with pipe flow calculations is presented in Chapter 7.

Part II (Chapters 8 through 19) presents consistent and reliable loss coefficient data on flow configurations most common to piping systems. Experimental test data and published formulas from worldwide sources are examined, integrated, and arranged into widely applicable equations—a valuable resource in this computer age. The results are also presented in straightforward tables and diagrams. The processes used to select and develop loss coefficient data for the various flow configurations are presented so the user can judge the merits of the results and the researcher can identify areas where further research is needed.

Friction factor, the main element of surface friction loss, is presented in Chapter 8 as an adjunct to quantifying the various features that contribute to head loss.

The flow configurations presented in Chapters 9 through 14 (entrances, contractions, expansions, exits, orifices, and flow meters) all exhibit some degree of flow contraction and/or expansion. As such, they have been treated as a family; where sufficient data for any one particular configuration were lacking, they were augmented by sufficient data in another.

Elbows, pipe bends, coils, and miter bends are presented in Chapter 15. The intricacies of converging and diverging flow through pipe junctions (tees) are presented in Chapter 16. Pipe joints are covered in Chapter 17, and valve information is offered in Chapter 18. The internal geometry of threaded (screwed) pipe fittings is discontinuous, creating additional pressure loss; and they are covered separately in Chapter 19.

Part III (Chapters 20 through 23) examines flow phenomena that can affect the performance of piping systems. Cavitation, when local pressure falls below the vapor pressure of a liquid, is studied in Chapter 20. Chapter 21 provides a brief depiction of flow-induced vibration in piping systems; water hammer and column separation are investigated. Situations where temperature rise in a flowing liquid may be of interest are presented in Chapter 22. Flow behavior in horizontal openings at low flow rates is evaluated in Chapter 23.

The book's nomenclature was selected so that it would be familiar to engineers worldwide. The book employs two systems of units: the English gravitational system (often called the U.S. Customary System or USCS) and the International System (or SI for Système International). Conversions between and within the two systems are provided in the appendix.

This book represents industrial experience gained working together at Aerojet General Corporation, Liquid Rocket Engine Test Division, and later, working separately at General Electric Company, Nuclear Energy Division, and at Westinghouse Electric Corporation, Oceanic Division. We are indebted to the many engineering colleagues who helped shape our experience in the field of fluid flow. We especially appreciate Dr. Phillip G. Ellison's helpful comments and suggestions.

We acknowledge the understanding and support of our wives, Bel and Joan.

Donald C. Rennels
Hobart M. Hudson

NOMENCLATURE

Symbol	Definition	Units	
		English	International System (SI)
	Roman Symbols		
A	Area	ft^2	m^2
a	Acceleration	ft/s^2	m/s^2
a	Acoustic velocity	ft/s	m/s
B	Bulk modulus	lb/in^2	N/m^2
C	Coefficient	Dimensionless	
c_p	Specific heat at constant pressure	Btu/lb-°F	J/kg-°C (N-m/kg-°C)
c_v	Specific heat at constant volume	Btu/lb-°F	J/kg-°C (N-m/kg-°C)
D	Diameter	ft	m
d	Diameter	in	mm
E	Modulus of elasticity	lb/in^2	N/m^2
E	Mechanical energy (per unit time, i.e., power)	ft-lb/s	N-m/s
e	Absolute roughness	in	mm
F	Factor	Dimensionless	
F	Force	lb	N ($kg\text{-}m/s^2$)
f	Friction factor (Darcy)	Dimensionless	
G	Mass flow rate per unit area	$lb/s\text{-}ft^2$	$kg/s\text{-}m^2$
g	Acceleration of gravity	ft/s^2	m/s^2
H	Head	ft	m
h	Enthalpy	Btu/lb	J/kg (N-m/kg)
J	Mechanical equivalent of heat	ft-lb/Btu	N-m/J (=1)
K	Loss coefficient (i.e., total pressure loss coefficient)	Dimensionless	
L	Length	ft	m
l	Length	in	mm
ln	Natural logarithm	Dimensionless	
log	Base-10 logarithm	Dimensionless	
M	Mach number	Dimensionless	
m	Mass[a]	slug ($lb_f\text{-}s^2/ft$)	kg

(*Continued*)

Symbol	Definition	Units	
		English	International System (SI)
m	Molecular weight	lb/mol_{lb}	kg/mol_{kg}
$\dot{m}$	Mass flow rate[a]	slug/s (lb_f-s/ft)	kg/s
m	Moisture content	Dimensionless	
N_{Fr}	Froude number	Dimensionless	
N_{Re}	Reynolds number	Dimensionless	
n	Number of mols	Dimensionless	
n	Ellipse major/minor axis ratio	Dimensionless	
P	Pressure	lb/ft^2	N/m^2 (pascal)
p	Pressure	lb/in^2	N/cm^2
p	Pitch	in	mm
Q	Volumetric flow rate	ft^3/s	m^3/s
Q	Heat flux	Btu/s	J/s (N-m/s)
q	Volumetric flow rate	gal/min	—
R	Individual gas constant	ft-lb/lb-°R[a]	N-m/kg-K
$\bar{R}$	Universal gas constant	ft-lb/mol-°R[a]	N-m/mol-K
R_P	Pressure ratio	Dimensionless	
R	Radius	ft	m
r	Radius	in	mm
T	Absolute temperature	°R	K
t	Common temperature	°F	°C
t	Time	s	s
t	Thickness	in	mm
U	Internal energy	Btu/lb	N-m/kg
u	Local velocity	ft/s	m/s
V	Average velocity	ft/s	m/s
V	Volume	ft^3	m^3
v	Specific volume	ft^3/lb	m^3/kg
W	Weight flow rate	lb/h	N/h
w	Weight	lb	N
$\dot{w}$	Weight flow rate	lb/s	N/s
x	Horizontal distance	ft	m
Y	Expansion factor	Dimensionless	
y	Radial location of local velocity	in	mm
y	Vertical distance	ft	m
Z	Elevation	ft	m
z	Compressibility factor	Dimensionless	
	Greek Symbols		
α	Bend angle or diffuser included angle	degrees	degrees
β	Diameter ratio	Dimensionless	
γ	Ratio of specific heats c_p/c_v	Dimensionless	
Δ	Finite difference (prefix)	Dimensionless	
ε	Absolute roughness	ft	m
θ	Momentum correction factor	Dimensionless	
λ	Jet contraction ratio	Dimensionless	
μ	Absolute (dynamic) viscosity	$lb\text{-}sec/ft^2$	$N\text{-}sec/m^2$ (Pascal-sec)
v	Kinematic viscosity	ft^2/sec	m^2/sec
π	pi (3.14159 . . .)	Dimensionless	
ρ_m	Mass density	$slug/ft^3$ ($lb_f\text{-}sec^2/ft^4$)	kg/m^3

Symbol	Definition	Units: English	Units: International System (SI)
ρ_w	Weight density	lb_f/ft^3	N/m^3
σ	Uncertainty	%	%
ϕ	Kinetic energy correction factor	Dimensionless	
ψ	Angle	degrees	degrees
ω	Acentric factor	Dimensionless	
	Subscripts		
o	Orifice or nozzle throat	Not defined	
1	Inlet or upstream	Not defined	
2	Outlet or downstream	Not defined	
a	Atmosphere	Not defined	
b	Velocity profile function exponent	Not defined	
b	Bend	Not defined	
c	Critical state	Not defined	
r	Reduced value	Not defined	
t	Total	Not defined	
x	Component in x-direction	Not defined	
y	Component in y-direction	Not defined	
z	Component in z-direction	Not defined	
	Superscripts		
$'$	Absolute value or derivative	Not defined (e.g., f')	
$\bar{}$	Average of initial and final values	Not defined (e.g., $\bar{x}$)	
$\dot{}$	Time derivative (rate)	Not defined (e.g., $\dot{w}$)	

[a] See Section 1.1 in Chapter 1, "Fundamentals," for the treatment of these units. There are instances identified in the text where lb_m is used instead of lb_f to simplify formulas for use with the English system and SI.

ABBREVIATION AND DEFINITION

Btu	British thermal unit	min	minutes
cP	centipoise	mol	moles
ft	feet	kg	kilograms
g	grams	m	meters
h	hours	mm	millimeters
in	inches	N	newtons
J	joules	P	poise
lb	pounds	s	seconds

PART I

METHODOLOGY

PROLOGUE

Part I of this work consists of Chapters 1 through 7. These chapters, with the exception of Chapters 5–7, establish the basic "rules of the road," so to speak.

Chapter 1, "Fundamentals," discloses the systems of units that are used throughout the book, nomenclature and meanings of fluid properties, important dimensionless ratios, equations of state, and expositions of flow velocity and flow regimes.

Chapter 2, "Conservation Equations," elaborates on the conservation equations, that is, conservation of mass, of momentum and of energy. The general energy equation, head loss, and grade lines are treated under conservation of energy.

Chapter 3, "Incompressible Flow," expounds on how the particulars of incompressible flow (i.e., flow of liquids) became known through the breakthroughs of Julius Weisbach (head loss formula, 1845), Osborne Reynolds (the Reynolds number, 1883), and Ludwig Prandtl (boundary layers and the smooth pipe friction factor formula, 1904–1929). Johann Nikuradse's artificially roughened pipe experiments provided data (1933) to flesh out Prandtl's smooth pipe friction factor formula and Theodor von Kármán's complete turbulence formula (1930). Discrepancies between Nikuradse's artificially roughened pipe data and data on commercial pipe were resolved by Cyril F. Colebrook and Cedric M. White (1937). Colebrook published a semirational formula for random roughness (1939) that is still used today.

Chapter 4, "Compressible Flow," gives several ways to calculate head loss in compressible flow in pipes using approximate formulas derived from incompressible flow formulas. It culminates in giving theoretical formulas for compressible flow using either the Mach number or absolute pressure. While the formulas are complicated enough to resist explicit solution, ways are given to solve them by trial-and-error methods.

Chapter 5, "Network Analysis," gives methods to solve distribution of flow in networks. Chapter 6, "Transient Analysis," provides methods for solving flow problems whose flow rates are not constant. Chapter 7, "Uncertainty Analysis," gives methods for estimating the probable error or uncertainty in predicting pressure drop and flow rate.

Pipe Flow: A Practical and Comprehensive Guide, First Edition. Donald C. Rennels and Hobart M. Hudson.

1

FUNDAMENTALS

In this chapter we consider the fundamentals concerning fluid flow systems, such as the systems of units employed in this work, the physical properties of fluids, and the nature of fluid flow.

1.1 SYSTEMS OF UNITS

This book employs two systems of units: the U.S. Customary System (or USCS) and the International System (or SI, for Système International). The latter is based on the metric system, a system devised in France during the French Revolution in the late 1700s, but uses internationally standardized physical constants. Conversions between the systems may be found in Appendix C.

The USCS is virtually indistinguishable from the English gravitational system. There is some confusion in regard to the differences. Some authors imply that in USCS the slug is basic and the pound is derived, while others hold that the pound is basic and the slug is derived. In the English gravitational system the latter is assumed. For general engineering use it does not matter which is basic, because both systems agree that there is the slug for mass, the pound for force, the foot for length, and the second for time. This is all that need concern us in this work. The SI, derived from the metric system and having a much shorter pedigree, is consequently much more standardized.

Much confusion has resulted from the use in both English and metric systems of the same terms for the units of force and mass. To help eliminate the ambiguity owing to this double use the following treatment has been adopted.

The equation relating force, mass, and acceleration is

$$F = ma, \tag{1.1}$$

where F, m, and a are defined in the nomenclature. In SI the unit of mass, the kilogram, is basic. The unit of force is derived by means of the equation above and is given a unique name, the newton. Mass is never referred to by force units and vice versa. In the English gravitational system (which predates the USCS) and the USCS, a similar set of units is available and familiar to engineers, but it is not uniformly used. The unit of force, the pound, can be considered to be basic and the unit of mass derived by means of the relation above. It is often called the slug. While the slug is not often used, its insertion here need not pose any inconvenience. Where mass units are called for they may be easily obtained from the pound-force unit by the use of Equation 1.1. By use of these conventions any fundamental equation given in this book may be used with either SI or English units.

It should be noted that Equation 1.1 returns, in the English gravitational system, a mass with units of lb_f-s^2/ft. This is not easily recognizable, so the engineering community has somewhat arbitrarily chosen the term "slug" to name the mass instead of lb_f-s^2/ft. Similarly, in SI, the force that comes out of the equation has units of kg-m/s^2, and this force has been given the name "newton." The equation does not contain a factor that transforms lb_f-s^2/ft to "slugs" or kg-m/s^2 to "newtons."

Pipe Flow: A Practical and Comprehensive Guide, First Edition. Donald C. Rennels and Hobart M. Hudson.

We knowingly or unknowingly assume that there is an implicit conversion factor that changes the names of these units. This factor for SI is N/(kg-m/s^2)/(kg), and in the English gravitational system it is lb$_f$/slug-ft/s^2. If you call these conversion factors "C_g," Equation 1.1 becomes:

$$F \text{ newtons} = [C_g, \text{N/(kg-m/s}^2)][m \text{ kg}] \times [a(\text{m/s}^2)],$$

or

$$F \text{ lb}_\text{f} = [C_g, \text{lb/(slug-ft/s}^2)][m \cdot \text{slug}] \times [a(\text{ft/s}^2)].$$

The numeric value of the conversion factor is 1.000, so it does not change the number obtained, but only the name of the number. This may be the reason many writers subscript the g with a c to obtain g_c, when a is the acceleration of gravity.

Unfortunately the modern engineer must deal with mixed units and nomenclature used in some current practice and remaining from past practice. Conversions are offered in Appendix C that can help the user to work with mixed units. (Some secondary equations are given in which the units are mixed for the convenience of users of the English gravitational system. These equations and the units they require will be clearly indicated in the text.) Appendix C gives the important base units and derived units used here as well as the most frequently used conversions between systems.

1.2 FLUID PROPERTIES

Understanding the subject of pressure loss in fluid flow requires an understanding of the fluid properties that cause it. The principal concepts of interest in pressure loss due to flow are pressure, density, velocity, energy, and viscosity. Of secondary interest are temperature and heat.

1.2.1 Pressure

Pressure: The force per unit area exerted by a fluid on an arbitrarily defined boundary or surface, usually the walls of the conduit in which the fluid is flowing, or its cross section. Pressures are measured and quoted in different ways. A picture of pressure relationships can be gained from a diagram such as that of Figure 1.1, in which are shown two typical pressures, one above, and the other below, atmospheric pressure.*

Absolute Pressure: The pressure measured with respect to a datum of absolute zero pressure in which there are no fluid forces imposed on the boundary.†

* In the English system of units, pressure p is expressed in pounds per square inch, or psi.

† Absolute pressure is often expressed as psia in the English system.

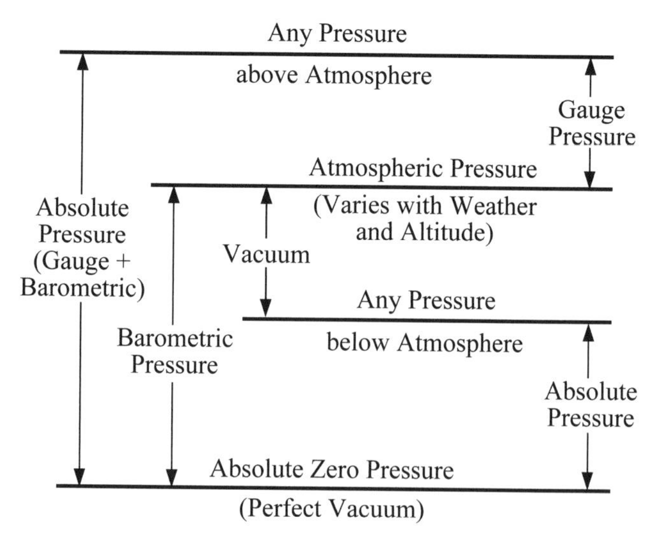

FIGURE 1.1. Pressure relationships.

Atmospheric Pressure: The absolute pressure of the local atmosphere.

Standard Atmospheric Pressure: The absolute pressure of the standard atmosphere at mean sea level. Standard atmospheric pressure, or one atmosphere, is 14.696 lb/in^2, 760.0 mm of mercury, 1.01325×10^5 N/m^2 (pascals), or 1.01325 bar.

Barometric Pressure: A barometer is an instrument used to measure atmospheric pressure by using water, air, or mercury. Thus atmospheric pressure is often called barometric pressure.

Critical Pressure: The pressure of a pure substance at its *critical state*; where the density of the saturated liquid is the same as the density of the saturated vapor. At pressures higher than the critical, a liquid may be heated from a low temperature to a very high one without any discontinuity indicating a change from the liquid to vapor phase. Values of critical pressure for selected gases are given in Appendix D.3.

Differential Pressure: The calculated or measured difference in pressure between any two points of interest.‡

Gauge Pressure: The pressure measured with respect to local atmospheric pressure. This is the pressure read by the common pressure gauge whose detecting element is a coil of flattened tube. (Sometimes this pressure is relative to standard atmospheric pressure. The reader is advised to determine which datum is used in other works or information sources.)§

‡ Differential pressure is often expressed as psid in the English system.

§ Gauge pressure is often expressed as psig in the English system.

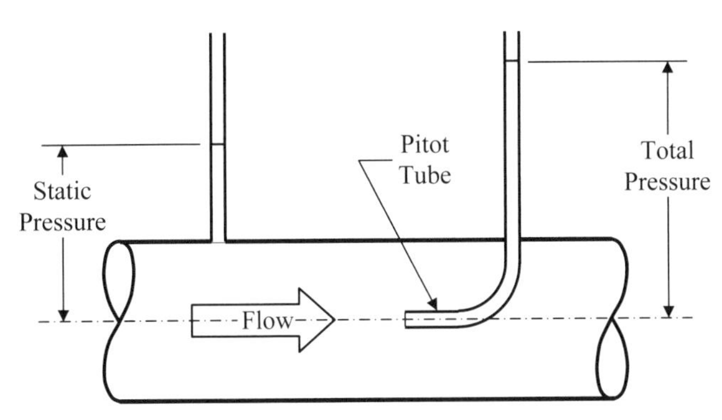

FIGURE 1.2. Total and static pressure.

Total Pressure: The pressure resulting from a moving fluid being brought to rest isentropically (without loss), as, for example, against a blunt object. (The kinetic energy of motion is converted to pressure when the fluid is brought to rest.) Total pressure is also known as stagnation pressure and pitot pressure (see Fig. 1.2).

Static Pressure: The pressure in a moving fluid before it is brought to rest. A pipe wall tap samples static pressure (see Fig. 1.2).

Vacuum: A pressure below local atmospheric pressure; often expressed as a negative pressure with respect to standard atmospheric pressure.

Vapor Pressure: The absolute pressure of a pure vapor in equilibrium with its liquid phase.

1.2.2 Density

Mass Density: The amount of material contained in a unit volume, measured in terms of its mass.

Weight Density: The amount of material contained in a unit volume, measured in terms of the force (weight) standard gravity exerts on the contained mass.

Specific Volume: The volume occupied by a unit mass or weight of material. (Which means it must be inferred from the units used. In the English system they will be ft^3/lb[force]; in SI they will be m^3/kg[mass].)

1.2.3 Velocity

Velocity: The speed of motion of a fluid with respect to a uniform datum. In pressure drop considerations it is usually used loosely with no direction implied. However, in impulse-momentum considerations, direction is an essential part of the measurement.

Average Velocity: A derived speed of a moving fluid whose various regions are not moving at the same speed but which accounts for the mass flux over the cross section of interest through which the fluid is moving.

Local Velocity: The actual speed of a moving fluid at a particular point of interest.

1.2.4 Energy

Energy (Work Energy): A measure of the ability of a substance to do or absorb work. It is usually measured in foot-pounds or newton-meters. (Newton-meters is also known as joules in SI.) Energy may exist in five forms: (1) potential, owing to a substance's elevation above an arbitrary datum; (2) pressure, which is a measure of a fluid's ability to lift some of itself to a level above an arbitrary datum or propel some of itself to a velocity; (3) kinetic, which resides in a substance's speed or velocity; (4) heat, which ultimately is a measure of the kinetic energy of the molecules of a substance; and (5) work. Work, in the case of fluid flow, is actually an effect of pressure moving some resistance. The work may be added to or subtracted from a fluid to change the status of the other four forms of energy. Pressure energy is sometimes called flow work because of its role in transferring work from one end of a conduit to another. Heat is considered separately below.

1.2.5 Viscosity

Viscosity: The resistance offered by a fluid to relative motion, or shearing, between its parts.

Absolute Viscosity: The frictional or shearing force per unit area of relatively moving surfaces per unit velocity for a unit separation of the surfaces. It is also called coefficient of viscosity and dynamic viscosity.

Kinematic Viscosity: Absolute viscosity per unit mass per unit volume of the flowing fluid. (A fluid's kinematic viscosity is its absolute viscosity divided by its mass density.)

1.2.6 Temperature

Temperature: In most fluid flow problems, temperature will refer simply to warmth (or lack of it), such as is perceived by our sense of touch and will be used to establish other fluid properties such as density and viscosity. It is usually measured on a somewhat arbitrary scale. The English system commonly uses the Fahrenheit scale, devised by the fifteenth-century German physicist Gabriel Fahrenheit [1]. It is based on the lowest temperature he could attain with a salt and ice mixture (assigned a value of 0°F) and human body temperature (to which he tried to assign a value of 96°F). This did not work out well and he ended up assigning 32°F to the melting point of ice and 212°F to the boiling point of water. The SI temperature scale (the Kelvin scale) is an absolute scale using the centigrade degree. The centigrade scale was devised by Swedish astronomer Anders Celsius in 1742 and incorporated into the metric system adopted in France at the close of the French Revolution [1]. On this scale—officially called Celsius since 1948—the melting point and boiling point of water at standard atmospheric pressure were assigned values of 0 and 100°C respectively.

Absolute Temperature: Temperature measured from absolute zero. It was noted in the late 1700s by the French physicist Jacques Charles (1746–1823) that gases expand and contract in direct proportion to their temperature changes. On a suitably chosen scale their volumes are thus directly proportional to their temperatures. The extrapolated temperature of zero volume according to the kinetic theory of gases is also the point at which molecular activity—and hence heat content—vanishes. No lower temperature is possible and so this temperature is called absolute zero. Two temperature scales based on this zero point are in common use. One, utilizing the Fahrenheit degree, is called the Rankine scale; temperatures on this scale are marked °R. The other, utilizing the Celsius degree, is called the Kelvin scale and its temperatures are marked K. The temperature 0°F corresponds to 459.67°R, and 0°C is identical to 273.15 K. Absolute zero is thus −459.67°F or −273.15°C.

Critical Temperature: The temperature of a pure substance at its *critical state*, above which its gas phase cannot be liquefied by the application of pressure, because at the critical temperature the latent heat of vaporization vanishes and the liquid cannot be distinguished from the gas. Values of critical temperature for selected gases are given in Appendix D.3.

1.2.7 Heat

Heat (Heat Energy): Heat is the measure of thermal energy contained in a substance. In fluid flow problems generally only sensible heat (i.e., heat obvious to the sense of touch or yielding a change in temperature) is of interest. It can be measured in the same units as work energy and indeed is interchangeable with energy. Usually heat is measured in units related to the heat–temperature relationship of water. In the English system the unit is the British thermal unit (usually abbreviated Btu). In SI it is the kilocalorie. The conversion to mechanical energy is the mechanical equivalent of heat. Its value is given in Appendix C.1.

Specific Heat: The measure of the change of heat capacity of a unit weight or mass of a substance for a unit change of temperature. It is almost always expressed in heat units, that is, Btu or kilocalories. The units of specific heat are thus Btu/lb-R (or Btu/lb-F) and kcal/kg-K (or kcal/kg-C).

1.3 IMPORTANT DIMENSIONLESS RATIOS

Researchers have devised many dimensionless ratios in order to describe the behavior of physical processes. The most important to us in analyzing pressure drop in fluid systems are described in the succeeding sections.

1.3.1 Reynolds Number

Named for the British engineer Osborne Reynolds (1842–1912), the Reynolds number is the ratio of momentum forces to viscous forces. It is extremely important in quantifying pressure drop in fluids flowing in closed conduits. It is given by:

$$N_{\mathrm{Re}} = \frac{VD\rho_w}{\mu g} = \frac{\dot{w}D}{\mu g A} \quad \text{(English)}, \tag{1.2a}$$

$$N_{\mathrm{Re}} = \frac{VD\rho_m}{\mu} = \frac{\dot{m}D}{\mu A} \quad \text{(SI)}. \tag{1.2b}$$

1.3.2 Relative Roughness

This quantity, as with the Reynolds number above, is extremely important in finding pressure drop in fluids flowing in pipes. It is rarely, if ever, assigned a symbol; but for illustration here let it be called R_{R}. It is defined as:

$$R_{\mathrm{R}} = \frac{\varepsilon}{D},$$

where ε is the absolute roughness of the pipe inner wall and D is the pipe inside diameter. (In practice it is usually just called ε/D.)

1.3.3 Loss Coefficient

The loss coefficient, or resistance coefficient, is the measure of pressure drop in fluid systems. It is defined as:

$$K = f\frac{L}{D}, \quad (1.3)$$

where:
K = loss coefficient measured in *velocity heads*,
f = Darcy friction factor,
L = length of pipe stretch for which the resistance coefficient applies, and
D = inside diameter of the pipe stretch.

More will be said about f and K in Chapters 3 and 8.

1.3.4 Mach Number

Named for the Czech physicist Ernst Mach (1838–1916), the Mach number is the ratio of the local fluid velocity u to the acoustic velocity A. It is very useful in describing compressible flow phenomena. It is given by:

$$M = \frac{u}{\mathrm{A}}. \quad (1.4a)$$

The average velocity V is usually substituted when the flow is in a conduit and the velocity profile is fairly flat. With this convention, the equation becomes:

$$M = \frac{V}{\mathrm{A}}. \quad (1.4b)$$

1.3.5 Froude Number

The *Froude number* N_{Fr} specifies the ratio of inertia force to gravity force on an element of fluid. It is named for William Froude, an English engineer and naval architect (1810–1879), who, in the later half of the nineteenth century, pioneered in the investigation of ship resistance by use of models. The Froude number is used in the investigation of similarity between ships and models of them. In this role, it is defined as the ratio of the velocity of a surface wave and the flow velocity. Our interest is in its application to pipe flow where the pipe is not flowing full. In this context it is expressed as:

$$N_F = \frac{V}{\sqrt{gD}} = \frac{V}{\sqrt{2gR}}, \quad (1.5)$$

where V is the characteristic velocity, g is the acceleration of gravity, D is the pipe diameter, and R is the pipe radius. The Froude number, unlike the Reynolds number, is independent of viscosity and so it applies to inviscid flow analysis.

1.3.6 Reduced Pressure

Reduced pressure, along with reduced temperature (described below), is useful in quantifying departures from the ideal state in gases. Reduced pressure is given by:

$$P_r = \frac{P}{P_c},$$

where P is the pressure of interest and P_c is critical pressure.

1.3.7 Reduced Temperature

As with reduced pressure described above, reduced temperature helps to reduce the state point of most gases to a common base, making it possible to quantify departures of most gases from the ideal equation describing the relationship between pressure, temperature, volume, and quantity of substance (the equation of state, described below). Reduced temperature is given by:

$$T_r = \frac{T}{T_c},$$

where T is the temperature of interest and T_c is critical temperature.

1.4 EQUATIONS OF STATE

This section presents various equations which describe the physical properties of fluids—principally the fluid's density as a function of pressure and temperature.

1.4.1 Equation of State of Liquids

An "equation of state of liquids" is not commonly expressed. This is because in usual engineering fluid-flow problems, the volume properties of the liquid are scarcely affected by changes in temperature or pressure in the flow path. Where their properties are significantly affected it is customary (because it is easiest and sufficiently accurate) to break the problem into small enough segments wherein the properties may be considered to be constant. Where this approach is not satisfactory, as, for instance, when dealing with liquids at pressures above the critical pressure, equations of state of liquids are available in the literature. Attention is directed to the works by Reid et al. [2] and Poling et al. [3], produced a quarter-century apart, which reflect the growth in information available in the literature on this subject.

1.4.2 Equation of State of Gases

Because gases exhibit large changes in volume, pressure, or temperature for comparable changes in one or both of the remaining of these three important variables, it has been necessary to formulate a workable expression relating them. The expression is called the equation of state. Two-variable relationships were discovered by Robert Boyle (1627–1691) and by Jacques Charles (1746–1823) and Joseph Gay-Lussac (1778–1850), which were soon combined into the perfect gas law:

$$PV = mRT, \tag{1.6a}$$

where m = mass of the gas, V is the volume, and R is the individual gas constant; or

$$PV = n\bar{R}T, \tag{1.6b}$$

where n = number of mols of gas considered and $\bar{R}$ is the universal gas constant. (In the English system Eq. 1.6a is usually written $PV = wRT$, where w = weight, lb, and the R used is expressed in weight units.)

Equation 1.6 adequately describes real gas behavior when pressure is low with respect to the critical pressure and temperature is high with respect to the critical temperature. However, with increasing pressure or decreasing temperature, or both, this relation departs increasingly from real gas behavior. A coefficient can be added to account for the departure, called the *compressibility factor*:

$$PV = zmRT, \tag{1.7}$$

where z is a function of the temperature and pressure of the gas. Dutch physicist Johannes van der Waals (1837–1923) noted that when z is plotted versus reduced pressure, that is, actual pressure divided by the critical pressure, for constant reduced temperature, that is, actual temperature divided by the critical temperature, the plotted points for any given reduced temperature for most gases fall into a narrow band [4]. If a line is faired through each band for each reduced temperature, a chart called a *compressibility chart* is obtained. A plot of this kind was published by L. C. Nelson and E. F. Obert in 1954 [5]. An example is shown in Figure 1.3 [3].* Many attempts have been made to find an analytic function, an equation of state, to describe this behavior, with varying success. Most of these "real gas" equations of state are limited in range of applicability. Two particularly attractive equations (solutions for z), suitable for wide ranges of pressure and temperature, the Redlich–Kwong equation and the Lee–Kesler equation, are described in Appendix D. Scores more are described by Poling et al. [3]. The utility of these equations is illustrated in Chapter 4, "Compressible Flow."

* Large charts of the compressibility factor are available. One is reprinted by Poling et al [3]. Where more precision is desired, a computer program, called *MIPROPS*, which calculates many fluid properties, including density, viscosity, entropy, and acoustic velocity, was published by the National Bureau of Standards (now the National Institute of Standards and Technology) and is available from the Department of Commerce.

1.5 FLUID VELOCITY

As stated in Section 1.2, velocity (so called; more accurately it would be called speed) is usually considered to be uniform over the cross section of flow. In reality, it is not. The fluid in contact with the conduit wall must be at zero velocity, and velocity ordinarily increases toward the center. The assumption of uniform velocity immensely simplifies fluid flow calculations. There is an inaccuracy introduced by this assumption, but, fortunately, it usually does not affect the confidence level of fluid flow computations. The inaccuracies can be quantified and will be considered in the following chapter.

Another assumption that is usually made is that the velocity is one-dimensional, that is, that radial components of flow velocities are inconsequential. Inaccuracies introduced by this assumption are small and are absorbed by the loss coefficients.

1.6 FLOW REGIMES

In the study of fluid flow it has long been recognized that there are two distinct kinds of flow or flow regimes. The first is characterized by preservation of layers or laminae in the flow stream. This kind of flow is called *laminar* or *streamline* flow. In cylindrical conduits the layers are cylindrical, the local velocities are strictly parallel to the conduit axis, and they vary parabolically in velocity from zero at the wall to a maximum at the center. The second is characterized by destruction and mixing of the layers seen in laminar flow, and the local motions in the fluid are chaotic or turbulent. This kind of flow is thus appropriately called *turbulent* flow. In circular conduits the axial velocity distribution is more nearly uniform than it is in laminar flow, although local velocity at the pipe wall is still zero. Laminar and turbulent flow velocity profiles are illustrated in Figure 1.4. Because their effects will be treated in the following chapter you need to know that these two types of flow exist.

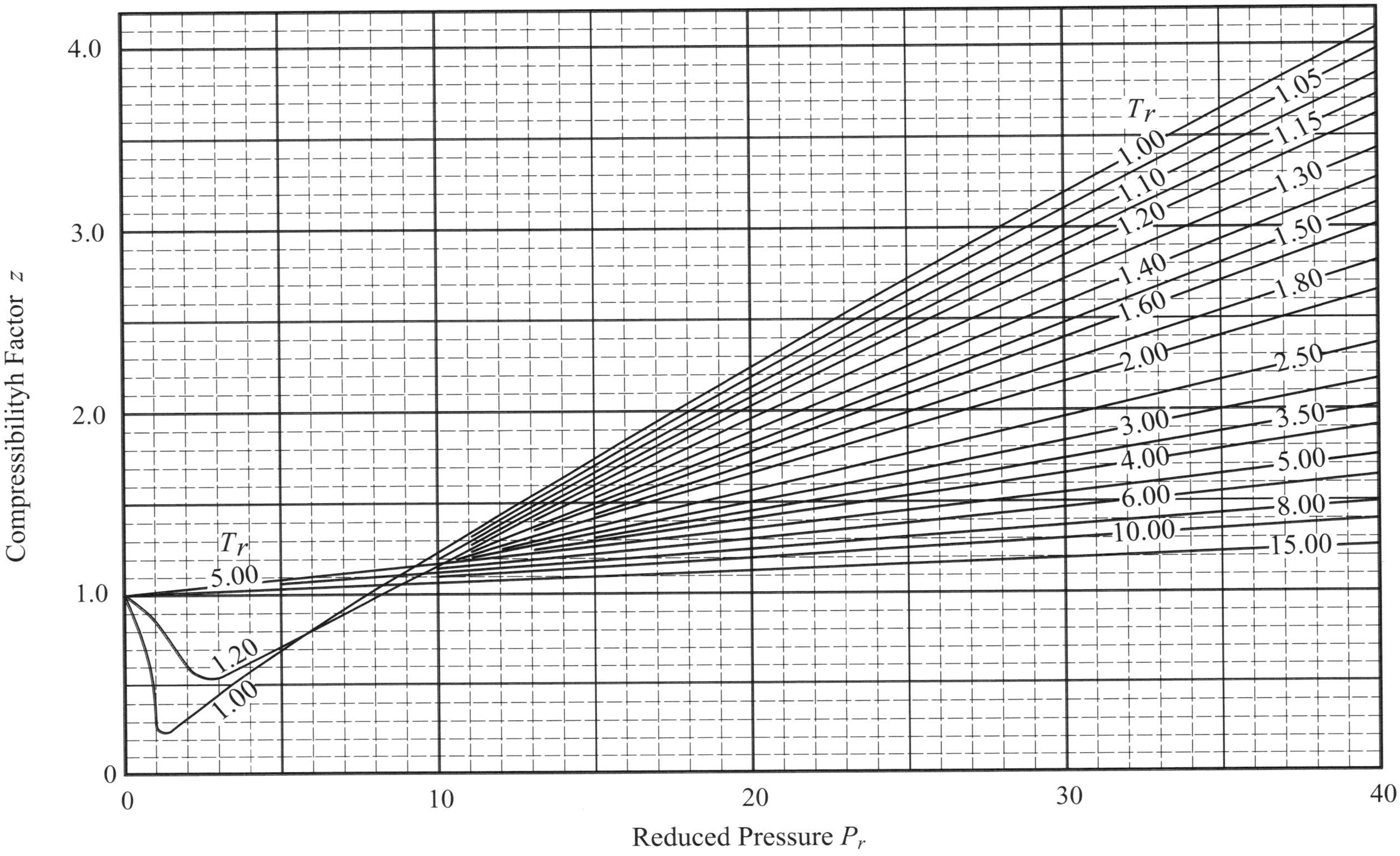

Compressibilityh Factor z

Reduced Pressure P_r

FIGURE 1.3. (a) Generalized compressibility factor.

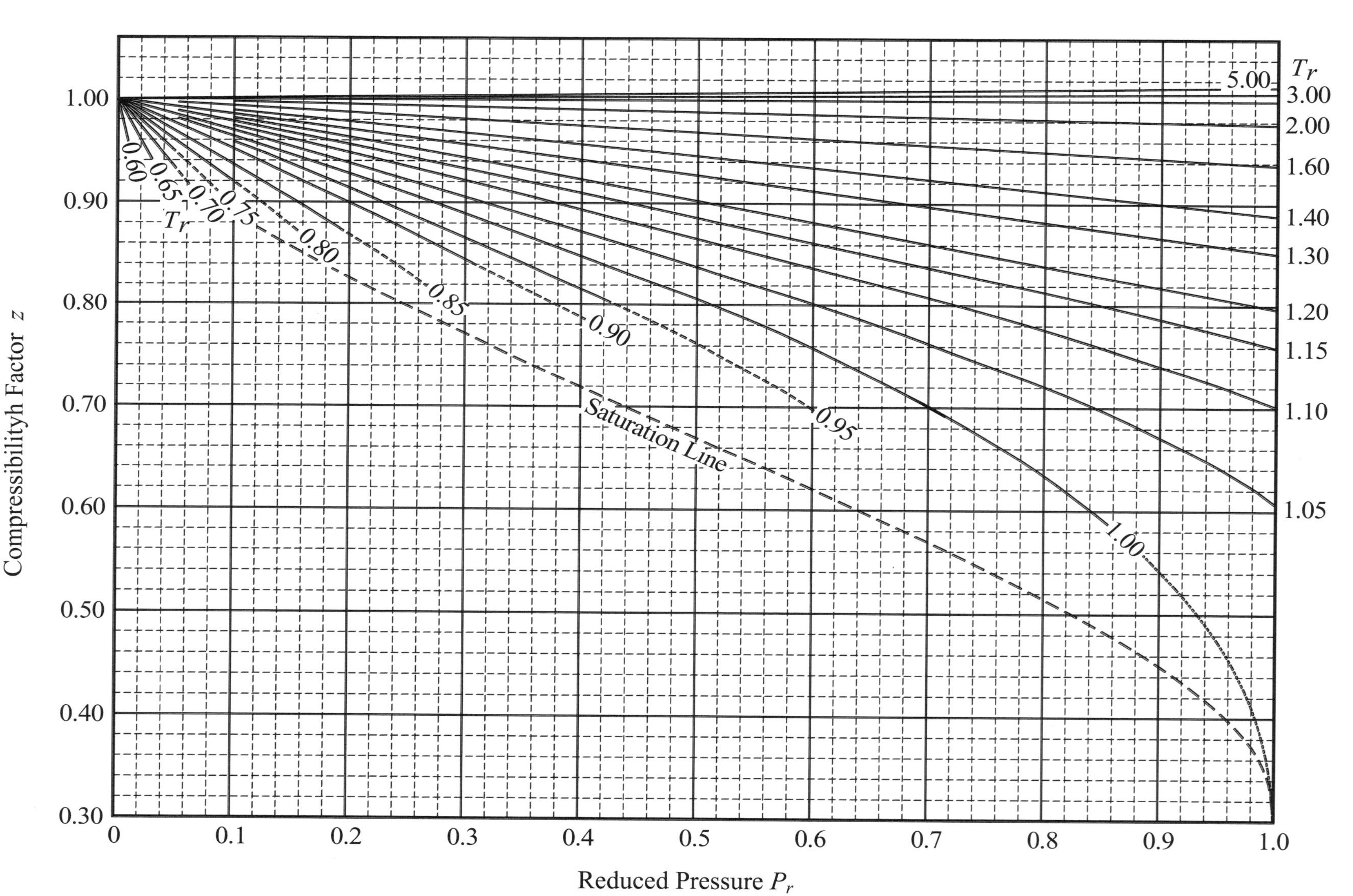

FIGURE 1.3. (*Continued*) (b) Generalized compressibility factor—subcritical range.

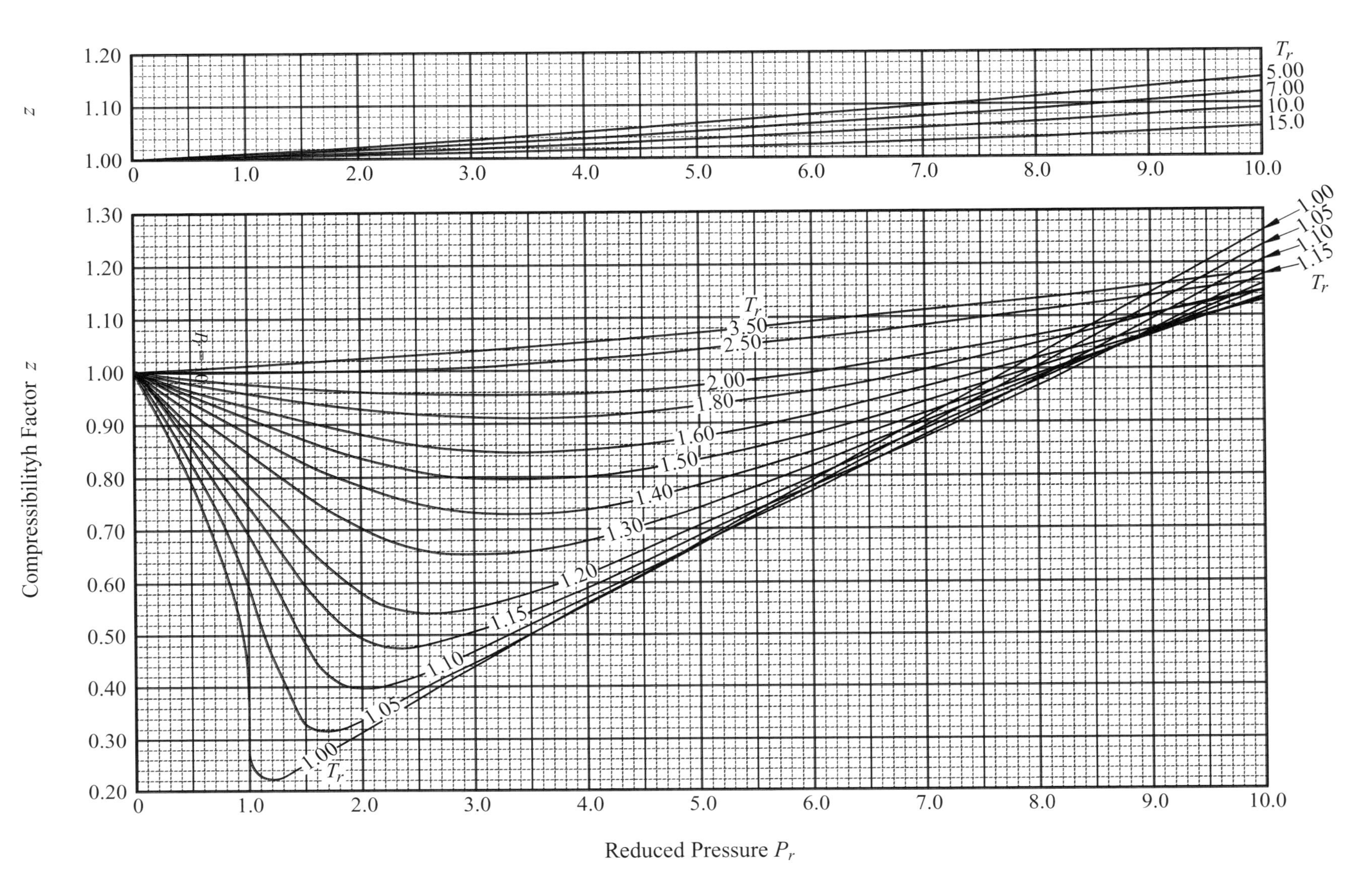

FIGURE 1.3. (*Continued*) (c) Generalized compressibility factor—pressure range to $P_r = 10$.

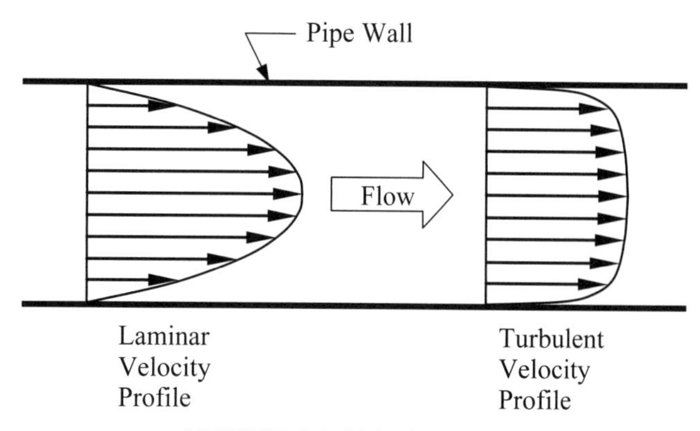

FIGURE 1.4. Velocity profiles.

REFERENCES

1. Graham, L., Heat, thermometry, in *Encyclopedia Britannica*, Vol. 8, 15th ed., Encyclopedia Britannica, 1978, Macropedia, p. 706.
2. Reid, R. C., J. M. Prausnitz, and T. K. Sherwood, *The Properties of Gases and Liquids*, 3rd ed., McGraw-Hill, 1977, p. 81.
3. Poling, B. E., J. M. Prausnitz, and J. P. O'Connell, *Properties of Gases and Liquids*, 5th ed., McGraw-Hill, 2001.
4. Baumeister, T., ed., *Marks' Standard Handbook for Mechanical Engineers*, 8th ed., McGraw-Hill, 1978, pp. 4–17.
5. Nelson, L. C. and E. F. Obert, Generalized pvT properties of gases, *Transactions of the American Society of Mechanical Engineers*, 76, 1954, 1057.

FURTHER READING

This list includes books and papers that may be helpful to those who wish to pursue further study.

Asimov, I., *Understanding Physics*, Vol. 1. Dorset Press, 1966.

Bedford, R. E., Thermometry, in *Encyclopedia Britannica*, Vol. 18, 15th ed., Encyclopedia Britannica, 1978, p. 322.

Fox, R. W., P. J. Pritchard, and A. T. McDonald, *Introduction to Fluid Mechanics*, 7th ed., John Wiley & Sons, 2008.

Streeter, V. L. and E. B. Wylie, *Fluid Mechanics*, 7th ed., McGraw-Hill, 1979.

Munson, B. R., D. F. Young, and T. H. Okiishi, *Fundamentals of Fluid Mechanics*, 3rd ed., John Wiley & Sons, 1998.

2

CONSERVATION EQUATIONS

This chapter will consider the equations for conservation of mass, energy and momentum, velocity profiles, and correction factors for momentum and energy. In general, the English gravitational system uses weight flow rate ($\dot{w}$), and the International System of Units (SI) uses mass flow rate ($\dot{m}$).

2.1 CONSERVATION OF MASS

The continuity equation is simply a statement that there is as much fluid flowing out of a system under consideration as there is flowing into it. It assumes that mass is conserved and that fluid is not being stored or released from storage within the system. The equations for weight rate of flow and mass rate of flow are:

$$\dot{w} = AV\rho_w, \tag{2.1a}$$

$$\dot{m} = AV\rho_m. \tag{2.1b}$$

When the continuity equation holds, the inlet flow rate is equal to the outlet flow rate, so that

$$A_1V_1(\rho_w)_1 = A_2V_2(\rho_w)_2, \tag{2.2a}$$

$$A_1V_1(\rho_m)_1 = A_2V_2(\rho_m)_2. \tag{2.2b}$$

These equations are expressions of the *continuity equation*.

In these equations it is customary to assume that the velocity profile is flat, that is, the velocity in the fluid flowing in a conduit is the same everywhere in the cross section. The velocity that accounts for all the weight flux (or mass flux) across the cross section of the conduit is the *average velocity*.

The velocity profile is, of course, *not* flat across the cross section! Does this assumption therefore cause an error in the continuity equation? No, because we use the same relation to define the average velocity as to determine the weight flux through the cross section. The same cannot be said, however, for the momentum flux or the energy flux as we shall discover in the next sections.

2.2 CONSERVATION OF MOMENTUM

The momentum equation is a statement that a fluid stream, as it relates to fluid flow when acted upon by external forces whose sum is not zero, must acquire a change in velocity. The amount of this force may be found by use of the momentum equation. It is thus an application of Newton's second law of motion (Eq. 1.1).

Consider an axisymmetric reducing flow passage as illustrated in Figure 2.1. Assume that velocity distribution is uniform at any cross section of the stream tube. P_1A_1 is the axial force acting on the fluid in the control volume owing to absolute pressure P_1 acting over area A_1; P_2A_2 is the axial force owing to absolute pressure P_2 acting over area A_2; and F is the apparent residual force owing to the diminishing stream pressure acting over the axial projection of the outer control volume boundary and to the frictional resistance on the surface of

Pipe Flow: A Practical and Comprehensive Guide, First Edition. Donald C. Rennels and Hobart M. Hudson.

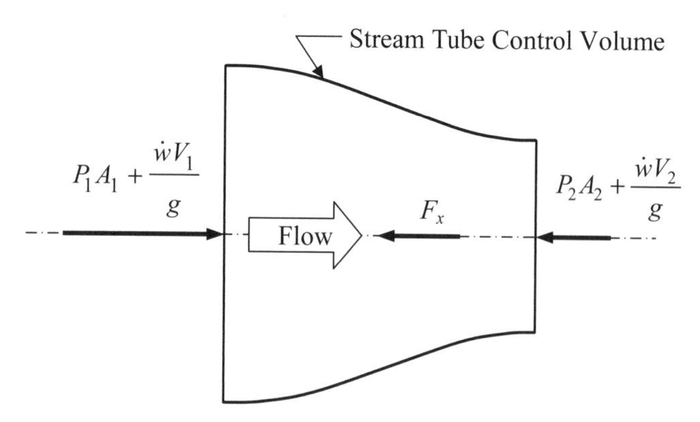

FIGURE 2.1. Axisymmetric reducing flow passage.

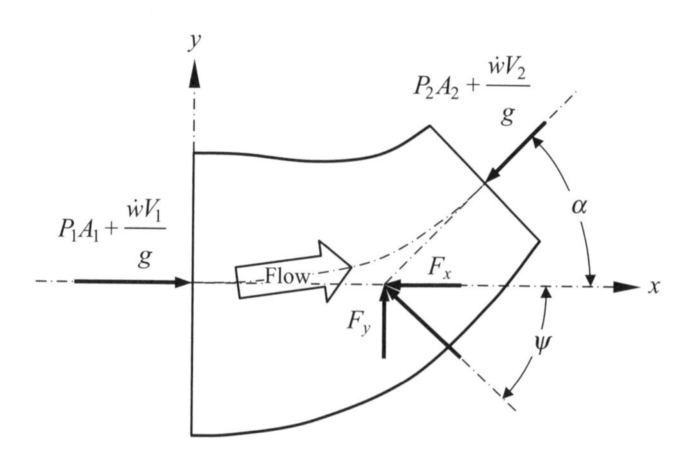

FIGURE 2.2. Nonaxisymmetric reducing flow passage.

the stream tube. The terms $\dot{w}V_1/g$ and $\dot{w}V_2/g$ are the entering fluid momentum and exiting fluid momentum, respectively.

The sum of these axial forces is:

$$\Sigma F = P_1A_1 - P_2A_2 - F_x.$$

The sum of the forces is equal to the change in the momentum of the fluid between the inlet and outlet of the control volume:

$$\Sigma F = \frac{\dot{w}V_1}{g} - \frac{\dot{w}V_2}{g} = \frac{\dot{w}}{g}(V_1 - V_2),$$

$$\Sigma F = \dot{m}V_1 - \dot{m}V_2 = \dot{m}(V_1 - V_2).$$

Combining the axial force equation with the change in momentum equations gives:

$$F_x = P_1A_1 - P_2A_2 + \frac{\dot{w}}{g}(V_1 - V_2),$$

$$F_x = P_1A_1 - P_2A_2 + \dot{m}(V_1 - V_2).$$

In this derivation, an axisymmetric stream tube shape was chosen so that only axial forces need be considered. Because both force and velocity are vector quantities, that is, they include both quantity and direction, the momentum equation can be written for each of the three orthogonal directions:

$$F_x = (P_1A_1)_x - (P_2A_2)_x + \frac{\dot{w}}{g}(V_1 - V_2)_x \quad \text{or}$$
$$F_x = (P_1A_1)_x - (P_2A_2)_x + \dot{m}(V_1 - V_2)_x,$$

$$F_y = (P_1A_1)_y - (P_2A_2)_y + \frac{\dot{w}}{g}(V_1 - V_2)_y \quad \text{or}$$
$$F_y = (P_1A_1)_y - (P_2A_2)_y + \dot{m}(V_1 - V_2)_y,$$

$$F_z = (P_1A_1)_z - (P_2A_2)_z + \frac{\dot{w}}{g}(V_1 - V_2)_z \quad \text{or}$$
$$F_z = (P_1A_1)_z - (P_2A_2)_z + \dot{m}(V_1 - V_2)_z.$$

Usually a nonaxisymmetric stream tube lies in a single plane so that an analysis in two directions is sufficient. For the stream tube shown in Figure 2.2 the momentum equations become:

$$F_x = P_1A_1 - P_2A_2\cos\alpha + \frac{\dot{w}}{g}(V_1 - V_2\cos\alpha) \quad \text{or}$$
$$F_x = P_1A_1 - P_2A_2\cos\alpha + \dot{m}(V_1 - V_2\cos\alpha),$$

$$F_y = -P_2A_2\sin\alpha - \frac{\dot{w}}{g}V_2\sin\alpha \quad \text{or}$$
$$F_y = -P_2A_2\sin\alpha - \dot{m}V_2\sin\alpha.$$

The angle ψ describing the orientation of F is:

$$\psi = \arctan\frac{F_y}{F_x}.$$

2.3 THE MOMENTUM FLUX CORRECTION FACTOR

Up to this point it has been assumed that velocity distribution in the fluid has been uniform across a plane normal to the direction of flow, when in fact it never is (Section 1.5). An assessment of the error incurred by this assumption in the momentum equation is in order. The total momentum at a given cross section of the stream tube is, assuming a flat velocity profile,

$$\dot{m}V = (AV\rho)V = AV^2\rho,$$

where V is the average fluid velocity. In an infinitely thin cylinder centered on the pipe center, this becomes the following differential equation,

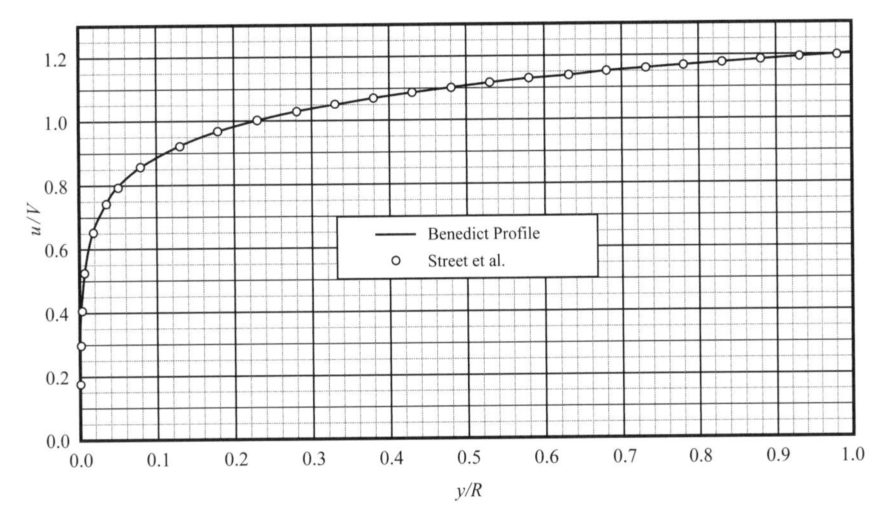

FIGURE 2.3. Plot of fully turbulent velocity profile for $f = 0.024$.

$$u d\dot{m} = u^2 \rho dA,$$

where u is the local velocity. If we integrate this differential equation over the total cross sectional area A where the fluid velocity is *not* uniform throughout, we will arrive at a value that is *not* equal to $\dot{m}V$. We need to introduce a correction factor:

$$\int u d\dot{m} = \theta \dot{m} V, \tag{2.3}$$

where θ is the momentum flux correction factor. For an axisymmetric velocity distribution the mass flow is:

$$d\dot{m} = u\rho dA = u\rho(2\pi r dr),$$

$$\dot{m} = \int d\dot{m} = 2\pi\rho \int u r dr, \tag{2.4}$$

where r is the radius from the center of the pipe to the local velocity. The momentum flux is given by:

$$u d\dot{m} = u^2 \rho dA = u^2 \rho(2\pi r dr),$$

$$\int u d\dot{m} = 2\pi\rho \int u^2 r dr. \tag{2.5}$$

Combining Equations 2.3–2.5, we obtain:

$$2\pi\rho \int u^2 r dr = \theta 2\pi\rho V \int u r dr,$$

or

$$\theta = \frac{1}{V} \frac{\int u^2 r dr}{\int u r dr}. \tag{2.6}$$

Ludwig Prandtl, Johann Nikuradse, and Theodor von Kármán, during the period from 1926 to 1932, determined an equation for the velocity profile in pipe flow. From that equation Robert P. Benedict [1] shows that the velocity profile can be expressed as*:

$$\frac{u}{V} = 1 + 3.75\sqrt{\frac{f}{8}} + 2.5\sqrt{\frac{f}{8}} \ln \frac{y}{R}. \tag{2.7}$$

The plot of this equation is shown in Figure 2.3. It will be seen that the slope of the curve is not zero at the pipe centerline. About this, Hunter Rouse [2] says that "[these equations] do not give a zero slope of the velocity distribution curve at the center line. This is a defect in the formulas, which, from a practical viewpoint, is nevertheless of little significance. The equations actually portray the true velocity distribution in the central region of the flow very well, although they were derived for the region near the wall."

Street et al. [3] give the following formulas for velocity profile and the resulting average velocity:

$$\frac{u}{V^*} = 5.75 \log_{10} \frac{y}{\varepsilon} + 8.5,$$

$$\frac{V}{V^*} = 5.75 \log_{10} \frac{R}{\varepsilon} + 4.75,$$

$$\frac{1}{\sqrt{f}} = 2.0 \log_{10} \frac{d}{\varepsilon} + 1.14.$$

* The development of this equation is given in Appendix F.

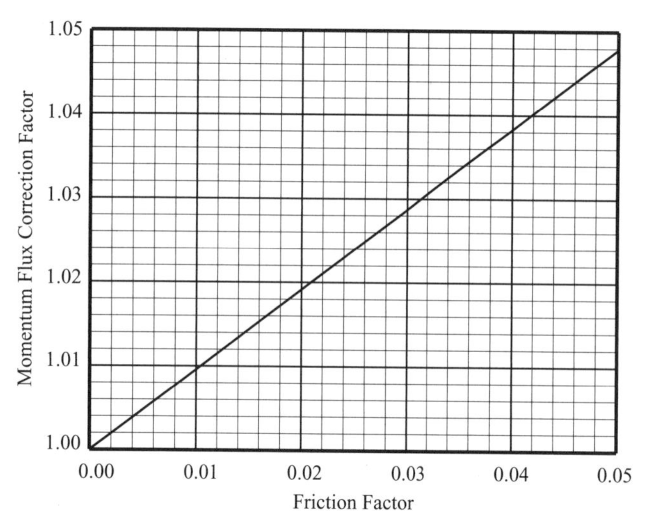

FIGURE 2.4. Momentum flux correction factor versus friction factor (for turbulent flow).

V^* is the "friction velocity," $V^* = \sqrt{\tau_0 / \rho_m}$, where τ_0 is the wall shear stress and ρ_m is the mass density (in either the English gravitational system or SI). By combining these three equations the following equation is obtained*:

$$\frac{u}{V} = \frac{\log_{10}\dfrac{y}{R} + \dfrac{1}{2\sqrt{f}} + 0.607231}{\dfrac{1}{2\sqrt{f}} - 0.044943}. \tag{2.8}$$

When Equation 2.8 is evaluated and compared with Equation 2.7, the difference is scarcely discernible. With either of these equations, performing the indicated integrations and ratio in Equation 2.6, the momentum flux correction factor is found to be:

$$\theta = 1 + 0.9765 f.$$

A plot of this equation (for turbulent flow) is shown in Figure 2.4.

With a friction factor of 0.04, θ is about 1.038. Because most friction factors encountered in engineering work are less than 0.04, the error attendant to assuming a flat velocity profile is therefore usually negligible. Laminar flow, however, is an exception. Here the velocity profile is parabolic, and performing the indicated integrations and subsequent divisions yields $\theta = 1.333$, a value which cannot be ignored. Other exceptions occur where the velocity profile is badly distorted, such as at the efflux of a conical expander.

* The development of this equation is given in Appendix F.

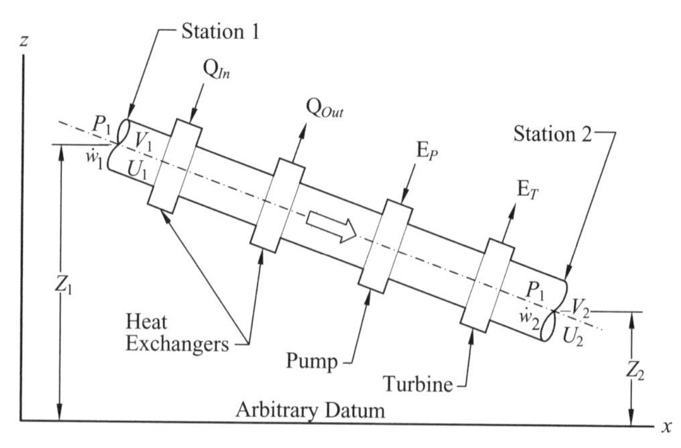

FIGURE 2.5. Energy fluxes.

2.4 CONSERVATION OF ENERGY

The energy equation is of paramount importance in our mathematical model of fluid flow losses. It accounts for the various energy changes within a flow system, or a portion of interest, and enables us to formulate a mathematical relationship that will provide consistently accurate predictions of pressure drop within it. The energy equation presents few difficulties once these energies have been identified.

As its name implies, the energy equation rests on the law of conservation of energy. This law, when applied to the steady flow of any real fluid, states that the rate of flow of energy entering a system is equal to that leaving the system. Figure 2.5 shows a hypothetical flow system with the fluid properties and circumstances and the energy fluxes affecting the energy balance.

In order to relate the energy inflows and outflows in a system it is necessary to put them in common units. It is convenient for this discussion to express energy in work units such as foot-pounds or newton-meters, and unit energies in terms of foot-pounds per pound of fluid, or newton-meters per newton. From Figure 2.5 it is seen that five kinds of energy flux must be considered: potential, pressure, kinetic, heat, and work.

2.4.1 Potential Energy

Every unit of fluid lifted above an arbitrary datum required a certain amount of work to lift it there. If the unit of fluid quantity is pounds (or newtons), the work required (in a uniform gravity field) is its weight times the height it was lifted, ft-lb (or N-m). Thus the unit energy is ft-lb/lb or ft (or N-m/N or m), equal numerically and dimensionally to its elevation Z above the datum. This is called the *elevation* or *potential head*.

2.4.2 Pressure Energy

Pressure is commonly expressed as force per unit area—for example, lb/in^2, lb/ft^2, or N/m^2 (pascals). If the fluid's pressure is divided by its weight density, its potential for doing work is expressed in potential energy terms. Consistent units will eliminate mixed unit problems. Thus lb/ft^2 and N/m^2 yield:

$$P/\rho_w = (\text{lb/ft}^2)/(\text{lb/ft}^3) = \text{ft},$$

$$P/\rho_m = (\text{N/m}^2)/(\text{N/m}^3) = \text{m}.$$

As an example, a fluid under pressure P can be lifted in a manometer to a height P/ρ_w or P/ρ_m. This is called the *pressure head.*

2.4.3 Kinetic Energy

The simple equations of motion show that in the absence of air or other resistance any body dropped from one elevation to another lower elevation acquires a velocity equal to the square root of twice the product of the elevation difference and the acceleration of gravity, that is,

$$V = \sqrt{2g\Delta Z}.$$

Conversely, any body moving with velocity V can, if the velocity can be directed upward, attain a height of:

$$\Delta Z = V^2/2g. \tag{2.9}$$

A fluid's energy of motion is thus $V^2/2g$ ft-lb/lb or simply ft (or N-m/N or m). This is called the *velocity head.* The symbol is H_{KE}.

In the hypothetical flow system shown in Figure 2.5 we might assume that every molecule of fluid is passing through the conduit, at any one cross section, at the same velocity. In such a case the fluid's average velocity would be the same as that of any particle of the flow, and its kinetic energy would be accurately described by Equation 2.9, where V is the fluid's average velocity. A real fluid, however, never flows in quite this fashion. At the wall of the conduit its velocity always approaches zero and it increases to a maximum at the center of the conduit for fully developed flow. The kinetic energies of its parts vary depending on their locations in the cross section. Because the square of the average is not the same as the average of the squares, a correction factor ϕ must be included if the average velocity is used to calculate the kinetic energy of the flowing fluid:

$$H_{KE} = \phi V^2/2g.$$

The correction factor will be treated in more detail in a later section, but suffice it to say now that ϕ is required to measure *precisely* the kinetic energy of the fluid.

2.4.4 Heat Energy

The English physicist James Prescott Joule (1818–1889) showed conclusively in experiments conducted between 1843 and 1850 that heat is equivalent to work. The physical constant relating the two is denoted here by the symbol J. To convert common heat units (Btu/lb or kcal/kN) to specific work units (ft or m) the heat units are multiplied by J in the proper units. Because transferred heat flux Q is usually calculated in heat units and the energy equation is usually set up with work units, it is convenient to convert the heat units to work units:

$$JQ = \left(\frac{\text{ft-lb}}{\text{Btu}}\right)\left(\frac{\text{Btu}}{\text{s}}\right) = \frac{\text{ft-lb}}{\text{s}},$$
$$= \left(\frac{\text{N-m}}{\text{kcal}}\right)\left(\frac{\text{kcal}}{\text{s}}\right) = \frac{\text{N-m}}{\text{s}}.$$

The units in the foregoing expression, now in work units per unit time, must be further converted to potential energy units:

$$\frac{JQ}{\dot{w}} = \frac{\text{ft-lb/s}}{\text{lb/s}} = \frac{\text{ft-lb}}{\text{lb}} = \text{ft}, \tag{2.10a}$$

$$= \frac{\text{N-m/s}}{\text{N/s}} = \frac{\text{N-m}}{\text{N}} = \text{m} \tag{2.10b}$$

Internal heat energy, that is, heat energy possessed by the fluid upon entering the flow system or leaving it, like transferred heat, is usually expressed in heat units; but unlike transferred heat it is treated on a per-unit-weight basis or a per-unit-mass basis. (For this discussion let us continue to treat the individual terms of the general energy equation on a per-unit-weight basis.) Internal heat energy, or simply internal energy, denoted by the symbol U, is converted to potential energy units as follows:

$$JU = \left(\frac{\text{ft-lb}}{\text{Btu}}\right)\left(\frac{\text{Btu}}{\text{lb}}\right) = \text{ft},$$

$$JU = \left(\frac{\text{N-m}}{\text{kcal}}\right)\left(\frac{\text{kcal}}{\text{N}}\right) = \text{m}, \quad \text{or}$$

$$= \frac{\text{joules}}{\text{N}} = \frac{\text{N-m}}{\text{N}} = \text{m}.$$

2.4.5 Mechanical Work Energy

The mechanical work done on the fluid in the flow system by a pump and, as in the case of heat flux, the work done by the fluid in a turbine must be expressed in power units, or work per unit time, to maintain dimensional homogeneity in the energy equation. These units may be converted to potential energy units as they were in the case of heat flux (Eq. 2.10a and 2.10b):

$$\frac{E_P}{\dot{w}} = \frac{\text{ft-lb / s}}{\text{lb/s}} = \frac{\text{ft-lb}}{\text{lb}} = \text{ft}$$
$$= \frac{\text{N-m/s}}{\text{N/s}} = \frac{\text{N-m}}{\text{N}} = \text{m}.$$

The same conversion also applies to turbine work, E_T.

The mechanical work energy is often called "flow work," because without flow there is no work performed. In the case of the pump, flow work is added to the flow, and in the case of the turbine, flow work is subtracted from the flow.

2.5 GENERAL ENERGY EQUATION

Having defined the energy fluxes in the hypothetical flow system in common units, we may now write the energy balance:

$$\frac{P_1}{(\rho_w)_1} + \frac{\phi_1 V_1^2}{2g} + Z_1 + JU_1 + \frac{JQ_1}{\dot{w}} + \frac{E_P}{\dot{w}} = \frac{P_2}{(\rho_w)_2} + \frac{\phi_2 V_2^2}{2g} + Z_2 + JU_2 + \frac{JQ_2}{\dot{w}} + \frac{E_T}{\dot{w}}. \tag{2.11a}$$

Equation 2.11a is set up for weight units in either the English gravitational system (where lb_f is basic) or SI (where the kilogram mass is basic) but using newtons as the force unit. For SI in mass units, the equation is:

$$\frac{P_1}{(\rho_m)_1 g} + \frac{\phi_1 V_1^2}{2g} + Z_1 + \frac{JU_1}{g} + \frac{JQ_1}{\dot{m}g} + \frac{E_P}{\dot{m}g} = \frac{P_2}{(\rho_m)_2 g} + \frac{\phi_2 V_2^2}{2g} + Z_2 + \frac{JU_2}{g} + \frac{JQ_2}{\dot{m}g} + \frac{E_T}{\dot{m}g}. \tag{2.11b}$$

As shown in Chapter 1, the units of ρ_m, m, and $\dot{m}$ are changed to force units when multiplied by g, and this entity may not be easily recognized by the user. For this reason a conversion factor called C_g may be inserted into the conversion to change the name of the entity. This factor for SI is N/(m/s^2)/(kg), and if you call it "C_g," Equation 1.1 becomes:

$$F\ \text{Newtons} = [C_g\ \text{N/(m/s}^2\text{/kg)}][m\ \text{kg} \times a(\text{m/s}^2)].$$

With this convention, each term in the SI General Energy Equation has the units of meters.

Other forms of energy, such as chemical, electric, or atomic, may need reckoning in a particular flow problem. Their inclusion should present no difficulties if they are treated as the five forms shown here have been.

The first three terms on each side of Equation 2.11a and 2.11b are called the Bernoulli terms, after Swiss mathematician Daniel Bernoulli (1700–1782), and are referred to as heads—P/ρ is called the pressure head, $\phi V^2/2g$ is called the velocity head, and Z is called the elevation or potential head.

2.6 HEAD LOSS

The general energy equation as given above (Eq. 2.11a and 2.11b) is valid for any real fluid. There is, however, an observation that should be made here. Consider the most elementary flow system: a horizontal pipe of constant cross section, without pump or turbine, and without external heat transfer, carrying a fluid from one end to the other. Let us also assume that changes of fluid pressure or temperature do not affect the fluid density during its passage through the flow system. (This kind of flow is called *incompressible flow* and it is very closely approximated by the flow of most liquids.) By the continuity equation (Eq. 2.2a and 2.2b), the average velocity does not change; therefore the $\phi V^2/2g$ terms are equal on both sides of Equation 2.11 and may be dropped. The elevation does not change from one side of the equation to the other, so the Z terms may be dropped. Without pump or turbine work the $E/\dot{w}$ terms may be dropped. Without external transferred heat the $JQ/\dot{w}$ terms may be dropped. This leaves only the P/ρ terms and the JU terms. Collect the JU terms and lump them into one term called ΔJU; the resulting equation is:

$$\frac{P_1}{\rho_1} - \frac{P_2}{\rho_2} = \Delta JU.$$

Again, as in Equation 2.11, ρ is either ρ_w or ρ_m, depending on the units chosen. The pressure head change is equal to the thermal energy term, ΔJU! In this illustration, we could have included the other Bernoulli or head terms and shown that ΔJU is equal to the change in total head. Appropriately enough, the change is called *head loss*, or H_L. In the general energy equation, where there is external heat transfer, only a portion of ΔJU is owing to head loss. But since we have observed that in incompressible flow the thermal terms usually do not affect the fluid density appreciably, we may drop the thermal terms altogether except for the portion that accounts for the loss of head, that is, H_L. Then we may write a simplified energy equation:

$$\frac{P_1}{\rho}+\frac{\phi_1 V_1^2}{2g}+Z_1+\frac{E_P}{\dot{w}}=\frac{P_2}{\rho}+\frac{\phi_2 V_2^2}{2g}+Z_2+\frac{E_T}{\dot{w}}+H_L, \tag{2.12}$$

where ρ is either ρ_w or ρ_m, depending on the units chosen, as in Equation 2.11. Head loss is not a loss of total energy; it is a loss of useful mechanical energy by conversion of mechanical energy to heat energy. This energy is seldom recoverable, and, because in the study of pressure drop in liquid systems the heat energy is usually of no interest, the head loss term represents the loss of useful energy. (It would be an exceptional case indeed where this lost heat energy could be partially recovered, say, by a low temperature, low pressure organic vapor turbine system, or a heating system.)

When a compressible fluid is flowing these generalizations cannot be made because there are significant conversions of heat energy to mechanical energy. Still, however, there are simplifications that can be made to make the general energy equation appear less formidable. These will be introduced in a later section (Section 2.8). Head loss will be treated in detail in the following chapters.

2.7 THE KINETIC ENERGY CORRECTION FACTOR

In Section 2.3 it was noted that the kinetic energy term requires a correction factor if the velocity profile is not flat and the energy is computed from the average velocity V. The value of the correction factor is important if an accurate energy balance is to be obtained. The expression for the kinetic energy correction factor may be derived in very much the same fashion as the momentum correction factor was. The total kinetic energy flux at a given cross section of the stream tube is:

$$\int u^2 d\dot{m}=\phi V^2 \dot{m}, \tag{2.13}$$

where ϕ is the kinetic energy correction factor, V is the average velocity, and u is the local velocity. For an axisymmetric velocity distribution in a circular duct, the mass flow is given by Equation 2.4:

$$\dot{m}=\int d\dot{m}=2\pi\rho\int urdr. \qquad \text{(2.4, repeated)}$$

The local kinetic energy flux is:

$$u^2 d\dot{m}=u^2(u\rho dA)=u^3\rho 2\pi rdr.$$

The total kinetic energy flux may be found by integrating along the radius:

$$\int u^2 d\dot{m}=2\pi\rho\int u^3 rdr. \tag{2.14}$$

Combining Equations 2.13, 2.4, and 2.14 yields:

$$2\pi\rho\int u^3 rdr=\phi V^2 2\pi\rho\int urdr,\ \phi=\frac{1}{V^2}\frac{\int u^3 rdr}{\int urdr}. \tag{2.15}$$

Robert P. Benedict [1] gives the following equation for velocity profile:

$$\frac{u}{V}=1+3.75\sqrt{\frac{f}{8}}+2.5\sqrt{\frac{f}{8}}\ln\frac{y}{R}. \qquad \text{(2.7, repeated)}$$

Using this equation, by performing the integrations indicated in Equation 2.15, Benedict obtains the following equation for the energy correction factor. It is (with coefficients rounded to four decimal places):

$$\phi=1+2.9297f-1.5537f^{3/2}.$$

A plot of this equation (for turbulent flow) is given in Figure 2.6.

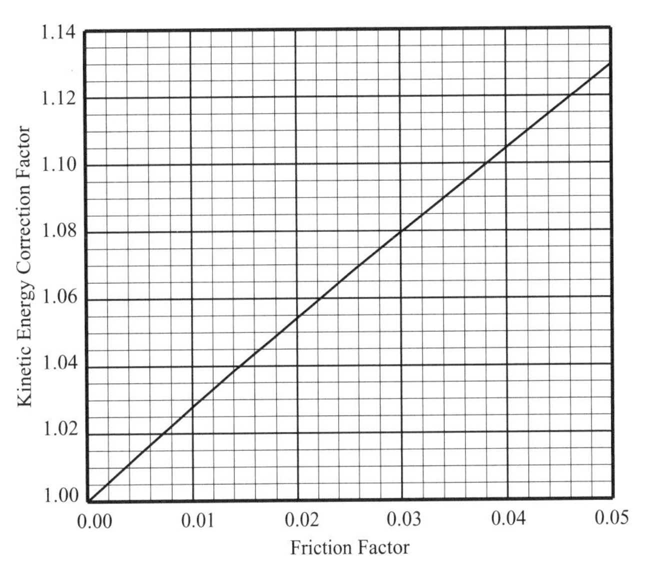

FIGURE 2.6. Kinetic energy correction factor versus friction factor (for turbulent flow).

In laminar flow, where the velocity profile is parabolic and is not a function of friction factor, the evaluation of Equation 2.15 may be accomplished analytically to show that $\phi = 2.000$. When analyzing a laminar flow system, it is important therefore to include ϕ. Turbulent flow, however, is present throughout the operating range of most modern piping systems and consideration of the kinetic energy correction factor is much less important as will be seen in the following section.

2.8 CONVENTIONAL HEAD LOSS

By convention the kinetic energy correction factor ϕ is dropped in engineering computations because its value is close to 1. The head loss term in the incompressible general energy equation is defined by ignoring the ϕ coefficient so that the equation becomes:

$$\frac{P_1}{\rho_1}+\frac{V_1^2}{2g}+Z_1+\frac{E_P}{\dot{w}}=\frac{P_2}{\rho_2}+\frac{V_2^2}{2g}+Z_2+\frac{E_T}{\dot{w}}+(H_L)_C, \quad (2.16)$$

where, as in Equation 2.11, ρ is either ρ_w or ρ_m, depending on the units chosen. Notations $(H_L)_C$ and $(H_L)_E$ will be used momentarily to distinguish the conventional value from the exact value. By solving Equations 2.16 and 2.12 simultaneously, conventional head loss is seen to be:

$$(H_L)_C=(H_L)_E+(\phi_2-1)\frac{V_2^2}{2g}-(\phi_1-1)\frac{V_1^2}{2g}.$$

It is evident that conventional head loss equals exact head loss when there is no change in flow area and thus, inherently, $V_2 = V_1$ and $\phi_2 = \phi_1$. When there is contraction of the flow passage as shown in Figure 2.7a, the contraction causes V_2 to exceed V_1 while flattening of the velocity profile causes ϕ_2 to approach 1 more closely than ϕ_1 does, so that $(\phi_1 - 1)$ exceeds $(\phi_2 - 1)$. Thus for a contraction the two effects tend to cancel, minimizing the difference between conventional and exact head losses. In Figure 2.7b, illustrating flow through an enlargement, again it is seen that the changes in velocity and kinetic energy correction factor are opposite, tending to minimize the difference between conventional and exact head losses. Finally it should be noted that head loss values are founded on or supported by experimental data, the evaluation of which is based upon the omission of the ϕ term in the velocity head. The net result of these effects is to markedly decrease the adverse influence of the uniform velocity assumption on fluid flow computations.

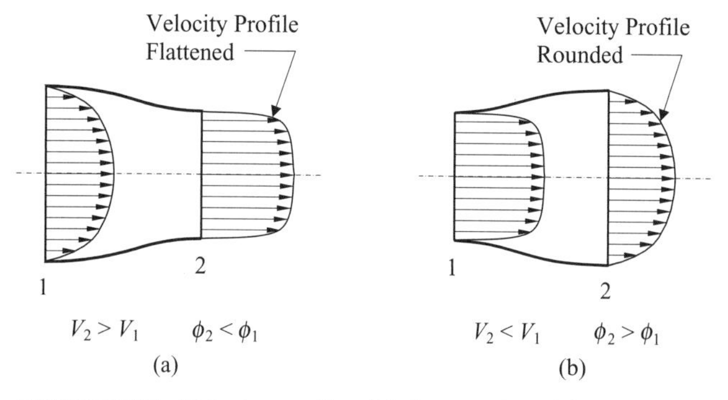

FIGURE 2.7. Velocity profiles. (a) Contraction. (b) Expansion.

2.9 GRADE LINES

It is helpful in visualizing the head loss process and the terms used in describing it if the various terms of the energy equation are plotted on the ordinate of a graph with length of the flow passage plotted on the abscissa. Figure 2.8 shows an example for flow through a pipe with an upward slope and a change in diameter.

The top line, variously called the *Energy Line*, *Energy Grade Line*, or *Total Head Line* (though "Total Useful Head Line" might be more appropriate), represents the sum of the elevation, pressure, and velocity heads. A pitot probe inserted in the flow would cause a column of the flowing fluid to rise in a manometer to that line as shown. If a pump or turbine were placed in the line as shown in Figure 2.5 there would be an appropriate rise or fall of the energy line representing the energy added to or subtracted from the flow.

The line below it represents the *Piezometric Head Line* or *Hydraulic Grade Line*. It is everywhere lower than the energy line by the value $V^2/2g$ or the velocity head, and it is the line to which a static pressure tap (or piezometer) will cause a column of the flowing fluid to rise.

Note in Figure 2.8 that the energy grade line dips at the sudden enlargement of the pipe owing to a loss of mechanical energy cause by turbulence downstream of the enlargement. The energy grade line downstream is positioned closer to the hydraulic grade line, reflecting the lower flow velocity due to the increase in pipe cross section. Note also that the hydraulic grade line rises rather abruptly downstream of the enlargement, indicating that not all of the kinetic energy difference before and after the enlargement is lost, but some is recovered and converted to pressure energy. Finally, note that energy and hydraulic grade lines are parallel as long as the pipe cross section remains constant, and that both lines slope downward to the right (in the direction of

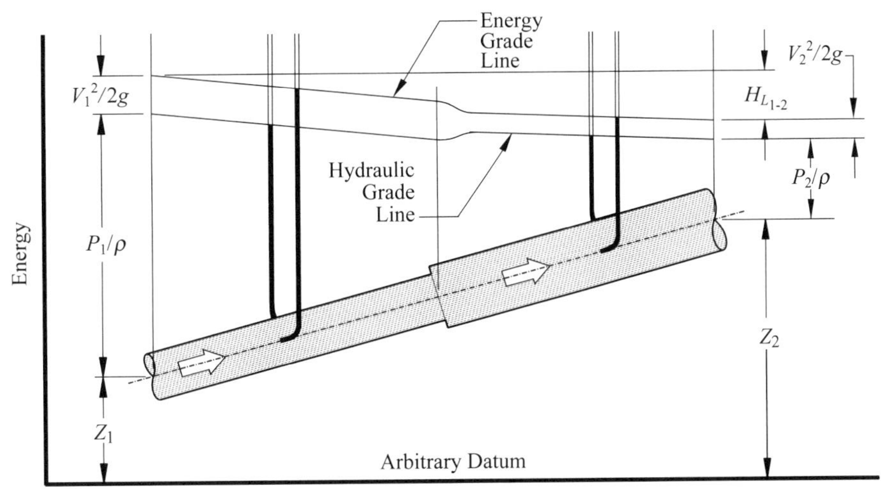

FIGURE 2.8. Grade lines.

flow), more steeply for the smaller, higher velocity pipe, as pipe friction converts mechanical energy to unavailable heat energy.

The following generalizations may be deduced:

1. The energy line for a real fluid will always slope downward in the direction of flow except where mechanical energy is added by a pump.
2. The vertical drop in the energy line represents the loss of total head or mechanical energy.
3. The energy line and the hydraulic grade line are coincident and lie in the free surface of a body of liquid at rest (as, for instance, in a reservoir).

REFERENCES

1. Benedict, R. P., *Fundamentals of Pipe Flow*, John Wiley & Sons, 1980.
2. Rouse, H., ed., *Engineering Hydraulics, Proceedings of the Fourth Hydraulics Conference, Iowa Institute of Hydraulic Research, June 12–15, 1949*, John Wiley & Sons, 1949.
3. Street, R. L., G. Z. Watters, and J. K. Vennard, *Elementary Fluid Mechanics*, 7th ed., John Wiley & Sons, 1996.

FURTHER READING

This list includes books that may be helpful to those who wish to pursue further study.

Sears, F. W., *Principles of Physics I: Mechanics, Heat and Sound*, Addison-Wesley Press, 1947.

Vennard, J. K., *Elementary Fluid Mechanics*, 4th ed., John Wiley & Sons, 1961.

Streeter, V. L. and E. B. Wylie, *Fluid Mechanics*, 7th ed., McGraw-Hill, 1979.

Munson, B. R., D. F. Young, and T. H. Okiishi, *Fundamentals of Fluid Mechanics*, 3rd ed., John Wiley & Sons, 1998.

3

INCOMPRESSIBLE FLOW

This chapter will explore kinds of flow, such as laminar and turbulent, the development of understanding of pressure losses in incompressible flow and the equations to describe them, and sources of pressure loss.

3.1 CONVENTIONAL HEAD LOSS

As was established in the previous chapter, if the fluid pressure and temperature do not appreciably affect the fluid density, the thermal terms of the general energy equation may be dropped, excepting one term called head loss (H_L). The head loss term designates the mechanical energy (embodied in the Bernoulli terms P/ρ, $\phi V^2/2g$, and Z) that is converted to thermal energy due to frictional resistance to flow. The resulting equation very closely describes the flow of most liquids. Further, when we neglect the kinetic energy correction factor, we obtain the conventional general energy equation:

$$\frac{P_1}{(\rho_w)_1}+\frac{V_1^2}{2g}+Z_1+\frac{E_P}{\dot{w}}=\frac{P_2}{(\rho_w)_2}+\frac{V_2^2}{2g}+Z_2+\frac{E_T}{\dot{w}}+H_L.$$

Now, for convenience and simplicity, assume that the flow line is level so that Z_1 equals Z_2, and that there is no pump or turbine. Then the energy equation is simplified to:

$$\frac{P_1}{(\rho_w)_1}+\frac{V_1^2}{2g}=\frac{P_2}{(\rho_w)_2}+\frac{V_2^2}{2g}+H_L$$

or

$$H_L=\frac{P_1}{(\rho_w)_1}-\frac{P_2}{(\rho_w)_2}+\frac{V_1^2-V_2^2}{2g}.$$

For ordinary liquids under ordinary conditions, as indicated above, change of specific weight is so modest for flow-induced temperature and pressure changes that $(\rho_w)_1$ may be equated to $(\rho_w)_2$. Making this simplification,

$$H_L=\frac{P_1-P_2}{\rho_w}+\frac{V_1^2-V_2^2}{2g}. \qquad (3.1)$$

Remember that H_L is a loss in the general energy equation; therefore, it represents a drop in the *Energy Grade Line*. Because there is no change in flow velocity when there is no change in flow area, this equation shows that pressure loss is directly proportional to head loss in a level constant area flow duct.

3.2 SOURCES OF HEAD LOSS

As was seen in Chapter 2, head loss amounts to a conversion of available mechanical energy to unavailable heat energy. Two principal sources of this conversion may be identified: (1) surface friction and (2) induced turbulence due to fittings and other changes in the flow path, such as valves.

Pipe Flow: A Practical and Comprehensive Guide, First Edition. Donald C. Rennels and Hobart M. Hudson.

3.2.1 Surface Friction Loss

Effort is required to cause a fluid to flow through a conduit. Whenever there is relative motion between two bodies in contact there is frictional resistance, and fluid flow in conduits is no exception.

The problem of a rational treatment of surface friction has been under investigation since at least the late 1700s. Some of the early experimenters recognized there are two flow regimes—one in which flow moves on in a tranquil, quiescent fashion, and one in which the flow is chaotic. The former has been named *streamline* or *laminar* flow, because the various axial layers of the fluid remain intact as the flow proceeds. The latter has been named *turbulent* flow, because layers in the flow conduit do not remain intact but are constantly being mixed due to turbulence, that is, chaotic motions in the flow.

Gradual progress in understanding surface friction started with the recognition that friction loss—at least for the turbulent regime—is approximately proportional to the square of the flow velocity. But the first rational formulation of pressure loss in flow of fluids in conduits was found for the laminar regime. A similar formulation for the turbulent region was not far behind.

3.2.1.1 Laminar Flow While most of the early researchers experimented in the turbulent regime, two very successful ones experimented in the laminar regime. Gotthilf Hagen (1797–1884), a German hydraulic engineer, published in 1839 a paper quantifying pressure loss in laminar flow. Independently, Jean Poiseuille (1797–1869), a French physician hoping to quantify flow losses in blood vessels, working at the same time, discovered the same governing relations. He published his work in 1841. The law governing pressure drop in laminar flow that they found is now called the Hagen–Poiseuille law in their honor [1,2]. It is:

$$H_L = \frac{32\mu LV}{\rho_w D^2}.$$

Although Hagen observed a transition in which his tranquil flow became chaotic or turbulent, he did not succeed in his attempts to understand why it happened.

3.2.1.2 Turbulent Flow One of the earlier fluid flow experimenters was Antoine de Chézy (1718–1798), a French hydraulic engineer. Chézy, in his analyses for the Yvette River aqueduct project in France in about 1770, made use of the fact that head loss—for aqueducts—is approximately proportional to the square of the flow velocity [3]. His formula was:

$$V = C\sqrt{RS},$$

where

R = hydraulic radius of conduit or channel,

S = slope of conduit = H_L/L,

L = conduit length, and

$C = V/(RS)^{1/2}$ from observation of other channels.

While Chézy's formula was developed for open channel flow, it is noteworthy that it can be applied to pipe flow as well. His equation can be rearranged and expressed as:

$$H_L = \frac{1}{C^2}\frac{L}{R}V^2.$$

Although Chézy's formula is dimensionally homogeneous, and Chézy recognized that C changed from channel to channel, it does not appear that he knew how it changed. It is ironic that Chézy's work on this formula was not published until 1897 in the United States by Clemens Herschel [3].

Gaspard Riche de Prony (1755–1839), another French engineer (famous for the Prony brake), published a formula in 1804 which may be expressed as [3]:

$$H_L = \left(\frac{a}{V} + b\right)\frac{L}{D}\frac{V^2}{2g},$$

where a and b are dimensionless coefficients. However, the equation is not dimensionally homogeneous. Prony believed that the formula for pressure loss in pipes was a power series in V, and his formula above was a first approximation using the second and third terms.

Henry Darcy (1803–1858), yet another French engineer, in 1857 proposed the following formula for smooth pipes [4]:

$$H_L = \left(a\frac{L}{D} + b\frac{L}{D^3}\right)V + \left(b\frac{L}{D} + c\frac{L}{D^2}\right)V^2,$$

in which a, b, and c are dimensionless coefficients. The first term was dropped for rough pipes and the coefficients changed somewhat. However, Darcy's formula was not dimensionally homogeneous either.

Julius Weisbach (1806–1871), a German engineer, published his formula in 1845 [5]:

$$H_L = f\frac{L}{D}\frac{V^2}{2g}. \quad (3.2)$$

Weisbach was the first to write a dimensionally homogeneous formula for surface friction pressure drop incorporating a dimensionless "friction factor" and the

2g divisor. His formula was so successful that it is still the formula in modern use. Darcy is usually given credit or joint credit for the formula, and the friction factor f in Weisbach's formula is usually called the *Darcy friction factor*.* In fact, though, Darcy's contribution was not the formula but the recognition that fluid resistance depends on the type and condition of the boundary material. However, neither Weisbach nor any of the other pioneers mentioned had any rational basis for the proportionality factors, or "friction factors," in their equations. The group fL/D is called the resistance (or loss) coefficient and given the symbol K. More will be said about f and K in Chapters 4 and 8.

3.2.1.3 Reynolds Number The next breakthrough came when Osborne Reynolds (1842–1912), a British engineer, showed in 1883 that the transition between laminar, or streamline, flow and turbulent flow occurs at a fairly definite value of a dimensionless number he had developed. The number named after him is the Reynolds number [6]:

$$N_{Re} = \frac{VD\rho_w}{g\mu} = \frac{\dot{w}\rho_w}{g\mu A} \quad \text{(English)} \qquad \text{(1.2a, repeated)}$$

$$N_{Re} = \frac{VD\rho_m}{\mu} = \frac{\dot{m}\rho_m}{\mu A} \quad \text{(SI).} \qquad \text{(1.2b, repeated)}$$

The Reynolds number is the ratio of momentum forces to viscous forces in the flow. It is now known that when the viscous forces predominate, the flow is laminar; when the momentum forces predominate, the viscous flow breaks down and becomes turbulent.

3.2.1.4 Friction Factors

The Laminar Flow Friction Factor Using Reynolds' new dimensionless number, a friction factor for use in Weisbach's friction head equation may be found for the laminar flow regime. Grouping the variables of the Reynolds number in the Hagen–Poiseuille law and grouping the remaining variables as in Weisbach's equation yields the following formula for *laminar* friction factor:

$$f = \frac{64}{N_{Re}}. \qquad (3.3)$$

The laminar friction factor is a function of Reynolds number alone, and is independent of any other factor.

* This nomenclature is necessary to distinguish it from other friction factors in use, especially the Fanning friction factor, which is one-quarter the Darcy friction factor.

The Turbulent Flow Friction Factor It was not until the early 1930s that the friction factor for turbulent flow was reduced to a rational basis. Ludwig Prandtl (1875–1953), a researcher at the University of Göttingen in Germany, through his work on velocity distribution, showed that the formula for friction factor for turbulent flow in smooth pipes should take the form [7]:

$$\frac{1}{\sqrt{f}} = A \log N_{Re}\sqrt{f} - B \quad \text{or} \quad \frac{1}{\sqrt{f}} = A \log_{10}(CN_{Re})\sqrt{f},$$

where A, B, and C are constants. At that time, Johann Nikuradse (1894–1979), an engineer on Prandtl's laboratory staff at Göttingen, was experimenting with flow in artificially roughened pipes. His research provided data to define the constants in Prandtl's equation [8,9]:

$$\frac{1}{\sqrt{f}} = 2\log_{10}\frac{N_{Re}\sqrt{f}}{2.51} \quad \text{or} \quad \frac{1}{\sqrt{f}} = -2\log_{10}\frac{2.51}{N_{Re}\sqrt{f}}. \qquad (3.4)$$

At the same time, Theodor von Kármán (1881–1963), a Hungarian engineer working as a professor at the University of Göttingen, also used Nikuradse's data to determine that for rough pipes flowing with complete turbulence, the friction factor is independent of Reynolds number and is equal to [10]:

$$\frac{1}{\sqrt{f}} = 2\log_{10} 3.7\frac{D}{\varepsilon} \quad \text{or} \quad \frac{1}{\sqrt{f}} = -2\log_{10}\frac{\varepsilon}{3.7D}. \qquad (3.5)$$

Nikuradse's work, published in 1933, showed that, for pipes roughened on the inside circumference with uniform sand grains, the friction factor followed the Hagen–Poiseuille law up to the critical Reynolds number, then rose up to the smooth pipe friction factor formulated by Prandtl, followed it down for a range of Reynolds numbers, then rose up again to meet the friction factor for turbulent flow formulated by von Kármán (Fig. 3.1).

His results, while duplicating results with *smooth* pipe and with *rough* pipe at high Reynolds numbers, unfortunately did not duplicate results with commercial pipe at intermediate Reynolds numbers. Friction factor results with commercial pipe, for increasing Reynolds numbers, yielded friction factors that followed the Hagen–Poiseuille law until rising abruptly at the critical Reynolds number ($N_{Re} \approx 2100$), then *declined* gradually in the transition zone to fair into the complete turbulence friction factor. Nikuradse's friction factor, on the other hand, after the abrupt rise at the critical Reynolds number, rose only to the smooth pipe line and *followed it* before *rising* to fair into the complete turbulence factor (compare Figs. 3.1 and 3.2).

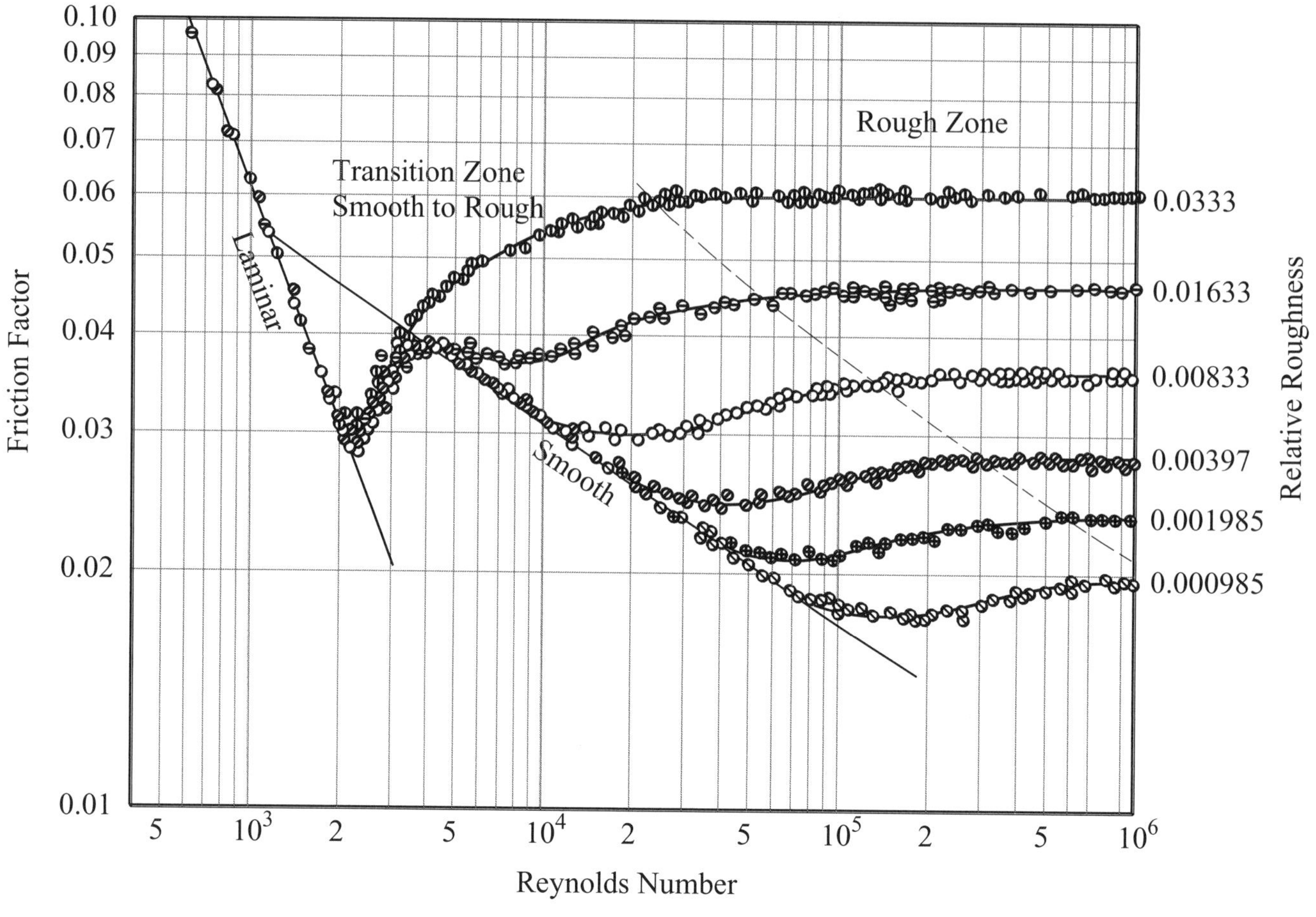

FIGURE 3.1. Nikuradse's uniform sand grain results [8].

Two British scientists, Cyril F. Colebrook (1910–1997) and Cedric M. White (1898–1993), showed experimentally in 1937 [11] that Nikuradse's results were due to the uniformity of roughness in his pipes. Artificially roughened pipes with *non*uniform sand grains duplicated very well the behavior of commercial pipes. Prandtl's boundary layer theory, which held that there is always a laminar flow boundary layer that thins as Reynolds number increases, explained the reason for the difference. In the critical zone, general laminar flow breaks up into a turbulent core and a laminar boundary layer next to the pipe wall. At this point in Nikuradse's pipes the uniform sand grains remained submerged in the boundary layer and were hidden to the turbulent core flow, and the friction factor was the same as for smooth pipe. With increasing Reynolds number, the boundary layer thinned until the sand grains began to emerge, and the friction factor transitioned from the smooth pipe value to the rough pipe value. With commercial pipe, on the other hand, the largest of the various sizes of protuberances were never submerged in the laminar boundary layer after the critical Reynolds number was passed and so the flow behaved somewhat like that for rough pipe from the beginning. For this reason the friction factor remained higher than for smooth pipe, but because many of the smaller protuberances remained submerged, it declined on increasing Reynolds number somewhat like that for smooth pipe until the rough pipe value was reached.

Upon a serendipitous suggestion by White, Colebrook [12] proposed an empirical combination of the Prandtl and von Kármán formulas (Eqs. 3.4 and 3.5) obtained by *inverting and adding* the arguments of the logarithms. The resulting expression modeled very accurately commercial pipe behavior in the turbulent regime. Their formula, published in 1939, is:

$$\frac{1}{\sqrt{f}} = -2\log_{10}\left(\frac{\varepsilon}{3.7D} + \frac{2.51}{N_{\text{Re}}\sqrt{f}}\right). \tag{3.6}$$

(Note that Prandtl's smooth pipe formula is slightly out of context when used in this formula; nevertheless, we

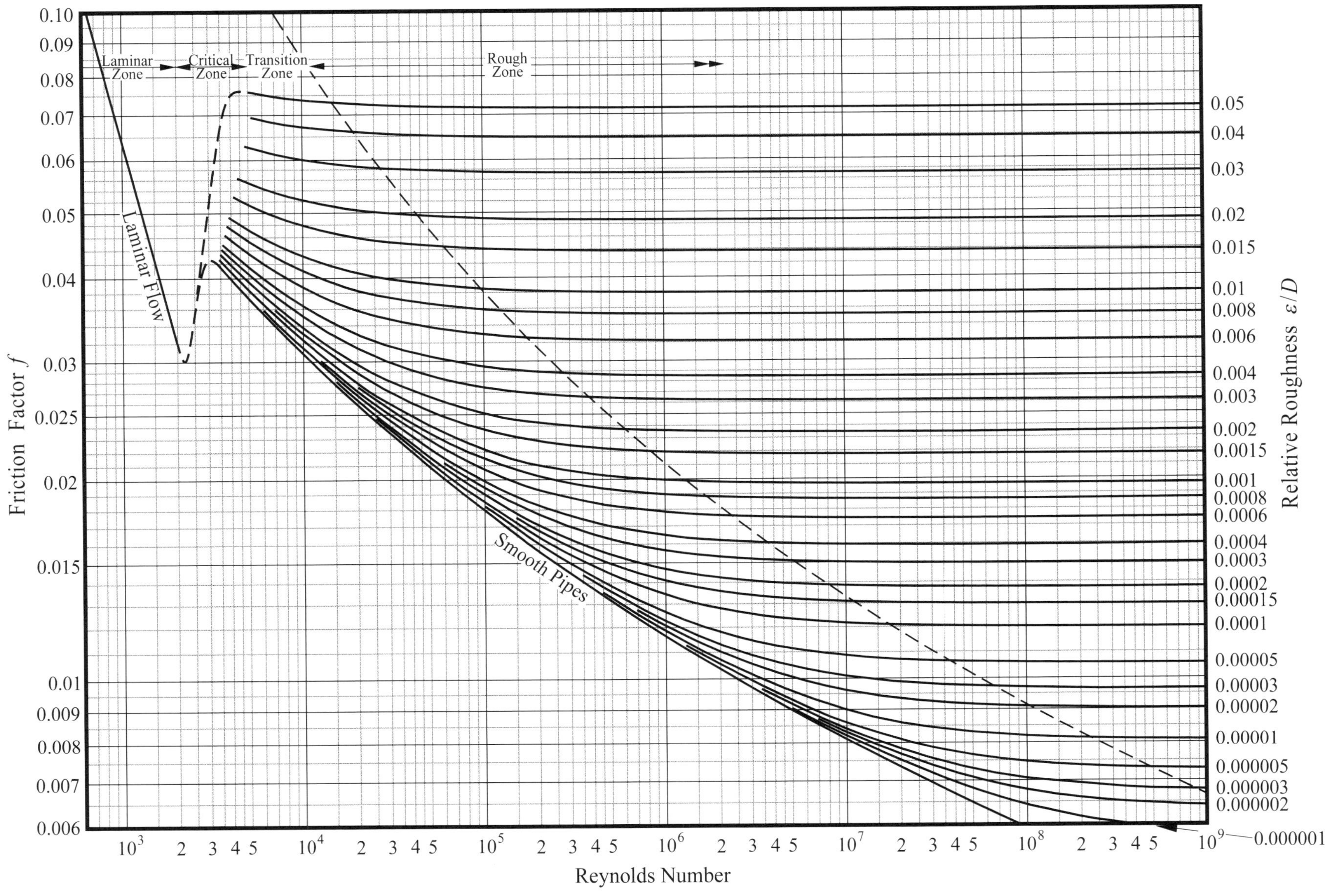

FIGURE 3.2. Friction factor versus Reynolds number and relative roughness for commercial pipe (after Moody [13]).

will call it the "Prandtl term" because of its origin.) Soon afterwards (1944), Lewis F. Moody [13] published a design chart based on the Colebrook–White formula. The formula and chart (Fig. 3.2) have been so successful that they are still in use today. The chart is popularly known as the "Moody Chart." See also the diagram in Part II, Chapter 8.

3.2.2 Induced Turbulence

The second source of pressure losses in pipe flow, in contrast to pipe friction, is *induced turbulence*. These losses are often referred to as *local losses* or sometimes as *minor losses*, although they are usually far from minor. When turbulence in excess of that normally present in the flow is caused by the flow passage shape, the energy resident in the turbulence is not usually recovered as mechanical energy and is consequently converted to heat. As shown in Chapter 2, mechanical energy converted to heat is described in the energy equation by the head loss term, H_L.

A particular solution to the local loss problem antedates the Darcy–Weisbach equation; this is the Borda sudden expansion loss. Jean-Charles de Borda, by reasoning, predicted in 1766 the head loss due to a sudden expansion. In modern terms his prediction may be written as:

$$H_L = \frac{V_1^2}{2g}\left(1 - \frac{A_1}{A_2}\right)^2.$$

This equation is often called the Borda–Carnot equation. Experiments proved Borda to be correct for the turbulent flow case.

Note that the head loss in Borda's equation is proportional to $V_1^2/2g$ times a *geometry-dependent constant*. This arrangement has been found to be generally true in subsequent pressure loss work. If we denote the geometry-dependent constant as K, the general case of induced turbulence head loss may be written as:

$$H_L = K\frac{V^2}{2g}. \tag{3.7}$$

K is known as the *resistance coefficient* or *loss coefficient*. As shown in Chapter 2, $V^2/2g$ is the *velocity head*, so that K is the head loss measured in velocity heads.

Making use of the fact that $V^2/2g$ is the velocity head, the Weisbach equation (Eq. 3.2) may also be written using the loss coefficient K. Factoring the $V^2/2g$ term from the equation shows that Equation 3.7 may be used to describe pipe friction when:

$$K = f\frac{L}{D}. \tag{1.3, repeated}$$

Equation 3.1 may be rearranged as:

$$\frac{P_1 - P_2}{\rho_w} = H_L - \frac{V_1^2 - V_2^2}{2g}.$$

Using the identity $A_1V_1 = A_2V_2$ we may write:

$$\frac{P_1 - P_2}{\rho_w} = H_L - \frac{V_1^2}{2g}\left[1 - \left(\frac{A_1}{A_2}\right)^2\right].$$

This is valid when there is a change in the flow area. Substituting Equation 3.7 for H_L and rearranging gives:

$$P_1 - P_2 = \frac{V_1^2\rho_w}{2g}\left[K_1 - 1 + \left(\frac{A_1}{A_2}\right)^2\right]. \tag{3.8}$$

The loss coefficient for induced turbulence—"local losses"—is ordinarily based on the inlet size, which controls the inlet velocity. With that convention, the loss coefficient is subscripted with a 1, denoting the inlet, and the velocity in the equation must also be the inlet velocity, as shown. If it is desired to base the loss coefficient on the outlet velocity, then the last three steps yield:

$$P_1 - P_2 = \frac{V_2^2\rho_w}{2g}\left[K_2 - \left(\frac{A_2}{A_1}\right)^2 + 1\right].$$

Using the identity $\dot{w} = AV\rho_w$, we can write:

$$V^2 = \frac{\dot{w}^2}{A^2\rho_w^2}.$$

Substituting this into Equation 3.8 yields:

$$P_1 - P_2 = \frac{\dot{w}^2}{2gA_1^2\rho_w}\left[K_1 - 1 + \left(\frac{A_1}{A_2}\right)^2\right].$$

If the loss coefficient is based on the outlet size, the pressure drop equation becomes:

$$P_1 - P_2 = \frac{\dot{w}^2}{2gA_2^2\rho_w}\left[K_2 + 1 - \left(\frac{A_2}{A_1}\right)^2\right].$$

If the inlet and outlet areas are the same ($A_1 = A_2$), the formula reduces to:

$$P_1 - P_2 = K\frac{\dot{w}^2}{2gA^2\rho_w}.$$

Unfortunately, this form is often used in practice even if the inlet and outlet areas are different.

While the head loss for induced turbulence is slightly dependent on the surface roughness, usually—unlike pipe friction—the feature geometry is by far the most important, and, after that, the Reynolds number. Part II gives K for a number of important pipe fittings and arrangements.

3.2.3 Summing Loss Coefficients

Pressure losses in incompressible flow are additive. If a piping stretch has various contributors to the overall pressure loss with different areas, it will be convenient to have a formula for the overall pressure loss in terms of the characteristics of the individual pressure loss contributors. We may write:

$$\Delta P_{OA} = \Delta P_a + \Delta P_b + \Delta P_c + \ldots + \Delta P_n.$$

Substituting for the individual ΔPs results in:

$$\Delta P_{OA} = \left(K\frac{\dot{w}^2}{2gA^2\rho_w} \right)_a + \left(K\frac{\dot{w}^2}{2gA^2\rho_w} \right)_b + \left(K\frac{\dot{w}^2}{2gA^2\rho_w} \right)_c + \ldots + \left(K\frac{\dot{w}^2}{2gA^2\rho_w} \right)_n.$$

The Ks themselves are based on different areas, so we cannot add them. However, if we factor out $\dot{w}^2/2gA\rho_w$ we get:

$$\Delta P_{OA} = \frac{\dot{w}^2}{2gA^2 w}\left(K_a\frac{A^2}{A_a^2} + K_b\frac{A^2}{A_b^2} + K_c\frac{A^2}{A_c^2} + \ldots + K_n\frac{A^2}{A_n^2} \right).$$

Each K, as modified by the ratio of the square of the ratio of the "standardized" area (A) to the actual area (A_a, A_b, A_c, or A_n), *can* be added. We say that the resistance coefficients are *referred* to the standardized area (that is, A). The general form is:

$$K = K_a\frac{A^2}{A_a^2} \quad \text{or} \quad K = K_a\frac{d^4}{d_a^4}, \tag{3.9}$$

where
K = standardized K (usable in the ΔP formula with A_1),
K_a = actual K (usable with A_a in the ΔP formula),
A = standardized area,
A_a = actual area,
d = standardized diameter, and
d_a = actual diameter.

Usually the "standardized" area is the area of the most important feature, usually pipe, in the stretch being considered.*

REFERENCES

1. Hagen, G. H. L., Ueber die Bewegung des Wassers in agen cylindrischen Rohren, *Poggendorfs Ann. Phys. Chem.*, 46, 1839, 423. (Translated as "On the flow of water in narrow cylindrical pipes.")
2. Poiseille, J. L. M., Recherches experimentales sur le mouvement des liquids dans les tubes tres petits diameters, *C. R. Acad. Sci.*, 1841. (Translated as "Experimental research on the flow of liquids through pipes of very small diameters.")
3. Rouse, H. and S. Ince, *History of Hydraulics, Iowa Institute of Hydraulic Research*, State University of Iowa, 1957.
4. Darcy, H. P. G., Recherches experimentales relatives aux mouvement del'eau dans les tuyaux, *Mem. Acad. Inst. Imp. Fr.*, 15, 1858, 141. (Translated as "Experimental research on the flow of water in pipes.")
5. Weisbach, J., Lehbuch der Ingenieur—und Maschinen-Mechanik, Braunschwieg, 1845.
6. Reynolds, O., An experimental investigation of the circumstances which determine whether the motion of water will be direct of sinuous, and the laws of resistance in parallel channels, *Philos. Trans. R. Soc. Lond.*, 1883.
7. Prandl, L., Ergeb. Aerodyn. Versuchanst. Gottingen. Series 3, 1927. (Translated as "Reports of the Aerodynamic Research Institute at Gottingen.").
8. Nikuradse, J., Laws of flow in rough pipes (in German), *Forsch.-Arb Ing.-Wesen*, 361, 1933.
9. Prandl, L., Neuere ergebnisse der turbulenzforschung, *Z. VDI*, 77, 1933, 105. (Translated as "Recent results of turbulence research.").
10. Von Kármán, T., Mechanische Ähnlichkeit und Turbulenz, *Nachrichten von der Gesellschaft der Wissenschofen zu Gőttingen*, 1930, Fachgruppe 1, Mathematik, no. 5, pp. 58–76. ("Mechanical similitude and turbulence," Tech. Mem. N.A.C.A., no. 611, 1931.).
11. Colebrook, C. F. and C. M. White, Experiments with fluid friction in roughened pipes, *Proc. R. Soc. Lond.*, 161, 1937, 367–381.
12. Colebrook, C. F., Turbulent flow in pipes, with particular reference to the transition region between the smooth and rough pipe laws, *J. Inst. Civ. Eng.*, 11, 1938–1939, 133–156.
13. Moody, L. F., Friction factors for pipe flow, *Trans. Am. Soc. Mech. Eng.*, 66, 1944, 671–684.

* The summing of loss coefficients is restated in Section 5.1 as series flow.

FURTHER READING

This list includes books and papers that may be helpful to those who wish to pursue further study.

Vennard, J. K., *Elementary Fluid Mechanics*, 4th ed., John Wiley & Sons, 1961.

Booth, R. and N. Epstein, Kinetic energy and momentum factors for rough pipes, *Can. J. Chem. Eng.*, 47, 1969, 515–517.

Streeter, V. L. and E. B. Wylie, *Fluid Mechanics*, 7th ed., McGraw-Hill, 1979.

Benedict, R. P., *Fundamentals of Pipe Flow*, John Wiley & Sons, 1980.

Guislain, S. J., How to make sense of friction factors in fluid flow through pipe, *Plant Engineering*, June 12, 1980, pp. 134–140.

Lamont, P. A., Pipe flow formulas compared with the theory of roughness, *Am. Water Works Assoc. J.*, 73(5), 1981.

Olujić, Ž., Compute friction factors fast for flow in pipes, *Chemical Engineering*, December 14, 1981, pp. 91–93.

Roberson, J. A. and C. T. Crowe, *Engineering Fluid Mechanics*, 3rd ed., Houghton Mifflin Company, 1985.

Munson, B. R., D. F. Young, and T. H. Okiishi, *Fundamentals of Fluid Mechanics*, 3rd ed., John Wiley & Sons, 1998.

4

COMPRESSIBLE FLOW

This chapter deals with finding the pressure drop in ducts flowing a compressible fluid. Six methods of finding the pressure drop are offered, ranging from approximate methods of varying accuracy to analytical methods with absolute accuracy within the assumptions made. Units used are mainly English gravitational units, but International System (SI) units can be substituted if dimensional homogeneity is maintained. (English gravitational units use lb_f, ft, and s as basic. This system is very nearly the same as the United States customary system (USCS) units, which use slug, ft, and s as basic. In either the USCS system or the English gravitational system, $lb_f = \text{slugs} \times g$, where g is the acceleration of gravity in ft/s^2.)

4.1 PROBLEM SOLUTION METHODS

As pointed out in Chapter 2, in incompressible flow without heat exchanger input or output the thermal terms may be equated to head loss. In compressible flow, however, such is not the case because there are significant conversions of heat energy to mechanical energy. Some methods for finding the pressure loss for a compressible fluid flowing in a duct are given below. Horizontal ducts are considered because the density of the flowing compressible fluid is usually low enough to neglect its effect on static pressure with modest changes in elevation.

It should be noted that each of the methods outlined below require the flow path to have a *constant cross-sectional area*. Procedures are given in Section 4.3.1.5 or at the end of Section 4.3.2 on how to handle changes in cross-sectional area.

Compressible flow may be treated in six principal fashions:

1. If the pressure drop is small compared to the system pressure, variation in flowing fluid density with changing system pressure may be ignored and the pressure drop found by incompressible flow formulas. This technique works well for pressure drops below about 10% of the inlet pressure, and it works for either isothermal flow or adiabatic flow. The formula is given in Section 4.2.1 below.
2. For pressure drops up to about 40% of the inlet pressure, and for resistance coefficient $K = 10$ or greater (or for smaller drops if K is smaller), incompressible flow formulas work fairly well if the fluid properties are determined at the average of inlet and outlet conditions. This also works with isothermal or adiabatic flow. The formula for this method is given early in Section 4.2.2 below, with an error chart for the adiabatic case.
3. For pressure drops up to about 40% of the inlet pressure, and for resistance coefficient $K = 6$ or greater, incompressible formulas work fairly well if the inlet and outlet pressures and their average are used in determining the pressure drop. This method is similar to the average properties method described above but takes more into account to

Pipe Flow: A Practical and Comprehensive Guide, First Edition. Donald C. Rennels and Hobart M. Hudson.

estimate the pressure drop. The formula for this method is given in the later part of Section 4.2.2 below, with an error chart for the adiabatic case.

4. Incompressible flow formulas may be used with "expansion factors," correction factors to account for compressible flow behavior. The correction factors are generally presented in the form of charts giving the factor ("Y") as a function of the relative pressure drop ($\Delta P/P_{\text{inlet}}$). While convenient, this method suffers from the disadvantage of requiring a different set of expansion factors for each value of the isentropic exponent γ. This is not a serious disadvantage for low pressure flow for many gases, but at higher pressures and/or lower temperatures the isentropic exponent varies considerably from its usual, ambient condition value. This technique is valid only for adiabatic flow in constant-area ducts. Formulas and charts are given in Section 4.2.3 below.
5. The ideal equation for compressible flow with friction may be used directly, using Mach number as a parameter. Indeed, this is the method used to determine the expansion factors used in method (4) above. The disadvantage of this method is that it usually requires a dedicated computer program to use, but for those who do not consider this to be a disadvantage they are presented. (The equations are considered to be ideal because the velocity profile is assumed to be flat and therefore θ and ϕ are unity, and the fluid's properties are assumed to be constant. The viscosity is also assumed to be constant so that friction factor does not change.) The equation for isothermal flow is different from the one for adiabatic flow. These formulas are shown in Sections 4.3.1 and 4.4 below.
6. The ideal equation for compressible flow with friction, using static pressures and temperatures at the inlet and outlet as parameters, is presented. This method is mathematically similar to the method for adiabatic flow in (5) above but the Mach number is not used. This method is given in Section 4.3.2 below.

4.2 APPROXIMATE COMPRESSIBLE FLOW USING INCOMPRESSIBLE FLOW EQUATIONS

This section demonstrates the use and accuracy of three approximate methods.

4.2.1 Using Inlet or Outlet Properties

This method works fairly well for pressure drops that are below 10% of the inlet pressure.

Neglecting elevation head Z, the energy equation for *incompressible* flow in a duct is:

$$\frac{P_1 - P_2}{\rho_w} = H_L - \frac{V_1^2 - V_2^2}{2g}, \qquad \text{(3.1, rearranged)}$$

where

P_1 = inlet static pressure, lb/ft^2,
P_2 = outlet static pressure, lb/ft^2,
ρ_w = flowing fluid specific weight, lb/ft^3,
H_L = loss of head, ft-lb/lb,
V_1 = inlet velocity, ft/s,
V_2 = outlet velocity, ft/s, and
g = acceleration of gravity, 32.1740 ft/s^2.

By multiplying both sides of the equation by ρ_w we may write:

$$P_1 - P_2 = \rho_w\left(H_L - \frac{V_1^2 - V_2^2}{2g}\right).$$

Noting that $\rho_w = 1/v$ we may change this to:

$$P_1 - P_2 = \frac{1}{v}\left(H_L - \frac{V_1^2 - V_2^2}{2g}\right). \qquad (4.1)$$

When we substitute the formula for H_L, which is

$$H_L = f\frac{L}{D}\frac{V^2}{2g} = K\frac{V}{2g}, \qquad \text{(3.2, repeated)}$$

into Equation 4.1 we obtain:

$$P_1 - P_2 = \frac{1}{v}\left(K\frac{V^2}{2g} - \frac{V_1^2 - V_2^2}{2g}\right). \qquad (4.2)$$

If we refer all the resistance coefficients K in the stretch of duct we are considering to the area at point 1 and sum the terms, we make the head loss term specific, and making this distinction changes Equation 4.2 to:

$$P_1 - P_2 = \frac{1}{v}\left(K_1\frac{V_1^2}{2g} - \frac{V_1^2}{2g} + \frac{V_2^2}{2g}\right).$$

Factor out $V_1^2/2g$ and we obtain:

$$P_1 - P_2 = \frac{V_1^2}{2gv}\left(K_1 - 1 + \frac{V_2^2}{V_1^2}\right). \qquad (4.3)$$

Because $AV\rho_w = \dot{w}$, which for uniform flow is constant throughout the duct, we may write:

$$V^2 = \frac{\dot{w}^2}{A^2\rho_w^2} = \frac{\dot{w}^2 v^2}{A^2}. \qquad (4.4)$$

If we substitute Equation 4.4 for V^2 in Equation 4.3, it becomes:

$$P_1 - P_2 = \frac{\dot{w}^2 v}{A_1^2 2g}\left(K_1 - 1 + \frac{A_1^2}{A_2^2}\right). \qquad (4.5)$$

Because we have assumed that the specific volume changes negligibly, and noting that A_1 must equal A_2, then the last two terms in the parenthetical expression drop out and the equation becomes:

$$P_1 - P_2 = K\frac{\dot{w}^2 v}{A^2 2g}. \qquad (4.6)$$

This equation is valid for incompressible flow. If it is used for compressible flow we should write it as:

$$P_1 - P_2 \approx K\frac{\dot{w}^2 v}{A^2 2g}. \qquad (4.7)$$

4.2.2 Using Average of Inlet and Outlet Properties

There are two methods presented here that use the average of the inlet and outlet properties of the flowing fluid. Specific volume is the best property to average. The first method is called the simple method because it assumes that the fluid specific volume is constant throughout the flow path. The second method is called the comprehensive method, and it is similar to the simple method except that it accounts for the effect of specific volume on velocity head.

In using the formulas for these methods, it must be recognized that they work satisfactorily only if the duct area is constant, or if the change in area is gradual, as in a fabricated reducer or expander. However, fabricated reducers and expanders are better described in Chapters 10 and 11. Using the formulas there for K and using the technique shown in Section 4.3.1.5 or at the end of Section 4.3.2 is encouraged.

4.2.2.1 Simple Average Properties The general energy equation for this problem was given in Section 4.2.1:

$$\frac{P_1 - P_2}{\rho_w} = H_L - \frac{V_1^2 - V_2^2}{2g}, \qquad (3.1, \text{rearranged})$$

and transformed to:

$$P_1 - P_2 = \frac{\dot{w}^2 v}{A_1^2 2g}\left(K_1 - 1 + \frac{A_1^2}{A_2^2}\right). \qquad (4.5, \text{repeated})$$

This is valid for incompressible flow. If we use it for compressible flow, using the *average* specific volume instead of the specific volume at the inlet or outlet, using $A = A_1 = A_2$ and $K = K_1 = K_2$, the equation should be written as:

$$P_1 - P_2 \approx K\frac{\dot{w}^2 \bar{v}}{A^2 2g}, \qquad (4.8)$$

where

$$\bar{v} = \frac{v_1 + v_2}{2}.$$

Equation 4.8 is the simple formula for pressure drop using the average specific volume. In order to implement the equation it is necessary to know the temperatures at the inlet and outlet as well as the respective pressures in order to determine the inlet and outlet fluid specific volumes. For Equation 4.8 (constant-area duct), the temperature at one end may be estimated from the temperature at the other using the following approximate equation:

$$\frac{T_2}{T_1} \approx F_{PR}\left(\frac{P_2}{P_1}\right)^{(\gamma-1)/\gamma}, \qquad (4.9)$$

where

T_1 = duct inlet absolute temperature,
T_2 = duct outlet absolute temperature,
P_1 = duct inlet absolute pressure,
P_2 = duct outlet absolute pressure, and
$F_{PR} \approx a + b(P_2/P_1) + c(P_2/P_1)^2$.

In this equation, F_{PR} is a parabolic fit of the ratio of T_2/T_1 in adiabatic flow with friction as determined by the equations in Section 4.3.1 to the temperature ratio T_2/T_1 in a nonflow adiabatic process. The product of F_{PR} and the nonflow adiabatic temperature ratio $(T_2/T_1)^{(\gamma-1)/\gamma}$ yields an approximate value of the adiabatic flow temperature ratio. It is then possible to determine the values of inlet and outlet specific volumes from gas or steam tables, or by the ideal gas equation $v = RT/P$.

The constants in Equation 4.9 and its limits of applicability are given in Table 4.1, in which ΔP is $P_1 - P_2$.

The error between the results using Equation 4.8 and using the ideal theoretical formula given in Section

TABLE 4.1. Constants and Limits for F_{Pr}

γ	a	b	c	Limits
1.4	1.650	−1.090	0.4412	For $20 \le K \le 100$: $\Delta P/P_1 \le 0.4$
1.3	1.496	−0.8209	0.3265	For $3 \le K < 20$: $\Delta P/P_1 \le 0.3$

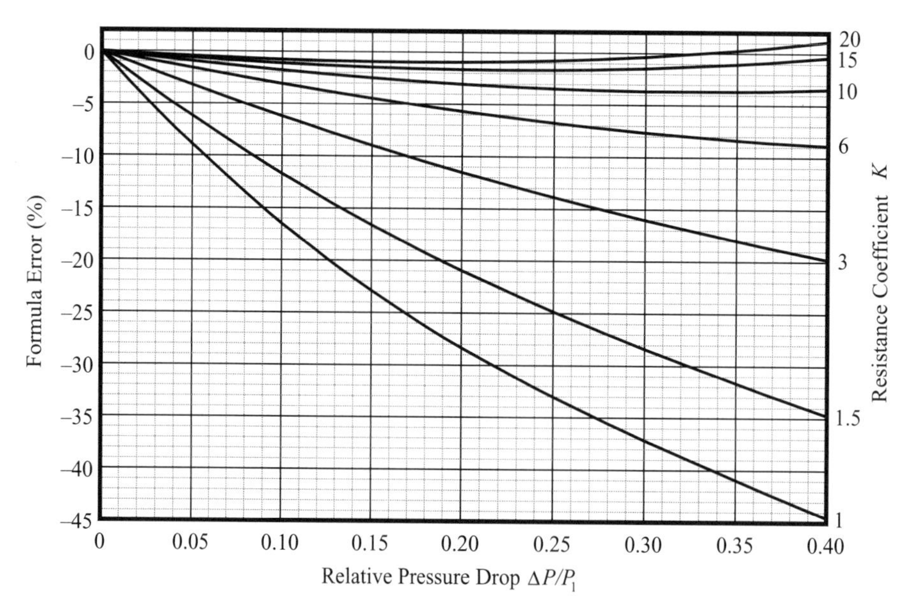

FIGURE 4.1. Error in simple average properties formula.

4.3.1, as a function of $\Delta P/P_1$ for several values of K, is shown in Figure 4.1. The chart assumes no error in the correction factor F_{PR}. Note that the method gives good results for higher values of K, and gives poor results at lower values of K, especially with increasing pressure ratio $\Delta P/P_1$.

4.2.2.2 Comprehensive Average Properties This method is similar to the simple method, but it does not assume constant specific volume throughout the flow path. The unknown fluid temperature, and hence the specific volume $\bar{v}$, may be estimated using the same relations given in the preceding section (Eq. 4.9).

Defining $\bar{v}$ as $\bar{v} = (v_1 + v_2)/2$, we can then modify the general case for the pressure drop (Eq. 4.2) to:

$$P_1 - P_2 \approx \frac{1}{\bar{v}}\left(K\frac{V^2}{2g} - \frac{V_1^2}{2g} + \frac{V_2^2}{2g}\right), \qquad \text{(4.2, modified)}$$

and by the application of Equation 4.4, $V^2 = \dot{w}v/A^2$, using the v applicable to each station, the equation becomes:

$$P_1 - P_2 \approx \frac{1}{\bar{v}}\left(K\frac{\dot{w}^2\bar{v}^2}{2gA^2} - \frac{\dot{w}^2v_1^2}{2gA^2} + \frac{\dot{w}^2v_2^2}{2gA^2}\right).$$

Factoring out $\dot{w}^2\bar{v}^2/2gA^2$ yields:

$$P_1 - P_2 \approx \frac{\dot{w}^2\bar{v}}{2gA^2}\left(K_1 - \frac{v_1^2}{\bar{v}^2} + \frac{v_2^2}{\bar{v}^2}\right). \qquad (4.10)$$

The expected error between the results using the comprehensive average properties formula (Eq. 4.10) and using the ideal theoretical formula given in Section 4.3.1 is shown in Figure 4.2. The chart assumes no error in Equation 4.9. When compared to the previous simple average properties method, this method gives fairly good results for higher values of K, and gives poor, but improved, results at lower values of K with increasing pressure ratio $\Delta P/P_1$.

For some problems the influence of elevation may not be negligible. For these, the ΔP due to elevation difference can be added:

$$P_1 - P_2 \approx \frac{\bar{v}\dot{w}^2}{2gA^2}\left(K - \frac{v_1^2}{\bar{v}^2} + \frac{v_2^2}{\bar{v}^2}\right) + \frac{1}{\bar{v}}(Z_2 - Z_1).$$

4.2.3 Using Expansion Factors

As in all the other pressure drop equations in Chapter 4, the expansion factor method must be used for constant-area ducts. Ordinarily it is also reserved for adiabatic flow and it will be so treated here.

The formula for pressure drop with incompressible flow in a horizontal pipe was given in Section 4.2.1:

$$P_1 - P_2 = K\frac{\dot{w}^2 v}{2gA^2}. \qquad \text{(4.6, repeated)}$$

Using the specific weight ρ_w it is:

$$P_1 - P_2 = K\frac{\dot{w}^2}{2gA^2\rho_w}, \qquad (4.11)$$

where

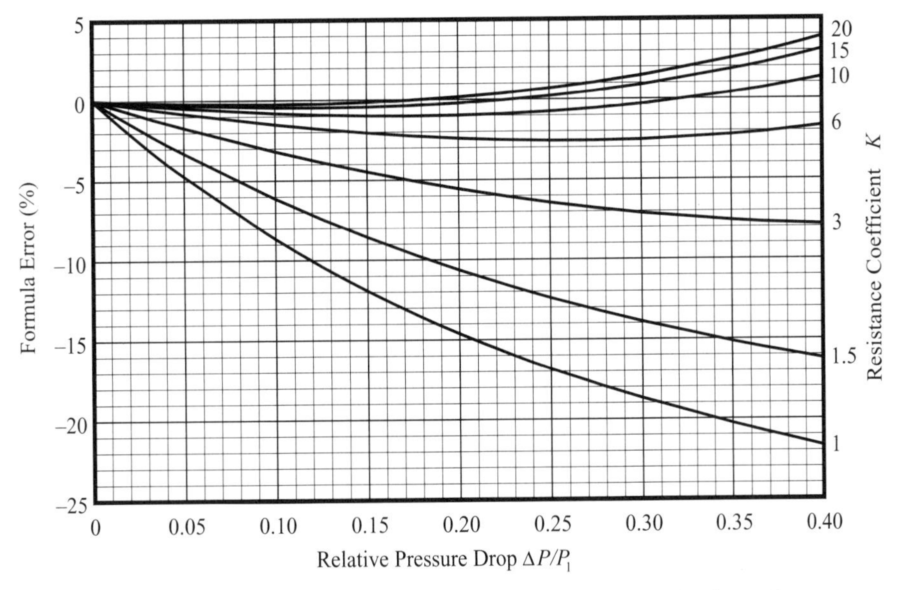

FIGURE 4.2. Error in comprehensive average properties formula.

$\dot{w}$ = weight flow rate of fluid through the duct, lb/s, and
A = duct cross-sectional area, ft^2.

The comparable equation for incompressible flow rate in a constant-area duct, as given by many fluid mechanics texts, is:

$$\dot{w} = 0.525d^2\sqrt{\frac{\Delta p \rho_w}{K}}, \tag{4.12}$$

where
d = duct diameter, inches,
Δp = static pressure drop, lb/in^2 = $p_1 - p_2$,
ρ_w = flowing fluid specific weight, lb/ft^3, and
0.525 = lumped conversion constants
$= \pi\sqrt{2g}/48 = 0.525021$.

(The symbol ρ_w used here for specific weight is often called weight density or simply density, and the symbol ρ, without the subscript, is often used for it. Care should be taken to distinguish between specific weight, or weight density, and mass density.) The fluid mechanics texts mentioned above adapt Equation 4.12 to compressible flow by adding an expansion factor Y:

$$\dot{w} = 0.525Yd^2\sqrt{\frac{\Delta p(\rho_w)_1}{K}} = 0.525Yd^2\sqrt{\frac{\Delta p}{Kv_1}}, \tag{4.13}$$

where
$(\rho_w)_1$ = fluid density at the *inlet*, lb/ft^3, and
v_1 = fluid specific volume at the *inlet*, ft^3/lb.

The comparable adaptation to Equation 4.11, using the same Y, is:

$$P_1 - P_2 = \frac{K\dot{w}^2}{Y^2 2gA^2(\rho_w)_1}. \tag{4.14}$$

The value of Y may be computed from:

$$Y = \sqrt{\frac{K\dot{w}^2}{(P_1 - P_2)2gA^2(\rho_w)_1}}. \tag{4.15}$$

To calculate Y utilizing the theoretical, ideal equation, by way of a computer program using d and returning p_1, p_2, V_1 and $\dot{w}$, let us make use of the relation $AV\rho_w = \dot{w}$ (Eq. 2.1a) to calculate $(\rho_w)_1$:

$$(\rho_w)_1 = \frac{\dot{w}}{AV_1}\ (\text{lb/ft}^3).$$

Also:

$$A = \frac{\pi d^2}{4(144)}\ (\text{with } d \text{ inches, } A \text{ ft}^2).$$

The program solution returns $p_1 - p_2 = \Delta p$ in lb/in^2, so:

$$P_1 - P_2 = 144\Delta p.$$

Substituting these into Equation 4.15, we obtain:

$$Y = \sqrt{\frac{K\dot{w}^2}{144\Delta p 2gA^2}\frac{AV}{\dot{w}}} = \sqrt{\frac{K\dot{w}V_1}{144\Delta p 2g}\frac{4(144)}{\pi d^2}}$$
$$= \sqrt{\frac{4K\dot{w}V_1}{2\pi g\Delta p d^2}} = \sqrt{\frac{2}{\pi g}}\sqrt{\frac{K\dot{w}V_1}{\Delta p d^2}}$$
$$= 0.14067\sqrt{\frac{K\dot{w}V_1}{\Delta p d^2}}.$$

Charts of expansion factor Y versus $\Delta p/p_1$ (or $\Delta P/P_1$, because $\Delta p/p_1 = \Delta P/P_1$) prepared this way are given in Figure 4.3, for $\gamma = 1.4$, and in Figure 4.4, for $\gamma = 1.3$.

Diatomic gases, such as nitrogen, oxygen, and hydrogen, typically have a ratio of specific heats (γ) close to 1.4, and polyatomic gases, such as carbon dioxide, methane, and steam, have a ratio of specific heats of about 1.3. If γ is not constant, you may not want to use these charts. You may need to determine the average value of γ to determine if it is close to the chart value. To utilize the charts, known values of Δp and p_1 may be used to find $\Delta p/p_1$; from this Y may be read from the chart, and, using Equation 4.13, the flow rate $\dot{w}$ may be found. If $\dot{w}$ is known but Δp is not, then it must be estimated, and the resulting $\dot{w}$ must be compared with the known value. From these, a correction in Δp may be found and the process repeated until the flow rates match.

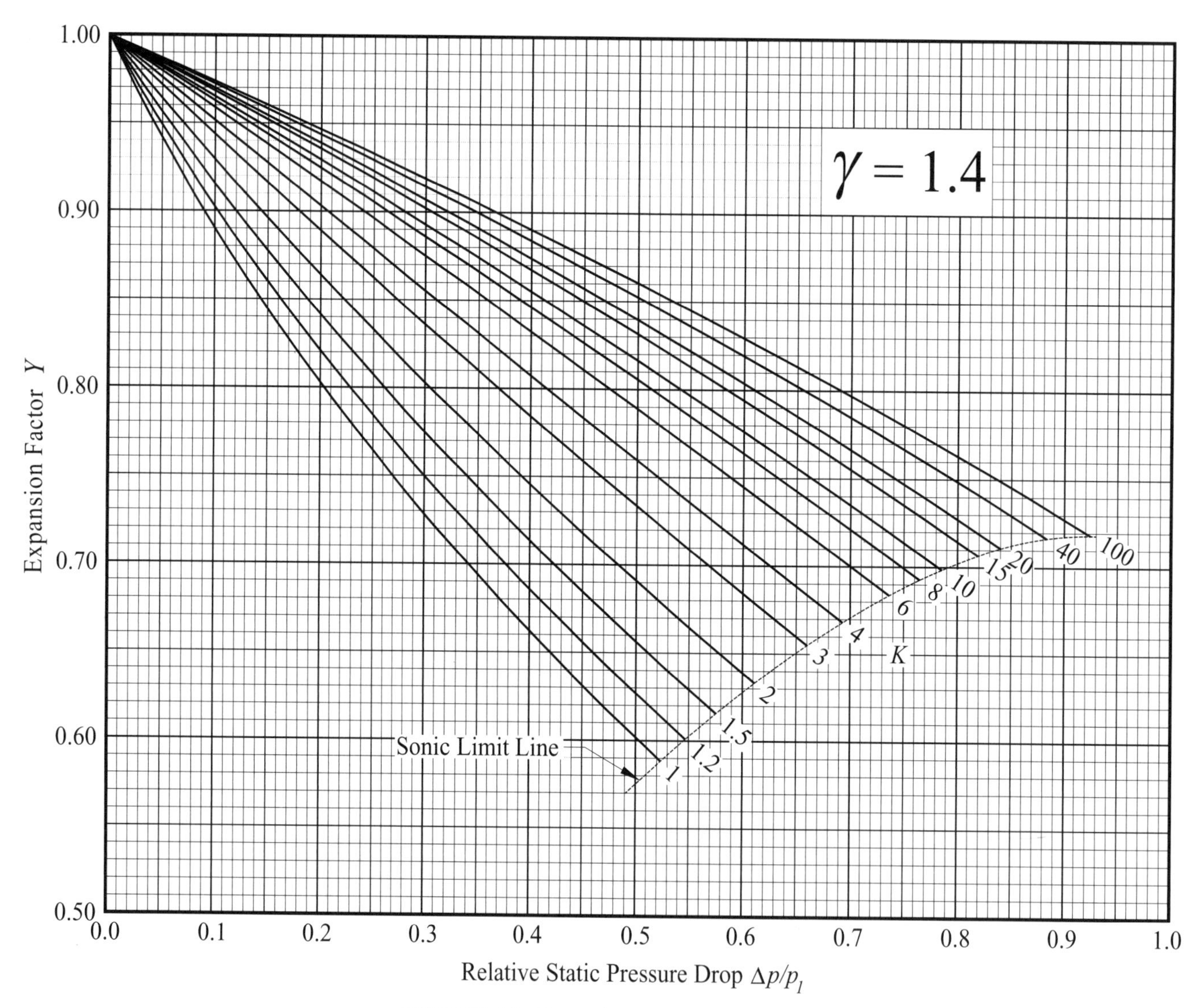

FIGURE 4.3. Expansion factor for $\gamma = 1.4$.

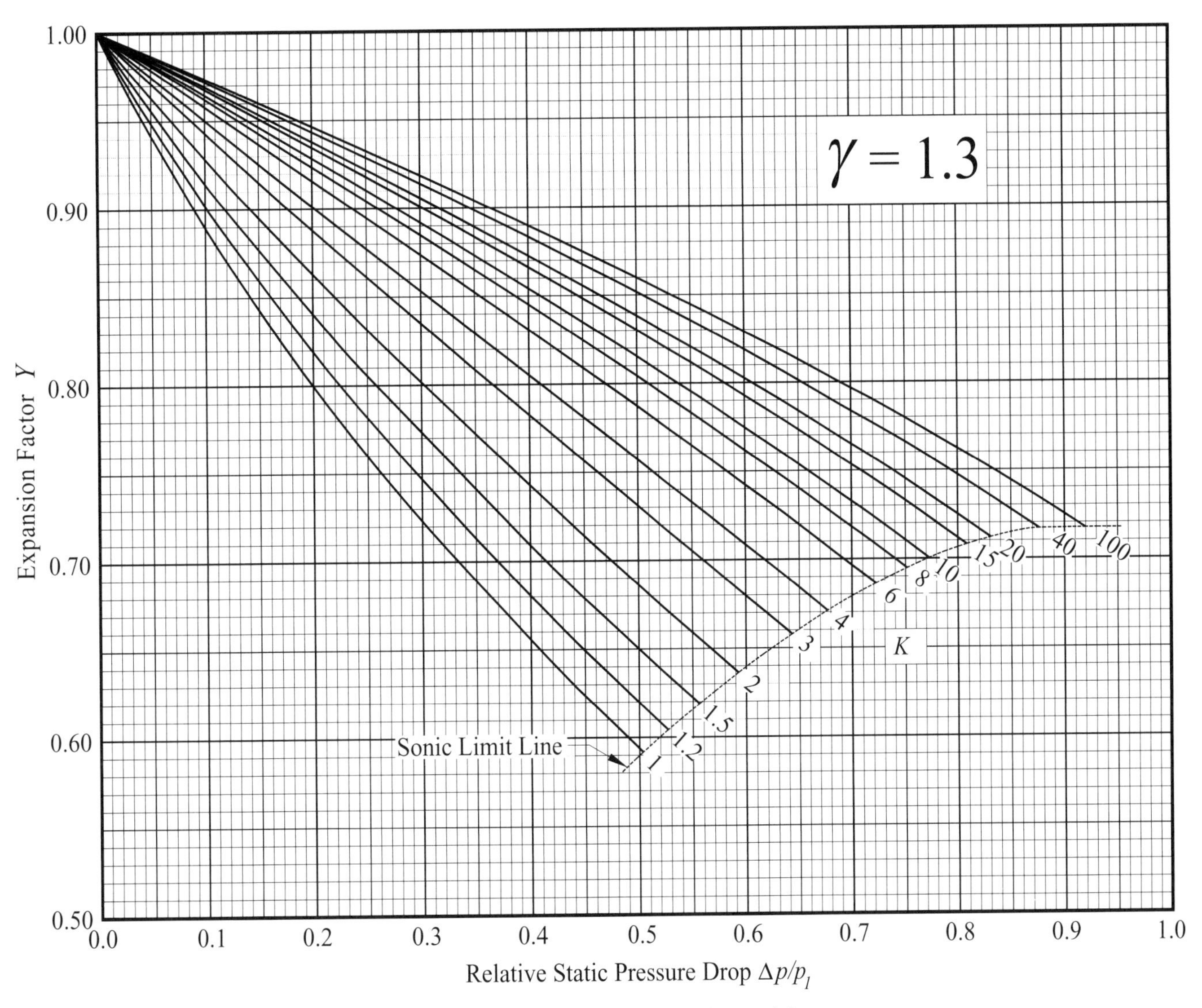

FIGURE 4.4. Expansion factor for $\gamma = 1.3$.

4.3 ADIABATIC COMPRESSIBLE FLOW WITH FRICTION: IDEAL EQUATION

Two theoretical equations are presented for pressure drop in compressible flow in constant-area ducts. One uses Mach number as the variable and the other uses pressure. Methods are offered for trial-and-error solutions to equations that cannot be solved explicitly.

4.3.1 Using Mach Number as a Parameter

Street et al. [1] give the following relation for a constant-area duct flowing a gas *with sonic velocity at the exit*:*

$$f_{\text{ave}}\frac{L_{\max}}{D} = \frac{1-M^2}{\gamma M^2} + \frac{\gamma+1}{2\gamma}\ln\left[\frac{(\gamma+1)M^2}{2\left(1+\frac{\gamma-1}{2}M^2\right)}\right] \equiv K_{\text{limit}}, \tag{4.16}$$

where

f_{ave} = average Darcy friction factor along the duct,
$L_{\max}$ = maximum attainable duct length with M at the inlet, ft (or m),
D = duct diameter, ft (or m),
γ = ratio of specific heats of flowing gas, and
M = Mach number of the gas flow *at the duct inlet.*

In the development of this equation, f is assumed to be a constant, and f_{ave} is taken as a reasonable value for f.

* Asher H. Shapiro also gives this equation.

In actuality, of course, since fluid temperature changes continuously along the duct, the fluid viscosity also changes, and then so does the Reynolds number—resulting in a varying friction factor. But it turns out that the variation is modest enough to be easily handled by using the average friction factor.

Notice that $f_{ave}L_{max}/D$ in the equation above is in the form of $K = fL/D$, so that it may be called K_{limit}, indicating that if K exceeds K_{limit}, L will exceed L_{max}, which is forbidden. In so doing, however, we must remember that K is a symbol for friction and induced turbulence (or local) losses, but in this context the local losses must be generated without any changes of flow cross section. This restriction nearly eliminates any local losses except those exhibited by bends.

The following solution techniques are very difficult if attempted by hand. A computer program incorporating the techniques and formulas given makes the solutions easy. The same may be said of the compressibility factor equations presented in Appendix D, especially the Lee–Kesler compressibility factor equation. If a computer program is not available to make the desired calculations and the compressibility factor may be neglected because the system pressures are modest and the fluid static temperature is reasonably higher than the critical temperature, then a spreadsheet may be employed to find the correct Mach number from Equation 4.16. The technique is described at the end of Section 4.3.2. Both Equation 4.16 and Equation 4.24 are relatively easy to solve because both have only one unknown—M_{inlet} in Equation 4.16 and T_2 in Equation 4.24.

Area changes should be handled as given in Section 4.3.1.5 or at the end of Section 4.3.2.

4.3.1.1 Solution when Static Pressure and Static Temperature Are Known

Equation 4.16 may be used to find the L_{max} of the duct if the essential duct data are available: flow rate, inlet static pressure, inlet static temperature, duct diameter, friction factor, and gas ratio of specific heats, molecular weight, and compressibility factor. The Mach number of a gas flowing in a duct (assuming a flat velocity profile) is:

$$M = \frac{v}{\text{A}} = \frac{V}{\text{A}}. \qquad \text{(1.4b, repeated)}$$

Using m for molecular weight in the proper units, the equation for the acoustic velocity A is:

$$\text{A} = \sqrt{\gamma g z \frac{\bar{R}}{m} T} \ (\bar{R} \text{ in weight units [English]}), \qquad (4.17\text{a})$$

$$\text{A} = \sqrt{\gamma z \frac{\bar{R}}{m} T} \ (\bar{R} \text{ in mass units [SI]}). \qquad (4.17\text{b})$$

In these two equations, m is not mass, but molecular weight; m is in lb/mol$_{lb}$ for English units, and is in kg/mol$_{kg}$ for SI units. The compressibility factor z may be evaluated using one of the formulas found in Appendix D, or, alternatively, found from a chart of z as a function of reduced pressure and reduced temperature, such as the Nelson–Obert chart. Utilizing Equations 1.3b, 4.17a, and 4.17b and V from $AV\rho_m = \dot{m}$ or $AV\rho_w = \dot{w}$ (Eq. 2.1a and 2.1b), we may write:

$$M = \frac{V}{\text{A}} = \frac{\dot{w}}{A\rho_w \text{A}} = \frac{\dot{w}}{AP}\sqrt{\frac{z\bar{R}T}{\gamma g m}} \ \text{(English)}, \qquad (4.18\text{a})$$

$$M = \frac{V}{\text{A}} = \frac{\dot{m}}{A\rho_m \text{A}} = \frac{\dot{m}}{AP}\sqrt{\frac{z\bar{R}T}{\gamma m}} \ \text{(SI)}. \qquad (4.18\text{b})$$

(In these formulas, care must be taken to distinguish between acoustic velocity A and flow area A; and while $\dot{m}$ is mass flow rate, m is molecular weight.*) Using this Mach number, evaluated at the duct inlet, L_{max} becomes immediately available from Equation 4.16.

Equation 4.16 may not be violated. The length of the duct may not exceed L_{max} where sonic velocity ($M = 1$) occurs at the exit. However, if the length of the duct is less than L_{max} as given by Equation 4.16, then the exit Mach number will be less than unity. This is the most frequently encountered case.

Consider a gas receiver discharging through a round duct of known length L_{line} to a lower pressure region and suppose that the pressure conditions are such that the discharging gas exits from the duct at subsonic velocity (see Fig. 4.5). Assume that friction factor f and diameter D are constant. If we know the flowing conditions at one end—either end—of the duct (flow rate, duct diameter, pressure, and temperature) we may find the Mach number M there using Equation 4.18 and then use Equation 4.16 to find the $(fL/D)_{limit}$ or K_{limit} at that end of the duct. (By Eq. 1.3, this can be called K_{limit} at that end. Remember that because f and D are constant, K in this context is simply length with a constant coefficient.) Note that since the flow exits from the duct subsonically, this K_{limit} includes a virtual length of duct at which the flow would attain sonic velocity (provided that the pressure at the virtual outlet is low enough). Now, because f/D is constant, K is proportional to L so that we can write:

$$(K_1)_{limit} = K_{line} + (K_2)_{limit}.$$

* Strictly speaking, kg/kg mol is molecular mass and lb$_f$/lb mol is molecular weight, but the term "molecular weight" is often used in both SI and the English system. The molecular mass and the molecular weight are numerically equal.

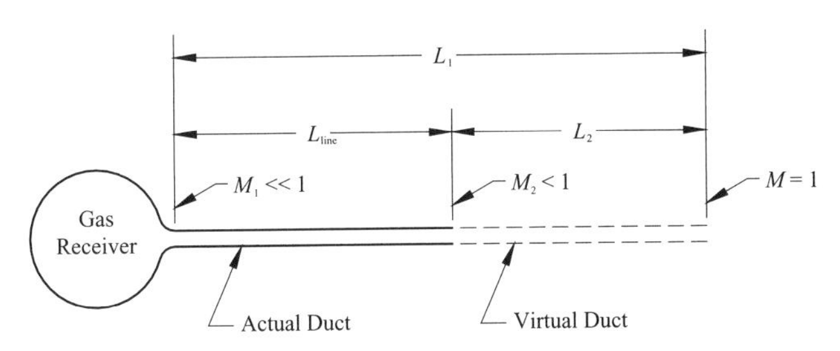

FIGURE 4.5. Subsonic constant-area gas flow duct.

In Figure 4.5, $(K_1)_{limit}$ corresponds to L_1, and $(K_2)_{limit}$ corresponds to L_2, the virtual portion of the duct. Knowing the line resistance coefficient K_{line} and limit resistance coefficient $(K\ )_{limit}$ at one end of the duct enables us to find the limit resistance coefficient at the other end of the duct. Then, since K_{limit} is associated with M at that end by Equation 4.16, we may find M at that end by solving the equation.

It must be noted that Equation 4.16 cannot be solved explicitly. A satisfactory solution can, however, be obtained using a trial-and-error method. Such a solution is undesirably tedious by hand, but it is easy using a programmable calculator or a computer. A solution technique is shown in Appendix E, Section E.2, suitable for implementation in a computer program. A technique using a spreadsheet is given in Section 4.5 for another explicitly unsolvable equation with one variable which is also applicable to Equation 4.16.

Once the unknown Mach number is found, the accompanying pressure and temperature may be found. The static pressure, in terms of the local Mach number and the static pressure P_* at the location where Mach number is unity (that is, where velocity is sonic), is given by:

$$\frac{P}{P_*} = \frac{1}{M}\sqrt{\frac{\gamma+1}{2\left(1+\frac{\gamma-1}{2}M^2\right)}}. \tag{4.19}$$

Taking the ratio of the expression evaluated for $M = M_1$ to that for $M = M_2$ yields:

$$\frac{P_1}{P_2} = \frac{M_2}{M_1}\sqrt{\frac{1+M_2^2(\gamma-1)/2}{1+M_1^2(\gamma-1)/2}}, \tag{4.20}$$

from which the desired pressure is easily found. The static temperature is available similarly from:

$$\frac{T}{T_*} = \frac{\gamma+1}{2\left(1+\frac{\gamma-1}{2}M^2\right)}. \tag{4.21}$$

T_* is the static temperature when the accompanying Mach number is unity. The ratio of the inlet and outlet static temperatures is thus:

$$\frac{T_1}{T_2} = \frac{1+M_2^2(\gamma-1)/2}{1+M_1^2(\gamma-1)/2}, \tag{4.22}$$

from which the desired temperature is easily found.

The foregoing relationships are useful if the static pressure and static temperature at one end of the duct are known. If one or the other of the static values is not known, but the corresponding total value is known (and this is often, if not usually, the case) these equations may still be solved, but account must be made for the divergence between total and static values. For instance, if a gas in a pressurized vessel is allowed to escape to atmosphere through a duct and it attains sonic velocity at the end of the conduit, the static pressure at the outlet end of the duct may be as low as half its total pressure and static temperature may be as low as 80% of its total temperature.

There are three cases in which the required static values are not all known: (1) static pressure and total temperature are known; (2) total pressure and total temperature are known; and (3) total pressure and static temperature are known. These will be considered in order. We must make use of the following relationships:

$$T = \frac{T_t}{1+M^2(\gamma-1)/2},$$

$$P = \frac{P_t}{\left[1+M^2(\gamma-1)/2\right]^{\gamma/(\gamma-1)}},$$

where T, P, T_t, P_t, and M are local values, T and P are static values, and T_t and P_t are total values.

Now, in order to simplify the equations, let us recast the equations for Mach number (Eq. 4.18a and 4.18b) in the following form:

$$M = B\sqrt{T}/P, \tag{4.23}$$

where

$$B = \frac{\dot{m}}{A}\sqrt{\frac{z\bar{R}}{\gamma m}} \text{ (for SI)},$$

$$B = \frac{\dot{w}}{A}\sqrt{\frac{z\bar{R}}{\gamma g m}} \text{ (for English)}.$$

4.3.1.2 Solution when Static Pressure and Total Temperature Are Known Now, if *static pressure* and *total temperature* are known, substitute the expression for static temperature T, in terms of total temperature T_t, in place of T; then:

$$M = \frac{B}{P}\sqrt{\frac{T_t}{1 + M^2(\gamma - 1)/2}}.$$

This equation is a quadratic in M^2 whose solution is:

$$M^2 = \frac{\sqrt{1 + 2(\gamma - 1)(B\sqrt{T_t}/P)^2} - 1}{\gamma - 1}.$$

Note the similarity of the expression $B\sqrt{T_t}/P$ in this equation to that for Mach number M in Equation 4.23. They are identical except that the one above contains T_t while Equation 4.23 contains simply T. Let us therefore call the expression (and similar expressions utilizing the available temperature and pressure, whether they be static or total) "core Mach number," M_{core}, because of its similarity to the simple expression for Mach number based on static values, and because it is the "core" of the expression for Mach number when values other than static values are utilized. Then, for the *static pressure* and *total temperature* case, we may write:

$$M^2 = \frac{\sqrt{1 + 2(\gamma - 1)M_{core}^2} - 1}{\gamma - 1}.$$

This M^2 may now be substituted into Equation 4.16 to find the $f_{ave}L_{max}/D$ or K_{limit}, and from thence to find the Mach number at the other end of the duct. Using Equations 4.19 through 4.22 in the preceding section, the unknown pressures and temperatures may be found at both ends of the pipe.

4.3.1.3 Solution when Total Pressure and Total Temperature Are Known

If *total pressure* and *total temperature* are known at one end of the duct, the expressions for static pressure in terms of total pressure and static temperature in terms of total temperature may be substituted into Equation 4.23 to obtain:

$$M^2 = M_{core}^2\left[1 + M^2(\gamma - 1)/2\right]^{(\gamma+1)/(\gamma-1)},$$

where M_{core} is again defined as the result of evaluating Equation 4.23 with the available temperature and pressure (total pressure and total temperature in this case) instead of with strictly the static values.

The foregoing formula presents a problem—it is another of those equations that cannot be solved explicitly. As described in Section 4.3.1.1 regarding Equation 4.16, a satisfactory solution can be obtained using a trial-and-error method with a programmable calculator or computer. In this case, however, the solution is a little more complicated than that for Equation 4.16. There are some added constraints that must be observed, but within these constraints the solution for M^2 is easily found, and having found it the unknown pressures and temperatures can be found by applying Equations 4.16 through 4.22 in Section 4.3.1.1. A trial-and-error solution technique applicable to the equation above is described in Section 4.3.2 using a spreadsheet. A computer program solution technique is also shown in Appendix E, Section E.3.

4.3.1.4 Solution when Total Pressure and Static Temperature Are Known

If *total pressure* and *static temperature* are known at one end of the duct, the expression for static pressure in terms of total pressure may be substituted into Equation 4.23 to obtain:

$$M^2 = M_{core}^2\left[1 + M^2(\gamma - 1)/2\right]^{2\gamma/(\gamma-1)},$$

where M_{core} is again defined as the result of evaluating Equation 4.18 with the available temperature and pressure (total pressure and static temperature in this case) instead of with strictly the static values.

The detailed derivation and solution technique are given in Appendix E, Section E.4. The derivation is very similar to that for the *total pressure/total temperature* case in Section 4.3.1.3 above, and the solution technique and caveats are identical. A solution technique using a spreadsheet is also described in Section 4.5 below.

4.3.1.5 Treating Changes in Area

While Street et al.'s adiabatic flow with friction equation is valid only for constant-area ducts, transitions between pipe sizes are as easy to handle as they are in incompressible flow calculations. For an expansion or contraction, it may be assumed that the change of area is negotiated isentropically (that is, without losses) so that the total pressure at the inlet and outlet of the fitting are the same. (Total temperature will always be the same in adiabatic flow.) Then another calculation may be made on the new size of pipe, using the total conditions at the adjacent end of the previous run of pipe. To account for the local loss in the fitting, the equivalent length of the fitting's resistance coefficient *in the smaller pipe* should be determined, whether it is upstream or downstream, and added to that stretch's length. It is recommended that the smaller pipe be used for determining the energy loss because loss coefficients of contractions and expansions found in Part II are based on the velocity in (or diameter of) the smaller pipe.

For calculating pressure drop in valves, it is recommended that a commercial valve sizing program be used, which will check for choking in the valve. If the equations given above are used to write a computer program running under the Microsoft Windows operating system, that program and the valve program may be run concurrently. The user may switch back and forth between both programs by pressing Alt+Tab, thereby avoiding having to reenter much of the same data in the compressible flow pressure drop program when switching back to it.

If such a program is unavailable to the user, it is very important to consider area reductions and increases within the valve where choking may occur. Such choking can have a profound effect on the energy losses within the valve.

4.3.2 Using Static Pressure and Temperature as Parameters

Richard Turton [2] gives the following equations:

$$f\frac{L}{D}=\frac{\gamma-1}{2\gamma}\left(\frac{P_1^2T_2^2-P_2^2T_1^2}{T_2-T_1}\right)\left(\frac{1}{P_1^2T_2}-\frac{1}{P_2^2T_1}\right)-\frac{\gamma+1}{\gamma}\ln\frac{P_1T_2}{P_2T_1}, \quad (4.24)$$

$$G=\sqrt{\frac{2\gamma}{\gamma-1}\frac{gP_1^2P_2^2m}{z\bar{R}}\left(\frac{T_2-T_1}{P_2^2T_1^2-P_1^2T_2^2}\right)}, \quad (4.25)$$

where

f = Darcy friction factor,
L = duct length, ft or m,
D = duct diameter, ft or m,
γ = gas ratio of specific heats, dimensionless,
P_1 = duct stretch inlet static pressure, lb/ft^2 or N/m^2,
P_2 = duct stretch outlet static pressure, lb/ft^2 or N/m^2,
T_1 = duct stretch inlet static temperature, °R or °K,
T_2 = duct stretch outlet static temperature, °R or °K,
m = molecular weight of flowing gas, lb/mol$_{lb}$ or kg/mol$_{kg}$,
G = gas weight velocity, lb/s-ft^2 or gas mass velocity kg/s-m^2,
g = acceleration of gravity, ft/sec^2, *used for English units only*,
z = gas compressibility factor, dimensionless, and
$\bar{R}$ = universal gas constant, ft-lb/mol$_{lb}$-°R or N-m/mol$_{kg}$-°K.

(As in Eq. 4.16, fL/D may be called K. The equation for G in the reference lacks g and z. The g is added to obtain dimensional homogeneity *when using English units*, and the z is added to account for real gas density.)

In Turton's equations, it is imperative to observe that area is constant, and that P is *static* absolute pressure, and T is *static* absolute temperature. *A consistent set of units must be used, either USCS or SI.* The gas flow cannot be choked unless it is at the *end* of the duct.

The friction factor in Equation 4.24 is assumed to be constant along the length of the pipe, which is tantamount to assuming fully developed turbulent flow throughout, and constant fluid viscosity. Because of these assumptions, the average f should be used. Solution of these equations may be accomplished provided sufficient input data are available.

The most likely potential variables are P_1 and T_1, P_2 and T_2, G, and f. Length and diameter of the duct could be variables, but probably will be specified in the design problem. A typical problem might be solved by estimating f, solving Equation 4.24 for the unknown value (usually T_2), then solving Equation 4.25 for flow rate G. (The weight or mass flow rate may be found by multiplying G by the duct area.) Then the estimate of f can be checked; if it is different from the assumed f, the procedure should be repeated. Since the dependence of f on G is usually quite weak, the iterative procedure will converge rapidly.

Because Equation 4.24 cannot be solved explicitly, it must be solved using a trial-and-error method. (See Street et al.'s Mach number-based equation [Section 4.3]). The solution, however, can be considerably more difficult than that for Equation 4.16. While Equation 4.16 ordinarily has one unknown variable, Equation 4.24 can have two—for instance, P_2 and T_2 if P_1 and T_1 are known (or P_1 and T_1 if P_2 and T_2 are known). However, the equation is given for the situation where P_1 and P_2 are known, so that the only unknown variable is T_2. (You may refer to Appendix E, Section E.1.2 for a solution technique using a computer program for a similar problem that can be adapted to this one.) Because Turton's equation and Street et al.'s equation can be equated, and both accurately model the same phenomenon, with the same input data, the results must be the same (and are). The same warnings are valid: for instance, Mach number at the outlet cannot exceed unity, and there may be no solution because the chosen variables result in a supersonic velocity at the outlet. If the inlet and outlet static values of pressure and/or temperature are not known in every instance while the total values are, the equations given for the relationships between static and total values can be used in the trial-and-error solution process. For solution by spreadsheet, see the suggested method in Section 4.5 below. When two unknowns exist, multiple solutions can be plotted to help solve the final variable.

Where the adiabatic compressible flow negotiates a change in area, the formulas above do not apply. This eventuality may be treated as discussed in Section

4.3.1.5, "Treating Changes in Area." For those wishing to work with static pressures as in Turton's formula above, a formula relating area change with pressure change for isentropic flow (reversible adiabatic flow) in terms of static pressures follows:

$$\dot{w} = A_2\sqrt{\frac{2g\gamma P_1[(P_2/P_1)^{2/\gamma}-(P_2/P_1)^{(\gamma+1)/\gamma}]}{v_1(\gamma-1)[1-(A_2/A_1)^2(P_2/P_1)^{2/\gamma}]}}\ \text{(for English)},$$

$$\dot{m} = A_2\sqrt{\frac{2\gamma P_1[(P_2/P_1)^{2/\gamma}-(P_2/P_1)^{(\gamma+1)/\gamma}]}{v_1(\gamma-1)[1-(A_2/A_1)^2(P_2/P_1)^{2/\gamma}]}}\ \text{(for SI)},$$

where
$\dot{w}$ = weight flow rate, lb/s, *for English units*,
$\dot{m}$ = mass flow rate, kg/s, *for SI units*,
A_1 = inlet area, ft^2 or m^2,
A_2 = outlet area, ft^2 or m^2,
g = acceleration of gravity, ft/s^2, *used for English units only*,
P_1 = inlet static pressure, lb/ft^2 or N/m^2,
P_2 = outlet static pressure lb/ft^2 or N/m^2,
γ = ratio of specific heats (dimensionless), and
v_1 = inlet specific volume, ft^3/lb or m^3/kg.

This formula will give the pressure change across an area change for *isentropic flow*—that is, with no flow losses. This is equivalent to assuming that the total pressure downstream of the area change is the same as that upstream. To account for the head loss in the area change, add a length of pipe to the upstream or downstream pipe (preferably the smaller pipe), which has a K (i.e., fL/D) equal to the K of the fitting with the head loss.

It may be noted that the equation given above cannot be solved explicitly. However, if a computer program is not available to execute the recommendations given in Section 4.3.1.5 above, the equation can be solved by a trial-and-error technique described below.

Rearrange the equation by squaring and factoring out the known constants:

$$\frac{\dot{w}^2 v_1(\gamma-1)}{2g\gamma P_1 A_2^2} = \frac{[(P_2/P_1)^{2/\gamma}-(P_2/P_1)^{(\gamma+1)/\gamma}]}{[1-(A_2/A_1)^2(P_2/P_1)^{2/\gamma}]}\ \text{(for English)},$$

$$\frac{\dot{m}^2 v_1(\gamma-1)}{2\gamma P_1 A_2^2} = \frac{[(P_2/P_1)^{2/\gamma}-(P_2/P_1)^{(\gamma+1)/\gamma}]}{[1-(A_2/A_1)^2(P_2/P_1)^{2/\gamma}]}\ \text{(for SI)}.$$

The group of known constants on the left side of the equation must be evaluated for comparison with the trial evaluation of the group on the right side of the equation. Separate evaluation of $(A_2/A_1)^2$, $2/\gamma$ and $(\gamma+1)/\gamma$ will also aid in the solution. Using a spreadsheet, the right-hand group may be easily evaluated and the value of the left-hand group subtracted from it. Evaluation with a complete range of possible pressure ratios is recommended, as $0.1 \le P_2/P_1 \le 1.0$. The calculated difference will in all probability bracket the final solution, which is when the calculated difference is zero. Another set of solutions with a very restricted range about zero may then be evaluated, which will show a good approximation of the correct value of P_2/P_1. A chart may be constructed in the spreadsheet to help visualize this value. Then an additional single line of calculation may be used to try, by experimental adjustment of P_2/P_1, to get a difference in the two sides of the equation as close to zero as possible, say 1×10^{-4} or less. When this is attained, a very nearly correct solution of P_2/P_1 will be revealed—closer than the precision of the calculated values. Knowing this will allow proceeding with solution of the pressure drop in the next section of pipe with the different area. If the value of P_2/P_1 is close to 1 then the value of K_2 of the area change fitting may be calculated from:

$$K_2 \approx K_1\left(\frac{A_2}{A_1}\right)^2.$$

4.4 ISOTHERMAL COMPRESSIBLE FLOW WITH FRICTION: IDEAL EQUATION

To obtain isothermal flow in a pipe the heat transferred out of the fluid through the pipe walls and the energy converted into heat by the friction process must be adjusted so that the temperature remains constant. Such an adjustment is approximated naturally in uninsulated pipes where velocities are low (well below the sonic) and where temperatures inside and outside the pipe are of the same order. Often the flow of gases in long pipelines may be treated isothermally.

Street et al. [1] give the following equation for isothermal compressible flow in a constant-area duct:

$$P_1^2 - P_2^2 = \frac{\dot{w}^2}{g}\frac{RT}{A^2}\left(2\ln\frac{P_1}{P_2} + f\frac{L}{D}\right)\ \text{(for English)}, \quad (4.26a)$$

$$P_1^2 - P_2^2 = \frac{\dot{m}^2 RT}{A^2}\left(2\ln\frac{P_1}{P_2} + f\frac{L}{D}\right)\ \text{(for SI)}, \quad (4.26b)$$

and

$$\dot{w} = Ag\sqrt{\frac{P_1^2 - P_2^2}{RT\left(2\ln\frac{P_1}{P_2} + f\frac{L}{D}\right)}} \quad \text{(for English)}, \qquad (4.27a)$$

$$\dot{m} = A\sqrt{\frac{P_1^2 - P_2^2}{RT\left(2\ln\frac{P_1}{P_2} + f\frac{L}{D}\right)}} \quad \text{(for SI)}, \qquad (4.27b)$$

where P_1 and P_2 are the inlet and outlet absolute static pressures, respectively.

Generally the solution of these equations must be accomplished by trial and error. Frequently the first term ($2\ln[P_1/P_2]$) of the parenthesis quantity is so small in comparison with *fl/d* that it may be neglected, thus allowing a preliminary direct solution. It is then much easier to obtain a value of P_2 and use it to approximate the first term of the parenthesis, shortening the trial-and-error process as a result. (A spreadsheet method similar to that given in Section 4.5 may be employed to make the solution relatively easy.)

Because the weight flow rate, $\dot{w}$, through the pipe line is constant and given by $\dot{w} = AV\rho_w$ at all sections, the Reynolds number, which is given by

$$N_{\text{Re}} = \frac{VD\rho_w}{\mu} = \frac{\dot{w}D}{\mu A}, \qquad (1.2, \text{repeated})$$

is also constant inasmuch as μ varies hardly at all if there is no temperature change. In English units μ is in lb_m/ft-s. (The viscosity does change with pressure, but the change is very small if the pressure change is small.) Therefore, with isothermal flow, the friction factor is virtually constant in the duct even though the velocity of the gas will increase and its density decrease as the pressure drops along the flow path.

An important limitation to these equations occurs for large pressure drops ($P_2 << P_1$). There is a point beyond which a reduction of *fl/d* (or *K*) is required for further reduction of P_2; therefore, if *K* is held constant, the pressure cannot drop below this point and thus the equation is applicable only between the pressure P_1 and the limiting value of P_2.

Equation 4.26 may be written in terms of Mach number, *M*:

$$\frac{M_1^2}{M_2^2} = 1 - \gamma M_1^2\left[2\ln\frac{M_2}{M_1} + f\frac{L}{D}\right]. \qquad (4.28)$$

Per Street et al., the limiting value of M_2 may be found by differentiating Equation 4.28 with respect to *l* and setting dP_2/dl equal to infinity. The result is:

$$M_2 = \sqrt{1/\gamma}.$$

Equations 4.26, 4.27, and 4.28 are therefore applicable only where $M_1 < M_2 \leq \sqrt{1/\gamma}$.

4.5 EXAMPLE PROBLEM: COMPRESSIBLE FLOW THROUGH PIPE

Using Richard Turton's Equations 4.24 and 4.25 and a spreadsheet, determine the flow of nitrogen gas through a 4″ schedule 40 pipe. In his original paper Turton states:

> This paper introduces two previously unpublished relationships for adiabatic frictional flow in pipes. These equations allow the direct evaluation of the flow rate for the situation when **upstream conditions and downstream pressure** are known.

Let us assume that the pipe is 100 ft long and its upstream pressure and temperature are 100 psia (14,400 lb/ft^2) and 530°R, respectively. The downstream pressure is 84.056 psia (12,104 lb/ft^2). The inside diameter of 4″ Sch 40 pipe is 4.026 in, and its cross-sectional flow area is 12.730 in^2 or 0.088405 ft^2. The molecular weight of nitrogen is 28.013 lb/mol$_{lb}$, and its ratio of specific heats is 1.400. The universal gas constant $\bar{R}$ is 1545.31 ft-lb/mol$_{lb}$°R. The equations are:

$$f\frac{L}{D} = \frac{\gamma - 1}{2\gamma}\left(\frac{P_1^2T_2^2 - P_2^2T_1^2}{T_2 - T_1}\right)\left(\frac{1}{P_1^2T_2} - \frac{1}{P_2^2T_1}\right) - \frac{\gamma + 1}{\gamma}\ln\frac{P_1T_2}{P_2T_1}, \qquad (4.24, \text{repeated})$$

$$G = \sqrt{\frac{2\gamma}{\gamma - 1}\frac{gP_1^2P_2^2m}{z\bar{R}}\left(\frac{T_2 - T_1}{P_2^2T_1^2 - P_1^2T_2^2}\right)}, \qquad (4.25, \text{repeated})$$

where

f = Darcy friction factor,*
L = duct length, ft,
D = duct diameter, ft
γ = gas ratio of specific heats
P_1 = duct inlet static pressure, lb/ft^2
P_2 = duct outlet static pressure, lb/ft^2
T_1 = duct inlet static temperature, °R
T_2 = duct outlet static temperature, °R,
m = molecular weight of flowing gas, lb/mol$_{lb}$,
G = gas weight velocity, lb/s-ft^2
g = acceleration of gravity, ft/s^2
z = gas compressibility factor, dimensionless, and
$\bar{R}$ = universal gas constant, ft-lb/mol$_{lb}$-°R.

* Having worked this problem before, we can "guess" that $f = 0.016466$.

To evaluate Equation 4.24 we must rearrange it as follows:

$$f(T_2)=\frac{\gamma-1}{2\gamma}\left(\frac{P_1^2T_2^2-P_2^2T_1^2}{T_2-T_1}\right)\left(\frac{1}{P_1^2T_2}-\frac{1}{P_2^2T_1}\right)-\frac{\gamma+1}{\gamma}\ln\frac{P_1T_2}{P_2T_1}-f\frac{L}{D},$$

where $f(T_2)$ reads "function of T_2" and is supposed to equal zero when the right value of T_2 is entered. This is where the trial and error comes in. We must try a range of values for T_2 to find the one that yields $f(T_2) = 0$. The spreadsheet is admirably suited to doing that.

The spreadsheet used to solve Equation 4.24 is shown in Figure 4.6. The cells above the double line delineating "Initial Search" results indicate input data and preliminary information used in the spreadsheet. For instance, the first cell after "P_1" has the formula =100*144, resulting in the 14,400 listed, and the cell after "$(P_1)^2$" has the formula =B4^2 because cell B4 is where 14,400 is stored. The cell under "1st Paren" has the formula =((E4*A13^2-E5)/(A13-B3))*G3 because it evaluates the term

$$\frac{\gamma-1}{2\gamma}\left(\frac{P_1^2T_2^2-P_2^2T_1^2}{T_2-T_1}\right),$$

using the value in cell E4 containing P_1^2 and the first trial T_2 in cell A13 to calculate $P_1^2T_2^2$, subtracts the value for $P_2^2T_1^2$ stored in cell E5, and divides the result by $T_2 - T_1$, which is stored in cell B3. The $ signs before the cell numbers and line numbers indicate that these cells have fixed contents (e.g., the contents of cell E5, $P_2^2T_1^2$, never changes and E5 nomenclature guarantees that it never changes. Cells that are intended to change from line to line lack the dollar sign.) The cell under "2nd Paren" similarly evaluates:

$$\left(\frac{1}{P_1^2T_2}-\frac{1}{P_2^2T_1}\right).$$

The columns labeled "1st Term" and "2nd Term" evaluate:

$$\frac{\gamma-1}{2\gamma}\left(\frac{P_1^2T_2^2-P_2^2T_1^2}{T_2-T_1}\right)\left(\frac{1}{P_1^2T_2}-\frac{1}{P_2^2T_1}\right)\ \text{and}\ \frac{\gamma+1}{\gamma}\ln\frac{P_1T_2}{P_2T_1},$$

while the column labeled $f(T_2)$ subtracts the latter from the former and also subtracts fL/D to yield the value of the function.

Next, the formulas in line 13 need to be copied into line 14 and beyond by selecting the first line and pasting it into a number of lines following. The formula in the first column increases the value of T_2 by 0.5°R in each subsequent row. When sufficient rows are added the value reported in the last column will change from positive to negative, indicating that the value of the function has passed through zero. Then another range of rows can be added as shown, with the increment of T_2 reduced drastically by, say, a factor of 10 as in the spreadsheet facsimile shown. This will pinpoint the value of T_2 that yields $f(T_2) = 0$ much more closely. Plots as shown of $f(T_2)$ versus T_2 may be constructed by the spreadsheet plot function to visualize where this occurs. Finally, a single line of calculations may be added to "zero in" on the desired result much more closely by changing T_2 by very small values.

From the first plot of results of the search (see Fig. 4.7), it can be seen that T_2 for $f(T_2) = 0$ lies between 528.2°R ≤ T_2 ≤ 528.4°R. From the second plot (see Fig. 4.8), it can be seen that 528.26°R ≤ T_2 ≤ 528.30°R. In the spreadsheet, 528.26°R may be entered on a new line and the following cell formulas copied to this new line. Then the 528.26 can be incremented in 0.01°R steps by hand until $f(T_2)$ crosses zero, and the resulting $T_2 \approx 528.270$ can be incremented by 0.001°R steps until $f(T_2)$ crosses zero again, and so on. Finally, the spreadsheet indicates that T_2 = 527.27804°R ± 0.000014°R, which is a whole lot closer than we need to know.

This calculation was duplicated to five significant figures using a dedicated Microsoft QuickBASIC 4.5 program from which we obtained the "guess" that $f = 0.016466$.

We can check these calculations using a hand-held calculator:

$$\frac{P_1^2T_2^2-P_2^2T_1^2}{T_2-T_1}=\frac{(14,400)^2(521.27804)^2-(12,104)^2(530)^2}{528.27804-530}=-9.707418\times10^{12},$$

$$\frac{\gamma-1}{2\gamma}\left(\frac{P_1^2T_2^2-P_2^2T_1^2}{T_2-T_1}\right)=0.142857(-9.707418\times10^{12})=-1.386774\times10^{12}\quad\text{(OK)},$$

$$\left(\frac{1}{P_1^2T_2}-\frac{1}{P_2^2T_1}\right)=\frac{1}{(14,400)^2528.27804}-\frac{1}{(12,104)^2530}=-3.749756\times10^{-12}\quad\text{(OK)},$$

$$\frac{\gamma+1}{\gamma}\ln\frac{P_1T_2}{P_2T_1}=\frac{1.4+1}{1.4}\ln\frac{14,400(528.27804)}{12,104(530)}=1.714286(1.185824)=0.2921794\quad\text{(OK)},$$

Solving Turton's Equation (4)						
P_1 =	14,400	lb/ft^2	$(P_1)^2$ =	2.07360E+08	$(\gamma-1)/2\gamma$ =	0.142857143
P_2 =	12,104	lb/ft^2	$(P_2)^2(T_1)^2$ =	4.11538E+13	$(\gamma+1)/\gamma$ =	1.714285714
γ =	1.4		$(P_2)^2(T_1)$ =	7.76486E+10		
f =	0.016466		P_2T_1 =	6.41512E+06		
L =	100	ft	z =	0.99859		
D =	0.3355	ft	fL/D =	4.907898659		
T_2	1st Paren	2nd Paren	1st Term	2nd Term	$f(T_2)$	
529.5	-4.85252E+12	-3.77082E-12	1.82980E+01	2.96140E-01	1.30939E+01	Initial Search
529.0	-2.41058E+12	-3.76221E-12	9.06912E+00	2.94521E-01	3.86670E+00	
528.5	-1.59661E+12	-3.75359E-12	5.99303E+00	2.92899E-01	7.92228E-01	
528.0	-1.18963E+12	-3.74495E-12	4.45512E+00	2.91277E-01	-7.44054E-01	
527.5	-9.45454E+11	-3.73629E-12	3.53249E+00	2.89653E-01	-1.66506E+00	
527.0	-7.82672E+11	-3.72762E-12	2.91750E+00	2.88027E-01	-2.27842E+00	
526.5	-6.66404E+11	-3.71893E-12	2.47831E+00	2.86400E-01	-2.71599E+00	
526.0	-5.79206E+11	-3.71022E-12	2.14898E+00	2.84771E-01	-3.04369E+00	
525.5	-5.11389E+11	-3.70150E-12	1.89290E+00	2.83141E-01	-3.29813E+00	
525.0	-4.57138E+11	-3.69276E-12	1.68810E+00	2.81509E-01	-3.50131E+00	
524.5	-4.12754E+11	-3.68400E-12	1.52059E+00	2.79875E-01	-3.66719E+00	
524.0	-3.75769E+11	-3.67523E-12	1.38104E+00	2.78240E-01	-3.80510E+00	
523.5	-3.44477E+11	-3.66644E-12	1.26300E+00	2.76604E-01	-3.92150E+00	
523.0	-3.17657E+11	-3.65763E-12	1.16187E+00	2.74966E-01	-4.02099E+00	
522.5	-2.94415E+11	-3.64881E-12	1.07426E+00	2.73326E-01	-4.10696E+00	
522.0	-2.74081E+11	-3.63997E-12	9.97644E-01	2.71685E-01	-4.18194E+00	
521.5	-2.56140E+11	-3.63111E-12	9.30072E-01	2.70042E-01	-4.24787E+00	
521.0	-2.40194E+11	-3.62223E-12	8.70040E-01	2.68398E-01	-4.30626E+00	
520.5	-2.25929E+11	-3.61334E-12	8.16357E-01	2.66752E-01	-4.35829E+00	
528.50	-1.59661E+12	-3.75359E-12	5.99303E+00	2.92899E-01	7.92228E-01	Fine Search
528.45	-1.54410E+12	-3.75273E-12	5.79458E+00	2.92737E-01	5.93942E-01	
528.40	-1.49487E+12	-3.75186E-12	5.60854E+00	2.92575E-01	4.08062E-01	
528.35	-1.44862E+12	-3.75100E-12	5.43377E+00	2.92413E-01	2.33458E-01	
528.30	-1.40509E+12	-3.75014E-12	5.26929E+00	2.92251E-01	6.91366E-02	
528.25	-1.36405E+12	-3.74927E-12	5.11420E+00	2.92088E-01	-8.57842E-02	
528.20	-1.32529E+12	-3.74841E-12	4.96774E+00	2.91926E-01	-2.32088E-01	
528.15	-1.28863E+12	-3.74754E-12	4.82919E+00	2.91764E-01	-3.70473E-01	
528.10	-1.25389E+12	-3.74668E-12	4.69794E+00	2.91602E-01	-5.01564E-01	
528.05	-1.22094E+12	-3.74581E-12	4.57341E+00	2.91439E-01	-6.25923E-01	
528.00	-1.18963E+12	-3.74495E-12	4.45512E+00	2.91277E-01	-7.44054E-01	
527.95	-1.15986E+12	-3.74408E-12	4.34260E+00	2.91115E-01	-8.56414E-01	
527.90	-1.13150E+12	-3.74322E-12	4.23544E+00	2.90952E-01	-9.63414E-01	
527.85	-1.10445E+12	-3.74235E-12	4.13326E+00	2.90790E-01	-1.06543E+00	
527.80	-1.07864E+12	-3.74149E-12	4.03573E+00	2.90627E-01	-1.16280E+00	
527.75	-1.05398E+12	-3.74062E-12	3.94253E+00	2.90465E-01	-1.25583E+00	
528.27804	-1.38677E+12	-3.74976E-12	5.20006E+00	2.92179E-01	-1.39575E-05	Final Search

FIGURE 4.6. Spreadsheet solution.

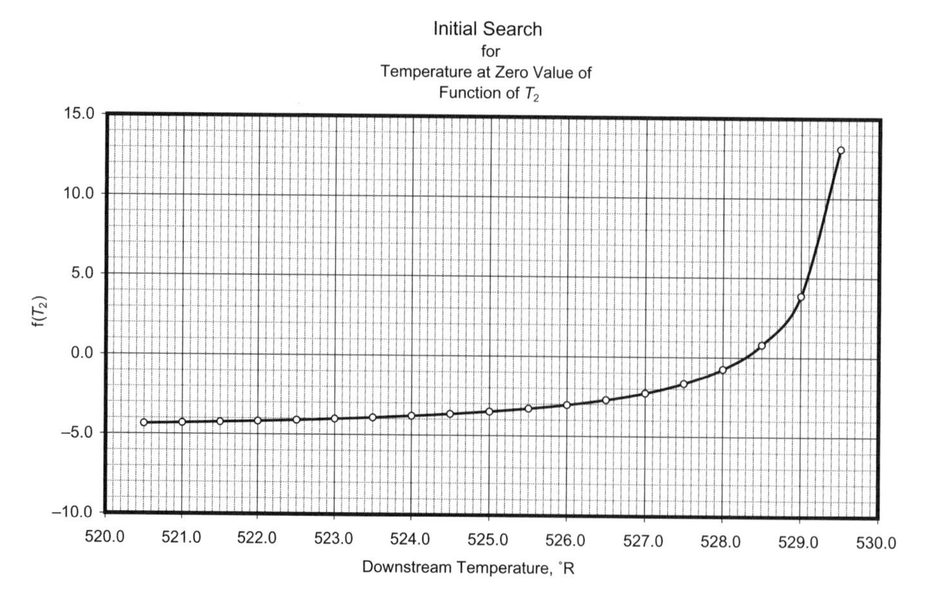

FIGURE 4.7. First plot.

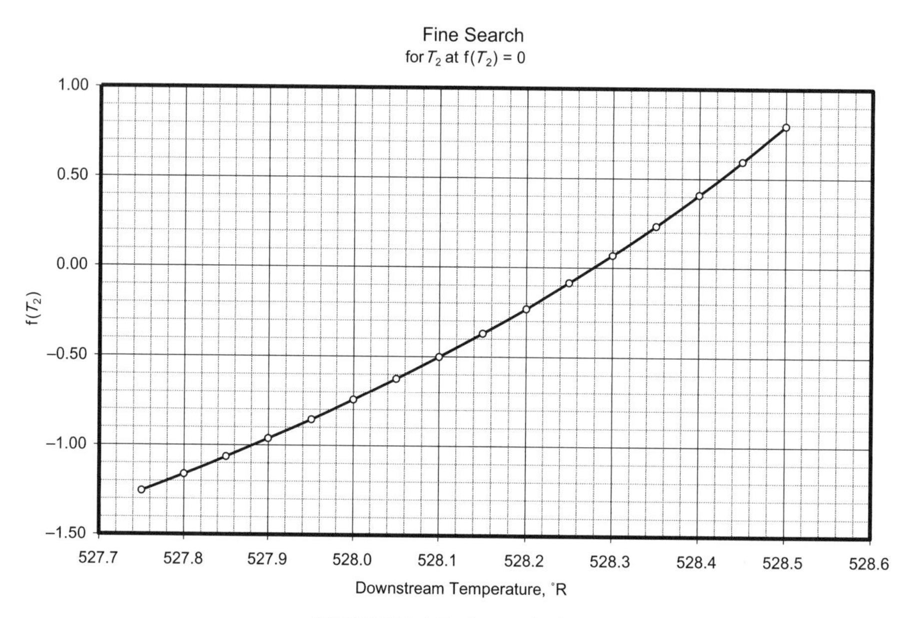

FIGURE 4.8. Second plot.

$$\frac{\gamma-1}{2\gamma}\left(\frac{P_1^2T_2^2-P_2^2T_1^2}{T_2-T_1}\right)\left(\frac{1}{P_1^2T_2}-\frac{1}{P_2^2T_1}\right)$$
$$=-1.386774\times10^{12}(-3.749756\times10^{-12})\quad\text{(OK)}$$
$$=5.20006\quad\text{(OK)},$$

$$\frac{\gamma-1}{2\gamma}\left(\frac{P_1^2T_2^2-P_2^2T_1^2}{T_2-T_1}\right)\left(\frac{1}{P_1^2T_2}-\frac{1}{P_2^2T_1}\right)-\frac{\gamma+1}{\gamma}\ln\frac{P_1T_2}{P_2T_1}-f\frac{L}{D}$$
$$=5.20006-0.2921794-4.907898659$$
$$=-1.39575\times10^{-5}\pm\text{round-off error}\quad\text{(OK)}.$$

Finally, using these results we can find the flow rate using Equation 4.25.

A portion of the contents of the root, namely

$$\frac{2\gamma}{\gamma-1}\left(\frac{T_2-T_1}{P_2^2T_1^2-P_1^2T_2^2}\right),$$

is the reciprocal of a portion of the first term in Equation 4.24, whose value is -1.386774×10^{12}, except that the terms in the denominator are reversed, changing the sign; the value is thus +1.386774 × 10^{12}. If we invert this we get +7.210980^{-13}. The remaining part of the root is:

$$\frac{gP_1^2P_2^2m}{z\bar{R}},$$

whose value is (including the compressibility factor of 0.99859 reported by the QuickBASIC program):

$$\frac{gP_1^2P_2^2m}{z\bar{R}} = \frac{32.1740(14{,}400)^2(12{,}104)^2 28.0134}{1545.31(0.99859)} = 1.77440\times10^{16}.$$

Now we can evaluate Equation 4.25:

$$G = \sqrt{\frac{2\gamma}{\gamma-1}\frac{gP_1^2P_2^2m}{z\bar{R}}\left(\frac{T_2-T_1}{P_2^2T_1^2-P_1^2T_2^2}\right)},$$

$$G = \sqrt{\frac{2\gamma}{\gamma-1}\frac{gP_1^2P_2^2m}{z\bar{R}}\left(\frac{T_2-T_1}{P_2^2T_1^2-P_1^2T_2^2}\right)}$$
$$= \sqrt{+7.210980\times10^{-13}(1.77440\times10^{16})}$$
$$= \sqrt{1.27952\times10^4} = 113.116\ \text{lb/s-ft}^2 \approx 113.12\ \text{lb/s-ft}^2.$$

The weight flow rate is the product of this weight velocity and the pipe cross-sectional area, which is 0.088405 ft^2:

$$\dot{w} = 113.116(0.088405) = 10.00002\ \text{lb/s}.$$

This weight velocity for a 4″ Sch 40 pipe is functionally identical to the 10.000 lb/s input into the QuickBASIC program (whose output values are rounded to five significant figure precision).

REFERENCES

1. Street, R. L., G. Z. Watters, and J. K. Vennard, *Elementary Fluid Mechanics*, John Wiley & Sons, 1996, pp. 608–609, 611–612.
2. Turton, R., A new approach to non-choking adiabatic compressible flow of an ideal gas in pipes with friction, *The Chemical Engineering Journal*, 30, 1985, 159–160.

FURTHER READING

This list includes books and papers that may be helpful to those who wish to pursue further study.

Shapiro, A. H., *The Dynamics and Thermodynamics of Compressible Flow*, Vol. 1, John Wiley & Sons, 1953.

Spotts, M. F., Numerical solution of complex equations, *Design News*, October 6, 1975, pp. 88–90.

Benedict, R. P., *Fundamentals of Pipe Flow*, John Wiley & Sons, 1980. (Chapters 3 and 8.)

Anderson, J., *Modern Compressible Flow, with Historical Perspective*, McGraw-Hill, 2003.

Bernard, P. S. and J. W. Wallace, *Turbulent Flow: Analysis, Measurement and Prediction*, John Wiley & Sons, 2003.

5

NETWORK ANALYSIS

A piping system is regularly made up of a number of piping elements that may be arranged in *series flow*, in *parallel flow*, or in *branching flow*. Series flow, where only one flow rate is involved, is rather straightforward. In parallel flow, two or more series combinations of piping elements diverge and then come together again downstream. In branching flow, a number of piping elements are combined in various series and parallel flow arrangements that, unlike parallel flow, do not necessarily come together again. Branching flow can be extremely complicated to analyze. The three types of flow network arrangements are examined in this chapter. Two example problems are presented to illustrate the setup and solution of pipe flow network arrangements.

Friction factor depends on pipe size, surface roughness, and Reynolds number, and can vary from pipe section to pipe section in parallel and branching flow networks. Differences in friction factor are usually large enough to justify extra computation. A simple approach is to first assign a common friction factor, say 0.020 or 0.030 to each pipe section. After initial network solution, compute Reynolds numbers and reassign friction factor in each pipe section using the Colebrook–White equation (Eq.8.3). One or two iterations may then be needed to obtain a satisfactory solution. Another approach is to assume fully turbulent flow friction factors (Eq.8.2) as a first approximation for each pipe section, followed by network solution and, perhaps, further iterations as described above.

Note that there is essentially one fundamental composition of fluid flow equations. It is suggested that each problem be worked out using the fundamental flow relationships presented in this book. The blind use of special formulae is not recommended. Calculated results can be converted to other units as appropriate after the flow network equations are solved. Unit conversions are available in Appendix C.2.

5.1 COUPLING EFFECTS

Calculation procedures for flow losses in piping systems are normally based on simple one-dimensional flow concepts using loss coefficients realized by experiment. The basic loss coefficients are for isolated components having sufficiently long inlet and outlet lengths of straight pipe to ensure that fully developed flow exists at the inlet to the component and redevelops again downstream of the component. However, when piping components are close coupled, interaction effects, or *coupling effects*, may appreciably affect their performance. Coupling effects may not completely disappear unless the components are separated by a spacer (a straight pipe section) of 30 pipe diameters or more. In practice, 4 or 5 pipe diameters may be sufficient to reduce coupling effects to a negligible level.

One method adopted to account for the close coupling of two pipe components is to multiply the sum of their loss coefficients by a correction factor. The correction factors are frequently less than unity; ignoring them often leads to an overestimation of pressure loss. An exception to this rule is a bend located upstream of a

Pipe Flow: A Practical and Comprehensive Guide, First Edition. Donald C. Rennels and Hobart M. Hudson.

diffuser where the nonuniform flow pattern at the diffuser entrance prevents normal pressure recovery. Likewise, a bend located downstream of a diffuser can cause premature flow separation from the diffuser wall and affect the symmetry of pressure recovery. Another exception is two or more elbows or bends in different planes (a twisted S-form).

Data for a number of close-coupled piping components are available in the literature; see Corp and Hartwell [1], Idel'chik [2], Murikami, et al. [3], and Miller [4].

5.2 SERIES FLOW

In series flow two or more piping elements are connected so that the fluid flows through one element and then another. Only one flow rate is involved. By reason, overall head loss for a number of piping elements connected in series is additive:

$$(H_L)_{\text{Overall}} = K_1 \frac{V_1^2}{2g} + K_2 \frac{V_2^2}{2g} + \ldots + K_N \frac{V_N^2}{2g},$$

where the subscript $_N$ denotes the Nth pipe element. Letting the subscript $_R$ denote a reference location that may or may not be an actual geometric location in the piping arrangement, the continuity relationships are:

$$A_1V_1 = A_2V_2 = \ldots = A_NV_N = A_RV_R.$$

When each term is expressed in terms of a common reference area and a common reference velocity, there results:

$$(H_L)_{\text{Overall}} = \left[K_1\left(\frac{A_R}{A_1}\right)^2 + K_2\left(\frac{A_R}{A_2}\right)^2 + \ldots + K_N\left(\frac{A_R}{A_N}\right)^2\right]\frac{V_R^2}{2g}.$$

By use of the continuity equations, the equation for head loss of a number of piping elements connected in series is:

$$(H_L)_{\text{Overall}} = \left[K_1/A_1^2 + K_2/A_2^2 + \ldots + K_N/A_N^2\right]\frac{\dot{w}^2}{2g\rho_w^2},$$

and the pressure loss equation is simply:

$$\Delta P_{\text{Overall}} = \left[K_1/A_1^2 + K_2/A_2^2 + \ldots + K_N/A_N^2\right]\frac{\dot{w}^2}{2g\rho_w}. \quad (5.1)$$

Equation 5.1 is the broad-spectrum equation for pressure loss of piping elements in series. When the resistance of all piping elements is based on the same reference area A the loss coefficients are directly additive and the pressure loss is given by Equation 5.2:

$$\Delta P_{\text{Overall}} = [K_1 + K_2 + \ldots + K_N]\frac{\dot{w}^2}{2g\rho_w A^2}. \quad (5.2)$$

5.3 PARALLEL FLOW

In parallel flow, two or more flow streams diverge and then converge downstream so that the flow divides among the pipe sections. The head loss is the same in every pipe section, and the individual flow rates are accumulative.* Overall resistance for a number N of piping sections connected in parallel is *not* a simple additive process.

To begin, the sum of the flow rates through each pipe section must equal the total flow into and out of the parallel network (the continuity principle):

$$\dot{w}_{\text{Total}} = \dot{w}_1 + \dot{w}_2 + \ldots + \dot{w}_N. \quad (5.3)$$

The pressure loss equation, or its equivalent, must be satisfied for each pipe section (the energy principle):

$$\Delta P_{\text{Overall}} = \left(\frac{K}{A^2}\right)_{\text{Overall}} \frac{\dot{w}_{\text{Total}}^2}{2g\rho_w}, \quad (5.4)$$

$$\Delta P_1 = \left(\frac{K}{A^2}\right)_1 \frac{\dot{w}_1^2}{2g\rho_w}, \quad (5.5)$$

$$\Delta P_2 = \left(\frac{K}{A^2}\right)_2 \frac{\dot{w}_2^2}{2g\rho_w}, \quad (5.6)$$

and

$$\Delta P_N = \left(\frac{K}{A^2}\right)_N \frac{\dot{w}_N^2}{2g\rho_w}, \quad (5.7)$$

where K in this case represents the sum of the individual flow elements within each pipe section. The algebraic sum of the head loss across each pipe section is the equal:

$$\Delta P_{\text{Overall}} = \Delta P_1 = \Delta P_2 = \ldots = \Delta P_N. \quad (5.8)$$

Substitution of Equations 5.4, 5.5, 5.6 and 5.7 into Equation 5.3 and taking into account Equation 5.8 gives:

$$\left(A/\sqrt{K}\right)_{\text{Overall}} = \left(A/\sqrt{K}\right)_1 + \left(A/\sqrt{K}\right)_2 + \ldots + \left(A/\sqrt{K}\right)_N.$$

* Vessels, reservoirs, or flow manifolds may be required at the divergent and convergent points to truly make the head loss the same in every pipe section when there are more than two parallel flow sections.

Inverting and squaring both sides gives:

$$(K/A^2)_{\text{Overall}} = \left[\frac{1}{\frac{1}{\sqrt{(K/A^2)_1}}+\frac{1}{\sqrt{(K/A^2)_2}}+\ldots+\frac{1}{\sqrt{(K/A^2)_N}}}\right]^2. \quad (5.9)$$

Equation 5.9 is the general equation for pressure loss of pipe sections in parallel. If the individual loss coefficients are based on one common reference area, the overall loss coefficient is:

$$K_{\text{Overall}} = \left[\frac{1}{\frac{1}{\sqrt{K_1}}+\frac{1}{\sqrt{K_2}}+\ldots+\frac{1}{\sqrt{K_N}}}\right]^2.$$

If all individual loss coefficients for each pipe section are equal to the same K, the overall loss coefficient is simply:

$$K_{\text{Overall}} = \frac{K}{N^2}.$$

The individual flow rates for each pipe section may be determined as:

$$\dot{w}_1 = \dot{w}_{\text{Total}}\sqrt{\frac{(K/A^2)_{\text{Overall}}}{(K/A^2)_1}},\ \dot{w}_2 = \dot{w}_{\text{Total}}\sqrt{\frac{(K/A^2)_{\text{Overall}}}{(K/A^2)_2}},\ldots,$$

$$\dot{w}_N\sqrt{\frac{(K/A^2)_{\text{Overall}}}{(K/A^2)_N}}.$$

If the individual loss coefficients for each pipe section are based on one common reference area, the flow rates are:

$$\dot{w}_1 = \dot{w}_{\text{Total}}\sqrt{\frac{K_{\text{Overall}}}{K_1}},\ \dot{w}_2 = \dot{w}_{\text{Total}}\sqrt{\frac{K_{\text{Overall}}}{K_2}},\ldots,$$

$$\dot{w}_N = \dot{w}_{\text{Total}}\sqrt{\frac{K_{\text{Overall}}}{K_N}}.$$

Finally, if all section loss coefficients are equal, the flow rates are simply:

$$\dot{w}_1 = \dot{w}_2 = \ldots = \dot{w}_N = \frac{\dot{w}_{\text{Total}}}{N}.$$

5.4 BRANCHING FLOW

A *branching network* is basically a number of piping sections combined in various series and parallel arrangements, but the branches do not necessarily come together again. Simple branching networks can often be solved by applying the methods given in Sections 5.2 and 5.3 for series and parallel flow. When the number of branches is large, the solution involves careful set up of the governing equations, and requires the aid of a computer program.

The solution of branching flow problems requires the application of *conservation of mass* and *conservation of energy* principles. In short, continuity and energy equations are set up to produce an arrangement of simultaneous equations that can be solved to calculate pressures and flow rates within the branching network.

The key step in setting up a particular branching flow problem is to create a flow schematic or an isometric diagram depicting the network. From this all flow junctions and terminal energy points can be identified and labeled; these points are called *nodes*. For diverging and converging flow junctions (tees), locate the node in the common channel. Indicate a flow direction for every branch, and identify and label its flow rate. If flow direction is uncertain and you are using a computer program for solution, enter the energy equation in the form of $P_A = P_B + CK\dot{w}|\dot{w}|$, where $|\dot{w}|$ is the absolute value of flow rate.

Complex branching flow problems require the simultaneous solution of a number of continuity and energy equations. Several general-purpose computational software programs can easily solve simultaneous equations. Spreadsheets, however, require complicated macros to solve a complex flow network and experience has shown that errors are prevalent. Consequently, if a spreadsheet is used, the solution should be carefully checked to show that it is in fact a correct solution.

Virtually all branching network problems can be solved using the methodology presented in this chapter. The possible arrangement of network flow problems is infinite. Two example problems are presented herein to illustrate the setup and solution of branching flow networks.

5.5 EXAMPLE PROBLEM: RING SPARGER

A ring sparger, as shown in Figure 5.1, is located above the test bay of a liquid rocket engine test stand to provide a shower of water in the event of a leak of rocket propellant or in the event of a fire. The sparger consists of two arm sections (headers) formed of 6″ steel

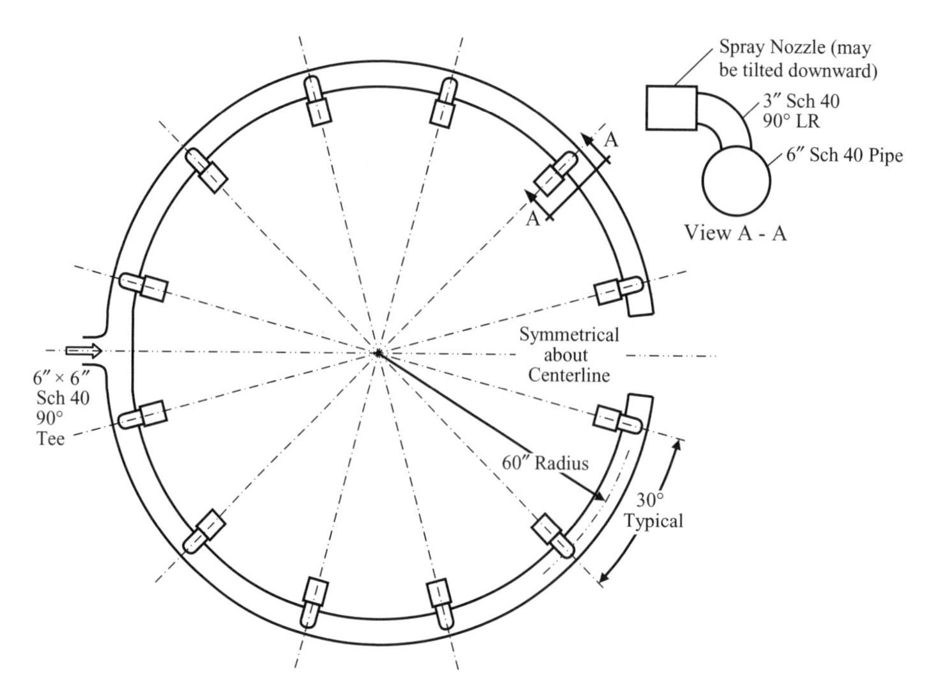

FIGURE 5.1. Ring sparger.

pipe. Each header contains six equally spaced spray nozzles mounted atop 3″ 90° elbows.

A network flow model of the ring sparger is developed in order to evaluate the uniformity of flow among the individual spray nozzles, to determine the pressure distribution within the header, and to determine the pressure at the inlet tee. As shown in Figure 5.2, only one arm is modeled because of flow symmetry about the inlet tee. The nodes are shown in Figure 5.2 as heavy dots and are numbered from 0 to 7.

5.5.1 Ground Rules and Assumptions

- The water temperature is 70°F.
- The total flow rate to the sparger is 6000 gpm.
- According to the manufacturer, the loss coefficient K_{Noz} of the flow nozzles is 10 in terms of velocity head in 3″ schedule 40 pipe.
- Node 0 is located at the inlet to the sparger. Nodes 1–6 are located within the sparger arm just upstream of the diverging branches (i.e., in the common channel). Node 7 is located inside the test bay outside the sparger and is at atmospheric pressure.
- Ignore coupling effects.
- Use Equation 16.18 to model pressure loss through the inlet tee. Assume that the radius ratio r/d of the branch edge equals 0.20.
- Use Equation 16.7 to model pressure loss due to diverging flow through run at Nodes 1 through 6.
- Use Equation 16.15 to model pressure loss due to diverging flow through branch at Nodes 1 through 6. Assume that the branch edge is sharp (radius ratio r/d equals zero).

5.5.2 Input Parameters

$\rho_w = 62.31$ lb/ft^3	Weight density of 70°F water
$\alpha' = 30°$	Angular separation of spray nozzles
$d_3 = 3.068$ in	Inside diameter of 3″ schedule 40 pipe
$d_6 = 6.065$ in	Inside diameter of 6″ schedule 40 pipe
$f_3 = 0.0173$	Friction factor for fully turbulent flow in 3″ elbow
$f_6 = 0.0149$	Friction factor for fully turbulent flow in 6″ pipe
$g = 32.174$ ft/s^2	Acceleration of gravity
$K_{Noz} = 10$	Loss coefficient of spray nozzles in terms of velocity in 3″ pipe
$K' = 0.57$	Coefficient for sharp-edged opening into 90° long radius (LR) elbow in terms of velocity in 3″ pipe
$p_7 = 0$ psig	Gage pressure within test bay
$q_{Spray} = 6000$ gpm	Total water flow rate to ring sparger
$r_{Ring} = 60$ in	Centerline radius of ring sparger
$rd_{ffb} = 0.20$	Rounding radius r/d of branch edge of inlet tee

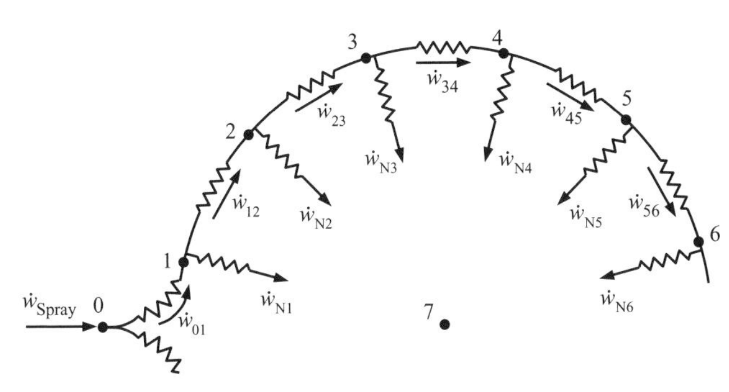

FIGURE 5.2. Ring sparger network flow diagram.

5.5.3 Initial Calculations

$\alpha = \pi\alpha'/180$ $= 0.05236$ rad	Angular separation of spray nozzles in radians
$A_3 = (\pi/4)(d_3/12)^2$ $= 0.05134\ \text{ft}^2$	Flow area of 3″ schedule 40 pipe
$A_6 = (\pi/4)(d_6/12)^2$ $= 0.20063\ \text{ft}^2$	Flow area of 6″ schedule 40 pipe
$K_{Ell} = 0.223$	Loss coefficient of 3″ sch 40, 90° LR elbow (Table 15.2)
$K_{Tee} = 1.59 \cdot 2^2 + \left(1.18 - 1.84\sqrt{rd_{ffb}} + 16\, rd_{ffb}\right)2 - 1.68 + 1.04\sqrt{rd_{ffb}} - 1.16\, rd_{ffb} = 6.09$	Inlet tee pressure drop coefficient (Equation 16.18) for $\dot{w}_{Spray}/\dot{w}_{01} = 2$
$K' = 0.57$	Loss coefficient for sharp-edged entrance from Equation 9.3 for input into Equation 16.15
$K_{30^\circ} = f_6\alpha(r_{Ring}/d_6) + (0.10 + 2.4f_6)\sin(\alpha/2) + 6.6f_6\left[\sqrt{\sin(\alpha/2)} + \sin(\alpha/2)\right] / (r_{Ring}/d_6)^{4(\alpha/\pi)} = 0.13$	Loss coefficient of 30° header pipe segment (Equation 15.1)
$\dot{w}_{Spray} = (\rho_w \cdot q_{Spray})/448.83 = 833.0\ \text{lb/s}$	Total weight flow rate to ring sparger

5.5.4 Network Equations

5.5.4.1 Continuity Equations

$$\dot{w}_{Spray} = 2\dot{w}_{01} \quad \text{Node 0,} \tag{1}$$

$$\dot{w}_{01} = \dot{w}_{12} + w_{N1} \quad \text{Node 1,} \tag{2}$$

$$\dot{w}_{12} = \dot{w}_{23} + w_{N2} \quad \text{Node 2,} \tag{3}$$

$$\dot{w}_{23} = \dot{w}_{34} + w_{N3} \quad \text{Node 3,} \tag{4}$$

$$\dot{w}_{34} = \dot{w}_{45} + w_{N4} \quad \text{Node 4,} \tag{5}$$

$$\dot{w}_{45} = \dot{w}_{56} + w_{N5} \quad \text{Node 5,} \tag{6}$$

$$\dot{w}_{56} = \dot{w}_{N6} \quad \text{Node 6.} \tag{7}$$

5.5.4.2 Energy Equations

Flow through header pipe:

$$p_0 = p_1 + \frac{\dot{w}_{01}^2}{288 g \rho_w A_6^2}\left(K_{Tee} + \frac{K_{30^\circ}}{2}\right), \tag{8}$$

$$p_1 = p_2 + \frac{\dot{w}_{12}^2}{288 g \rho_w A_6^2}\left(1.62 - 0.98\frac{\dot{w}_{01}}{\dot{w}_{12}} - 0.64\frac{\dot{w}_{01}^2}{\dot{w}_{12}^2} + 0.03\frac{\dot{w}_{12}^6}{\dot{w}_{01}^6} + K_{30^\circ}\right), \tag{9}$$

$$p_2 = p_3 + \frac{\dot{w}_{23}}{288 g \rho_w A_6^2}\left(1.62 - 0.98\frac{\dot{w}_{12}}{\dot{w}_{23}} - 0.64\frac{\dot{w}_{12}^2}{\dot{w}_{23}^2} + 0.03\frac{\dot{w}_{12}^6}{\dot{w}_{23}^6} + K_{30^\circ}\right), \tag{10}$$

$$p_3 = p_4 + \frac{\dot{w}_{34}^2}{288 g \rho_w A_6^2}\left(1.62 - 0.98\frac{\dot{w}_{23}}{\dot{w}_{34}} - 0.64\frac{\dot{w}_{23}^2}{\dot{w}_{34}^2} + 0.03\frac{\dot{w}_{34}^6}{\dot{w}_{23}^6} + K_{30^\circ}\right), \tag{11}$$

$$p_4 = p_5 + \frac{\dot{w}_{45}^2}{288 g \rho_w A_6^2}\left(1.62 - 0.98\frac{\dot{w}_{34}}{\dot{w}_{45}} - 0.64\frac{\dot{w}_{34}^2}{\dot{w}_{45}^2} + 0.03\frac{\dot{w}_{45}^6}{\dot{w}_{34}^6} + K_{30^\circ}\right), \tag{12}$$

$$p_5 = p_6 + \frac{\dot{w}_{56}^2}{288 g \rho_w A_6^2}\left(1.62 - 0.98\frac{\dot{w}_{45}}{\dot{w}_{56}} - 0.64\frac{\dot{w}_{45}^2}{\dot{w}_{56}^2} + 0.03\frac{\dot{w}_{56}^6}{\dot{w}_{45}^6} + K_{30^\circ}\right). \tag{13}$$

Flow through spray nozzles:

$$p_1 = p_7 + \frac{\dot{w}_{N1}^2}{288 g \rho_w A_3^2}\left[\left(0.81 - 1.13\frac{\dot{w}_{01}}{\dot{w}_{N1}}\right)\frac{d_3^4}{d_6^4} + 1.00 + 1.12\frac{d_3}{d_6} - 1.08\frac{d_3^3}{d_6^3} + K_{Entr} + K_{Ell} + K_{Noz}\right], \tag{14}$$

$$p_1 = p_7 + \frac{\dot{w}_{N2}^2}{288 g \rho_w A_3^2}\left[\left(0.81 - 1.13\frac{\dot{w}_{12}}{\dot{w}_{N2}}\right)\frac{d_3^4}{d_6^4} + 1.00 + 1.12\frac{d_3}{d_6} - 1.08\frac{d_3^3}{d_6^3} + K_{Entr} + K_{Ell} + K_{Noz}\right], \tag{15}$$

$$p_1 = p_7 + \frac{\dot{w}_{N3}^2}{288 g \rho_w A_3^2}\left[\left(0.81 - 1.13\frac{\dot{w}_{23}}{\dot{w}_{N3}}\right)\frac{d_3^4}{d_6^4} + 1.00 + 1.12\frac{d_3}{d_6} - 1.08\frac{d_3^3}{d_6^3} + K_{Entr} + K_{Ell} + K_{Noz}\right], \tag{16}$$

TABLE 5.1. Calculated Flow Rate

	q_{Spray} ($\dot{w}_{Spray}$)	q_{N1} ($\dot{w}_{N1}$)	q_{N2} ($\dot{w}_{N2}$)	q_{N3} ($\dot{w}_{N3}$)	q_{N4} ($\dot{w}_{N4}$)	q_{N5} ($\dot{w}_{N5}$)	q_{N6} ($\dot{w}_{N6}$)
Flow rate, gpm (lb/s)	6000 (833.0)	481 (66.8)	490 (68.1)	499 (69.3)	506 (70.2)	511 (70.9)	513 (71.2)

$$p_1 = p_7 + \frac{\dot{w}_{N4}^2}{288 g \rho_w A_3^2}\left[\left(0.81 - 1.13\frac{\dot{w}_{34}}{\dot{w}_{N4}}\right)\frac{d_3^4}{d_6^4} + 1.00 + 1.12\frac{d_3}{d_6} - 1.08\frac{d_3^3}{d_6^3} + K_{Entr} + K_{Ell} + K_{Noz}\right], \tag{17}$$

$$p_1 = p_7 + \frac{\dot{w}_{N5}^2}{288 g \rho_w A_3^2}\left[\left(0.81 - 1.13\frac{\dot{w}_{45}}{\dot{w}_{N5}}\right)\frac{d_3^4}{d_6^4} + 1.00 + 1.12\frac{d_3}{d_6} - 1.08\frac{d_3^3}{d_6^3} + K_{Entr} + K_{Ell} + K_{Noz}\right], \tag{18}$$

$$p_1 = p_7 + \frac{\dot{w}_{N6}^2}{288 g \rho_w A_3^2}\left[\left(0.81 - 1.13\frac{\dot{w}_{56}}{\dot{w}_{N6}}\right)\frac{d_3^4}{d_6^4} + 1.00 + 1.12\frac{d_3}{d_6} - 1.08\frac{d_3^3}{d_6^3} + K_{Entr} + K_{Ell} + K_{Noz}\right], \tag{19}$$

5.5.5 Solution

There are 19 equations and 19 unknowns ($p_0, p_1, p_2, p_3, p_4, p_5, p_6, \dot{w}_{01}, \dot{w}_{12}, \dot{w}_{23}, \dot{w}_{34}, \dot{w}_{45}, \dot{w}_{56}, \dot{w}_{N1}, \dot{w}_{N2}, \dot{w}_{N3}, \dot{w}_{N4}, \dot{w}_{N5}$, and $\dot{w}_{N6}$). Thus the equations can be solved simultaneously. Several general-purpose computational software programs are capable of solving simultaneous equations.* Because of the repetitive nature of the equations, and because the branches "come together" at a common node, the solution may also be obtained by compiling a Basic or Fortran computer program, or by employing a spreadsheet program.

Calculated flow rates and header pressures are given in Tables 5.1 and 5.2.

A check of Reynolds numbers revealed that flow is fully turbulent throughout the ring sparger. Thus there was no need to adjust pipe friction factors.

The nozzle flow rates are fairly uniform so adequate spray coverage is assured. The inlet pressure p_0 is within the pressure capacity of the water supply system and the internal pressure of the sparger is well within the pressure capacity of schedule 40 steel pipe.

* The simultaneous equations in the example problems in this chapter were solved using Mathcad (PTC Corporation, Needham, MA), a software program used in engineering and other areas of technical computing. Be aware that some versions may limit the number of simultaneous equations that can be solved to fifty or less. Another computational software program, Mathematica (Wolfram Research, Champaign, IL), is also capable of solving simultaneous equations. Both programs have extensive capabilities and have first-rate user's manuals.

TABLE 5.2. Calculated Header Pressure

	p_0	p_1	p_2	p_3	p_4	p_5	p_6
Pressure, psig	80.7	34.6	36.3	37.4	39.1	40.1	40.7

Note that header pressures and nozzle flows increase along the length of the header pipe in this particular manifold design. Header pressure increases because velocity head is converted to pressure head within the sparger arm as flow is removed through the branches, and this increase in pressure is greater than the drop in pressure due to friction and local loss between the header branches. If friction and local losses between the branches were increased so that they overcome the conversion of velocity head into pressure head, header pressure and nozzle flows would decrease along the length of the header.

5.6 EXAMPLE PROBLEM: CORE SPRAY SYSTEM

Two independent core spray systems are designed to cool the core of a nuclear reactor in the event of a loss of coolant accident (LOCA), in which a break is postulated in any steam or liquid line that forms part of the reactor coolant pressure boundary. The core spray pump draws water from a suppression chamber located below the reactor vessel and injects it into the reactor vessel through spargers (pipe manifolds with spray nozzles) located above the reactor core as shown in Figures 5.3 and 5.4. To enhance plant safety, the core spray system contains redundant pumps, valves, spargers, and so on, and can draw water from an alternate source, but these features are not shown for simplicity. A minimum flow bypass line protects the pump when it operates at or near shutoff head.

Vessel pressure and core flow begin to decrease and water level begins to drop immediately after the start of the postulated LOCA event. The core spray pump is initiated at the beginning of the blowdown period and the injection valve is opened when the pressure in the reactor vessel reaches 120 psia. The core spray system continues to provide cooling water to the top of the core as the vessel pressure drops to or near atmospheric. The purpose of this analysis is to:

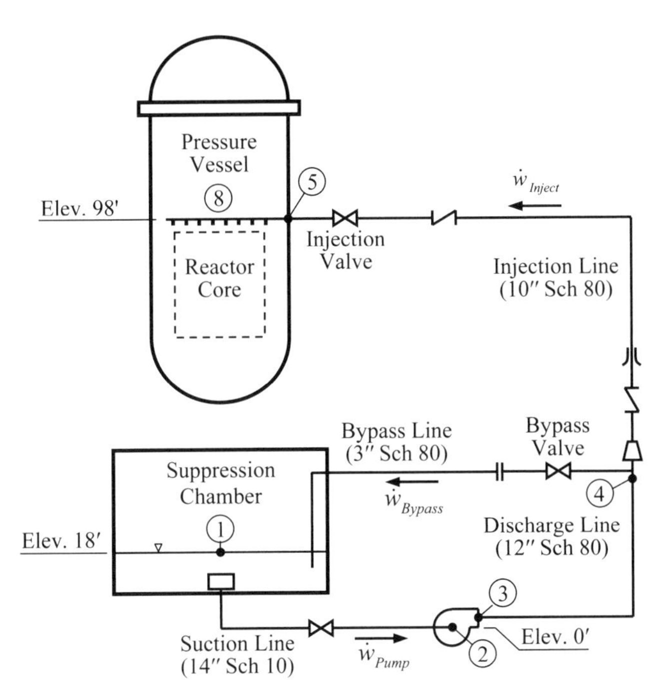

FIGURE 5.3. Core spray system.

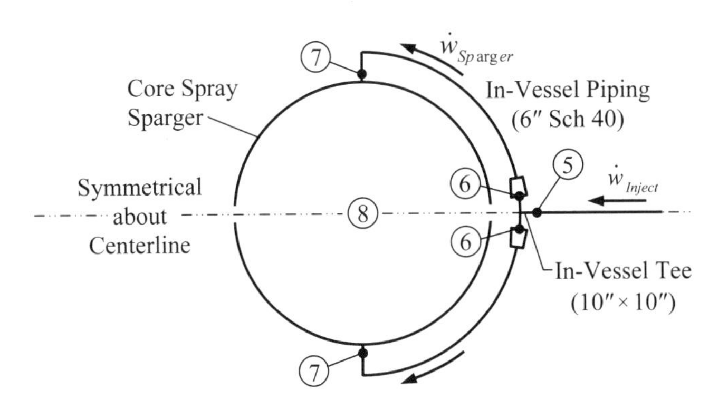

FIGURE 5.4. In-vessel piping and spargers.

1. Demonstrate that bypass flow rate is at least 15% of rated flow when the *bypass valve is open* and the *injection valve is closed* during initial operation of the core spray pump.
2. Determine core spray injection flow rate when the *injection valve is opened*, the *bypass line valve remains open*, and the *vessel pressure progressively drops from 120 to 14.7 psia*
3. Determine core spray injection flow rate when the *injection valve is opened*, the *bypass line valve is closed*, and the *vessel pressure progressively drops from 120 to 14.7 psia.*

Pipe surface conditions can deteriorate over time. The uncertainties of pipe pressure drop calculations are increased by the somewhat unpredictable change in pipe surface roughness, and therefore in friction factor, due to the effects of age and usage.* As an exercise, perform this analysis assuming two different pipe surface roughness: (1) new, clean steel pipe and (2) moderately corroded steel pipe.

5.6.1 New, Clean Steel Pipe

Calculate core spray system flow rate assuming new, clean steel pipe during initial operation.

5.6.1.1 Ground Rules and Assumptions

- The design, or rated, flow rate of the core spray system is 4000 gpm.
- The absolute roughness e of the pipe is equal to 0.00180 in.
- The suppression chamber pressure p_1 is 14.70 psia, the water temperature is 120°F, and the water level is minimum.
- The strainer in the pump suction line may fill with debris during the LOCA event; assume a loss coefficient of 6.0 based on a "dirty" strainer.
- A curve fit of the manufacturer's head versus flow rate curve of the centrifugal pump provides the following quadratic equation:

$$\Delta p = 5.16 \cdot \rho_w \frac{NO^2}{NT^2} - 0.063 \frac{NO}{NT} \dot{w}_{Pump} - \frac{0.0021}{\rho_w} \dot{w}_{Pump}^2 + \frac{0.0015}{\rho_w^2} \frac{NT}{NO} \dot{w}_{Pump}^3,$$

  where the test speed NT of the pump was 3540 rpm. The operating speed NO is 3600 rpm during the LOCA event.
- The diameter ratio of the high beta flow nozzle in the 10″ injection line is 0.520.
- The diameter ratio of the sharp-edged orifice in the 3″ bypass line is 0.460.
- The sparger pressure drop was measured as 16 psid at a test flow rate of 2000 gpm in 80°F water.
- For simplicity, Nodes 3 and 4 are located at the same elevation as Node 2 in the network flow model.
- At Nodes 4 and 5, assume the radius ratio r/d of the branch edges equals 0.10 for the diverging flow tees.
- At Node 4, use Equation 16.7 for pressure loss for diverging flow through run, and use Equation 16.15 for pressure loss for diverging flow through branch.

* See Section 8.6 for a discussion of age and usage on pipe-carrying capacity.

- At Node 5, use Equation 16.18 for pressure loss for diverging flow from branch into run.

5.6.1.2 *Input Parameters* All loss coefficients are in terms of velocity in their respective pipe sections unless otherwise noted.

$\beta_o = 0.460$ — Diameter ratio d_o/d_3 of orifice in 3″ bypass line

$\beta_{FN} = 0.520$ — Diameter ratio d_T/d_{10} of flow nozzle in 10″ injection line

$\Delta p_{Test} = 16$ psid — Sparger test pressure drop

$\rho_w = 61.71$ lb/ft^3 — Density of water at 120°F during postulated LOCA (Appendix A)

$\rho_{Test} = 62.22$ lb/ft^3 — Density of water at 80°F sparger test condition (Appendix A)

$\mu = 1.168 \cdot 10^{-5}$ lb-s/ft^2 — Viscosity of water at 120°F (Appendix A)

$A_3 = 0.04587$ ft^2 — Flow area of 3″ schedule 80 pipe (Appendix B)

$A_6 = 0.2006$ ft^2 — Flow area of 6″ schedule 40 pipe (Appendix B)

$A_{10} = 0.4987$ ft^2 — Flow area of 10″ schedule 80 pipe (Appendix B)

$A_{12} = 0.7056$ ft^2 — Flow area of 12″ schedule 80 pipe (Appendix B)

$A_{14} = 0.9940$ ft^2 — Flow area of 14″ schedule 10 pipe (Appendix B)

$d_3 = 2.900$ in — Inside diameter of 3″ schedule 80 pipe (Appendix B)

$d_6 = 6.065$ in — Inside diameter of 6″ schedule 40 pipe (Appendix B)

$d_{10} = 9.562$ in — Inside diameter of 10″ schedule 80 pipe (Appendix B)

$d_{12} = 11.374$ in — Inside diameter of 12″ schedule 80 pipe (Appendix B)

$d_{14} = 13.500$ in — Inside diameter of 14″ schedule 10 pipe (Appendix B)

$Elev_1 = 18$ ft — Elevation of minimum water level in suppression chamber

$Elev_2 = 0$ ft — Elevation of core spray pump suction inlet

$Elev_5 = 98$ ft — Elevation of core spray sparger

$f_3 = 0.0175$ — Friction factor for fully turbulent flow in 3″ schedule 80 pipe for e of 0.00180 in (Equation 8.2 or Table 15.3)

$f_6 = 0.0149$ — Friction factor for fully turbulent flow in 6″ schedule 40 pipe for e of 0.00180 in (Equation 8.2 or Table 15.2)

$f_{10} = 0.0136$ — Friction factor for fully turbulent flow in 10″ schedule 80 pipe for e of 0.00180 in (Equation 8.2 or Table 15.3)

$f_{12} = 0.0131$ — Friction factor for fully turbulent flow in 12″ schedule 80 pipe for e of 0.00180 in (Equation 8.2 or Table 15.3)

$f_{14} = 0.0127$ — Friction factor for fully turbulent flow in 14″ schedule 10 pipe for e of 0.00180 in (Equation 8.2 or Table 15.1)

$g = 32.174$ ft/s^2 — Acceleration of gravity

$K_{CkValve} = 1.20$ — Loss coefficient of swing check valves in 10″ injection line

$K_{Exit3} = 1.00$ — Loss coefficient of 3″ bypass line exit into suppression pool (Section 12.1)

$K_{FN} = 0.563$ — Loss coefficient of flow nozzle in injection line in terms of the velocity in the nozzle constriction (Diagram 14.1)

$K_{LREll3} = 0.218$ — Loss coefficient for fully turbulent flow in 3″ schedule 80, 90° LR elbow (Table 15.3)

$K_{LREll10} = 0.183$ — Loss coefficient for fully turbulent flow in 10″ schedule 80, 90° LR elbow (Table 15.3)

$K_{LREll12} = 0.179$ — Loss coefficient for fully turbulent flow in 12″ schedule 10, 90° LR elbow (Table 15.1)

$K_{LREll14} = 0.177$ — Loss coefficient for fully turbulent flow in 14″ schedule 10, 90° LR elbow (Table 15.1)

$K_o = 2.05$ — Loss coefficient of sharp-edged orifice in bypass line in terms of velocity in the orifice constriction (Diagram 13.2)

$K_{SREll6} = 0.275$ — Loss coefficient for fully turbulent flow in 6″ schedule 40, 90° short radius (SR) elbow (Table 15.2)

$K_{Strainer} = 6.0$ — Loss coefficient of "dirty" strainer

$K_{Valve3} = 0.20$ — Loss coefficient of open gate valve in bypass line (input 10^{15} for closed valve)

$K_{Valve10} = 0.20$ — Loss coefficient of open gate valve in injection line (input 10^{15} for closed valve)

$K_{Valve14} = 0.20$ Loss coefficient of open gate valve in pump suction line

$L_{1,2} = 40$ ft Pump suction line straight pipe length

$L_{3,4} = 9$ ft Pump discharge line straight pipe length

$L_{4,5} = 145$ ft Injection line straight pipe length

$L_{6,7} = 12$ ft In-vessel line straight pipe length

$L_{4,1} = 60$ ft Bypass line straight pipe length

$NO = 3600$ rpm Operating speed of pump during LOCA

$NT = 3540$ rpm Test speed of pump

$p_1 = 14.7$ psia Suppression chamber pressure

$p_8 = 120$ to 14.7 psia Reactor vessel pressure—progressively decreasing to atmospheric pressure during LOCA

$q_{Test} = 2000$ gpm Sparger test flow rate

$rd = 0.10$ Radius ratio r/d of branch edge of diverging tees at Nodes 4 and 5

5.6.1.3 *Initial Calculations*

Loss coefficient of bypass line orifice in terms of velocity in 3″ pipe:

$$K_{Orifice} = \frac{K_o}{\beta_o^4} = \frac{2.05}{0.460^4} = 45.78.$$

Loss coefficient of injection line flow nozzle in terms of velocity in 10″ pipe:

$$K_{FlowNoz} = \frac{K_{FN}}{\beta_{FN}^4} = \frac{0.563}{0.520^4} = 7.694.$$

Calculate K' for input into Equation 16.15:

$$K' = 0.57 - 1.07rd_T^{1/2} - 2.13rd_T + 8.24rd_T^{3/2} - 8.84rd_T^2 + 2.90rd_T^{5/2} = 0.204.$$

Calculate loss coefficient of sparger based on test data:

$$\dot{w}_{Test} = \frac{\rho_{Test} q_{Test}}{448.83} = \frac{62.22 \times 2000}{448.83} = 277.25 \text{ lb/s},$$

$$K_{Sparger} = \frac{288\, g \rho_{Test} A_6^2 \Delta p_{Test}}{\dot{w}_{Test}^2} = \frac{288 \times 32.174 \times 62.31 \times 0.2006^2 \times 16}{277.25^2} = 4.83.$$

5.6.1.4 *Adjusted Parameters**

After initial solution of the simultaneous equations, a check of Reynolds number revealed that flow is not fully turbulent throughout the system. Accordingly, pipe friction factors were adjusted upward using the Colebrook–White formula (Equation 8.3):

$f_3 = 0.0180$ Adjusted friction factor in 3″ schedule 80 pipe

$f_6 = 0.0153$ Adjusted friction factor in 6″ schedule 40 pipe

$f_{10} = 0.0140$ Adjusted friction factor in 10″ schedule 80 pipe

$f_{12} = 0.0137$ Adjusted friction factor in 12″ schedule 80 pipe

$f_{14} = 0.0134$ Adjusted friction factor in 14″ schedule 10 pipe

Also, elbow loss coefficients were adjusted upward in direct proportion to friction factor increase per Equation 15.2:

$K_{LREll3} = 0.225$ 3″ schedule 80, 90° LR elbow

$K_{SREll6} = 0.282$ 6″ schedule 40, 90° SR elbow

$K_{LREll10} = 0.189$ 10″ schedule 80, 90° LR elbow

$K_{LREll12} = 0.187$ 12″ schedule 80, 90° LR elbow

$K_{LREll14} = 0.187$ 14″ schedule 10, 90° LR elbow

5.6.1.5 *Network Flow Equations*

Continuity equations:

$$\dot{w}_{Pump} = \dot{w}_{Inject} + \dot{w}_{Bypass} \quad \text{Node 4,} \tag{1}$$

$$\dot{w}_{Sparger} = \dot{w}_{Inject}/2 \quad \text{Node 5.} \tag{2}$$

Energy equations:

$$p_1 = p_2 + \frac{K_{Strainer} + K_{Valve\,14} + 4 \cdot K_{LREll\,14} + f_{14}\dfrac{L_{1,2}}{d_{14}/12} + 1}{288 \cdot g \cdot \rho_w \cdot A_{14}^2} \dot{w}_{Pump}^2 + \frac{\rho_w}{144}(Elev_2 - Elev_1), \tag{3}$$

$$p_2 = p_3 - 5.16 \cdot \rho_w \frac{NO^2}{NT^2} + 0.063 \frac{NO}{NT} \dot{w}_{Pump} - \frac{0.0021}{\rho_w} \dot{w}_{Pump}^2 + \frac{0.0015}{\rho_w^2} \frac{NT}{NO} \dot{w}_{Pump}^3, \tag{4}$$

* Friction factors and elbow loss coefficients will vary slightly as a function of valve lineup and vessel pressure from the representative values shown here.

$$p_3 = p_4 + \frac{3 \cdot K_{LREll10} + f_{12}\dfrac{L_{3,4}}{d_{12}/12}}{288 \cdot g \cdot \rho_w \cdot A_{12}^2}\dot{w}_{Pump}^2, \quad (5)$$

$$p_4 = p_5 + \frac{1.62 - 0.98\dfrac{\dot{w}_{Pump}}{\dot{w}_{Inject}} - 0.64\dfrac{\dot{w}_{Pump}^2}{\dot{w}_{Inject}^2} + 0.03\dfrac{\dot{w}_{Inject}^6}{\dot{w}_{Pump}^6}}{288 \cdot g \cdot \rho_w \cdot A_{10}^2}\dot{w}_{Inject}^2 + \frac{K_{FlowNoz} + K_{Valve\,10} + 2 \cdot K_{CkVakve} + 6 \cdot K_{LREll\,12} + f_{10}\dfrac{L_{4,5}}{d_{10}/12}}{288 \cdot g \cdot \rho_w \cdot A_{10}^2}\dot{w}_{Inject}^2 + \frac{\rho_w}{144}(Elev_5 - Elev_2), \quad (6)$$

$$p_5 = p_6 + \frac{1.59\dfrac{\dot{w}_{Inject}^2}{\dot{w}_{Sparger}^2} + \left(1.18 - 1.84\sqrt{rd} + 1.16rd\right)\dfrac{\dot{w}_{Inject}}{\dot{w}_{Sparger}} - 1.68 + 1.04\sqrt{rd} - 1.16rd}{288 \cdot g \cdot \rho_w \cdot A_8^2}\dot{w}_{Sparger}^2, \quad (7)$$

$$p_6 = p_7 + \frac{K_{SREll6} + f_6\dfrac{L_{6,7}}{d_6/12} + 1 - \dfrac{A6^2}{A8^2}}{288 \cdot g \cdot \rho_w \cdot A_6^2}\dot{w}_{Sparger}^2, \quad (8)$$

$$p_7 = p_8 + \frac{K_{Sparger}}{288 \cdot g \cdot \rho_w \cdot A_6^2}\dot{w}_{Sparger}^2, \quad (9)$$

$$p_4 = p_1 + \frac{\left(0.81 - 1.13\dfrac{\dot{w}_{Pump}}{\dot{w}_{Bypass}}\right)\dfrac{d_3^4}{d_{12}^4} + 1.00 + 1.12\dfrac{d_3}{d_1} - 1.08\dfrac{d_3^3}{d_1^3} + K_{9,3}}{288 \cdot g \cdot \rho \cdot A_3^2}\dot{w}_{Bypass}^2 + \frac{K_{Orifice} + K_{Valve\,3} + 4 \cdot K_{LREll3} + K_{Exit} + f_3\dfrac{L_{4,1}}{d_3/12}}{288 \cdot g \cdot \rho \cdot A_3^2}\dot{w}_{Bypass}^2 + \frac{\rho}{144}(Elev_1 - Elev_2). \quad (10)$$

5.6.1.6 Solution There are 10 equations and 10 unknowns ($p_2, p_3, p_4, p_5, p_6, p_7, \dot{w}_{Pump}, \dot{w}_{Inject}, \dot{w}_{Sparger}$ and $\dot{w}_{Bypass}$). A general-purpose computational software program was used to perform the simultaneous solution of equations 1 through 10. After initial solution, Reynolds number was calculated for each pipe section. Pipe friction factors, and elbow loss coefficients, were adjusted accordingly and the solution was repeated to improve the accuracy of the results. The following three cases were investigated:

1. Initial operation of the core spray system was simulated by opening the bypass line valve ($K_{Valve3} = 0.20$) and closing the injection line valve ($K_{Valve10} = 10^{12}$). For this valve lineup, bypass flow rate of 624 gpm exceeded 15% of rated core spray flow.
2. The bypass valve remained open ($K_{Valve3} = 0.20$), the injection valve was opened ($K_{Valve10} = 0.20$), and vessel pressure was progressively decreased from 120 to 14.7 psia to simulate core spray injection during the postulated LOCA.
3. In this case, the bypass valve was closed ($K_{Valve3} = 10^{10}$), the injection valve was opened ($K_{Valve10} = 0.20$), and vessel pressure was progressively decreased from 120 to 14.7 psia to simulate core spray injection during the postulated LOCA.

Calculated flow rates are shown in Table 5.3. Core spray injection flow rate as a function of vessel pressure is shown in Figure 15.5.

The calculated results show that the core spray system will indeed deliver at least 4000 gpm of cooling water to the top of the reactor core during the postulated LOCA event with the bypass valve open. Closing the bypass valve will increase injection flow by over 300 gpm.

Net positive suction head (NPSH), another important aspect of core spray system performance, is evaluated as an example problem in Chapter 20.

5.6.2 Moderately Corroded Steel Pipe

Calculate core spray system flow rate assuming moderately corroded steel pipe. The analysis is basically the same as for new, clean steel pipe except as indicated below.

5.6.2.1 Ground Rules and Assumptions

- The absolute roughness e of the pipe is equal to 0.0130 in.*
- The remaining ground rules and assumption are the same as above for new, clean steel pipe.

5.6.2.2 Input Parameters No changes are necessary.

* In practice, core spray system piping is kept full with demineralized water. It is a closed system so that original oxygen content is soon depleted. In addition, surveillance tests are performed periodically. Assuming that the absolute roughness increases to 0.0130 in is very conservative.

TABLE 5.3. Core Spray System Flow during a Postulated LOCA (New, Clean Steel Pipe—e = 0.00180 in)

Injection valve	Closed	Open					Open				
Bypass valve	Open	Open					Closed				
Vessel pressure, psia	120	120	90	60	30	14.7	120	90	60	30	14.7
Q_{Pump}, gpm (lb/s)	624 (86)	4622 (635)	4905 (674)	5163 (710)	5400 (742)	5514 (758)	4474 (615)	4768 (656)	5038 (693)	5287 (727)	5407 (743)
Q_{Bypass}, gpm (lb/s)	624 (86)	486 (67)	459 (63)	430 (59)	399 (55)	382 (53)	0	0	0	0	0
Q_{Inject}, gpm (lb/s)	0	4136 (569)	4446 (611)	4733 (651)	5001 (688)	5132 (706)	4474 (615)	4768 (656)	5038 (693)	5287 (727)	5407 (744)

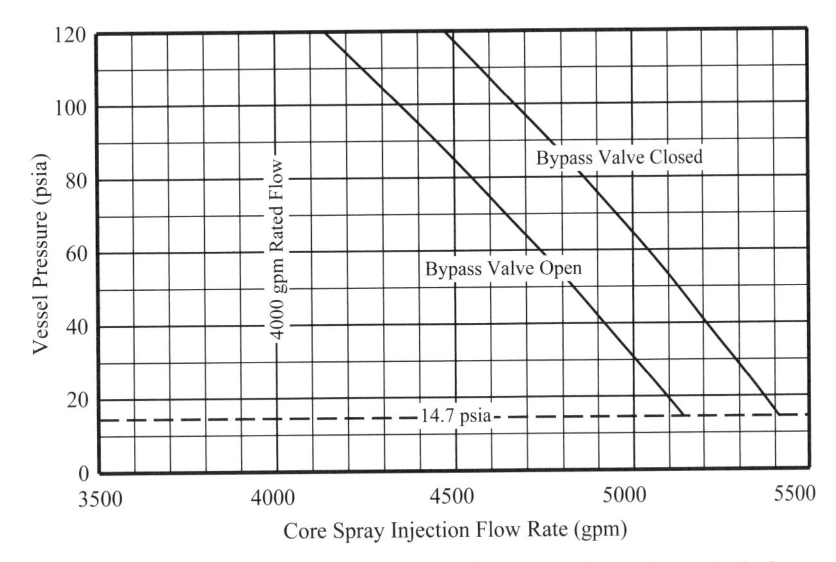

FIGURE 5.5. Core spray injection flow rate versus vessel pressure (new, clean steel pipe—e = 0.00180 in).

5.6.2.3 Adjusted Parameters* After initial solution of the simultaneous equations, a check of Reynolds number revealed that flow is not fully turbulent throughout the system. Accordingly, pipe friction factors were adjusted upward using the Colebrook–White formula (Equation 8.3).

$f_3 = 0.0296$ Adjusted friction factor in 3″ schedule 80 pipe

$f_6 = 0.0240$ Adjusted friction factor in 6″ schedule 40 pipe

$f_{10} = 0.0213$ Adjusted friction factor in 10″ schedule 80 pipe

$f_{12} = 0.0204$ Adjusted friction factor in 12″ schedule 80 pipe

$f_{14} = 0.0197$ Adjusted friction factor in 14″ schedule 10 pipe

* Friction factors and elbow loss coefficients will vary slightly as a function of valve lineup and vessel pressure from the representative values shown here.

Also, elbow loss coefficients were adjusted upward in direct proportion to friction factor increase per Equation 15.2.

$K_{LREll3} = 0.396$ 3″ schedule 80, 90° LR elbow

$K_{SREll6} = 0.437$ 6″ schedule 40, 90° SR elbow

$K_{LREll10} = 0.307$ 10″ schedule 80, 90° LR elbow

$K_{LREll12} = 0.298$ 12″ schedule 80, 90° LR elbow

$K_{LREll14} = 0.292$ 14″ schedule 10, 90° LR elbow

5.6.2.4 Network Flow Equations No changes are necessary.

5.6.2.5 Solution Calculated flow rates are shown in Table 5.4. Core spray injection flow rate as a function of vessel pressure is shown in Figure 15.6.

The calculated results show that the core spray system will still deliver at least 4000 gpm of cooling water to the top of the reactor core during the postulated LOCA in the event the absolute roughness e of the pipe increases to 0.0130 inch.

TABLE 5.4. Core Spray System Flow during a Postulated LOCA (Moderately Corroded Pipe—e = 0.0130 in)

Injection valve	Closed	Open					Open				
Bypass valve	Open	Open					Closed				
Vessel pressure, psia	120	120	90	60	30	14.7	120	90	60	30	14.7
Q_{Pump}, gpm (lb/s)	605 (83)	4550 (625)	4831 (664)	5088 (699)	5323 (732)	5436 (747)	4393 (604)	4687 (644)	4955 (681)	5202 (715)	5322 (732)
Q_{Bypass}, gpm (lb/s)	605 (83)	475 (65)	450 (62)	424 (58)	395 (54)	380 (52)	0	0	0	0	0
Q_{Inject}, gpm (lb/s)	0	4075 (560)	4381 (602)	4664 (641)	4928 (678)	5056 (695)	4393 (604)	4687 (644)	4955 (681)	5202 (715)	5322 (732)

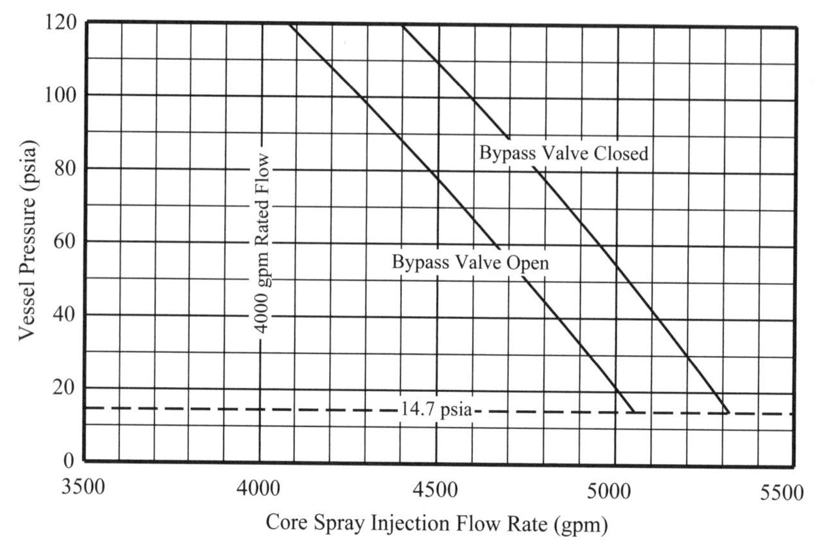

FIGURE 5.6. Core spray injection flow rate versus vessel pressure (moderately corroded steel pipe—e = 0.0130 in).

NPSH, another important aspect of core spray system performance, is evaluated as an example problem in Chapter 20.

REFERENCES

1. Corp, C. I. and H. T. Hartwell, Experiments on Loss of Head in U, S, and Twisted S Pipe Bends, Bulletin of the University of Wisconsin, Engineering Experiment Station Series No. 66, 1927, pp. 1–181.
2. Idel'chik, I. E., *Handbook of Hydraulic Resistance—Coefficients of Local Resistance and of Friction*, Gosudarstvennoe Energeticheskoe Izdatel'stvo, Moskva-Leningrad, 1960. (Translated from Russian; published for the U.S. Atomic Energy Commission and the National Science Foundation, Washington, D. C. by the Israel Program for Scientific Translations, Jerusalem, 1966.).
3. Murikami, M., Y. Shimuzu, and H. Shiragami, Studies on fluid flow in three-dimensional bend conduits, *Bulletin of Japan Society of Mechanical Engineers*, 12(54), 1969, 1369–1379.
4. Miller, D. S., *Internal Flow, a Guide to Losses in Pipe and Duct Systems*, The British Hydromechanics Research Association, 1971.

FURTHER READING

This list includes books and papers that may be helpful to those who wish to pursue further study.

Ito, T., H. Fukawa, H. Okamoto, and H. Hoshino, Analysis of Hydraulic Pipe Line Network Using and Electronic Computer, *Mitsubishi Technical Bulletin 46*, 1967.

Streeter, V. L., *Fluid Mechanics*, 6th ed., McGraw-Hill, 1975, pp. 556–568.

Vennard, J. K. and R. L. Street, *Elementary Fluid Mechanics*, 5th ed., John Wiley & Sons, 1975, pp. 424–439.

Ahuja, R. K., T. L. Magnanti, and J. Orlin, *Network Flows: Theory, Algorithms, and Applications*, Prentice Hall, 1993.

Jones, G. F., *Gravity-Driven Water Flow in Networks*, John Wiley & Sons, 2011.

6

TRANSIENT ANALYSIS

In the analysis of *unsteady flow*, or *transient flow*, processes, the *system* is abandoned in favor of a *control volume*. Attention is given to a fixed region in space, not to a fixed mass in motion. The control volume, defined by an imaginary closed boundary, is set up to surround the equipment under study. Both mass and energy may cross the boundary. Momentum or other properties may be involved, but herein we are considering only mass and energy. We are considering only *bulk flows*. Bulk flows are characterized by negligible *propagation effects*. Propagation effects can be thought of as "startup" effects where momentum or other properties play a role. They are generally of short duration. According to F. J. Moody [1]: "Propagation effects probably are not important in analyses when t_p [*propagation time*] is less than about 0.1 t_d [*bulk disturbance time*]."

In transient flow processes the principles of mass and energy conservation are of utmost importance. The rates at which mass and energy enter the control volume may not be the same as the rates of flow of mass and energy out of the control volume. Furthermore, the rates of flow may vary with time.

Transient analysis is a complex and wide-ranging topic. Books by F. J. Moody [1] and J. A. Fox [2] may be particularly helpful. There are many software programs that can solve a variety of hydraulic transient problems. The attempt here is to introduce the subject by describing the basic methodology and providing example problems that may be useful in their own right.

6.1 METHODOLOGY

It is virtually impossible to set up a completely general set of equations that suit all transient flow processes. Nonetheless, as a first step the control volume must be defined to fit the physical situation. From that, the instantaneous rate at which mass *enters* the control volume may be equated to those at which mass *leaves* the control volume and at which mass is *stored* within the control volume:

Total mass inflow rate = total mass outflow rate + total mass storage rate.

Similarly, the instantaneous rate at which energy *enters* the control volume may be equated to those at which energy *leaves* the control volume and at which energy is *stored* within the control volume:

Total energy inflow rate = total energy outflow rate + total energy storage rate.

Differential equations may be written from consideration of mass and energy principles to simulate the physical situation over a finite interval of time. It is convenient to choose a scheme in which the flow properties are continuous in space and time. Sometimes, however, fluid properties or physical arrangement may be discontinuous so that the problem must be split

Pipe Flow: A Practical and Comprehensive Guide, First Edition. Donald C. Rennels and Hobart M. Hudson.

into two or more parts. It is possible to integrate these equations when the relations between the several variables and time are known. Often closed-form integration is difficult or impossible. In that case, a computer program may be developed to perform the integration using time-step interval processes.*

Transient analysis involving an incompressible fluid may be straightforward, while that involving a compressible fluid may be complex. Several examples of setting up and solving transient flow problems are presented in the following sections. The examples provide solutions to practical problems. The principals demonstrated in the example problems can be applied to other transient flow problems.

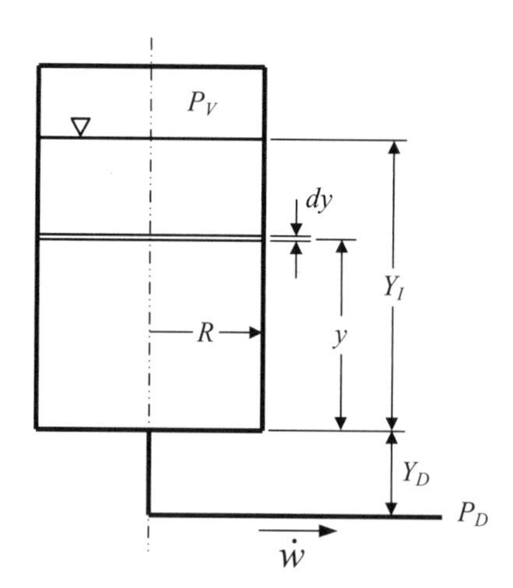

FIGURE 6.1. Drain from an upright cylindrical vessel.

6.2 EXAMPLE PROBLEM: VESSEL DRAIN TIMES

The draining of vessels presents interesting applications of the conservation of mass and energy principles. In these problems, the walls of the vessel and discharge piping make up the control volume. The driving force may be due to pressure as well as gravity.

6.2.1 Upright Cylindrical Vessel

Consider the upright cylindrical vessel of radius R in Figure 6.1. The initial liquid surface height is Y_I. Gravity and pressure draining occurs from a pipe outlet located a vertical distance Y_D from the bottom of the vessel.† The vessel may be pressurized at a constant value P_V and the pressure at the pipe discharge may be P_D. Determine the time t (in seconds) required to drain the vessel.

The change of liquid mass within the cylindrical vessel can be expressed as:

$$\dot{w} = \rho_w A_V \frac{dy}{dt}, \tag{6.1}$$

where ρ_w is the density of the fluid and A_V is the cross-sectional area of the vessel. The flow rate $\dot{w}$ through the drain line can be expressed as:

$$\dot{w} = \sqrt{\frac{2g\rho_w A_P^2}{K_P}[P_V - P_D + \rho_w(Y_D + y)]}, \tag{6.2}$$

where A_P is the flow area of the pipe and K_P is the total loss coefficient of the drain line.‡

Substitution of Equation 6.2 into Equation 6.1, letting $A_v = \pi R^2$, and rearranging yields:

$$dt = \frac{\pi R^2}{\sqrt{\frac{2gA_P^2}{K_P}}\sqrt{\frac{P_V - P_D}{\rho_w} + Y_D + y}} dy. \tag{6.3}$$

The integral form of Equation 6.3 within the limits $y = Y_I$ and $y = 0$ is:

$$\int dt = \frac{\pi R^2}{\sqrt{\frac{2gA_P}{K_P}}} \int_0^{Y_I} \frac{1}{\sqrt{C_{\Delta H} + y}} dy, \tag{6.4}$$

where

$$C_{\Delta H} = \frac{P_V - P_D}{\rho_w} + Y_D.$$

Integration of Equation 6.4 yields the general equation for drain time from an upright cylindrical vessel:

$$t = \frac{2\pi R^2}{\sqrt{\frac{2gA_P{}^2}{K_P}}}\left(\sqrt{C_{\Delta H} + Y_I} - \sqrt{C_{\Delta H}}\right). \tag{6.5}$$

* General purpose programming languages such as FORTRAN and BASIC, as well as technical calculation software such as Mathcad (PTC Corporation, Needham, MA) and Mathematica (Wolfram Research, Champaign, IL), can be used to perform time-step integration. Spreadsheet programs may also be used. See Section 6.4 for an example time-step integration problem solved using Mathcad.

† Y_D may be positive or negative. When the pipe outlet is located above the bottom of the vessel, Y_D is negative, and the equation for a partially drained upright cylindrical vessel is employed.

‡ Drain line losses typically include surface friction loss as well as pipe entrance, pipe exit, valve, and fitting losses (see Part II).

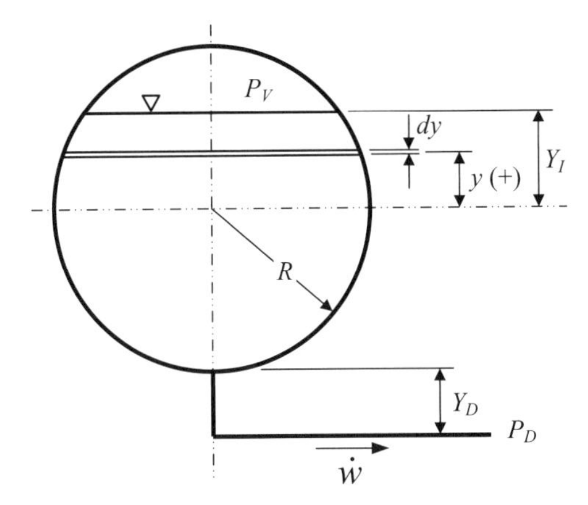

FIGURE 6.2. Drain from a spherical vessel.

The time to partially drain the vessel can be obtained by integrating Equation 6.4 within the limits $y = Y_I$ and $y = Y_{Int}$, where Y_{Int} is an intermediate height between Y_I and 0. Thus, the equation for a partially drained upright cylindrical vessel is:

$$t = \frac{2\pi R^2}{\sqrt{\frac{2gA_P^2}{K_P}}}\left(\sqrt{C_{\Delta H} + Y_I} - \sqrt{C_{\Delta H} + Y_{Int}}\right).$$

6.2.2 Spherical Vessel

A degree of complexity is added to the drain problem when considering a spherical vessel because the cross-sectional area of the vessel varies with height. A spherical vessel of radius R is shown in Figure 6.2. The initial liquid surface height Y_I may be located anywhere between +R and –R. Gravity and pressure draining occurs from a pipe outlet located a vertical distance Y_D from the bottom of the vessel. The vessel may be pressurized at a constant value P_V and the pressure at the pipe discharge may be P_D. Determine the time t (in seconds) required to drain the vessel.

As was the case for the cylindrical vessel, the change of liquid mass within the vessel can be expressed as:

$$\dot{w} = \rho_w A_V \frac{dy}{dt}, \qquad (6.1,\ \text{repeated})$$

where ρ_w is the density of the fluid. The cross-sectional area of the vessel A_V varies with height y as expressed below:

$$A_V = \pi\left(R^2 - y^2\right). \qquad (6.6)$$

The flow rate $\dot{w}$ through the drain line can be expressed as:

$$\dot{w} = \sqrt{\frac{2g\rho_w A_P^2}{K_P}\left[P_V - P_D + \rho_w\left(Y_D + R + y\right)\right]}, \qquad (6.7)$$

where A_P is the flow area of the pipe and K_P is the loss coefficient of the drain line. Substitution and rearranging of Equations 6.1, 6.6 and 6.7 yields:

$$dt = \frac{\pi\left(R^2 - y^2\right)}{\sqrt{\frac{2gA_P^2}{K_P}}\sqrt{\frac{P_V - P_D}{\rho_w} + Y_D + R + y}}\,dy. \qquad (6.8)$$

The integral form of Equation 6.8 within the limits $y = y_I$ and $y = -R$ is:

$$\int dt = \frac{\pi}{\sqrt{\frac{2gA_P^2}{K_P}}}\int_{-R}^{Y_I}\frac{R^2 - y^2}{\sqrt{C_{\Delta H} + R + y}}\,dy, \qquad (6.9)$$

where

$$C_{\Delta H} = \frac{P_V - P_D}{\rho_w} + Y_D.$$

Integration of Equation 6.9 yields the drain time for a spherical vessel:

$$t = \frac{2\pi R^2}{\sqrt{\frac{2gA_P^2}{K_P}}}\left(\sqrt{C_{\Delta H} + R + Y_I} - \sqrt{C_{\Delta H}}\right) - \frac{2\pi}{15\sqrt{\frac{2gA_P^2}{K_P}}}\left\{\begin{array}{l}[8(C_{\Delta H}+R)^2 - 4(C_{\Delta H}+R)Y_I + 3Y_I^2]\\ \sqrt{C_{\Delta H}+R+Y_I} - [8(C_{\Delta H}+R)^2 +\\ 4(C_{\Delta H}+R)R + 3R^2]\sqrt{C_{\Delta H}}\end{array}\right\}.$$

The time to *partially* drain the vessel can be obtained by integrating Equation 6.9 within the limits $y = Y_I$ and $y = Y_{Int}$, where Y_{Int} is an intermediate height between Y_I and $-R$. Thus, the equation for a partially drained spherical vessel is:

$$t = \frac{2\pi R^2}{\sqrt{\frac{2gA_P^2}{K_P}}}\left(\sqrt{C_{\Delta H} + R + Y_I} - \sqrt{C_{\Delta H} + R + Y_{Int}}\right) - \frac{2\pi}{15\sqrt{\frac{2gA_P^2}{K_P}}}\left\{\begin{array}{l}[8(C_{\Delta H}+R)^2 - 4(C_{\Delta H}+R)Y_I + 3Y_I^2]\\ \sqrt{C_{\Delta H}+R+Y_I} - [8(C_{\Delta H}+R)^2 +\\ 4(C_{\Delta H}+R)Y_{Int} + 3Y_{Int}^2]\sqrt{C_{\Delta H}+R+Y_{Int}}\end{array}\right\}.$$

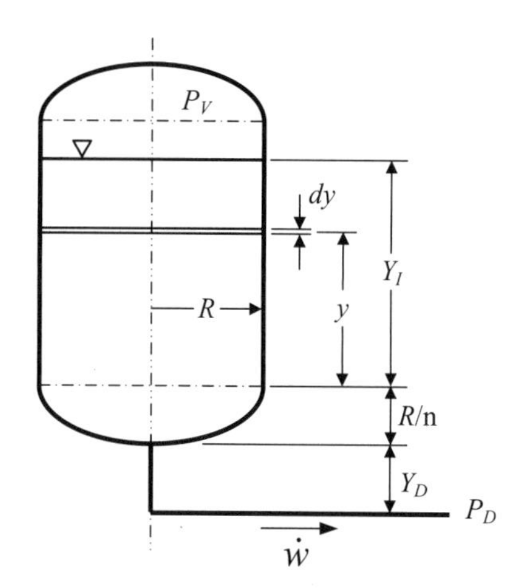

FIGURE 6.3. Drain from a cylindrical vessel with elliptical heads.

6.2.3 Upright Cylindrical Vessel with Elliptical Heads

Consider the cylindrical vessel with elliptical heads in Figure 6.3. The ratio of the major axis to the minor axis of the elliptical head is denoted by n. Note that n = 1 represents a hemispherical head. Assume that the initial liquid surface height Y_I is located within the cylindrical region of the vessel. Draining occurs from a pipe outlet located a vertical distance Y_D from the bottom of the vessel. The vessel may be pressurized at a constant value P_V and the pressure at the pipe discharge may be P_D. Determine the time t required to drain the vessel.

The preceding drain problems for cylindrical and spherical vessels are revisited in order to obtain solutions. As before, the change of liquid mass within the vessel can be expressed as:

$$\dot{w} = \rho_w A_V \frac{dy}{dt}, \qquad (6.1, \text{repeated})$$

where ρ_w is the weight density of the fluid. Because of discontinuity in geometry at the interface between the cylinder and the elliptical head, the problem is split into two parts. Accordingly, the total time t is the sum of the time t_{Cyl} to drain the cylindrical section, and the time t_{BHd} to drain the bottom head:

$$t = t_{Cyl} + t_{BH}.$$

In the case of the cylindrical region of the vessel, the cross-sectional area A_V remains constant with height. The flow rate $\dot{w}$ through the drain line can be expressed as:

$$\dot{w} = \sqrt{\frac{2g\rho_w A_P^2}{K_P}\left[P_V - P_D + \rho_w\left(Y_D + \frac{R}{n} + y\right)\right]}, \qquad (6.10)$$

where A_P is the flow area of the pipe and K_P is the loss coefficient of the drain line. Substitution and rearranging of Equations 6.1 and 6.10 and letting $A_V = \pi R^2$ yields:

$$dt = \frac{\pi R^2}{\sqrt{\frac{2gA_P^2}{K_P}}\sqrt{\frac{P_V - P_D}{\rho_w} + Y_D + \frac{R}{n} + y}}\, dy. \qquad (6.11)$$

To determine drain time for the cylindrical region of the vessel, we employ the integral form of Equation 6.11 within the limits $y = Y_I$ and $y = 0$:

$$\int dt = \frac{\pi R^2}{\sqrt{\frac{2gA_P^2}{K_P}}}\int_0^{Y_I} \frac{1}{\sqrt{C_{\Delta H} + \frac{R}{n} + y}}\, dy, \qquad (6.12)$$

where

$$C_{\Delta H} = \frac{p_V - p_D}{\rho_w} + Y_D.$$

Integration of Equation 6.12 yields the time t_{Cyl} to drain the cylindrical region of the vessel:

$$t_{Cyl} = \frac{2\pi R^2}{\sqrt{\frac{2gA_P^2}{K_P}}}\left(\sqrt{C_{\Delta H} + \frac{R}{n} + Y_I} - \sqrt{C_{\Delta H} + \frac{R}{n}}\right). \qquad (6.13)$$

In the case of the elliptical bottom head region of the vessel, the cross-sectional area varies with height y as expressed below:

$$A_V = \pi\left(R^2 - n^2 y^2\right). \qquad (6.14)$$

In this case, the flow rate through the drain line can be expressed as:

$$\dot{w} = \sqrt{\frac{2g\rho_w A_P}{K_P}\left[P_V - P_D + \rho_w\left(Y_D + \frac{R}{n} + y\right)\right]}. \qquad (6.15)$$

Substitution and rearrangement of Equations 6.1, 6.14 and 6.15 yields:

$$dt = \frac{\pi\left(R^2 - n^2 y^2\right)}{\sqrt{\frac{2gA_P^2}{K_P}}\sqrt{\frac{p_V - p_D}{\rho_w} + Y_D + \frac{R}{n} + y}}\, dy. \qquad (6.16)$$

To determine drain time t_{BHd} of the elliptical bottom head region of the vessel we employ the integral form of Equation 6.16 within the limits $y = 0$ and $y = -R/n$:

$$\int dt = \frac{\pi}{\sqrt{\frac{2gA_P^2}{K_P}}} \int_{-\frac{R}{n}}^{0} \frac{R^2 - n^2y^2}{\sqrt{C_{\Delta H} + \frac{R}{n} + y}} dy, \tag{6.17}$$

where

$$C_{\Delta H} = \frac{p_V - p_D}{\rho_w} + Y_D.$$

Integration of Equation 6.17 yields the time t_{BHd} to drain the elliptical bottom head region of the vessel:

$$t_{BH} = \frac{2\pi R^2}{\sqrt{\frac{2gA_P^2}{K_P}}} \left(\sqrt{C_{\Delta H} + \frac{R}{n}} - \sqrt{C_{\Delta P}} \right) - \frac{2\pi n^2}{15\sqrt{\frac{2gA_P^2}{K_P}}} \left\{ \begin{array}{l} \left(8\left(C_{\Delta H} + \frac{R}{n}\right)^2 \right) \sqrt{C_{\Delta H} + \frac{R}{n}} \\ - \left[8\left(C_{\Delta H} + \frac{R}{n}\right)^2 + \right. \\ \left. 4\left(C_{\Delta H} + \frac{R}{n}\right)\frac{R}{n} + 3\frac{R^2}{n^2} \right] \sqrt{C_{\Delta H}} \end{array} \right\}. \tag{6.18}$$

In conclusion, the results of Equations 6.13 (tCyl) and 6.18 (tBH) are added to obtain the total time t to drain a cylindrical vessel with elliptical heads when pressure, as well as gravity, provides the driving force. It is left to the reader to adapt these equations for the case when the vessel is partially drained.

6.3 EXAMPLE PROBLEM: POSITIVE DISPLACEMENT PUMP

Assume that a container with fixed volume V holds a perfect gas at an initial absolute pressure of P_1 and an initial absolute temperature of T_1. As shown in Figure 6.4, a positive displacement pump removes gas from the container at a constant volumetric flow rate D ft^3/s (or m^3/s) when the volume is measured at the pump inlet pressure and temperature. Assume no pressure drop in the discharge line and no appreciable storage in the discharge line and pump. Gravitation effects are slight and are neglected. The problem is developed in the English system of units. In the International System of Units (SI), substitute mass m for weight w.*

The walls of the container, pump and discharge line, and a section a-a across the end of the pump discharge line, form the boundary of the control volume in this problem. First assume no heat transfer through the boundary—then assume heat transfer.

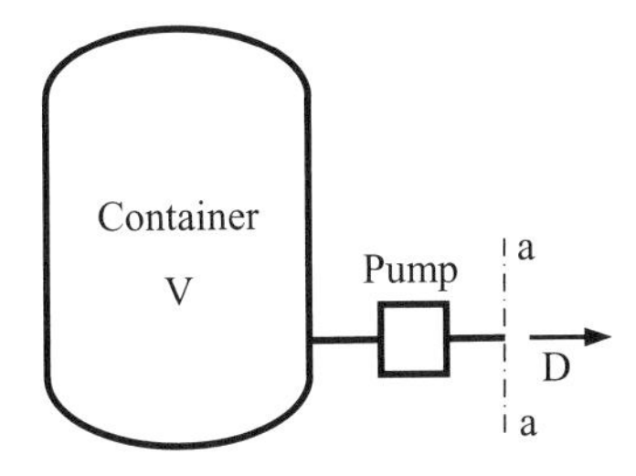

FIGURE 6.4. Positive displacement pump and container.

6.3.1 No Heat Transfer

The weight w of gas within the container at any given time is:

$$w = \frac{PV}{RT}, \tag{6.19}$$

where R is the individual gas constant. The rate at which this gas is removed from the container at a particular time is:

$$-\frac{dw}{dt} = \frac{P}{RT}\left(\frac{dV}{dt}\right)_{pump} = \frac{PD}{RT}. \tag{6.20}$$

Differentiation of Equation 6.19 yields:

$$\frac{dw}{dt} = \frac{\partial w}{\partial P}\frac{dP}{dt} + \frac{\partial w}{\partial T}\frac{dT}{dt} + \frac{\partial w}{\partial V}\frac{dV}{dt}, \tag{6.21}$$

and

$$\frac{\partial w}{\partial P} = \frac{V}{RT}, \frac{\partial w}{\partial T} = -\frac{PV}{RT^2}, \frac{\delta w}{\partial V} = \frac{P}{RT}. \tag{6.22}$$

Substituting Equation 6.22 into Equation 6.21 and rearranging gives:

$$\frac{dw}{dt} = \frac{V}{RT}\frac{dP}{dt} - \frac{PV}{RT^2}\frac{dT}{dt} + \frac{P}{RT}\frac{dV}{dt}.$$

In this problem there is no change of volume ($dV/dt = 0$). Therefore:

$$\frac{dw}{dt} = \frac{V}{RT}\frac{dP}{dt} - \frac{PV}{RT^2}\frac{dT}{dt}. \tag{6.23}$$

Substitution of Equation 6.20 into Equation 6.23 and rearrangement yields:

$$dt = \frac{V}{DT}dT - \frac{V}{DP}dP. \tag{6.24}$$

We employ the integral form of Equation 6.24:

$$\int_{t_1}^{t_2} dt = \frac{V}{D}\int_{T_2}^{T_1}\frac{dT}{T} - \frac{V}{D}\int_{P_2}^{P_1}\frac{dP}{P}.$$

* This example problem is an enlargement of a problem presented by Mackey et al. [3].

Integration yields:

$$t_2 - t_1 = \frac{V}{D}[(\ln P_1 - \ln P_2) - (\ln T_1 - \ln T_2)]$$
$$= \frac{V}{D}\left[\ln\frac{P_1}{P_2} - \ln\frac{T_1}{T_2}\right] = \frac{V}{D}\ln\frac{P_1T_2}{P_2T_1}.$$

Then,

$$\frac{w_1}{w_2} = \frac{P_1T_2}{P_2T_1} = e^{\frac{D(t_2-t_1)}{V}}. \quad (6.25)$$

In addition to the mass balance, there is an energy balance to satisfy. The rate at which the internal energy of the gas in the container is decreasing at a particular time must equal the rate at which energy is crossing the control volume. Assume that gravitational potential energy and kinetic energy are insignificant.

The rate at which the internal energy is decreasing is:

$$-dQ = -\frac{c_v V}{R}\frac{dP}{dt}.$$

The rate at which energy crosses the prescribed section a in the form of internal energy and flow work is:

$$dQ = \frac{c_p PD}{R}. \quad (6.26)$$

The energy balance is:

$$-\frac{c_v V}{R}\frac{dP}{dt} = \frac{c_p PD}{R}. \quad (6.27)$$

For constant specific heats, $\gamma = c_p/c_v$, the integral form of Equation 6.27 can be expressed as:

$$-\int_{P_1}^{P_2}\frac{dP}{P} = \frac{\gamma D}{V}\int_{t_1}^{t_2} dt.$$

Integration yields:

$$\ln P_1 - \ln P_2 = \frac{\gamma D}{V}(t_2 - t_1) \quad (6.28)$$

or

$$t_1 - t_2 = \frac{V(\ln P_2 - \ln P_1)}{\gamma D}.$$

Rearranging Equation 6.28 also yields:

$$\frac{P_1}{P_2} = e^{\frac{\gamma D(t_2-t_1)}{V}}.$$

6.3.2 Heat Transfer

Assume that there is heat transfer to the control volume at the instantaneous rate H Btu/s (or J/s) at a particular time. The same mass balance holds as in the case of no heat transfer:

$$\frac{w_1}{w_2} = \frac{P_1T_2}{P_2T_1} = e^{\frac{D(t_2-t_1)}{V}}. \quad (6.25, \text{repeated})$$

The rate at which the internal energy of the gas in the container is decreasing is:

$$-dQ = -\frac{c_v V}{R}\frac{dP}{dt} - H.$$

The rate at which energy crosses the prescribed section a in the form of internal energy and flow work is:

$$dQ = \frac{c_p PD}{R}. \quad (6.26, \text{repeated})$$

The new energy balance is:

$$H = \frac{c_v V}{R}\frac{dP}{dt} + \frac{c_p DP}{R}.$$

The integral form can be expressed as:

$$\int_{t_1}^{t_2}\left(\frac{H}{P} - \frac{c_v D}{R}\right)dt = \frac{c_p V}{R}\int_{P_1}^{P_2}\frac{dP}{P}.$$

Partial integration yields:

$$\frac{c_v V(\ln P_2 - \ln P_1)}{R} = \int_{t_1}^{t_2}\frac{H}{P}dt - \frac{c_p D(t_2 - t_1)}{R}. \quad (6.29)$$

For constant specific heats, $\gamma = c_p/c_v$, and rearranging, Equation 6.29 becomes:

$$t_1 - t_2 = \frac{V(\ln P_2 - \ln P_1)}{\gamma D} - \frac{R}{c_p D}\int_{t_1}^{t_2}\frac{H}{P}dt.$$

Then,

$$\frac{P_1}{P_2} = e^{\frac{\gamma D(t_2-t_1)}{V} - \frac{R}{Vc_p}\int_{t_1}^{t_2}\frac{H}{P}dt}.$$

It is possible to integrate the residual differential equation when the relations between heat transfer H and the several variables and time are known. A numerical or time-step integration may then be employed to obtain the solution.

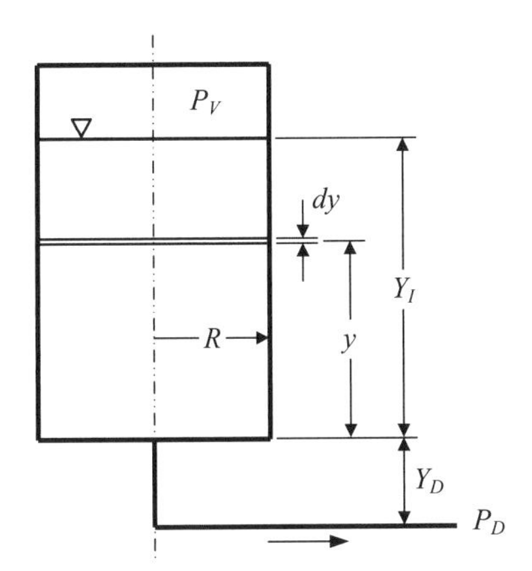

FIGURE 6.5. Drain from an upright cylindrical vessel (repeated).

6.4 EXAMPLE PROBLEM: TIME-STEP INTEGRATION

Numerical integration is the approximate computation of an integral using numerical techniques. The term is sometimes used to describe the numerical solution of differential equations using a *time-step integration* process. Herein the time-step process will be demonstrated by obtaining particular solutions to the differential equation developed in Section 6.2.1 representing drain time from an upright cylindrical vessel.

6.4.1 Upright Cylindrical Vessel Drain Problem

Consider the upright cylindrical vessel shown in Figure 6.5. The radius of the vessel is 6 ft. The vessel contains 70°F water at an initial height of 20 ft. The vessel pressure is constantly maintained at 100 lb/in^2. The drain line discharges to atmosphere from an outlet located a distance of 4 ft below the bottom of the vessel. The loss coefficient of the drain line is based on velocity in the 6″ schedule 40 drain pipe.

The input parameters are:

$R = 6$ ft	Vessel radius
$Y_I = 20$ ft	Initial height of water
$Y_D = 4$ ft	Vertical distance of drain line exit from bottom of vessel
P_V (= 100 lb/in^2) = 14,400 lb/ft^2	Vessel pressure
P_D (= 14.7 lb/in^2) = 2117 lb/ft^2	Atmospheric pressure
$A_P = 0.20063$ ft^2	Drain line flow area (6″ schedule 40 pipe)
$K = 5$	Drain line loss coefficient
$\rho_w = 62.31$ lb/ft^3	Density of 70°F water

6.4.2 Direct Solution

Equation 6.5 can be employed to directly solve for the drain time.

$$t = \frac{2\times\pi\times 6^2}{\sqrt{\dfrac{2\times 32.174\times 0.2006^2}{5}}}\left(\sqrt{\frac{14400-2117}{62.31}+4+20}-\sqrt{\frac{14400-2117}{62.31}+4}\right)$$

$$= 216.38 \text{ s.}$$

6.4.3 Time-Step Solution

Now calculate the drain time t using Equation 6.3, the integral form of Equation 6.5, assuming progressively smaller time steps Δt.*

Several programs are available to perform step-by-step integration. They are all similar in that they form a loop to repeatedly execute a calculation until a certain condition is met. The one shown below, uses a "while" loop, a programming feature found in Mathcad† as well as in other computational programs.

$y = Y_I$	Sets initial water level
$t = 0$	Sets transient start time at zero
$\Delta t = 10$ s (=1 s) (=0.1 s)	Sets time-step interval (three intervals will be used)
while $y > 0$	Sets condition to be met
$dy = ((\sqrt{2gA_P / K_P}\sqrt{((P_V - P_D)/\rho_w) + Y_D + y})/\pi R^2)\Delta t$	Repeatedly calculates Equation 6.3 (rearranged)
$t = t + \Delta t$	Resets time
$y = y - dy$	Resets water level; continues execution while $y > 0$
$t = 220$ s (=217 s) (=216.4 s)	The condition is met when $y \le 0$ (three solutions for three time step intervals)

* In this application, Δt is substituted for dt. When Δt *is* small, $\Delta t \approx dt$, and the closer the time-step interval is to zero the better the results are, provided the precision of the arithmetic (as in significant figures) does not start providing inaccurate answers.

† Mathcad is a computational software program used in engineering and other areas of technical computing. The Mathcad Professional version may be needed to provide the programming tools required to solve time-step problems.

Setting the time step at 10 seconds would likely provide a sufficiently accurate answer for this simple computation. Other computations, such as a transient analysis program that simulates a loss of coolant accident (LOCA) in a nuclear power plant, would use small time steps, and would simultaneously track pressure, temperature, phase change, fluid density, flow rate, and so on, in addition to water level.

REFERENCES

1. Moody, F. J., *Unsteady Thermofluid Mechanics*, John Wiley & Sons, 1990.
2. Fox, J. A., *Hydraulic Analysis of Unsteady Flow in Pipe Networks*, The MacMillan Press Ltd, 1984.
3. Mackay, C. O., W. N. Barnard, and F. O. Ellenwood, *Engineering Thermodynamics*, John Wiley & Sons, 1957.

FURTHER READING

This list includes books and papers that may be helpful to those who wish to pursue further study.

Streeter, V. L. and E. B. Wylie, *Hydraulic Transients*, McGraw-Hill, 1967.

Webb, S. W. and J. L. Caves, Fluid transient analysis in pipelines with nonuniform liquid density, *Journal of Fluids Engineering, Transactions of American Society of Mechanical Engineers*, 105, 1983, 423–428.

Blevins, R. D., *Applied Fluid Dynamics Handbook*, Van Nostrand Reinhold Company, 1984.

Nanayakkara, S. and N. D. Perrieira, Wave propagation and attenuation in piping systems, *Journal of Vibration, Acoustics, Stress, and Reliability in Design*, 108, 1986, 441–446.

Pejovic, S., A. P. Boldy, and D. Obradovic, *Guidelines to Hydraulic Transient Analysis*, Gower Technical Press, 1987.

Thorley, R. D. and C. H. Tiley, Unsteady and transient flow of compressible fluids in pipelines—A review of theoretical and some experimental studies, *Heat and Fluid Flow*, 8, 1987, 3–15.

Ellis, J., *Pressure Transients in Water Engineering: A Guide to Analysis and Interpretation of Behavior*, Thomas Telford Ltd., 2008.

7

UNCERTAINTY

Uncertainty is the probable range of error. The uncertainty associated with hydraulic analysis may be defined as the *statistical difference* between *calculated* pressure drop and *true* pressure drop. It may also be defined as the statistical difference between *calculated* flow rate and *true* flow rate.*

There is uncertainty associated with practically every variable involved in calculating pressure drop or flow rate. In the uncertainty analysis, it is assumed that the individual errors have an equal probability of being positive or negative; and further, that the probability density function describing the uncertainty is normally distributed. This is done for three reasons: first, these assumptions simplify the mathematical manipulations required; second, the existing knowledge regarding the uncertainty of the various errors does not allow a more sophisticated treatment; and finally, this allows convenient expression of the results. With these assumptions, the combined value of the various errors can be determined and expressed as ±1 (or 2 or 3) standard deviation(s) of the mean (or calculated) pressure drop or flow rate.

7.1 ERROR SOURCES

The predominant source of error in the calculation of pressure drop or flow rate is associated with the loss coefficients of the various elements within the flow system. The accuracy of loss coefficients is subject to dimensional and surface roughness differences, experimental and theoretical variations, and to uncertainties associated with modeling the loss coefficient over a wide range of variables.

Suggested 3-sigma values of uncertainty associated with the loss coefficients documented in Part II are given in Table 7.1.† The uncertainty values are expressed as percentage. They reflect the authors' judgment based on selecting, developing, and formulating the loss coefficients for the various flow elements.

The suggested uncertainty values for tees are for normal ranges of flow and diameter ratio. Higher values may be appropriate outside of these ranges. It should be noted that some loss coefficient values in junctions pass through zero so that uncertainty values on a percentage basis are unrealistic. Fixed uncertainty values may be appropriate under these conditions.

7.2 PRESSURE DROP UNCERTAINTY

The basic pressure drop equation for a pipe section can be expressed as:

$$P_1 - P_2 = \left(K_1 - 1 + \frac{A_1^2}{A_2^2} \right) \frac{\dot{w}^2}{2g\rho_w A_1^2} + \rho_w (Z_2 - Z_1). \quad (7.1)$$

In pressure drop analysis the values of flow rate, density, and flow area are established so that they contain little or no uncertainty. Assuming a normal distribution of error, the percentage uncertainty of calculated pressure

* There is somewhat of a difficulty here because we also have uncertainty associated with determining the true value. Perhaps we should say that the uncertainty is the difference between the *calculated* value and the *measured* value. Then that could encompass errors not only in calculating but also in measuring.

† If a data distribution is approximately normal, then 3 sigma (or three standard deviations) accounts for 99.73% of the data set.

Pipe Flow: A Practical and Comprehensive Guide, First Edition. Donald C. Rennels and Hobart M. Hudson.

TABLE 7.1. Suggested Uncertainty Values

	Three-Sigma Uncertainty (%)
Friction factor	
Laminar flow	5
Turbulent flow	
Smooth	5
Known surface roughness	10
Unknown surface roughness	>10
Entrances	
Sharp-edged	6
Round-edged	10
Bevel-edged	25
Through an orifice	
Sharp-edged	5
Round-edged	10
Thick-edged	15
Bevel-edged	25
Contractions	
Sudden	8
Rounded	15
Conical	15
Beveled	25
Smooth	10
Pipe reducer–contracting	20
Expansions	
Sudden	5
Conical diffuser	
$\alpha \leq 20°$	10
$20° < \alpha < 40°$	10–20
$40° < \alpha < 180°$	15
Stepped conical diffuser	20
Two-stage conical diffuser	20
Curved wall diffuser	20
Pipe reducer–expanding	20
Exits	
From a straight pipe	5
From a conical diffuser	15
From an orifice	
Sharp-edged	5
Round-edged	10
Thick-edged	15
Bevel-edged	25
From a smooth nozzle	5
Orifices	
Sharp-edged	5
Round-edged	10
Bevel-edged	25
Thick-edged	15
Multihole	20
Noncircular	20
Flow meters	
Flow nozzle	5
Venturi tube	15
Nozzle/Venturi	15
Bends	
Elbows and pipe bends	
≤90°	10
>90°	15
Coils	15
Miter bends	15
Tees	
Diverging flow through run	10
Diverging flow through branch	15
Diverging flow from branch	15
Converging flow through run	20
Converging flow through branch	10
Converging flow into branch	15
Joints	
Welds protrusion	25
Backing rings	15
Misalignment	25
Valves	
Specified	5
Estimated	20–100
Threaded fittings	
Reducers—contracting	20–50
Reducers—expanding	20
Elbows	30
Tees	30

Note: The designer may select higher uncertainty values based on the needs of his equipment, applicable codes, local engineering practice, and design margins to meet customer/hardware performance requirements..

drop in a pipe section composed of n different elements can be determined by:

$$\sigma_{dP} = \frac{\sqrt{\sum_{i=1}^{n} (\mathrm{N}_i \sigma_i K_i)^2}}{\sum_{i=1}^{n} \mathrm{N}_i K_i}, \tag{7.2}$$

where N_i is the number of like or similar elements that share a common loss coefficient K_i.* The percentage uncertainty σ_{dP} can be based on one, two, or three standard deviation(s).

* Treat surface friction as a single element unless there are size or surface roughness differences along the length of the pipe section.

7.3 FLOW RATE UNCERTAINTY

The basic flow rate equation for a pipe section can be expressed as:

$$\dot{w} = \sqrt{\frac{2g\rho_w A_1^2}{\left(K_1 - 1 + \frac{A_1^2}{A_2^2}\right)}[P_1 - P_2 + \rho_w(Z_1 - Z_2)]}. \quad (7.3)$$

In this case the values of pressure drop, density, and flow area are established so they contain little or no uncertainty. Assuming a normal distribution of error, the percentage uncertainty of calculated flow rate of a pipe section composed of n different elements can be determined by:

$$\sigma_{\dot{w}} = \frac{\sqrt{\sum_{i=1}^{n}(\mathrm{N}_i\sigma_i K_i)^2}}{2\sum_{i=1}^{n}\mathrm{N}_i K_i}, \quad (7.4)$$

where, as before, N_i is the number of like or similar elements that share a common loss coefficient K_i.* The percentage uncertainty $\sigma_{\dot{w}}$ can be based on one, two, or three standard deviation(s).

7.4 EXAMPLE PROBLEM: PRESSURE DROP

Water at 70 F is flowing at the rate of 125 lb/s through the 4″ schedule 40 pipe section shown in Figure 7.1. In accordance with Equations 7.1 and 7.2 and Table 7.1, calculate pressure drop in the pipe section within 3-sigma uncertainty assuming new, clean steel pipe.

* See previous note.

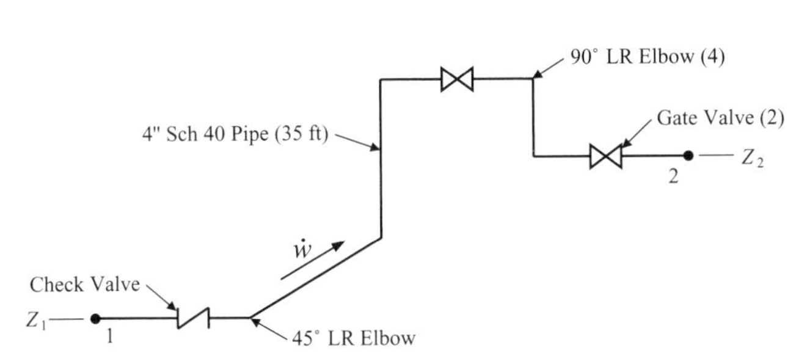

FIGURE 7.1. Four-inch pipe section.

7.4.1 Input Data

$$\dot{w} = 125 \text{ lb/s} \quad Z_1 = 500 \text{ ft} \quad Z_2 = 100 \text{ ft}$$

$$\rho_w = 62.30 \text{ lb/ft}^3 \text{ (Table A.1)}$$

$$\mu = 2.037\times10^{-5} \text{ lb-s/ft}^2 \text{ (Table A.1)}$$

4″ Schedule 40 Pipe (New, Clean Steel)

$$L = 35 \text{ ft} \qquad \varepsilon = 0.00015 \text{ ft (Table 8.1)}$$

$$D = 0.3355 \text{ ft (Table B.1)} \quad \text{A} = 0.0884 \text{ ft}^2 \text{ (Table B.1)}$$

$$f_\text{T} = 0.0163 \text{(initially assume fully turbulent flow) (Table 15.2)}$$

$$N_\text{Re} = \frac{\dot{w}D}{\mu g A} = 7.238\times10^5 \quad \text{(Eq. 1.2a)}$$

$$f = \left[2\log_{10}\left(\frac{\varepsilon}{3.7D} + \frac{2.41}{N_\text{Re}\sqrt{f_\text{T}}}\right)\right]^{-2} = 0.01702 \quad \text{(Eq. 8.3)}$$

$$f = \left[2\log_{10}\left(\frac{\varepsilon}{3.7D} + \frac{2.41}{N_\text{Re}\sqrt{f}}\right)\right]^{-2} = 0.01700$$

$$\mathrm{N}_1 = 1 \quad \sigma_1 = 10\% \quad K_1 = fL/D \quad \text{(Eq. 1.3)}$$

$$K_1 = 1.774$$

45° LR Elbow

$$\mathrm{N}_2 = 1 \quad \sigma_2 = 10\% \quad K_\text{T} = 0.144 \text{ (Table 15.2)}$$

$$K_2 = \frac{f}{f_\text{T}}K_\text{T} = 0.150$$

90° LR Elbow

$$\mathrm{N}_3 = 4 \quad \sigma_3 = 10\% \quad K_\text{T} = 0.211 \text{ (Table 15.2)}$$

$$K_3 = \frac{f}{f_\text{T}}K_\text{T} = 0.220$$

Check Valve

$$\mathrm{N}_4 = 1 \quad \sigma_4 = 5\% \quad K_4 = 1.20 \text{ (specified)}$$

Gate Valve

$$\mathrm{N}_5 = 2 \quad \sigma_5 = 5\% \quad K_5 = 0.20 \text{ (specified)}$$

7.4.2 Solution

$$\begin{aligned} K_{Total} &= N_1K_1 + N_2K_2 + N_3K_3 + N_4K_4 + N_5K_5 \\ &= 1\times1.774 + 1\times0.144 + 4\times0.211 + 1\times1.20 + 2\times0.20 \\ &= 4.404 \end{aligned}$$

$$\begin{aligned} (P_1 - P_2)_{\text{Mean}} &= \frac{K_{Total}\dot{w}^2}{2g\rho_w A^2} + \rho_w(Z_2 - Z_1) \\ &= \frac{4.404\times125^2}{2\times32.174\times62.30\times0.08814^2} + 62.30(5-0) \\ &= 2196 \text{ lb/ft}^2 + 312 \text{ lb/ft}^2 \\ &= 2508 \text{ lb/ft}^2 \end{aligned}$$

$$\begin{aligned} (p_1 - p_2)_{\text{Mean}} &= 15.25 \text{ lb/in}^2 + 2.16 \text{ lb/in}^2 \\ &= 17.41 \text{ lb/in}^2 \end{aligned}$$

$$\begin{aligned} \sigma_{dp} &= \frac{\sqrt{\begin{matrix}(N_1\sigma_1K_1)^2 + (N_2\sigma_2K_2)^2 + \\ (N_3\sigma_3K_3)^2 + (N_4\sigma_4K_4)^2 + (N_5\sigma_5K_5)^2\end{matrix}}}{N_1K_1 + N_2K_2 + N_3K_3 + N_4K_4 + N_5K_5} \\ &= \frac{\sqrt{\begin{matrix}(1\times10\times1.774)^2 + (1\times10\times0.144)^2 + \\ (4\times10\times0.211)^2 + (1\times5\times1.20)^2 + (2\times5\times0.20)^2\end{matrix}}}{4.404} \\ &= 4.743\% \end{aligned}$$

$$p_1 - p_2 = 15.25\left(1 \pm \frac{4.743}{100}\right) + 2.16$$

$$(p_1 - p_2)_{\text{Min}} = 16.69 \text{ lb/in}^2$$

$$(p_1 - p_2)_{\text{Max}} = 18.15 \text{ lb/in}^2$$

Based on the loss coefficient values and 3-sigma uncertainties assigned to the various piping elements, the predicted pressure drop in the pipe section ranges from 16.69 to 18.15 lb/in^2.

7.5 EXAMPLE PROBLEM: FLOW RATE

A 14-in pipeline connects two reservoirs as shown in Figure 7.2. Calculate the flow rate in the pipeline within 3-sigma uncertainty in accordance with Equations 7.3 and 7.4 and Table 7.1. Assume new, clean steel pipe and water temperature of 70°.

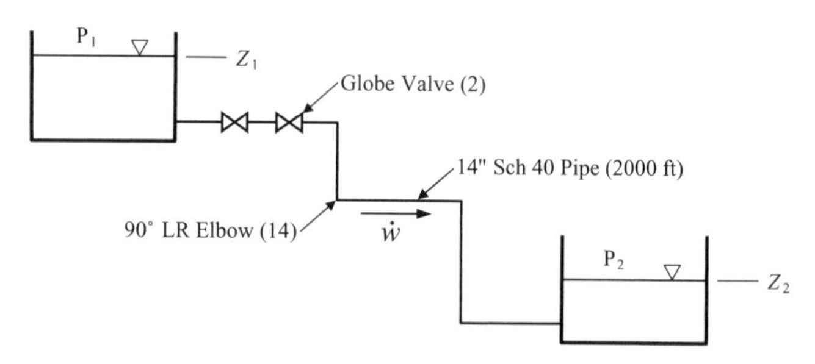

FIGURE 7.2. Fourteen-inch pipeline.

7.5.1 Input Data

$\rho_w = 62.30 \text{ lb/ft}^3$ (Table A.1)

$\mu = 2.037\times10^{-5} \text{ lb/s/ft}^2$ (Table A.1)

$Z_1 = 500$ ft

$Z_2 = 100$ ft

$P_1 = P_2 =$ Atmospheric

14″ Schedule 40 Pipe (New, Clean Steel)

$L = 2000$ ft $\quad \varepsilon = 0.00015$ ft (Table 8.1)

$D = 1.0937$ ft (Table B.1) $\quad A = 0.9394 \text{ ft}^2$ (Table B.1)

$f_T = 0.0127$ (initially assume fully turbulent flow) (Table 15.2)

$N_1 = 1 \quad \sigma_1 = 10\% \quad K_1 = \dfrac{fL}{D} = 23.23$ (Eq. 1.3)

Rounded Entrance

$N_2 = 1 \quad \sigma_2 = 10\% \quad K_2 = 0.10$ (Diagram 9.2 for $r/d = 0.24$)

Globe Valve

$N_3 = 2 \quad \sigma_3 = 5\% \quad K_3 = 3.50$ (specified)

90° LR Elbow

$N_4 = 14 \quad \sigma_4 = 10\% \quad K_T = 0.175$ (Table 15.2)

Exit

$N_5 = 1 \quad \sigma_5 = K_5 = 1.00$ (Section 12.1)

7.5.2 Solution

$$\begin{aligned}K_{Total} &= \mathrm{N}_1K_1 + \mathrm{N}_2K_2 + \mathrm{N}_3K_3 + \mathrm{N}_4K_4 + \mathrm{N}_5K_5\\ &= 1\times 23.23 + 1\times 0.10 + 2\times 3.50 + 14\times 0.175 + 1\times 1.00\\ &= 33.77\end{aligned}$$

$$\begin{aligned}\dot{w} &= \sqrt{\frac{2g\rho A^2}{K_{Total}}\rho(Z_1 - Z_2)}\\ &= \sqrt{\frac{2\times 32.174\times 62.30\times 0.9394^2}{33.77}\times 62.30(500-100)}\\ &= 1616\ \mathrm{lb/s}\end{aligned}$$

$$N_{\mathrm{Re}} = \frac{\dot{w}D}{\mu g A} = 2.870\times 10^6 \text{ (calculate Reynolds number and iterate on friction factor)}$$

$$f = \left[2\log_{10}\left(\frac{\varepsilon}{3.7D} + \frac{2.41}{N_{\mathrm{Re}}\sqrt{f_{\mathrm{T}}}}\right)\right]^{-2} = 0.01320 \quad \text{(Eq. 8.3)}$$

$$f = \left[2\log_{10}\left(\frac{\varepsilon}{3.7D} + \frac{2.41}{N_{\mathrm{Re}}\sqrt{f}}\right)\right]^{-2} = 0.01319$$

$$K_1 = \frac{fL}{D} = 24.13 \quad \text{(Eq. 1.3)}$$

$$K_4 = \frac{f}{f_{\mathrm{T}}}K_{\mathrm{T}} = 0.182 \quad \text{(Eq. 1.3)}$$

$$\begin{aligned}K_{Total} &= \mathrm{N}_1K_1 + \mathrm{N}_2K_2 + \mathrm{N}_3K_3 + \mathrm{N}_4K_4 + \mathrm{N}_5K_5\\ &= 1\times 24.13 + 1\times 0.10 + 2\times 3.50 + 14\times 0.182 + 1\times 1.00\\ &= 34.77\end{aligned}$$

$$\begin{aligned}\dot{w}_{\mathrm{Mean}} &= \sqrt{\frac{2g\rho A^2}{K_{Total}}\rho(Z_1 - Z_2)}\\ &= \sqrt{\frac{2\times 32.174\times 62.30\times 0.9394^2}{34.77}\times 62.30(500-100)}\\ &= 1592\ \mathrm{lb/s}\end{aligned}$$

$$\begin{aligned}\sigma_{\dot{w}} &= \frac{\sqrt{(\mathrm{N}_1\sigma_1K_1)^2 + (\mathrm{N}_2\sigma_2K_2)^2 + (\mathrm{N}_3\sigma_3K_3)^2 + (\mathrm{N}_4\sigma_4K_4)^2 + (\mathrm{N}_5\sigma_5K_5)^2}}{2(\mathrm{N}_1K_1 + \mathrm{N}_2K_2 + \mathrm{N}_3K_3 + \mathrm{N}_4K_4 + \mathrm{N}_5K_5)}\\ &= \frac{\sqrt{(1\times 10\times 24.13)^2 + (1\times 15\times 0.10)^2 + (2\times 5\times 3.50)^2 + (14\times 10\times 0.182)^2 + (1\times 6\times 1.00)^2}}{2\times 34.77}\\ &= 3.526\%\end{aligned}$$

$$\dot{w} = 1592\times\left(1\pm\frac{3.526}{100}\right)$$

$$\dot{w}_{Min} = 1536\ \mathrm{lb/s}$$

$$\dot{w}_{Max} = 1648\ \mathrm{lb/s}$$

Based on the loss coefficient values of the various piping elements and their assigned 3-sigma uncertainties, the predicted flow rate in the pipeline ranges from 1536 to 1648 lb/s.*

* One further iteration on friction factor resulted in predicted flow rate within the range of 1534–1646 lb/s.

PART II

LOSS COEFFICIENTS

PROLOGUE

As explained in Chapter 2, head loss represents a conversion of available mechanical energy to unavailable heat energy. The two principal sources of this conversion are: (1) surface friction and (2) induced turbulence due to pipe fittings and other changes in the flow path, such as flow meters and valves. The gradual process leading to understanding and quantifying surface friction was presented in Chapter 3. Its basic feature, friction factor, is presented in Chapter 8 as an adjunct to quantifying the various features that contribute to head loss. Induced turbulence, in the form of loss coefficients (or resistance coefficients), is dealt with in the remaining chapters of Part II.

Chapters 9 through 19 present rational and comprehensive investigations of pipe flow configurations commonly encountered by the professional engineer. Experimental test data and formulas for loss coefficients from worldwide sources are evaluated, integrated, and developed into widely applicable equations. The processes used to select and develop loss coefficient data for the various flow configurations are described so the reader can judge their merit and understand their limitations. The end results are presented in straightforward tables and diagrams located at the end of each chapter, where a user familiar with the work can quickly find them.

The flow configurations presented in Chapters 9 through 14—"Entrances," "Contractions," "Expansions," "Exits," "Orifices," and "Flow Meters"—all exhibit some degree of flow contraction and/or expansion. As such, they were treated as a family; they share semiempirical formulas that were rationally tailored to meet the specifics of the various flow configurations. Where sufficient data for a particular flow configuration were lacking, they were augmented by ample data in a related configuration.

Bends, tees, joints, and valves are treated in Chapters 15 through 18, respectively. The loss coefficient data in Part II are on the whole applicable to pipe components with butt weld, socket weld, flanged, or otherwise smooth-walled end connections. However, the internal geometry of threaded fittings is discontinuous, creating additional pressure loss, and they are covered separately in Chapter 19.

The loss coefficient data are independent of the kind of fluid as long as it is homogeneous and incompressible. The data are valid for turbulent flow conditions commonly encountered throughout the operating range of most industrial piping systems. The effect of Reynolds number on loss coefficients is mainly evident at its small values. The loss coefficient values are generally applicable to Reynolds numbers greater than 10^5, but they can be used with some loss of accuracy at lower Reynolds numbers in the turbulent flow regime.

In the case of laminar flow, the data can be used for rough estimates and only when the Reynolds number is greater than 100. In the case of compressible flow, they can be applied at Mach numbers up to approximately 0.3 with little or no loss of accuracy. They may be used at higher subsonic velocities up to about

Pipe Flow: A Practical and Comprehensive Guide, First Edition. Donald C. Rennels and Hobart M. Hudson.

Mach number 0.8 with decreased accuracy. In addition, the data can be applied to square passages or to rectangular passages of low aspect ratio with moderate loss of accuracy.

The loss coefficient always represents the number of velocity heads, $V^2/2g$, lost. The numerical value of any loss coefficient is intimately related to the inherent velocity in the associated pressure drop equation. In many cases the relationship is self-evident and the loss coefficient is simply labeled as K. In cases where there is a change in flow area a subscript is used to denote the relationship. For example, in the case of a contraction, K_2 indicates that the loss coefficient is related to the velocity at point 2, the downstream velocity. In the case of flow through tees, a subscript—two numbers separated by a comma—defines the flow path, and a sub-subscript defines the related velocity. For instance, $K_{1,2_2}$ indicates that pressure loss is from point 1 to point 2 and the loss coefficient is related to the velocity at point 2.

8

SURFACE FRICTION

The gradual process leading to understanding and quantifying surface friction was presented in Chapter 3. Surface friction and its main element, friction factor, are further considered in this chapter as an adjunct to quantifying the various features that contribute to head loss.

The loss coefficient due to surface friction (analogous to the loss coefficient due to local loss) is expressed as:

$$K = f\frac{L}{D}. \qquad \text{(1.3, repeated)}$$

Thus the product of the friction factor f and the geometric factor L/D (or l/d) represents the number of velocity heads lost due to surface friction. The friction factor under discussion here is that corresponding to fully developed velocity profiles that are encountered only after 20 or more pipe diameters downstream of a pipe inlet or other major disturbance. In practice, this condition is rarely met. However, satisfactory results are generally obtained ignoring this limitation.

8.1 FRICTION FACTOR

The relationship of friction factor to the Reynolds number and surface roughness has three distinct differing regions of application.

8.1.1 Laminar Flow Region

As identified in Chapter 3, an expression for laminar flow friction factor was developed by the mid-1800s:

$$f = \frac{64}{N_{\text{Re}}}. \qquad \text{(3.3, repeated)}$$

Note that f is a function of the pipe Reynolds number only. Protrusions on the pipe surface do not cause turbulence in laminar flow. For laminar flow, pipes of different surface roughness have the same friction factor for the same Reynolds number.

The upper range of Reynolds number for laminar flow is somewhat indefinite, being dependent upon several incidental conditions, and may be as high as 4000. However, such high values are of little practical interest, and the engineer may take the upper limit of laminar flow to be defined by a Reynolds number of 2100.

8.1.2 Critical Zone

For pipe Reynolds numbers between 2100 and 3000 to 4000, the friction factor can have large uncertainties and is highly indeterminate. Hence this region is called the *critical zone*. The flow in this zone may be laminar or turbulent (or an unsteady mix of both) depending on the pipe entrance, initial disturbances, and pipe roughness. This transition region from laminar to turbulent flow is accompanied by a considerable increase in friction factor, and thereby in pressure drop in the pipe. As a rule, however, pipes flowing significant amounts of fluid, and which have measurable pressure loss, have a Reynolds number much greater than 3000 or 4000. Notwithstanding, the engineer may on occasion have to

Pipe Flow: A Practical and Comprehensive Guide, First Edition. Donald C. Rennels and Hobart M. Hudson.

make a conservative selection of friction factor when pipe flow operates within the critical zone.

8.1.3 Turbulent Flow Region

Turbulent flow occurs more frequently in engineering applications, hence the greater interest in this flow region. Whereas the friction factor is independent of surface roughness in laminar flow, roughness is of fundamental importance in turbulent flow except in the case of smooth pipes.

8.1.3.1 Smooth Pipes For turbulent flow, if the surface roughness is very slight, as for glass tubes, drawn metal tubing, or so-called perfectly smooth pipes, the friction factor is essentially a function of Reynolds number only.* Analytical and experimental work in the early 1930s led to the following implicit formula for friction factor for turbulent flow in smooth pipes:

$$\frac{1}{\sqrt{f}} = -2\log_{10}\frac{2.51}{N_{\mathrm{Re}}\sqrt{f}}. \qquad \text{(3.4, repeated)}$$

Squaring and inverting yields:

$$f = \left[2\log_{10}\left(\frac{2.51}{N_{\mathrm{Re}}\sqrt{f}}\right)\right]^{-2}. \qquad (8.1)$$

8.1.3.2 Rough Pipes Also in the early 1930s, analytical and experimental work for rough pipe in the fully turbulent region, where friction factor is no longer a function of Reynolds number, resulted in the following formula:

$$\frac{1}{\sqrt{f}} = -2\log_{10}\frac{\varepsilon}{3.7D}, \qquad \text{(3.5, repeated)}$$

where ε (or e) is the *surface roughness*, or more aptly, the *absolute roughness*, of the pipe walls, and the ratio ε/D (or e/d) is termed the *relative roughness* of the pipe. Squaring and inverting Equation 3.5 yields:

$$f = \left[2\log_{10}\left(\frac{\varepsilon}{3.7D}\right)\right]^{-2}. \qquad (8.2)$$

8.2 THE COLEBROOK–WHITE EQUATION

If solutions to the equations for turbulent flow friction factor in smooth and rough pipes are plotted on a single graph, a noticeably sharp intersection between the two regions is apparent. In reality, Nikuradse's artificially roughened pipe results show a jump from the laminar friction factor directly to the smooth pipe friction factor followed by a gradual transition to the rough pipe friction factor (Fig. 3.1), whereas actual commercial pipes do not show this kind of jump. Instead, commercial pipes show a jump from the laminar friction factor to a point *above* the rough pipe friction factor, then, on increasing Reynolds numbers, the friction factor gradually settles down to the rough pipe number. Colebrook and White [1] showed experimentally that this behavior is due to commercial pipe's randomly sized roughness protuberances (as opposed to Nikuradse's uniform roughness imparted by the uniform sand grains he used to roughen his pipes), and their formula mimics this behavior, albeit not analytically, but empirically. Building on their own work and on the work of other researchers in the 1930s, Colebrook and White [2] developed an expression that bridged this intersection quite well:

$$\frac{1}{\sqrt{f}} = -2\log_{10}\left(\frac{\varepsilon}{3.7\,D} + \frac{2.51}{N_{\mathrm{Re}}\sqrt{f}}\right). \qquad \text{(3.6, repeated)}$$

It is worthy of note that, while the friction factor is not available explicitly in the Colebrook–White equation, solution by a trial-and-error method is very easy in a computer program. Squaring and inverting Equation 3.6 yields:

$$\begin{aligned} f &= \left[2\log_{10}\left(\frac{\varepsilon}{3.7D} + \frac{2.51}{N_{\mathrm{Re}}\sqrt{f}}\right)\right]^{-2} \\ &= \left[2\log_{10}\left(\frac{e}{3.7d} + \frac{2.51}{N_{\mathrm{Re}}\sqrt{f}}\right)\right]^{-2}. \end{aligned} \qquad (8.3)$$

If a guessed friction factor, say 0.02, is introduced on the right side of the equation and the equation solved for the friction factor on the left side, a better estimate of the friction factor is obtained. If this better estimate is substituted on the right side and the equation solved again, an even better estimate is obtained. After three to five iterations, the solved friction factor is accurate to four significant figures or better. If, however, the computation time consumed in this many iterations is burdensome, the number of iterations may be reduced to two or three, because the Colebrook–White equation itself is limited to an accuracy estimated by experimenters as ±3%. The desire for four significant figure accuracy is thus likely unwarranted.

* It should be noted that drawn metal tubing is assigned an absolute roughness value of 0.000060 in (see Table 8.1). All the same, drawn metal tubes effectively act as smooth pipes except at very high Reynolds number.

8.3 THE MOODY CHART

The various equations for friction factor, although suitably summarizing the data on pipe flow, were hardly suitable for engineering use in the days prior to the use of computers. It was opportune to introduce a composite plot for presentation of the friction factor.

As described in Chapter 3, in 1944 American engineer L.F. Moody [3] developed a composite plot of all regions of interest for presentation of the friction factor in a suitable form for engineering use. The chart is repeated at the end of this chapter for convenience (see Diagram 8.1). Other friction factor charts have been developed over the years. The "Moody Chart" is still the popular choice and is still in use today.

8.4 EXPLICIT FRICTION FACTOR FORMULATIONS

Because of Moody's work and the demonstrated applicability of the Colebrook–White equation over a wide range of Reynolds numbers and relative roughness, Equation 8.3 has become the accepted standard for calculating friction factor in the turbulent flow region. Clearly, however, it suffers from being an implicit equation in f and thus requires charts, tables, or successive approximations to extract the value of f. While the Moody Chart is sufficient for the numerical solution of specific engineering problems, cases arise where we need not only specific values but an explicit formulation for the friction factor. For example, in dealing with total losses in a system in which friction is only one of a number of factors, in handling a problem involving the solution of simultaneous equations, or in any problem where we want a direct analytical solution, we may need an explicit expression for friction factor as a function of the controlling variables.

Moody quickly recognized the need for an explicit equation for friction factor and was possibly the first to provide one. Since the end of the 1940s, many alternative explicit equations have been developed to avoid the iterative process inherent in the Colebrook–White equation. Several are offered herein. These approximations vary in the degree of accuracy depending on the complexity of their functional forms; the more complex ones usually providing friction factor estimates of higher accuracy. These formulas may be used on their own merit or may be used as the first guess in the Colebrook–White equation to reduce the number of necessary iterations.

8.4.1 Moody's Approximate Formula

In 1947, Moody [4] proposed the following approximate formulation for friction factor:

$$f \cong 0.0055\left[1+\left(20{,}000\frac{e}{d}+\frac{10^6}{N_{Re}}\right)^{1/3}\right].$$

Moody noted that the formula agrees with the Colebrook–White equation for f within an error of ±5% for values of N_{Re} from 4000 to 10^7, and for values of ε/D up to 0.01 or values of f up to 0.05.

8.4.2 Wood's Approximate Formula

In 1966, Wood [5] proposed an explicit formula which is valid for $N_{Re} > 10{,}000$ and for ε/D within 10^{-5} and 0.04:

$$f = 0.094\left(\frac{\varepsilon}{D}\right)^{0.025}+0.53\frac{\varepsilon}{D}+88\left(\frac{\varepsilon}{D}\right)^{0.44}N_{Re}^{1.62(\varepsilon/D)^{0.134}}.$$

Wood noted that the accuracy of the formula in the specified range is between −4% and +6%.

8.4.3 The Churchill 1973 and Swamee and Jain Formulas

Stuart Churchill [6] developed an empirical formula by substituting for Prandtl's implicit smooth pipe formula (Eq. 3.4) an explicit one proposed by Nikuradse in 1932—

$$f = \frac{1}{[1.8\log_{10}(N_{Re}/7)]^2}$$

—in the Colebrook–White equation to obtain:

$$f = \left\{2\log_{10}\left[\frac{\varepsilon}{3.7D}+\left(\frac{7}{N_{Re}}\right)^{0.9}\right]\right\}^{-2}.$$

Churchill's explicit formula was published in 1973. In 1976, Swamee and Jain [7] published an almost identical formula, in which the constant in the coefficient of the smooth pipe term was tweaked slightly (6.97 vs. 7), perhaps to obtain better accuracy. Their formula gives a friction factor within 3% of that from the Colebrook–White equation for ε/D from 0.000001 to 0.01 and for N_{Re} from 5000 to 10^8. If either the Churchill or the Swamee and Jain formula is used for the first guess in the Colebrook–White equation the number of cycles necessary to close for four significant figure accuracy is reduced significantly.

8.4.4 Chen's Formula

Chen [8] proposed an accurate formula encompassing all the normal ranges of N_{Re} and ε/D within the turbulent region:

$$f=\left\{-2\log_{10}\left[\frac{\varepsilon}{3.7065D}-\frac{5.0452}{N_{\mathrm{Re}}}\log_{10}\left(\frac{1}{2.8257}\left(\frac{\varepsilon}{D}\right)^{1/1098}+\frac{5.8506}{N_{\mathrm{Re}}^{0.8981}}\right)\right]\right\}^{-2}.$$

Chen published this explicit formula in 1979.

8.4.5 Shacham's Formula

By substituting $f = 0.03$ in the right hand side of Equation 8.3, and substituting the result in Equation 8.3 again, Shacham [9] devised the empirical formula

$$f=\left\{-2\log_{10}\left[\frac{\varepsilon}{3.7D}-\frac{5.02}{N_{\mathrm{Re}}}\log_{10}\left(\frac{\varepsilon}{3.7D}+\frac{14.5}{N_{\mathrm{Re}}}\right)\right]\right\}^{-2},$$

thus obtaining the effect of two iterations of Equation 8.3. Shacham published this explicit formula in 1980.

8.4.6 Barr's Formula

In 1981, Barr [10] proposed the following explicit formula:

$$f=\left\{-2\log_{10}\left[\frac{\varepsilon}{3.7\,D}+\frac{4.518\log_{10}\left(\frac{N_{\mathrm{Re}}}{7}\right)}{N_{\mathrm{Re}}\left(1+\frac{N_{\mathrm{Re}}^{0.52}}{29}\left(\frac{\varepsilon}{D}\right)^{0.7}\right)}\right]\right\}^{-2}.$$

8.4.7 Haaland's Formulas

In 1983, Haaland [11] proposed a variation in the effects of the relative roughness by the following expression:

$$f=\left\{-1.8\log_{10}\left[\frac{6.9}{N_{\mathrm{Re}}}+\left(\frac{\varepsilon}{3.7D}\right)^{1.11}\right]\right\}^{-2}. \qquad (8.4)$$

In deference to experiments using smooth pipes (as in natural gas pipelines) that showed that the transition from the smooth to the rough regime is much more abrupt than indicated by the Colebrook–White equation, Haaland also proposed the following formulation:

$$f=\left\{-\frac{1.8}{n}\log_{10}\left[\left(\frac{6.9}{N_{\mathrm{Re}}}\right)^{n}+\left(\frac{\varepsilon}{3.75D}\right)^{1.11n}\right]\right\}^{-2},$$

where with $n = 3$ the formulation gives values of f that are close to the completely abrupt transition between smooth and rough pipe flow as recommended by the American Gas Association [12].

8.4.8 Manadilli's Formula

In 1997, Manadilli [13] proposed the following explicit formula valid for N_{Re} ranging from 5235 to 10^8 and for any value of ε/D:

$$f=\left[-2\log_{10}\left(\frac{\varepsilon}{3.7D}+\frac{95}{N_{\mathrm{Re}}^{0.983}}-\frac{96.82}{N_{\mathrm{Re}}}\right)\right]^{-2}.$$

8.4.9 Romeo's Formula

Romeo et al. [14] proposed the following explicit formula in 2002:

$$f=\left[-2\log_{10}\left(\frac{\varepsilon}{3.7065D}-\frac{5.0272}{N_{\mathrm{Re}}}\mathrm{A}\right)\right]^{-2},$$

where

$$\mathrm{A}=\log_{10}\left\{\frac{\varepsilon}{3.827D}-\frac{4.567}{N_{\mathrm{Re}}}\log_{10}\left[\left(\frac{\varepsilon}{7.7918D}\right)^{0.9924}+\left(\frac{5.3326}{208.815+N_{\mathrm{Re}}}\right)^{0.9345}\right]\right\}.$$

8.4.10 Evaluation of Explicit Alternatives to the Colebrook–White Equation

With the exception of the early formulations by Moody and Wood, the explicit formulas accurately reproduce the implicit Colebrook–White equation.

In 2009, Yildirim [15] presented the results of a computer-based analysis of a number of explicit alternatives to the Colebrook–White equation. According to Yildirim's statistical analyses, the formulas by Chen, Barr, Haaland, and Romeo et al. are the most accurate of the formulas presented above.* Among these formulas, Haaland's formulation appears to be the most convenient one to use. Because of its accuracy and simplicity, Equation 8.4 is recommended for practical use as an explicit alternative to the Colebrook–White equation.

* The Chen, Barr, Haaland, and Romeo formulas had extreme values of mean relative error of less than 1.2% and extreme values of maximum relative errors of less than 4.7% for different ε/D values ranging from 1×10^{-6} to 5×10^{-2} and for N_{Re} values ranging from 4×10^{3} to 10^{8} for a 20×500 grid. Over most of the entire grid, the relative errors were less than one-half the extreme values.

8.5 ALL-REGIME FRICTION FACTOR FORMULAS

In 1977 Churchill [16] published a formula covering all flow regimes—laminar, critical, transition, and fully turbulent—and for any relative roughness. It is based on his 1973 formula and on earlier work with his collaborator, R. Usagi. It also incorporates the laminar flow friction factor formula of Hagen and Poiseuille, and a fit of the data of Wilson and Azad [17] in the critical zone. Churchill's formula is smooth and continuous.

8.5.1 Churchill's 1977 Formula

Churchill's 1977 all-regime friction factor formula is:

$$f_D = 8\left[\left(\frac{8}{N_{\text{Re}}}\right)^{12} + \frac{1}{(A+B)^{3/2}}\right]^{1/12}, \tag{8.5}$$

where

$$A = \left[2.457\ln\frac{1}{\left(\frac{7}{N_{\text{Re}}}\right)^{0.9} + \frac{0.27\varepsilon}{D}}\right]^{16} \quad \text{and} \quad B = \left(\frac{37{,}530}{N_{\text{Re}}}\right)^{16}.$$

In this formula Churchill used a friction factor that is one-eighth of the customary Darcy friction factor. The multiplier outside the brackets transforms Churchill's factor into the Darcy factor, hence the $_D$ subscript on f (which is used here only to emphasize that this work always uses the Darcy friction factor).

The laminar zone friction factor formula is recognizable in the first term within the brackets in Equation 8.5. The 1973 Churchill formula is recognizable as the expression for A, and the critical zone fit of Wilson and Azad is embodied as the expression for B.* Churchill's 1977 function is shown plotted against a backdrop of the Moody chart in Figure 8.1. The figure shows that Equation 8.5 yields excellent agreement with the Hagen–Poiseuille law, Nikuradse's results in the critical zone, and von Kármán's formula for complete turbulence, though there is some disagreement in the transition zone.

This formula (or the modification described below) is highly recommended when one is not sure if the Reynolds number is in the turbulent region.

* As noted in Section 8.3, the engineer must make conservative selection of friction factor when pipe flow operates within the critical zone.

8.5.2 Modifications to Churchill's 1977 Formula

A linearized Hoerl function curve fit for $1/\sqrt{f}$ for smooth pipes is:

$$\frac{1}{\sqrt{f}} = -2\log_{10}\left[0.883\frac{(\ln N_{\text{Re}})^{1.282}}{N_{\text{Re}}^{1.007}}\right].$$

This is a much better fit than Prandtl's proposed fit used by Churchill. When the argument for the logarithm is substituted in the Churchill formula for $(7/N_{\text{Re}})^{0.9}$, the resulting formula gives better fidelity to the Colebrook–White equation for $\varepsilon/D \le 0.002$ in the transition zone, and especially for $\varepsilon/D = 0$.

The modified Churchill formula resulting is:

$$f_D = \left[\left(\frac{64}{N_{\text{Re}}}\right)^{12} + \frac{1}{(A+B)^{3/2}}\right]^{1/12},$$

where

$$A = \left[0.8687\ln\frac{1}{\frac{0.883(\ln N_{\text{Re}})^{1.282}}{N_{\text{Re}}^{1.007}} + \frac{0.27\varepsilon}{D}}\right]^{16}$$

and

$$B = \left(\frac{13{,}269}{N_{\text{Re}}}\right)^{16}.$$

In Figure 8.1 it may be observed that the 1977 Churchill formula exhibits some upward "curl" in the transition region from the critical zone to a Reynolds number of about 10^5. Schroeder [18] makes a case that the Colebrook–White equation itself already predicts a higher than observed friction factor in this region. This curl can be largely eliminated by the addition of a subtractive term, $110\varepsilon/N_{\text{Re}}D$, to the Colebrook–White group in the Churchill equation. Then the Churchill equation becomes:

$$f_D = \left[\left(\frac{64}{N_{\text{Re}}}\right)^{12} + \frac{1}{(A+B)^{3/2}}\right]^{1/12}, \tag{8.6}$$

where

$$A = \left[0.8687\ln\frac{1}{\frac{0.883(\ln N_{\text{Re}})^{1.282}}{N_{\text{Re}}^{1.007}} + \frac{0.27\varepsilon}{D} - \frac{110\varepsilon}{N_{\text{Re}}D}}\right]^{16}$$

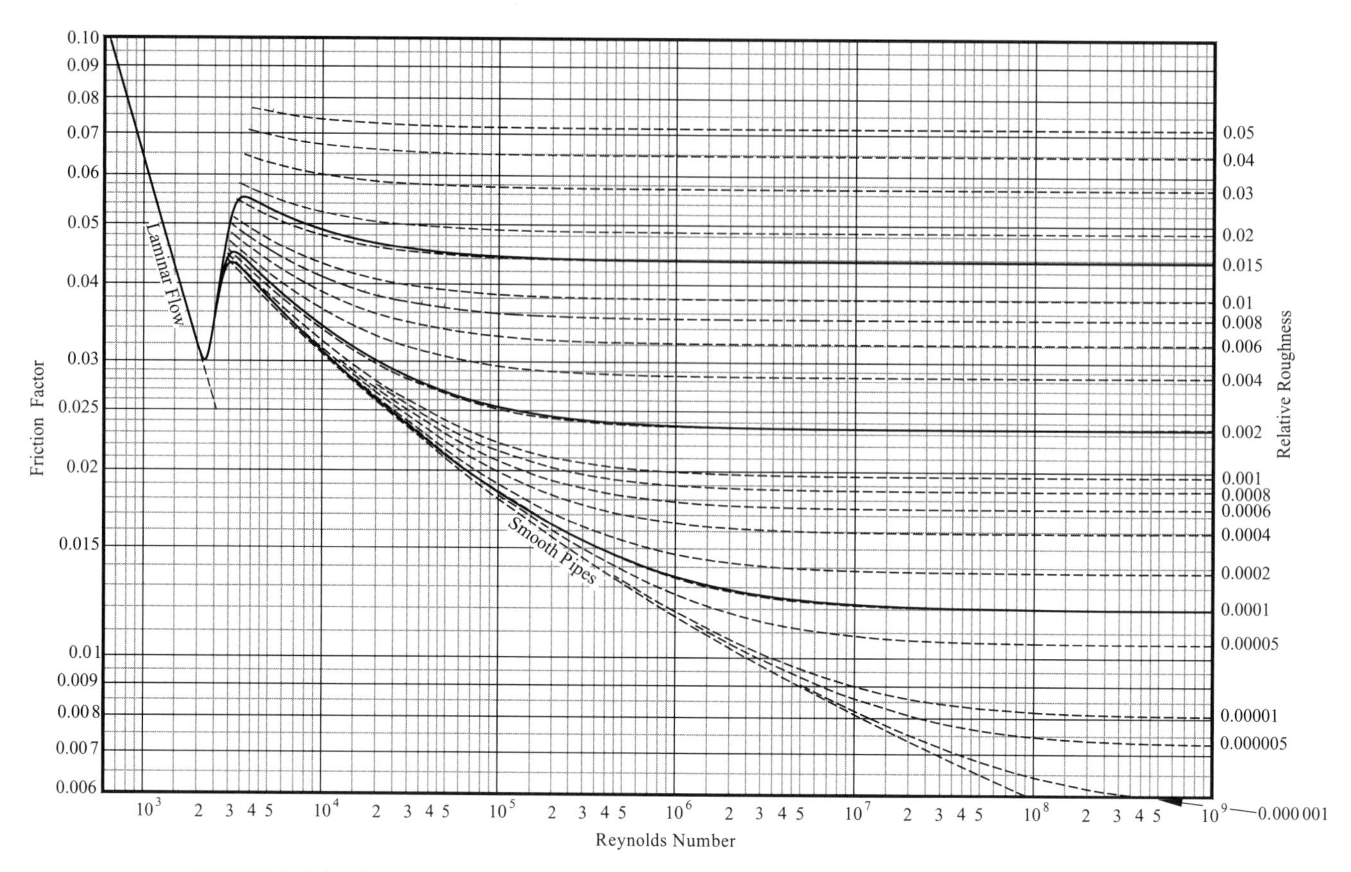

FIGURE 8.1. The Churchill friction factor equation at ε/D = 0.015, 0.002 and 0.0001 (Eq. 8.5).

and

$$B = \left(\frac{13,269}{N_{\mathrm{Re}}}\right)^{16}.$$

Figure 8.2 shows the results of this modification of Churchill's 1977 formulation. A case can be made that by incorporating the linearized Hoerl function curve fit and by eliminating the curl, the modified Churchill formula is equal to the more accurate of the explicit formulas evaluated in Section 8.4.

8.6 SURFACE ROUGHNESS

Surface roughness is defined as irregularities in the surface texture of the pipe inner wall. The degree of roughness is a function of the pipe material, its manufacturing process, and the environment to which it has been exposed. Establishing the correct surface roughness, or *absolute roughness*, of pipe is essential to reduce uncertainty in estimating friction factor necessary for calculating pressure loss in pipe.

8.6.1 New, Clean Pipe

Typical values of absolute roughness for new, clean pipe are given in Table 8.1. These values may be sufficient for initial operation, for piping systems that contain noncorrosive fluid, or for closely monitored piping systems that are cleaned as necessary. Aside from these conditions, corrosion and scale buildup can considerably increase the absolute roughness of pipe, resulting in significantly reduced carrying capacity.

8.6.2 The Relationship between Absolute Roughness and Friction Factor

The absolute roughness ε of new, clean carbon steel pipe may be taken as 0.000150 ft, the absolute roughness of moderately corroded carbon steel pipe with small depositions of scale may be taken as 0.00130 ft, and the absolute roughness of heavily corroded carbon steel pipe with large depositions of scale may be taken as 0.0100 ft. Using the Colebrook–White equation (Eq. 8.3), the friction factor of carbon steel pipe was calculated at these three surface roughness conditions as a function of pipe

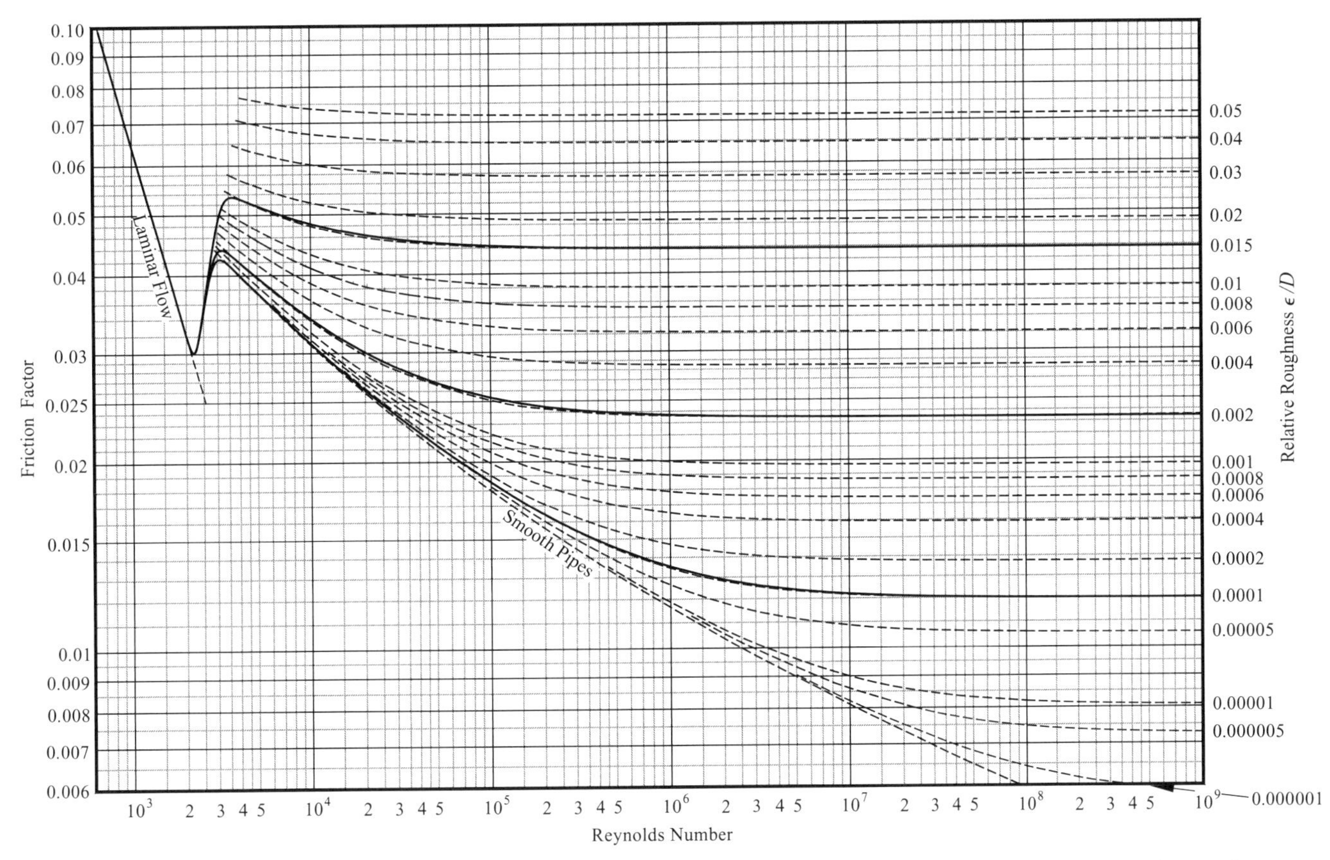

FIGURE 8.2. The Churchill friction factor equation at ε/D = 0.015, 0.002 and 0.0001, with reduced curl (Eq. 8.6).

TABLE 8.1. Typical Values of Absolute Roughness for New, Clean Pipe

	English System		International System of Units (SI)
Pipe or Lining Material	e (inch)	ε (feet)	e (mm), or ε (meter × 10^3)
Asbestos cement	0.000096	0.000008	0.0024
Carbon steel, commercial	0.0018	0.00015	0.045
Concrete, smoothed	0.012	0.0010	0.30
Concrete, ordinary	0.040	0.0033	1.0
Concrete, coarse	0.12	0.010	3.0
Glass tube	0.000060	0.0000050	0.0015
Iron, cast, uncoated	0.0102	0.00085	0.26
Iron, cast, asphalted	0.0048	0.00040	0.12
Iron, cast, cement lined	0.000096	0.000008	0.0024
Iron, cast, bituminous lined	0.000096	0.000008	0.0024
Iron, cast, centrifugally spun	0.00012	0.000010	0.030
Iron, galvanized	0.0060	0.00050	0.15
Iron, wrought	0.0022	0.00018	0.060
Fiberglass	0.00020	0.000010	0.005
Polyvinyl chloride (PVC) and plastic	0.000060–0.00024	0.0000005–0.000020	0.0015–0.0060
Stainless steel, commercial	0.0018	0.00015	0.045
Steel, riveted	0.036–0.36	0.0030–0.030	0.90–9.0
Tubing, drawn (aluminum, brass, copper, lead, etc.)	0.000060	0.0000050	0.0015
Wood stave	0.0072–0.036	0.00060–0.0030	0.18–0.90

It should be noted that relative roughness, e/d or ε/D, is dimensionless—it is important to ensure that absolute roughness and the internal diameter are in the same units.

TABLE 8.2. Friction Factor as a Function of Pipe Size, Condition, and Reynolds Number

	Friction Factor for Commercial Carbon Steel Pipe (Percent Increase)								
	New, Clean Pipe			Moderately Corroded Pipe			Heavily Corroded Pipe[a]		
	$\varepsilon = 0.000150$ feet			$\varepsilon = 0.00130$ feet			$\varepsilon = 0.0100$ feet		
Nominal Pipe Size (Schedule 40)	10^4	10^6	10^8	10^4	10^6	10^8	10^4	10^6	10^8
1″	0.0334	0.0227	0.0225	0.0478	0.0436	0.0436	*0.1112*	*0.1097*	*0.1097*
	—	—	—	(43%)	(92%)	(94%)	*(233%)*	*(383%)*	*(387%)*
2″	0.0322	0.0193	0.0190	0.0405	0.0346	0.0345	*0.0788*	*0.0768*	*0.0768*
	—	—	—	(26%)	(79%)	(82%)	*(145%)*	*(254%)*	*(304%)*
4″	0.0316	0.0168	0.0163	0.0363	0.0283	0.0282	0.0600	0.0571	0.0570
	—	—	—	(15%)	(68%)	(73%)	(90%)	(239%)	(250%)
8″	0.0312	0.0149	0.0141	0.0337	0.0235	0.0233	0.0480	0.0438	0.0437
	—	—	—	(8%)	(57%)	(65%)	(54%)	(193%)	(210%)
16″	0.0311	0.0137	0.0124	0.0324	0.0201	0.0198	0.0410	0.0353	0.0352
	—	—	—	(4%)	(47%)	(60%)	(32%)	(157%)	(183%)
32″	0.0310	0.0128	0.0109	0.0317	0.0173	0.0168	0.0363	0.0284	0.0283
	—	—	—	(2%)	(35%)	(54%)	(17%)	(122%)	(160%)

[a] Values in italics may be out of range of Equation 8.3.

size and Reynolds number. The calculated results, presented in Table 8.2, reveal that increase in surface roughness due to age and usage can result in significant increase in friction factor. It is evident that increased surface roughness becomes more important as Reynolds number increases, and becomes less important as pipe diameter increases. The calculated results in Table 8.2 provide an appreciation and understanding of these relationships.

8.6.3 Inherent Margin

Piping system design practice often provides pressure drop margin and this margin may be sufficient to accommodate increased pressure loss due to increased surface roughness.

1. At least initially, the absolute roughness of the pipe walls may be less than the assumed value.
2. Historical sources have often provided conservative loss coefficient data for pipe fittings.
3. A conservatively large pipe size is utilized; for example, a 10″ pipe size (or 11″, which is not available) may have been sufficient, but a 12″ pipe size is selected.
4. Actual equipment (pumps, valves, etc.) performance may exceed specified performance.
5. Surface friction accounts for only a portion of total system resistance.
6. The designer arbitrarily chose to increase absolute roughness in initial pressure loss calculations.

Nonetheless, the designer is advised to search the literature for pipe aging data specific to her or his application, and possibly add contingency depending on the circumstances, the scope of the work, and the need for being conservative in design.

8.6.4 Loss of Flow Area

The encrustation of pipe with scale, dirt, sludge, tubercules, or other foreign bodies results in a reduction of pipe diameter (and flow area) in addition to increase in surface roughness. By and large, the effect of loss of flow area is much less significant than increased surface roughness and may be ignored except in extreme cases.

8.6.5 Machined Surfaces

The absolute roughness of newly machined or otherwise manufactured surfaces (other than pipe and tubing) may be taken to be twice the root mean square (RMS) surface finish. Thus the absolute roughness of a surface machined to a finish of 250 micro-inch is taken to be 0.000500 in. There is no certain basis for applying this rule except that it appears to be reasonable and seems to work well. Take into account the possible effects of age and usage on machined surfaces just as you would for pipe.

8.7 NONCIRCULAR PASSAGES

The foregoing friction factor equations for circular pipes may be adapted to noncircular passages through the use of the *hydraulic diameter*. The hydraulic diameter, d_h, is an arbitrary definition of a value calculated so that the ratio of pressure forces acting over the flow area to the frictional forces acting along the wetted perimeter* is the same for circular and noncircular passages. It turns out that a multiplier of 4 is necessary to satisfy this definition. For example, applying the definition to a circular passage produces:

$$d_h = \frac{4\dfrac{\pi d^2}{4}}{\pi d} = d,$$

as should be the case. The calculated hydraulic diameter of several noncircular flow passages follow.

For a *square passage* of width w,

$$d_h = \frac{4w^2}{4w} = w.$$

* The wetted perimeter is the perimeter of the flow passage in contact with the fluid. The hydraulic diameter concept is particularly important in open channel flow calculations.

For a *rectangular passage* of width w and length l,

$$d_h = \frac{4wl}{2(w+l)} = \frac{2wl}{w+l}.$$

For a *slit* of width w and (infinite) length l,

$$d_h \approx \frac{4wl}{2l} = 2w.$$

For an *annulus* of outer diameter d_o and inner diameter d_i,†

$$d_h = \frac{4\dfrac{\pi}{4}\left(d_o^2 - d_i^2\right)}{\pi(d_o + d_i)} = d_o - d_i.$$

The hydraulic diameter d_h is substituted for d in Equation 1.3 (loss coefficient due to surface friction), as well as in the various friction factor equations and the "Moody Chart." The hydraulic diameter is also used when computing Reynolds number for noncircular flow passages.

† Additional correction factors are required to accurately calculate friction factor for flow through an annulus. These correction factors, available in the literature, differentiate between concentric and eccentric alignment of the annulus, and between laminar versus turbulent flow.

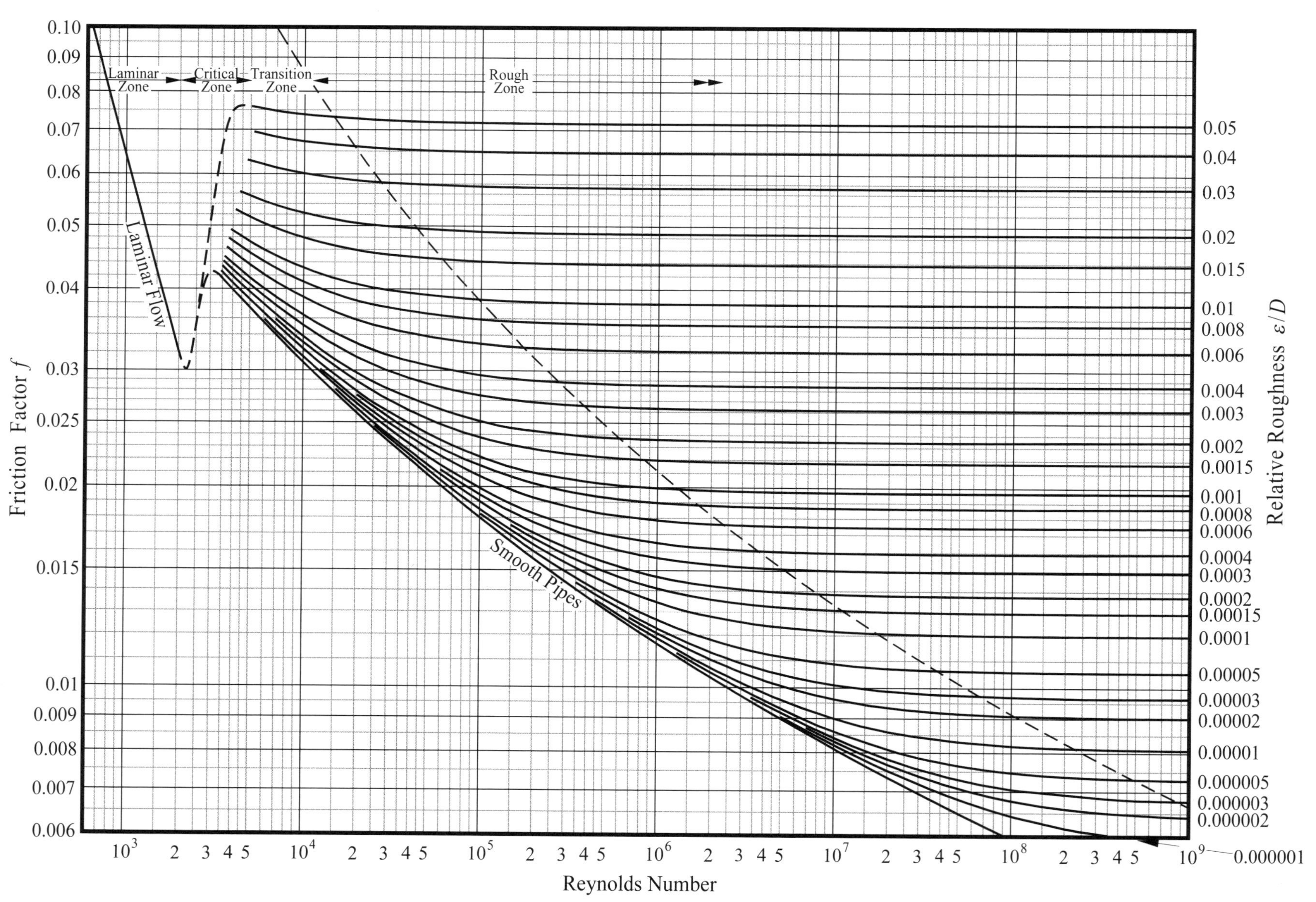

DIAGRAM 8.1. Friction factor versus Reynolds number and relative roughness for commercial pipe (after Moody [3]).

REFERENCES

1. Colebrook, C. F. and C. M. White, Experiments with fluid friction in roughened pipes, *Proceedings of the Royal Society of London*, 161, 1937, 367–381.
2. Colebrook, C. F., Turbulent flow in pipes, with particular reference to the transition region between the smooth and rough pipe laws, *Journal of the Institution of Civil Engineers*, 11, 1938–1939, 133–156.
3. Moody, L. F., Friction factors for pipe flow, *Transactions of the American Society of Mechanical Engineers*, 66, 1944, 671–684.
4. Moody, L. F., An approximate formula for pipe friction factors, *Transactions of the American Society of Mechanical Engineers*, 69, 1947, 1005–1006.
5. Wood, D. J., An explicit friction factor relationship, *Civil Engineering-ASCE*, 36(12), 1966, 60–61.
6. Churchill, S. W., Empirical expressions for the shear stress in turbulent flow in commercial pipe, *American Institute of Chemical Engineering Journal*, 19(2), 1973, 375–376.
7. Swamee, P. K. and A. K. Jain, Explicit equations for pipe-flow problems, *Journal of the Hydraulics Division, American Society of Civil Engineers*, 102(HY5), 1976, 657–664.
8. Chen, N. H., An explicit equation for friction factor, *American Institute of Chemical Engineering Journal*, 19(2), 1980, 229–230.
9. Shacham, M., An explicit equation for friction factor in pipe, *Industrial & Engineering Chemistry Fundamentals*, 19(2), 1980, 228–229.
10. Barr, D. I. H., Solutions to the Colebrook-White functions for resistance to uniform turbulent flow, *Proceeding of the Institute of Civil Engineers*, Part 2, 71, 1981, 529.
11. Haaland, S. E., Simple and explicit formulas for the friction factor in turbulent pipe flow, *Transactions of the ASME, Journal of Fluids Engineering*, 105, 1983, 89–90.
12. Uhl, A. E. et al., Steady Flow in Gas Pipelines, *Institute of Gas Technology Report No. 10, American Gas Association*, New York, 1956.
13. Manadilli, G., Replace implicit equations with sigmoidal functions, *Chemical Engineering*, 104(8), 1997, 187.
14. Romeo, E., C. Royo, and A. Monzon, Improved explicit equations for estimation of friction factor in rough and smooth pipes, *Chemical Engineering*, 86, 2002, 369–374.
15. Yildirim, G., Computer-based analysis of explicit approximations to the implicit Colebrook-White equation in turbulent flow friction factor calculation, *Advances in Engineering Software*, 40, 2009, 1183–1190.
16. Churchill, S. W., Friction-factor equation spans all fluid-flow regimes, *Chemical Engineering*, 84, 1977, 91–92.
17. Wilson, N. W. and R. S. Azad, A continuous prediction method for fully developed laminar, transitional and turbulent flows in pipes, *Journal of Applied Mechanics*, 42, 1975, 51–54.
18. Schroeder, D. W., *A Tutorial on Pipe Flow Equations*, Stoner Associates, Inc., Carlisle, Pennsylvania, August 16, 2001.

FURTHER READING

This list includes books and papers that may be helpful to those who wish to pursue further study.

Kemler, E., A study of the data on the flow of fluids in pipes, *Transactions of ASME*, 55, 1933, 7–32.

Colebrook, C. F. and C. M. White, The reduction of carrying capacity of pipes with age, *Journal of the Institution of Civil Engineers*, 10, 1937–1938, 99–118.

Freeman, J. R., Experiments on the Flow of Water in Pipes and Pipe Fittings, *1889 to 1893*. (Published by the *ASME* in a special volume in 1941.).

Tao, L. N. and W. F. Donovan, Through-flow in concentric and eccentric annuli of fine clearance with and without relative motion of the boundaries, *Transactions of the ASME*, 77, 1955, 1291–1301.

Rouse, H. and S. Ince, *History of Hydraulics*, Iowa Institute of Hydraulic Research, State University of Iowa, Iowa City, IA, 1957.

Lohrenz, J. and F. Kurata, A friction factor plot...for smooth circular conduits, concentric annuli, and parallel plates, *Industrial & Engineering Chemistry*, 52(8), 1960, 703–706.

U.S. Bureau of Reclamation, Friction Factors for Large Conduit Flowing Full, Engineering Monograph, No. 7, U.S. Department of the Interior, Washington, D.C., 1965.

Selander, W. N., *Explicit Formulas for the Computation of Friction Factors in Turbulent Pipe Flow*, Atomic Energy of Canada Limited, Chalk River, Ontario, Canada, 1978 (AECL-6354).

Guislain, S. J., How to Make Sense of Friction Factors in Fluid Flow Through Pipe, *Plant Engineering*, June 12, 1980, pp. 134–140.

Lamont, P. A., Pipe flow formulas compared with the theory of roughness, *American Water Works Association Journal*, 73(5), 1981.

Olujić, Ž., Compute Friction Factors Fast for Flow in Pipes, *Chemical Engineering*, December 14, 1981, pp. 91–93.

Obot, N. T., Determination of incompressible flow friction factor in smooth circular and noncircular passages: A generalized approach including validation of the nearly century old hydraulic diameter concept, *Transactions of the ASME, Journal of Fluids Engineering*, 110, 1988, 431–440.

Scaggs, W. F., R. P. Taylor, and H. W. Coleman, Measurement and prediction of rough wall effects on friction factor—Uniform roughness results, *Transactions of the ASME, Journal of Fluids Engineering*, 110, 1988, 385–391.

Taylor, R. P., W. F. Scaggs, and H. W. Coleman, Measurement and prediction of the effects of nonuniform surface roughness on turbulent flow friction coefficients, *Transactions of the ASME, Journal of Fluids Engineering*, 110, 1988, 380–383.

Farshad, F. F. and H. H. Rieke, Surface-roughness design values for modern pipes, *Society of Petroleum Engineers, Drilling & Completion*, SPE 89040, 2006, 212–215.

9

ENTRANCES

Pressure loss at the entrance into a straight pipe or passage of constant cross section is governed by several parameters: the distance from the pipe edge to the wall in which it is installed; the thickness of the inlet pipe edge; the angle at which the pipe is mounted into the wall; and, most definitely, rounding or beveling of the edge of the pipe inlet.

The entrance is a special form of contraction (see Chapter 10). A most important parameter for contractions is β, the ratio of downstream diameter d_2 to upstream diameter d_1. In the case of an entrance, however, the upstream diameter d_1 goes to infinity so that β goes to zero. With this key fact, loss coefficient equations developed in Chapter 10 for contractions can be adapted to various entrance configurations.*

It was long accepted that the loss coefficient of a sharp-edged flush-mounted entrance takes on a value of 0.50, or even as low a value as 0.43. Analyses supporting these values took into account expansion loss from measured or theoretical contraction factors, but did not account for loss due to the onrush of fluid into the pipe to form the contraction. Early test data seemed to support these values. This may have been because (1) the entrance edges may not have been truly sharp,† (2) test apparatus and test methods were not refined, and (3) of preconceived notions as to what the test results should be. In any case, it is demonstrated in Chapter 10 that the loss coefficient for a sharp-edged contraction—a flush-mounted, sharp-edged entrance in this case—can take on values higher than the long accepted values.‡ Accordingly, certain test results and formulas in this chapter are adjusted upwards to conform to a currently accepted value of 0.57.

* The loss coefficient equations developed in Chapter 13 for orifices can also be adapted to various entrance configurations as evident in Section 9.4.

† See discussion regarding edge sharpness of orifices in Section 13.3.1.

9.1 SHARP-EDGED ENTRANCE

The loss that arises from a sharp-edged entrance may be thought of as arising out of three effects. The first is the contraction of the main flow into the pipe and subsequent separation from the pipe wall leading up to a vena contracta. The second is the expansion loss of the main flow from the vena contracta to reattachment at the pipe wall. The third is the readjustment of the velocity profile downstream of the vena contracta and beyond the reattachment point.

9.1.1 Flush Mounted

A sharp-edged entrance is illustrated in Figure 9.1. The following expression was developed in Chapter 10 for the loss coefficient of a sharp-edged contraction in a straight pipe:

$$K_2 = 0.0696\left(1-\beta^5\right)\lambda^2 + (\lambda-1)^2, \qquad \text{(10.4, repeated)}$$

‡ The test data of Benedict, et al. [1] clearly show this.

Pipe Flow: A Practical and Comprehensive Guide, First Edition. Donald C. Rennels and Hobart M. Hudson.

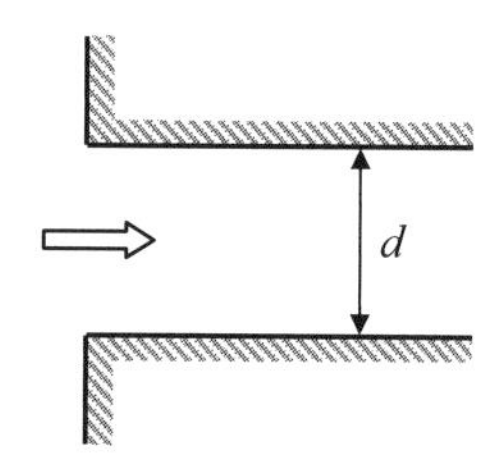

FIGURE 9.1. Flush-mounted sharp-edged entrance.

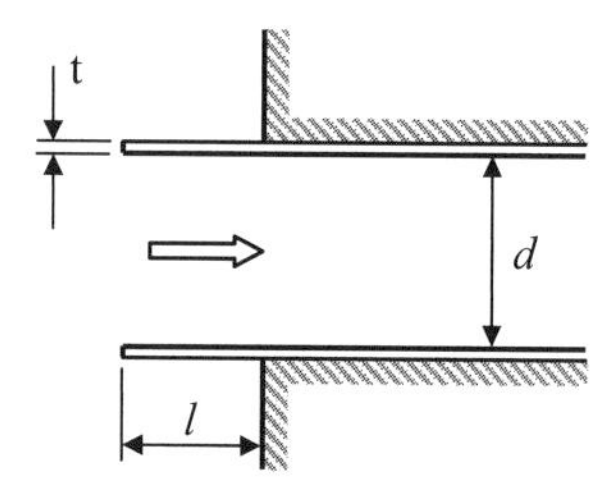

FIGURE 9.2. Sharp-edged entrance mounted at a distance.

where the diameter ratio $\beta = d_2/d_1$, and where the jet contraction coefficient λ is given by:

$$\lambda = 1 + 0.622(1 - 0.215\beta^2 - 0.785\beta^5). \quad (10.3, \text{repeated})$$

In the case of a flush-mounted pipe entrance, d_1 equals infinity, which means that β equals 0 and λ equals 1.622. Thus the loss coefficient for a flush-mounted sharp-edged entrance becomes:

$$K = 0.57.$$

As previously noted, the loss coefficient for a sharp-edged contraction—a flush-mounted, sharp-edged entrance in this case—can take on values higher than the oft-cited value of 0.50.

9.1.2 Mounted at a Distance

The loss coefficient K of the entry of a straight pipe extending a distance into a reservoir from its wall (see Fig. 9.2) is a function of the relative wall thickness t/d of the pipe, and on the relative distance l/d from the pipe edge to the wall.* In actuality, the effect of distance from the wall practically ceases at l/d equal to 0.5. It has long been accepted that K is maximum and equal to one velocity head for a pipe edge of infinitesimal thickness and an infinite distance from the wall. Its minimum value is created by a thick inlet edge, or by a pipe entry flush mounted to the wall, and, as worked out in the previous section, is equal to the value of 0.57.

* A short cylindrical tube, extending into a reservoir from its wall, is known as *Borda's mouthpiece*. If the wall of the tube is thin, or its inner edge is sharp, the contraction of the jet is found to be greater than in a jet from a sharp-edged orifice or from a flush mounted sharp-edged entrance. If the tube terminates outside the wall and its length is about equal to its diameter or less, the liquid in the reservoir will issue from the tube without touching its sides. Historically and academically of interest, in point of fact, Borda's mouthpiece has no practical value.

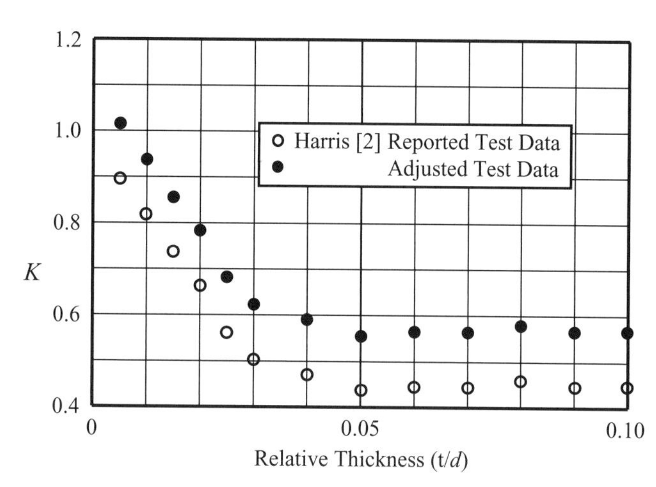

FIGURE 9.3. Pipe intake mounted at $l/d > 0.5$.

Test data reported by Harris [2] for an intake mounted at a relative distance l/d equal to 3.3 from the wall are shown in Figure 9.3. Harris' data are adjusted upwards by 0.12 to be consistent with a value of 0.57 at relative wall thickness t/d equal to or greater than 0.05. Note that this adjustment increases the minimum and maximum values above their long-held values.

A curve fit of Harris's adjusted data provides the following equation for the loss coefficient of a pipe intake mounted at a distance l/d equal to or greater than 0.5 as a function of pipe wall thickness to diameter ration t/d:

$$K = 1.12 - 22\frac{t}{d} + 216\left(\frac{t}{d}\right)^2 + 80\left(\frac{t}{d}\right)^3 \quad (t/d \le 0.05). \tag{9.1}$$

The results of Equation 9.1 are presented in Diagram 9.1 as a function of thickness-to-diameter ratio t/d ranging from 0 to 0.05.

9.1.3 Mounted at an Angle

The loss coefficient of a truly sharp-edged entrance mounted at an angle α from the wall (see Fig. 9.4) can be determined from Weisbach's formula [3], modified on the basis that the value at a 90° angle is 0.57 rather than 0.50:

$$K \approx 0.57 + 0.30\cos\alpha + 0.20\cos^2\alpha.$$

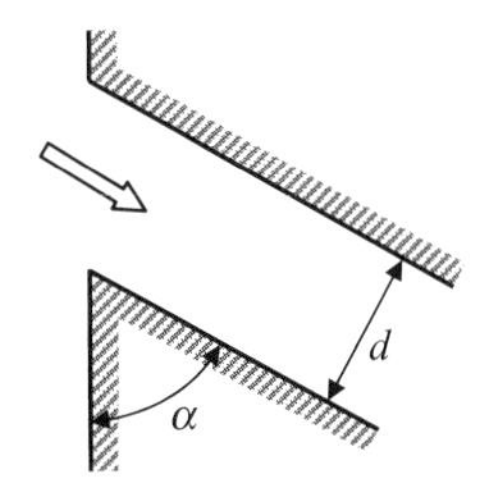

FIGURE 9.4. Mounted at an angle.

In practice, the edge may not be truly sharp and a value less than 0.57 may be substituted as the first term in Equation 9.2. The equation is not reliable at values of α less than 20°.

9.2 ROUNDED ENTRANCE

Rounding the inlet edge of a pipe entrance (see Fig. 9.5) streamlines the contraction of the main flow into the pipe and diminishes or prevents flow stream separation from the wall downstream of the entrance section so that the vena contracta is reduced or eliminated. Thus head loss is substantially reduced.

The following expression for the loss coefficient of a rounded contraction in a straight pipe was developed in Chapter 10 for the case where the rounding ratio r/d_2 is equal to or less than 1:

$$K_2 = 0.0696\left(1-0.569\frac{r}{d_2}\right)\left(1-\sqrt{\frac{r}{2}}\beta\right)(1-\beta^5)\lambda^2+(\lambda-1)^2 \quad (r/d_2 \le 1), \tag{10.6, repeated}$$

where the diameter ratio $\beta = d_2/d_1$, and where the jet contraction coefficient λ is given by:

$$\lambda = 1+0.622\left(1-0.30\sqrt{\frac{r}{d_2}}-0.70\frac{r}{d_2}\right)^4 (1-0.215\beta^2-0.785\beta^5). \tag{10.7, repeated}$$

In the case of a pipe entrance, d_1 goes to infinity so that β equals 0. Thus the loss coefficient of a flush-mounted rounded entrance simplifies to:

$$K = 0.0696\left(1-0.569\frac{r}{d}\right)\lambda^2+(\lambda-1)^2 \quad (r/d<1), \tag{9.2}$$

where the jet contraction coefficient λ is given by:

$$\lambda = 1+0.622\left(1-0.30\sqrt{\frac{r}{d}}-0.70\frac{r}{d}\right)^4.$$

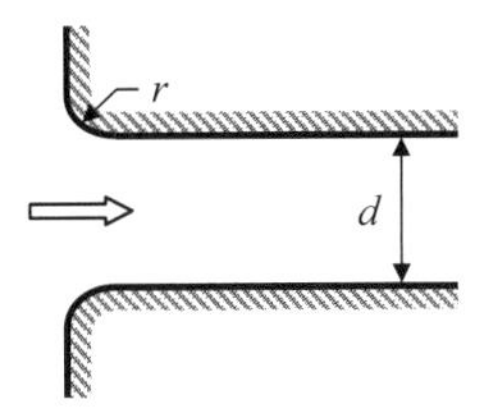

FIGURE 9.5. Flush-mounted rounded entrance.

For the case of a generously rounded entrance where r/d is equal to or greater than 1, the jet contraction ratio λ equals 1, and the loss coefficient for a flush-mounted rounded entrance becomes:

$$K = 0.03 \quad (r/d \ge 1).$$

Over the years a wide range of loss coefficient values for flush-mounted round-edged entrances has been reported by various authors. Early values may stem from test data such as those reported by Hamilton [4] in 1929. Hamilton's test results, shown in Figure 9.6, may have been influenced by preconceived notions that the loss coefficient of a sharp-edged entrance is 0.43 (see Harris [2]), and that full suppression of head loss takes place with rounding radius r greater than $0.14d$. It is true that a rounding radius r greater than about $0.14d$ prevents flow stream separation from the wall and, thereby, alleviates a significant expansion loss from a vena contracta to reattachment at the pipe wall. However, increase in rounding radius r beyond $0.14d$ continues to reduce head loss due to contraction of fluid into the pipe and reduces subsequent downstream readjustment of the velocity profile. Based on data presented in Chapter 10 for rounded contractions (see Diagram 10.1), reduction in head loss continues up to a rounding radius of $1.0d$. The results of Equation 9.2 are compared with Hamilton's test data in Figure 9.6.

Equation 9.2 is presented in Diagram 9.2 as a function of rounding ratio r/d ranging from 0 to 1.0. A useful curve fit of Equation 9.2 for $r/d \le 1.0$ is given by:

$$K = 0.57-1.07(r/d)^{1/2}-2.13(r/d)+8.24(r/d)^{3/2}-8.48(r/d)^2+2.90(r/d)^{5/2}. \tag{9.3}$$

Equation 9.3 is used in Section 16.1.2 to characterize the effect of rounding the edge of the branch to main channel connection of a tee.

9.3 BEVELED ENTRANCE

Beveling (or chamfering) the inlet edge of a pipe entrance, as shown in Figure 9.7, reduces the head loss. The important parameters are the nondimensional bevel length to diameter ratio, l/d and the bevel angle ψ.

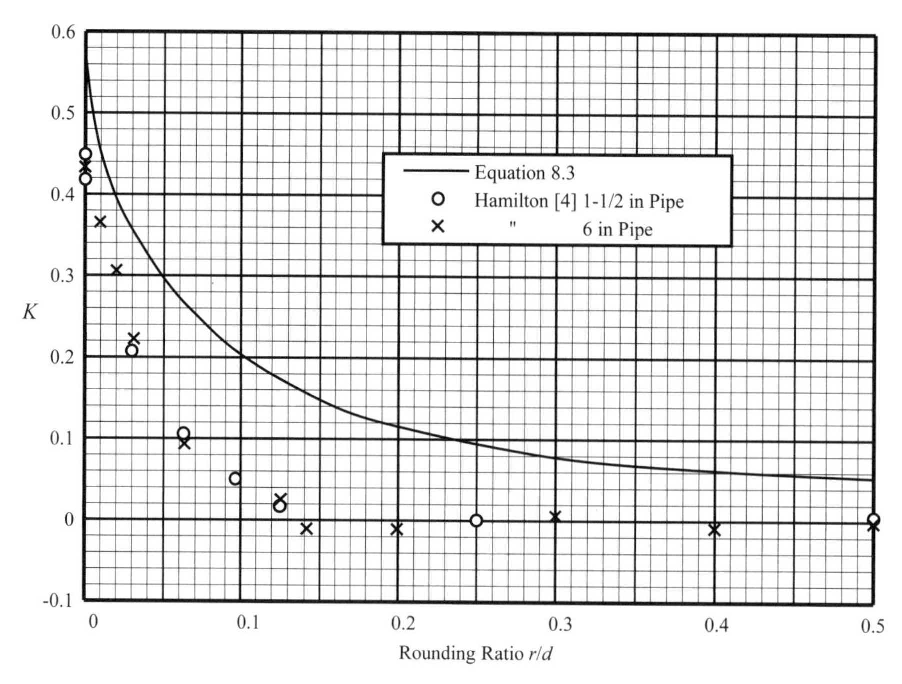

FIGURE 9.6. Comparison of Equation 8.3 with loss coefficient data from Hamilton [4].

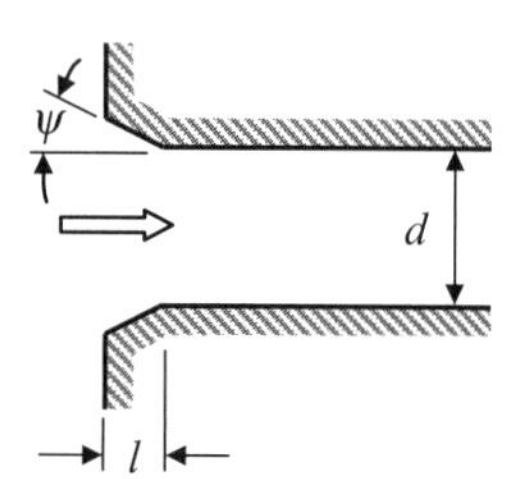

FIGURE 9.7. Flush mounted beveled entrance.

There is little or no credible data for a beveled entrance. The equations in this section are related to similar equations developed in Section 13.4 for beveled orifices. The following approximate equation was developed for a beveled entrance of length l and angle ψ:

$$K_2 \approx 0.0696\left(1 - C_b \frac{l}{d}\right)\lambda^2 + (\lambda - 1)^2, \qquad (9.4)$$

where the jet contraction coefficient λ is given by:

$$\lambda = 1 + 0.622\left[1 - 1.5C_b\left(\frac{l}{d}\right)^{\frac{1-\sqrt[4]{l/d}}{2}}\right],$$

and where C_b, a function of bevel angle ψ in degrees and bevel length to diameter ratio l/d, is given by:

$$C_b = \left(1 - \frac{\psi}{90}\right)\left(\frac{\psi}{90}\right)^{\frac{1}{1+l/d_2}}.$$

The results of Equation 9.4 are presented in Diagram 9.3 as a function of bevel length to diameter ratio l/d ranging from zero to one.

9.4 ENTRANCE THROUGH AN ORIFICE

Loss coefficient equations developed in Chapter 13 for various orifice configurations in a transition section can be adapted to represent a pipe entrance through an orifice from a reservoir by recognizing that d_1 is in effect equal to infinity so that d_o/d_1, or β, goes to zero.

Note that the loss coefficients (K_os) presented in this section are based on the velocity (or flow area) of the orifice restriction. When summing the loss coefficients in a piping stretch they must be transformed to the "standardized" area used in the ΔP formula; usually the pipe flow area (see Section 3.2.3):

$$K = K_a \frac{A^2}{A_a^2} \quad \text{or} \quad K = K_a \frac{d^4}{d_a^4}. \qquad \text{(3.8, repeated)}$$

9.4.1 Sharp-Edged Orifice

A sharp-edged orifice in an entrance section is illustrated in Figure 9.8. Equation 13.5 for a sharp-edged orifice in a transition section can be transformed into a pipe entrance. Because β is equal to zero, the jet velocity

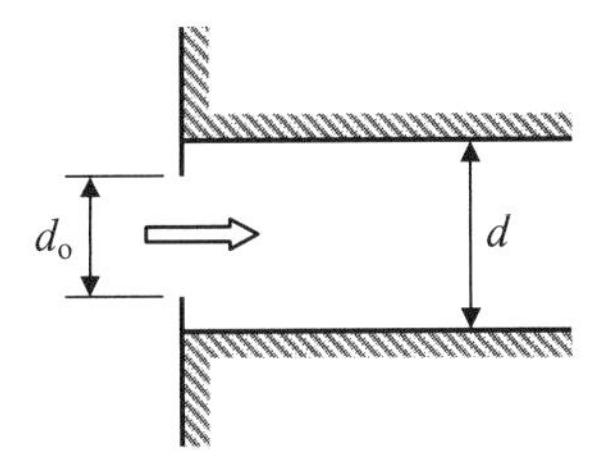

FIGURE 9.8. Entrance through a sharp-edged orifice.

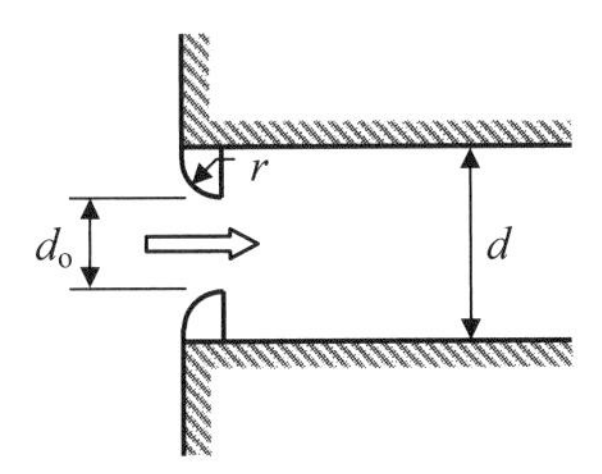

FIGURE 9.9. Entrance through a round-edged orifice.

ratio λ is equal to 1.622, and the loss coefficient turns out to be:

$$K_o = 0.183 + \left[1.622 - \left(\frac{d_o}{d}\right)^2\right]^2. \tag{9.5}$$

The results of Equation 9.5 are shown as the uppermost curves in Diagrams 9.4 and 9.5.

9.4.2 Round-Edged Orifice

A round-edged orifice in an entrance section is illustrated in Figure 9.9. Equation 13.8 for a round-edged orifice in a transition section can be transformed into a pipe entrance. Because β is equal to zero, the jet velocity ratio λ is equal to 1.622, and the loss coefficient equation turns out to be:

$$K_o = 0.0696\left(1 - 0.569\frac{r}{d_o}\right)\lambda^2 + \left(\lambda - \left(\frac{d_o}{d}\right)^2\right)^2 \quad (r/d_o \le 1) \tag{9.6}$$

where the jet contraction coefficient λ is given by:

$$\lambda = 1 + 0.622\left(1 - 0.30\sqrt{\frac{r}{d_o}} - 0.70\frac{r}{d_o}\right)^4.$$

For the case of a generously rounded orifice where $r/d_o > 1$, the jet contraction ratio $\lambda = 1$ and the loss coefficient becomes:

$$K_o = 0.030 + \left(1 - \left(\frac{d_o}{d}\right)^2\right)^2 \quad (r/d_o > 1).$$

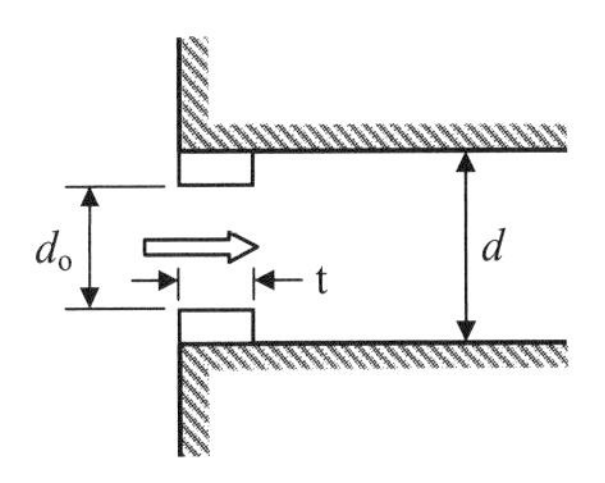

FIGURE 9.10. Entrance through a thick-edged orifice.

The results of Equation 9.6 are presented in Diagram 9.4 as a function of rounding ratio r/d ranging from zero to one.

9.4.3 Thick-Edged Orifice

A thick-edged orifice in an entrance section is illustrated in Figure 9.10. Equation 13.16 for a thick-edged orifice in a transition section can be transformed into a pipe entrance where the thickness t is less than or equal to $1.4d$:

$$K_o = 0.183 + C_{th}\left[1.622 - \left(\frac{d_o}{d}\right)^2\right]^2 + (1 - C_{th})\left\{0.387 + \left[1 - \left(\frac{d_o}{d}\right)^2\right]^2\right\} \quad (t/d_o < 1.4), \tag{9.7}$$

where C_{th} is given by:

$$C_{th} = \left[1 - 0.50\left(\frac{t}{1.4d}\right)^{2.5} - 0.50\left(\frac{t}{1.4d}\right)^3\right]^{4.5}.$$

For thickness t equal to or greater than $1.4d_o$, surface friction loss becomes significant and the loss coefficient can be determined from the following equation:

$$K_o = 0.57 + \left[1 - \left(\frac{d_o}{d}\right)^2\right]^2 + f_o\left(\frac{t}{d} - 1.4\right) \quad (t/d_o \ge 1.4).$$

The results of Equation 9.7 are presented in Diagram 9.5 as a function of thickness ratio t/d ranging from zero to 1.4.

9.4.4 Beveled Orifice

A beveled orifice in an entrance section is illustrated in Figure 9.11. Equation 9.4 for a beveled entrance can be transformed into an entrance through a beveled orifice by substituting $\lambda - (d_o/d)^2$ for $\lambda - 1$ in the last term of the equation:

$$K_o \approx 0.0696\left(1 - C_b\frac{l}{d_o}\right)\lambda^2 + \left[\lambda - \left(\frac{d_o}{d}\right)^2\right]^2, \tag{9.8}$$

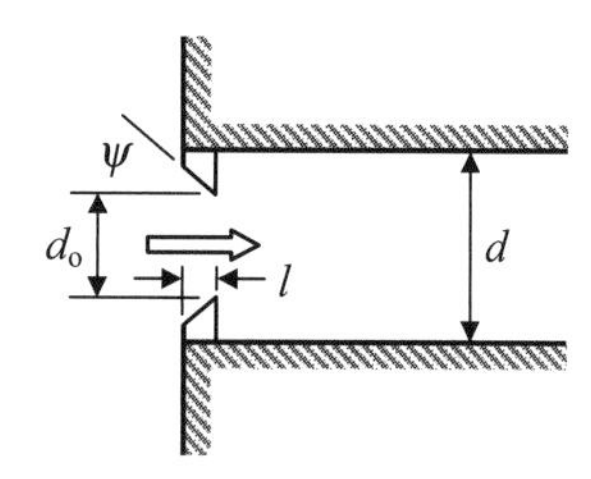

FIGURE 9.11. Entrance through a beveled orifice.

where the jet contraction coefficient λ is given by:

$$\lambda = 1 + 0.622\left[1 - C_b\left(\frac{l}{d_o}\right)^{\frac{1-\sqrt[4]{l/d_o}}{2}}\right],$$

and where C_b is given by:

$$C_b = \left(1 - \frac{\psi}{90}\right)\left(\frac{\psi}{90}\right)^{\frac{1}{1+l/d_o}}.$$

The results of Equation 9.8 for an entrance through a 45° beveled orifice are presented in Diagram 9.6 as a function of bevel length to diameter ratio l/d_o ranging from zero to one.

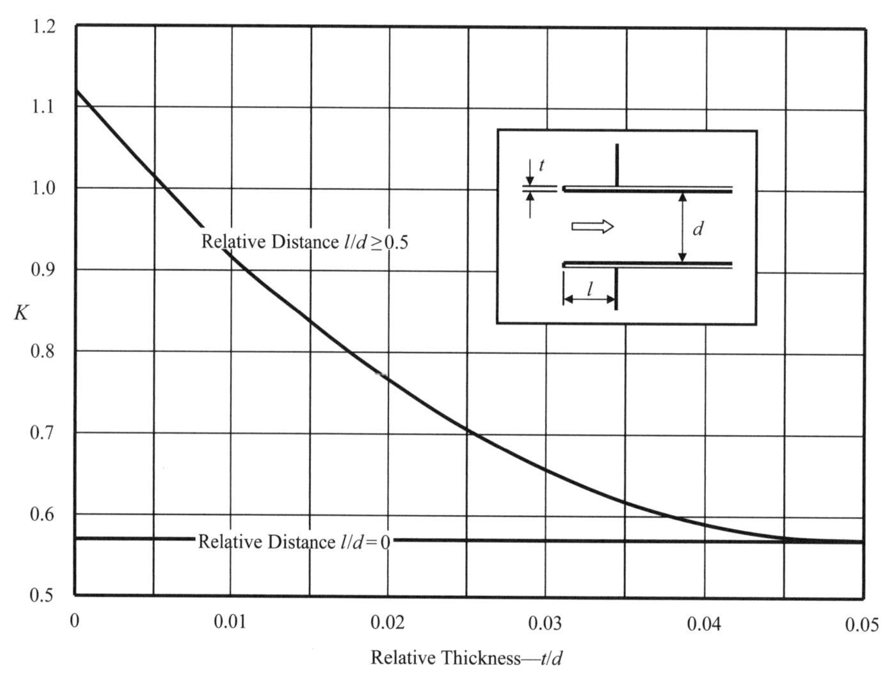

DIAGRAM 9.1. Loss coefficient K of a sharp-edged entrance mounted at a distance.

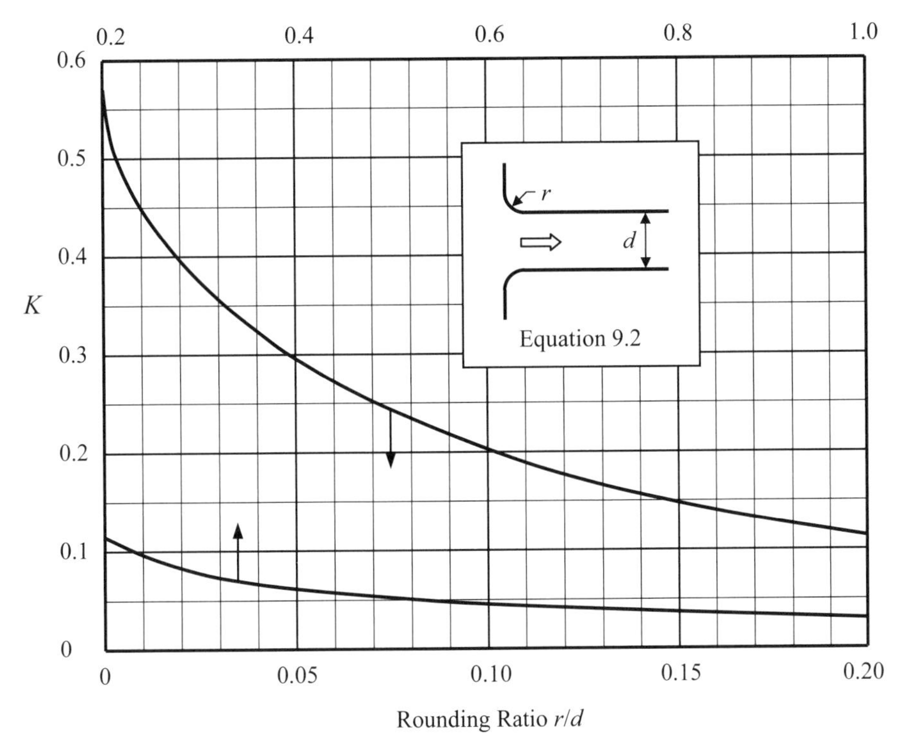

DIAGRAM 9.2. Loss coefficient K of a flush-mounted rounded entrance.

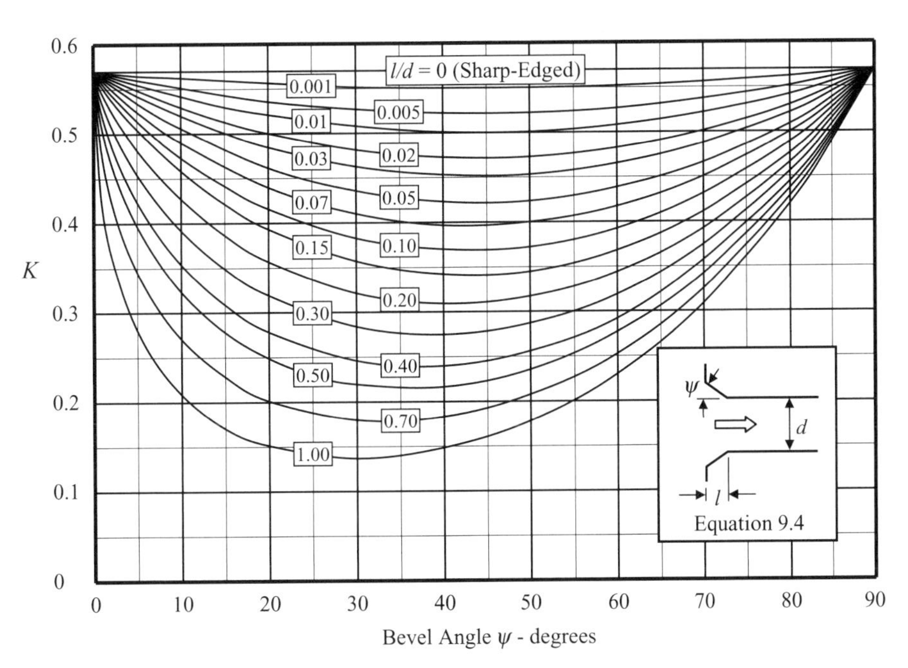

DIAGRAM 9.3. Loss coefficient K of a flush-mounted beveled entrance.

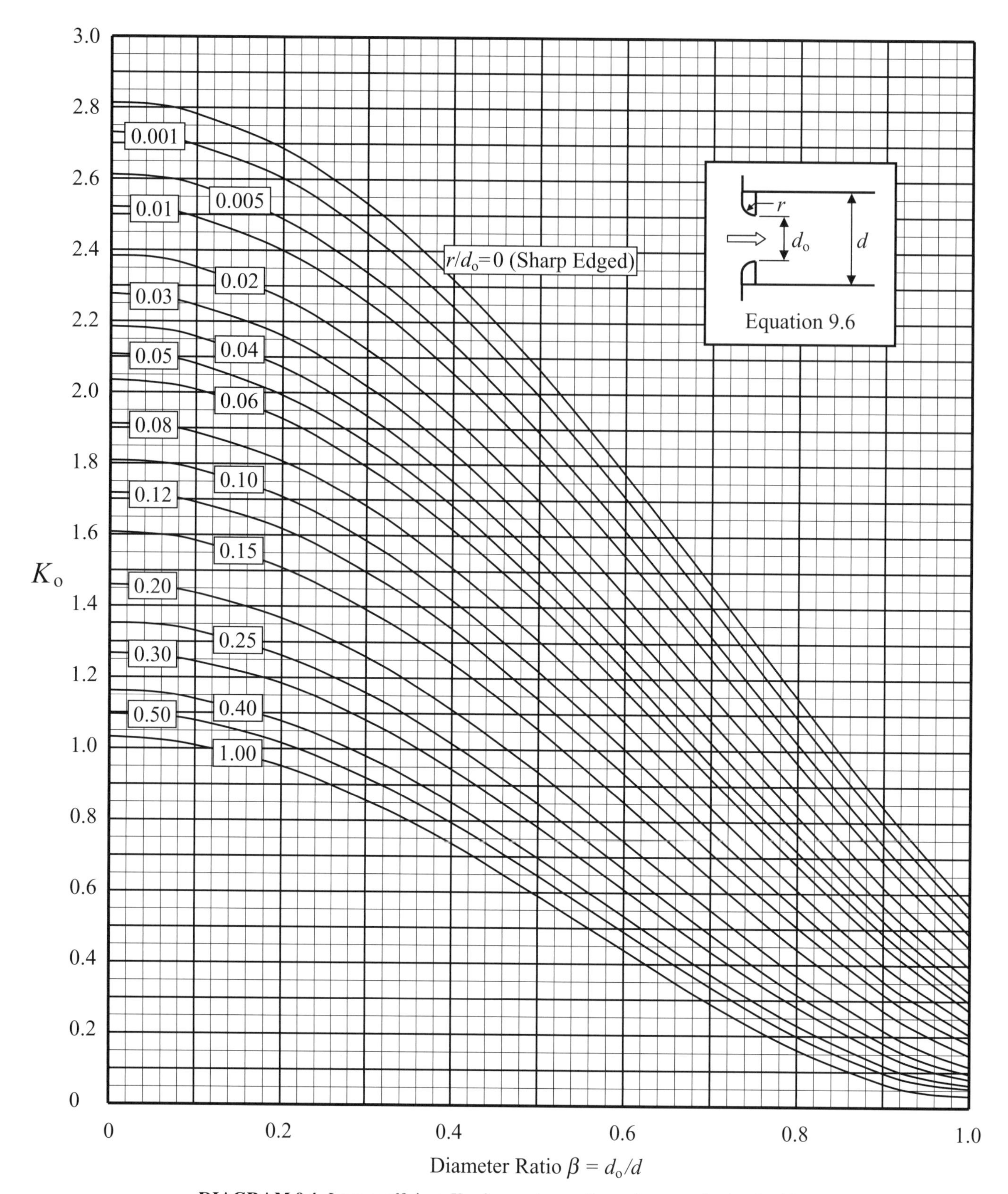

DIAGRAM 9.4. Loss coefficient K_o of an entrance through a round-edged orifice.

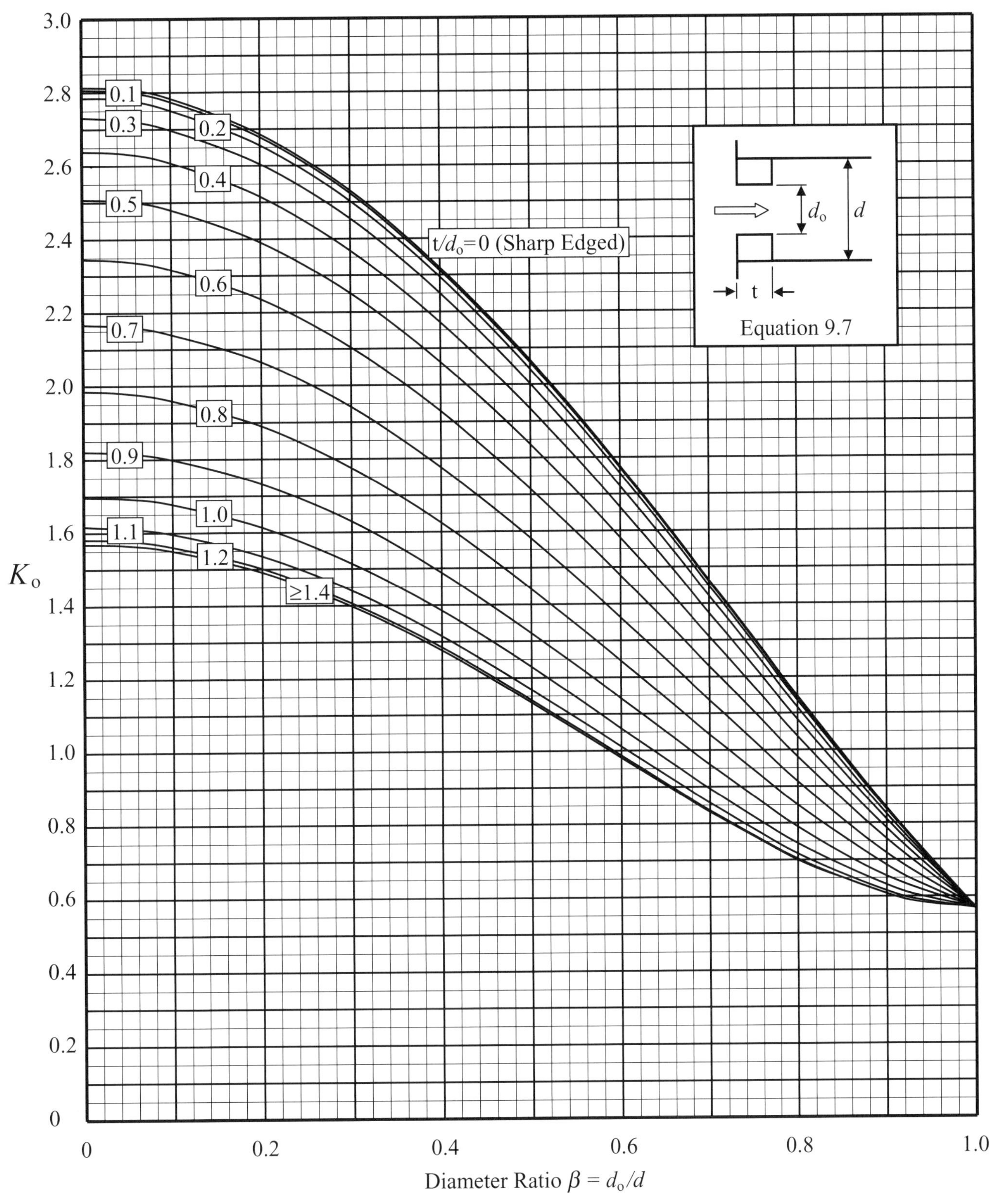

DIAGRAM 9.5. Loss coefficient K_o of an entrance through a thick-edged orifice.

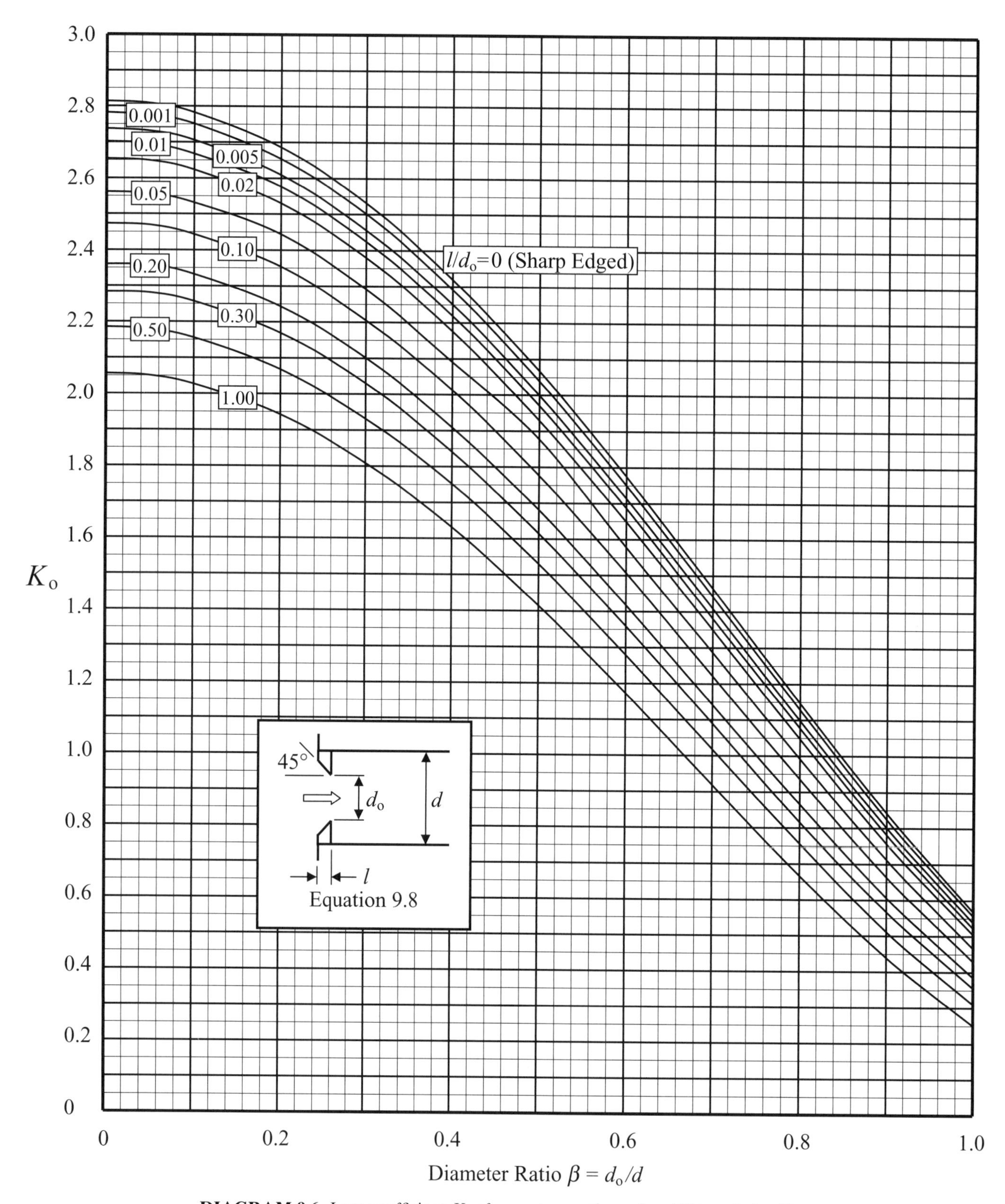

DIAGRAM 9.6. Loss coefficient K_o of an entrance through a 45° beveled orifice.

REFERENCES

1. Benedict, R. P., N. A. Carlucci, and S. D. Swetz, Flow losses in abrupt enlargements and contractions, *Transactions of the American Society of Mechanical Engineers, Journal of Engineering for Power*, 88, 1966, 73–81.
2. Harris, C. W., The Influence of Pipe Thickness on Re-Entrant Intake Losses, *University of Washington Engineering. Experiment Station*, Bulletin No. 48, November 1, 1928.
3. Weisbach, J., *Mechanics of Engineering*, translated by, E. B. Coxe, Van Nostrand Book Co., 1872.
4. Hamilton, J. B., Suppression of Pipe Intake Losses by Various Degrees of Rounding, *University of Washington Engineering Experiment Station*, Bulletin No. 51, November 15, 1929.

FURTHER READING

This list includes works that may be helpful to those who wish to pursue further study.

Rouse, H. and M. M. Hassan, Cavitation-Free Inlets and Contractions, *Mechanical Engineering*, 1933, pp. 213–216.

Deissler, R. G., Turbulent heat transfer and friction factor in the entrance region of smooth passages, *Transactions of the American Society of Mechanical Engineers*, 76, 1955, 1221–1233.

Ross, D., Turbulent flow in the entrance region of a pipe, *Transactions of the American Society of Mechanical Engineers*, 78, 1956, 915–923.

Campbell, W. D. and J. C. Slattery, Flow in the entrance of a tube, *Transactions of the American Society of Mechanical Engineers, Journal of Basic Engineering*, 88, 1963, 41–46.

Barbin, A. R. and J. B. Jones, Turbulent flow in the inlet region of a smooth pipe, *Transactions of the American Society of Mechanical Engineers, Journal of Basic Engineering*, 85, 1974, 29–34.

Wang, J.-S. and J. P. Tullis, Turbulent flow in the entry region of a rough pipe, *Transactions of the American Society of Mechanical Engineers, Journal of Fluids Engineering*, 96, 1974, 62–68.

Dong, W. and J. H. Lienhard, Contraction coefficients for Borda mouthpieces, *Transactions of the American Society of Mechanical Engineers, Journal of Fluids Engineering*, 108, 1986, 377–379.

Bullen, P. R., D. J. Cheeseman, L. A. Hussain, and A. E. Ruffell, The determination of pipe contraction pressure loss coefficients for incompressible turbulent flow, *International Journal of Heat and Fluid Flow*, 8(2), 1987, 111.

Bullen, P. R., D. J. Cheeseman, and L. A. Hussain, The effects of inlet sharpness on the pipe contraction loss coefficient, *International Journal of Heat and Fluid Flow*, 9(4), 1988, 431.

10

CONTRACTIONS

Flow through a *sudden* or *sharp-edged contraction* is shown in Figure 10.1. The flow *accelerates* as it approaches the contraction and the outer filaments adjacent to the wall achieve a high inward radial velocity of about the same order as the axial velocity. The high radially inward velocity causes the jet to contract and the flow stream to separate from the wall. The point of minimum cross-sectional flow area in the separated region is called the *vena contracta*. The jet subsequently *decelerates* and expands to fill the passage. Rounding, tapering, or beveling the entrance section reduces the high radially inward velocity and substantially reduces the head loss.

10.1 FLOW MODEL

Taking the total head loss H_2 of a contraction as the sum of the losses in the acceleration and deceleration regions, and treating them as a gradual contraction and a sudden expansion respectively,* gives:

$$H_2 = K_2\frac{V_2^2}{2g} = K_{acc}\frac{V_C^2}{2g} + \frac{(V_C - V_2)^2}{2g}, \qquad (10.1)$$

where V_C is the local velocity at the vena contracta and V_2 is the velocity in the downstream pipe. The first term on the right represents the gradual acceleration of the fluid to the vena contracta and K_{acc} is the loss coefficient for the acceleration portion of the flow. The second term represents the sudden expansion of the fluid stream downstream of the vena contracta.†

Re-arrangement of Equation 10.1 gives:

$$K_2 = K_{acc}\frac{V_C^2}{V_2^2} + \left(\frac{V_C}{V_2} - 1\right)^2.$$

The ratio V_C/V_2 can be expressed as the jet velocity ratio λ and the equation becomes:

$$K_2 = K_{acc}\lambda^2 + (\lambda - 1)^2. \qquad (10.2)$$

Undoubtedly, the universal velocity profile exists at the vena contracta as well as in the fully developed flow regions in the upstream and downstream pipes. Nonetheless, the simple assumption is made that the velocity profile is uniform at the vena contracta and in the pipes. Successful correlation with test data in the following sections and chapters validates this simplification.

Equation 10.2 allows for expressing the loss coefficient of various types of contractions (sharp edged, rounded, conical, etc.) by the use of suitable terms for λ and K_{acc} based on available data. Additionally, the jet velocity ratio λ can be used to determine the local velocity and, thereby, to estimate the local pressure at the vena contracta.

A term, the so-called beta ratio β, is used to describe contractions and expansions. For a pipe (or circular)

* This treatment was suggested by Vennard [1].

† The sudden expansion term derives from Equation 11.5, the Borda–Carnot equation.

Pipe Flow: A Practical and Comprehensive Guide, First Edition. Donald C. Rennels and Hobart M. Hudson.

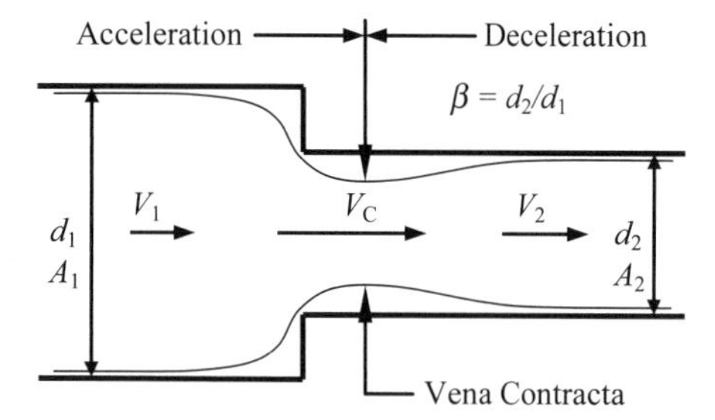

FIGURE 10.1. Sudden contraction.

contraction it is simply the ratio of the smaller diameter to the larger diameter:

$$\beta = d_2 / d_1.$$

In the case of a noncircular passage, an effective beta ratio can be calculated as a ratio of flow areas:

$$\beta = \sqrt{A_2 / A_1}.$$

Contraction losses are less sensitive to upstream conditions than expansion losses. A pointed velocity profile ahead of a contraction actually reduces the loss. Contraction losses are relatively insensitive to downstream conditions.

10.2 SHARP-EDGED CONTRACTION

Early measurements by Weisbach [2] accurately established the magnitude of the jet contraction coefficient C_C (ratio of jet contraction area A_C to area A_2) in free discharge water tests through sharp-edged, or sudden, contractions. They were found, for flows at high Reynolds numbers, to be dependent upon the area ratio A_2/A_1. Von Mises [3] analytically confirmed these experimental values for two-dimensional orifice flow. Kirchoff [4] gave a theoretical minimum jet contraction coefficient, at $A_2/A_1 = 0$, for a perfect liquid passing through a long slit or a circular opening as $C_C = \pi/(\pi + 2) = 0.611$. Weisbach's measurements as well as measurements by Freeman [5] in free discharge water tests of square ring nozzles* are shown in Table 10.1.

The free discharge data are represented in Figure 10.2 in the form of a jet velocity ratio λ which is simply the reciprocal of the jet contraction coefficient C_C. A curve fit of Weisbach's data† yields the jet velocity ratio as:

* "Square ring nozzles" or square-edged nozzles evolved from the mistaken belief that sharp edges increased the reach of fire nozzles.

† Freeman's data were not employed because the contraction was preceded by a conical converging section (or nozzle).

TABLE 10.1. Jet Contraction Coefficient

A_2/A_1	C_C (Weisbach)	C_C (Freeman)
0.0	0.617	–
0.1	0.624	0.632
0.2	0.632	0.644
0.3	0.643	0.659
0.4	0.659	0.676
0.5	0.681	0.696
0.6	0.712	0.717
0.7	0.755	0.744
0.8	0.813	0.784
0.9	0.892	0.890
1.0	1.000	1.000

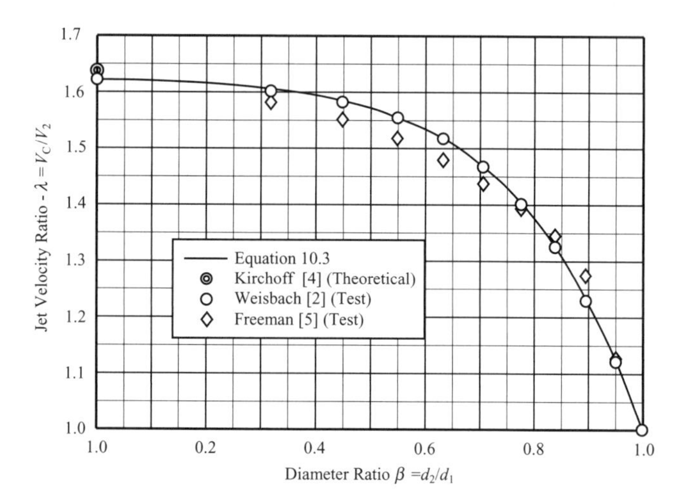

FIGURE 10.2. Jet velocity ratio curve fit.

$$\lambda = 1 + 0.622(1 - 0.215\beta^2 - 0.785\beta^5), \qquad (10.3)$$

where β is the ratio of the downstream diameter d_2 to the upstream diameter d_1. Equation 10.3 closely matches Weisbach's data as shown in Figure 10.2.

A wide range of loss coefficient values for sharp-edged contractions is found in the literature. Historically, the maximum value did not exceed 0.5. Benedict et al. [6] report experimental results that belie this notion. Benedict et al.'s test data were used to develop the following expression for the loss coefficient of sharp-edged contractions:

$$K_2 = 0.0696(1 - \beta^5)\lambda^2 + (\lambda - 1)^2, \qquad (10.4)$$

where $\beta = d_2/d_1$ and where the jet velocity ratio λ is given by:

$$\lambda = 1 + 0.622(1 - 0.215\beta^2 - 0.785\beta^5). \qquad (10.3, \text{repeated})$$

As evident in Figure 10.3, Equation 10.4 closely models Benedict et al.'s test data. This equation was developed in parallel with a similar expression for sharp-edged orifices in Chapter 13 (see Eq. 13.3). The only difference between the two equations is that the $(\lambda - 1)^2$ expansion term in Equation 10.4 has been replaced with a $(\lambda - \beta^2)^2$ expansion term in Equation 13.3. The acceleration term is the same in both cases. However, the vena contracta expands to the original flow area in the case of flow through an orifice in a straight pipe, whereas it expands to a new downstream flow area in the case of flow through a contraction.

Note that the maximum loss coefficient for a sharp-edged contraction can take on values above 0.5. This oft-quoted value was derived from a mean discharge coefficient of 0.815 assigned by Weisbach [2] in the mid-nineteenth century. The discharge coefficient is the actual flow divided by the ideal flow. The derivation is $K = 1/C_D^2 - 1 = 0.506$, where $C_D = 0.815$.

The relative magnitudes of the acceleration and deceleration portions of the total loss coefficient are shown as dotted lines in Figure 10.3. The loss produced by deceleration is noticeably greater than the loss produced by acceleration even though the decrease of velocity of the former is less than the increase of velocity of the latter. This is an example of the characteristic efficiency associated with acceleration and the inefficiency associated with deceleration in fluid flow.

The jet velocity ratio λ was developed from test data for free discharge from nozzles or orifices. It was not developed for internal (or confined) flow in piping systems. Although the source of λ is imperfect, it gives specific results that match test data for internal incompressible flow quite well.

10.3 ROUNDED CONTRACTION

The head loss of a contraction can be reduced by rounding the inlet edge of the entrance to the narrow section (see Fig. 10.4). Rounding diminishes or prevents flow stream separation from the wall downstream of the entrance section and thus substantially reduces the head loss.

The rounding contour may be the arc of a circle, or it may take the form of an ellipse, lemniscate, or other smoothly curved shape. For a circular inlet, the rounding radius r is simply the radius of the circle. In the case of an elliptical inlet contour, the rounding radius can be expressed as:

$$r = \sqrt[3]{r_1^2 r_2}, \tag{10.5}$$

where r_1 and r_2 are the semimajor (longitudinal) and semiminor (radial) axes, respectively.*

Observations indicate that a rounding radius r greater than about $0.14d_2$ prevents flow stream separation from the wall. Even so, further increase in r reduces loss due to acceleration of fluid into the contraction, as well as loss due to downstream readjustment of the velocity

* There is no theoretical basis for this relationship, but it is reasonable and it works quite well.

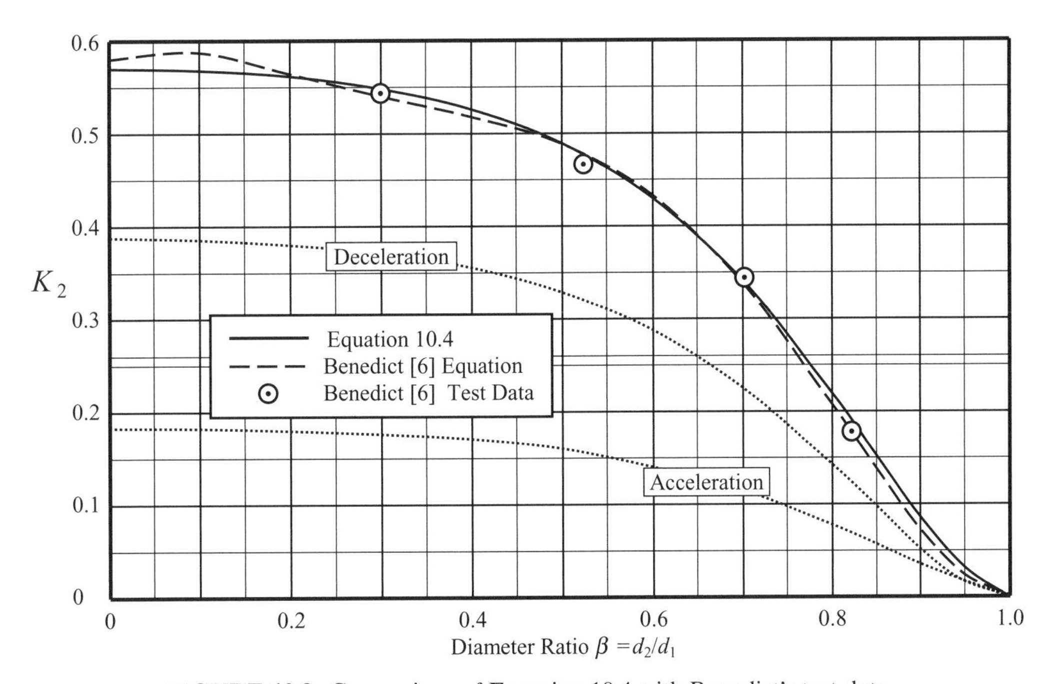

FIGURE 10.3. Comparison of Equation 10.4 with Benedict's test data.

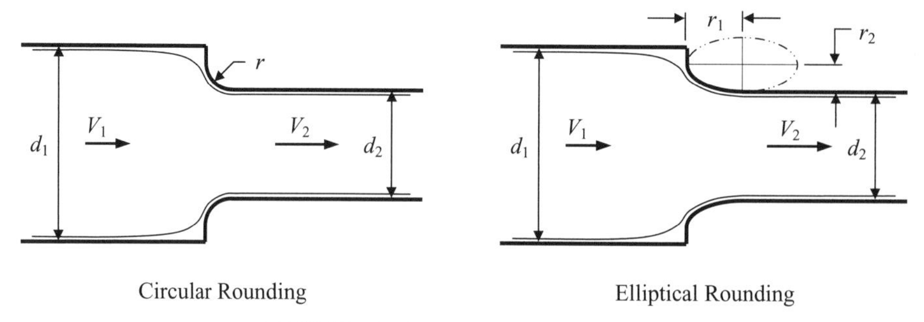

FIGURE 10.4. Rounded contraction.

profile. As the rounding radius r approaches $1.0d_2$ the head loss becomes minimal.

The following formulation for the loss coefficient of a rounded contraction was derived from a formulation for round-edged orifices that was developed in Section 13.3. The difference between the two formulations is in the sudden expansion term. As was the case for sharp-edged contractions, the sudden expansion term $(\lambda - \beta^2)^2$ for orifices has simply been replaced with $(\lambda - 1)^2$ for contractions. Thus, the following expression was developed for the loss coefficient of a rounded contraction for the case where the rounding ratio r/d_2 is equal to or less than 1:

$$K_2 = 0.0696\left(1 - 0.569\frac{r}{d_2}\right)\left(1 - \sqrt{\frac{r}{d_2}}\beta\right)(1-\beta^5)\lambda^2 + (\lambda - 1)^2 \quad (r/d_2 \le 1), \tag{10.6}$$

where the diameter ratio $\beta = d_2/d_1$, and where the jet contraction coefficient λ is given by:

$$\lambda = 1 + 0.622\left(1 - 0.30\sqrt{\frac{r}{d_2}} - 0.70\frac{r}{d_2}\right)^4 (1 - 0.215\beta^2 - 0.785\beta^5). \tag{10.7}$$

For the case of a generously rounded nozzle where r/d_2 is equal to or greater than 1, the jet contraction ratio λ equals 1 and the loss coefficient for a rounded contraction becomes:

$$K_2 = 0.030(1-\beta)(1-\beta^4) \quad (r/d_2 \ge 1). \tag{10.8}$$

Loss coefficients of rounded contractions can be determined from Diagram 10.1. The dashed line in Diagram 10.1 represents the boundary where full rounding cannot be achieved by simple circular rounding. In this case, an ellipse, lemniscate, or other noncircular curved shape may be employed to achieve a rounding radius ratio r/d_2 approaching, equal to, or greater than 1 (see Eq. 10.5). The parameters at which circular rounding is limited because of geometry restrictions are given by:

$$\beta_{\text{limit}} = \frac{1}{1 + 2r/d_2}$$

and

$$r/d_{2\text{limit}} = \frac{1/\beta - 1}{2}.$$

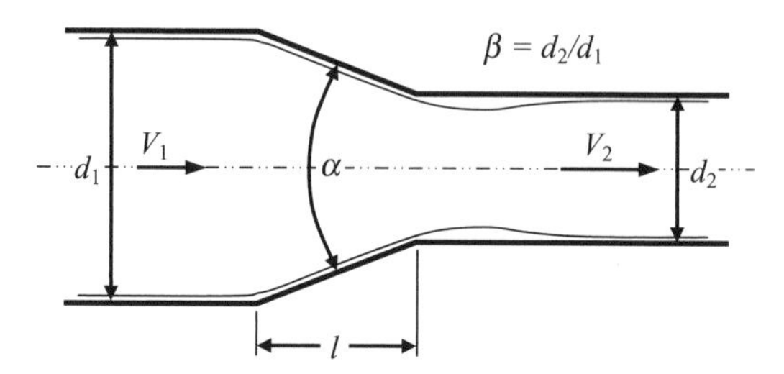

FIGURE 10.5. Conical contraction.

10.4 CONICAL CONTRACTION

Pressure loss in a contracting passage can be materially reduced by providing a gradually converging, conical section as shown in Figure 10.5.

The main geometric considerations of conical contractions are the diameter ratio $\beta = d_2/d_1$, the divergence angle α, and the length l of the conical section.* These variables are interrelated as follows:

$$l = \frac{d_1 - d_2}{2\tan(\alpha/2)} = \frac{d_2(1/\beta - 1)}{2\tan(\alpha/2)} \tag{10.9}$$

and

$$\alpha = 2\,\text{atan}\left(\frac{d_1 - d_2}{2l}\right) = 2\,\text{atan}\left(\frac{1/\beta - 1}{2l/d_2}\right). \tag{10.10}$$

Surface friction losses may be significant for long stretches at small included angle. The loss coefficient

* In the following equations α is generally expressed in radians; the modifications for using degrees is obvious.

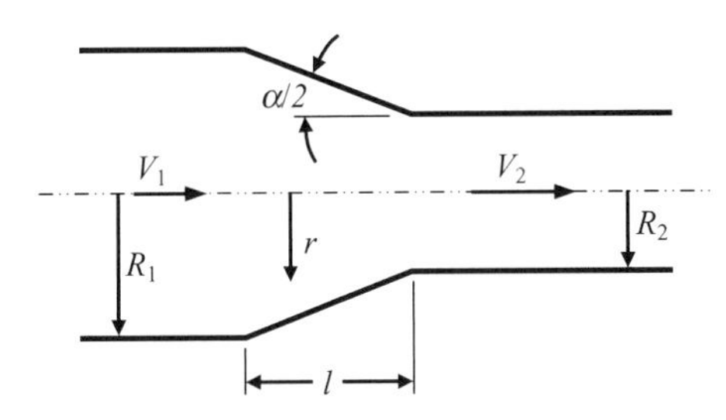

FIGURE 10.6. Surface friction loss.

equation for conical contractions may be conveniently written in the form of:

$$K_2 = K_{fr2} + K_{con2}, \qquad (10.11)$$

where K_{fr2} represents the surface friction loss and K_{con2} represents the local loss.

10.4.1 Surface Friction Loss

A theoretical equation for surface friction loss coefficient K_{fr2} in a conical contraction has been reported by several investigators including Levin and Claremont [7]. It is a classic equation of early hydraulic analysis. Referring to Figure 10.6, the head loss due to surface friction in terms of the velocity at point 2 can be expressed as:

$$dh_2 = f\frac{dl}{2r}\frac{u^2}{2g}. \qquad (10.12)$$

From geometry considerations:

$$dl = \frac{dr}{\sin(\alpha/2)}. \qquad (10.13)$$

The velocity profile along the length of the contraction can be expressed as:

$$u = V_2\left(\frac{R_2}{r}\right)^2. \qquad (10.14)$$

Substitution of Equations 10.13 and 10.14 into Equation 10.12 gives:

$$dh_2 = \frac{f}{2\sin(\alpha/2)}\frac{V_2^2}{2g}\frac{R_2^4}{r^5}dr. \qquad (10.15)$$

The integral form of Equation 10.15 within the limits $r = R_1$ to $r = R_2$ is:

$$dh_2 = \frac{fR_2^4}{2\sin(\alpha/2)}\frac{V_2^2}{2g}\int_{R_2}^{R_1}\frac{dr}{r^5}.$$

Integration yields

$$\Delta h_2 = \frac{fR_2^4}{2\sin(\alpha/2)}\frac{V_2^2}{2g}\left(\frac{1}{4R_2^4} - \frac{1}{4R_1^4}\right) = \frac{f_c\left(1 - R_2^4/R_1^4\right)}{8\sin(\alpha/2)}\frac{V_2^2}{2g}.$$

Recognizing that $\beta^4 = R_2^4/R_1^4$ yields:

$$\Delta h_2 = \frac{f\left(1-\beta^4\right)}{8\sin(\alpha/2)}\frac{V_2^2}{2g},$$

or

$$K_{fr2} = \frac{f\left(1-\beta^4\right)}{8\sin(\alpha/2)}, \qquad (10.16)$$

where the friction factor f is based on the relative roughness of the conical surface and the hydraulic diameter and Reynolds number at the cone exit. This formulation appears to adequately represent surface friction loss in a conical contraction. It is also employed to represent surface friction loss in diverging conical sections (Section 11.2) in bevel-edged entrance sections (Section 9.3) and in flow meters (Chapter 14).

Equation 10.16 is plotted as a function of diameter ratio β and divergence angle α in Diagram 10.2 for a friction factor f of 0.020. Because K_{fr2} is directly proportional to f, the surface friction loss coefficient can be determined by simple proportion for other values of f. It is evident that surface friction loss may generally be ignored at large included angles or at small stretches.

10.4.2 Local Loss

The coefficient of local loss can be determined as follows:

$$K_{con2} = 0.0696\sin(\alpha/2)(1-\beta^5)\lambda^2 + (\lambda-1)^2, \qquad (10.17)$$

where the jet contraction ratio λ is given by:

$$\lambda = 1 + 0.622(\alpha/180)^{4/5}\left(1 - 0.215\beta^2 - 0.785\beta^5\right). \qquad (10.18)$$

Equation 10.17 is compared to test data reported by Levin and Clermont [7]* in Figure 10.7. Note that at divergence angle α equal to 180°, Equation 10.17 evolves into the straightforward equation for a sharp-edged contraction (see Eq. 10.4).

* Levin and Clermont used the theoretical equation for surface friction in a conical contraction (Eq. 10.15) to remove surface friction loss from their test results. Thus their reported data are for local loss only.

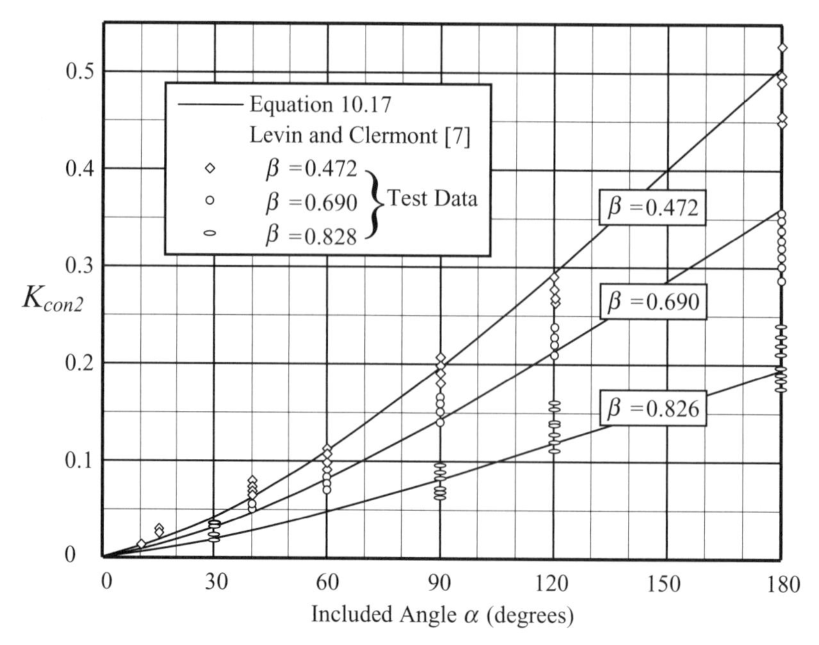

FIGURE 10.7. Comparison of Equation 10.17 with Levin and Clermont's data.

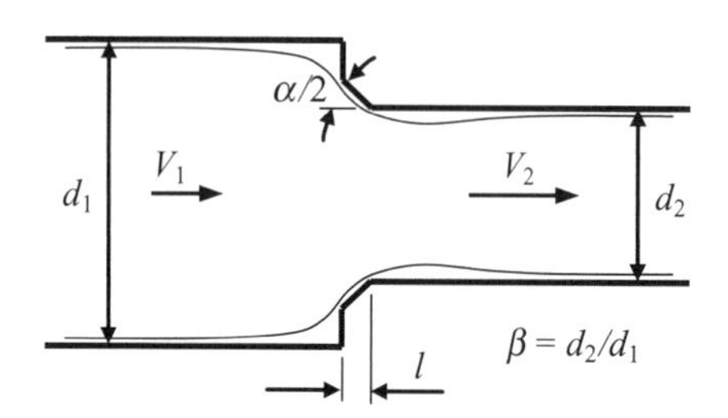

FIGURE 10.8. Beveled contraction.

In this flow configuration the sharpness of the corner or edge between the cone and the downstream passage becomes important. Equation 10.17 assumes the corners are absolutely sharp. If this is not the case, and the corners are generously rounded, rounding of the corners can be taken into account by measuring or estimating the rounding radius and incorporating the coefficients from the rounded contraction of Section 10.2 into the above equations.

Equation 10.17 is plotted as a function of diameter ratio β and divergence angle α in Diagram 10.3. Equation 10.17 is also plotted as a function of diameter ratio β and length ratio l/d_2 in Diagram 10.4.

10.5 BEVELED CONTRACTION

Beveling (or chamfering) the inlet edge of the entrance to the narrow section of a contraction, as illustrated in Figure 10.8, reduces the head loss. The important parameters are the nondimensional bevel length to diameter ratio l/d_2 and the included angle α. The equations developed in this section are related to similar equations for beveled entrances in Section 9.3 and bevel-edged orifices in Section 13.5. There are little or no credible data on beveled flow configurations so the equations are tentative. The beveled contraction transforms into sharp-edged inlets at $\alpha = 0°$ and at $\alpha = 180°$. A limit is reached as the bevel length increases to that of a conical contraction.

The loss coefficient of a contraction with a bevel of length l and included angle α can be tentatively determined as*:

$$K_2 \approx 0.0696[1 + C_B(\sin(\alpha/2) - 1)](1-\beta^5)\lambda^2 + (\lambda - 1)^2, \tag{10.19}$$

where the diameter ratio $\beta = d_2/d_1$, where the jet contraction coefficient λ is given by:

$$\lambda = 1 + 0.622\left[1 + C_B\left(\left(\frac{\alpha}{180}\right)^{4/5} - 1\right)\right] (1 - 0.215\beta^2 - 0.785\beta^5), \tag{10.20}$$

and where C_B is the ratio of bevel length l to the length of a conical contraction of corresponding diameter ratio and included angle.

* In the following equations α is generally expressed in radians; the modifications for using degrees are obvious.

With the aid of Equation 10.9, the ratio is determined as:

$$C_B = \frac{l}{d_2}\frac{2\beta\tan(\alpha/2)}{1-\beta}. \qquad (10.21)$$

Equation 10.19 assumes the corners of the bevel are absolutely sharp. Keep in mind that substantial rounding or chamfering may be applied to the edges of manufactured items. If such is the case, the loss coefficient may best be determined by treating the bevel as a rounded contraction, or as somewhere between a rounded contraction and a sharply edged bevel.

Beveled contraction loss coefficients for included angles α equal to 30°, 60°, 90°, 120°, and 150° can be approximately determined from Diagrams 10.5 through 10.9. The dashed lines in each diagram represent the boundary where the length of the bevel is limited by geometry. The parameters at which the bevel is limited are given by:

$$\beta_{\text{limit}} = \frac{1}{1+2\dfrac{l}{d_2}\tan(\alpha/2)}$$

and

$$\frac{l}{d_{2\text{limit}}} = \frac{1/\beta-1}{2\tan(\alpha/2)}.$$

As can be confirmed by comparison with Diagram 10.3, this lower limit is consistent with the loss coefficient of a conical contraction of corresponding divergence angle. Surface friction loss may become significant for long stretches at small included angle. It may be taken into account by utilizing Equation 10.16.

10.6 SMOOTH CONTRACTION

The resistance of contractions can be greatly reduced by providing a curvilinear transition section from the larger section to the smaller section (see Fig. 10.9). The entrance and exit contours may follow the arc of circles or other smooth curves. The convergent entrance section of a classical or Herschel Venturi tube is an example. The fluid stream does not separate from the walls and the losses are small and mainly due to surface friction.

The loss coefficient of a smooth contraction is effectively determined as wholly due to surface friction losses:

$$K_2 \approx K_{fr2} = \frac{f(1-\beta^4)}{8\sin(\alpha/2)}, \qquad (10.16, \text{repeated})$$

where the effective included angle α can be determined using Equation 10.10. The friction factor f is the friction factor as determined by the relative roughness of the surface of the cone, and the hydraulic diameter and Reynolds number at the cone exit.

10.7 PIPE REDUCER: CONTRACTING

Standard butt-weld pipe reducers, American National Standards Institute (ANSI) reducers, are commonly used to join pipe sections of different diameters (see Fig. 10.10). Industry standards define the length of butt-weld reducers but there are no standards regarding the dimensions of the straight and conical sections, or the curvature of the transition sections. As a rule, however, the fittings are generously rounded at the intersections of the conical and straight sections so that they resemble smooth contractions.

In any case, the losses are small and primarily due to surface friction. They may be simply accounted for by adding one-half the length of the reducer to the length of straight pipe attached at each end of the reducer. If actual dimensions are known and more accuracy is required, employ Equation 10.16 or Diagram 10.3 to estimate the loss coefficient more accurately. In the case of an eccentric reducer, use Equation 10.10 to calculate the equivalent divergence angle α for a concentric reducer and use that value in the appropriate equation or diagram.

Where little or no rounding is provided, as may be the case for large, specially constructed reducers, the losses are best evaluated as a conical contraction using Equation 10.11.

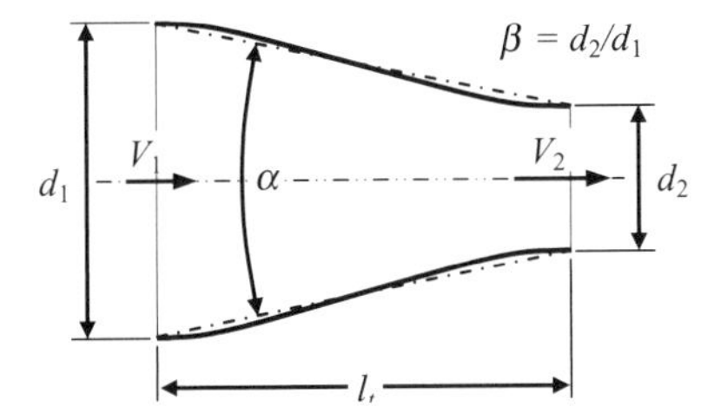

FIGURE 10.9. Smooth contraction.

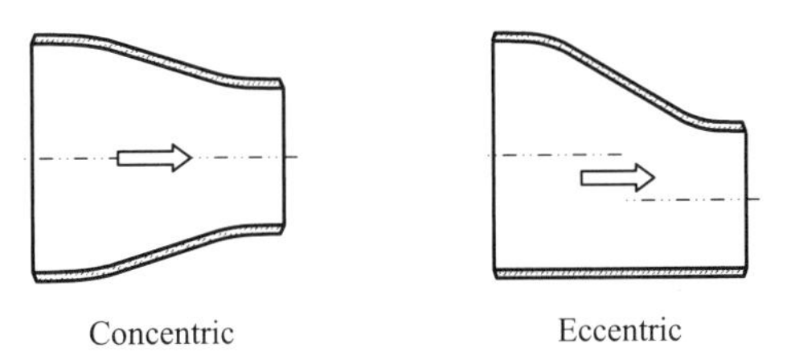

FIGURE 10.10. Welded pipe reducer—contracting.

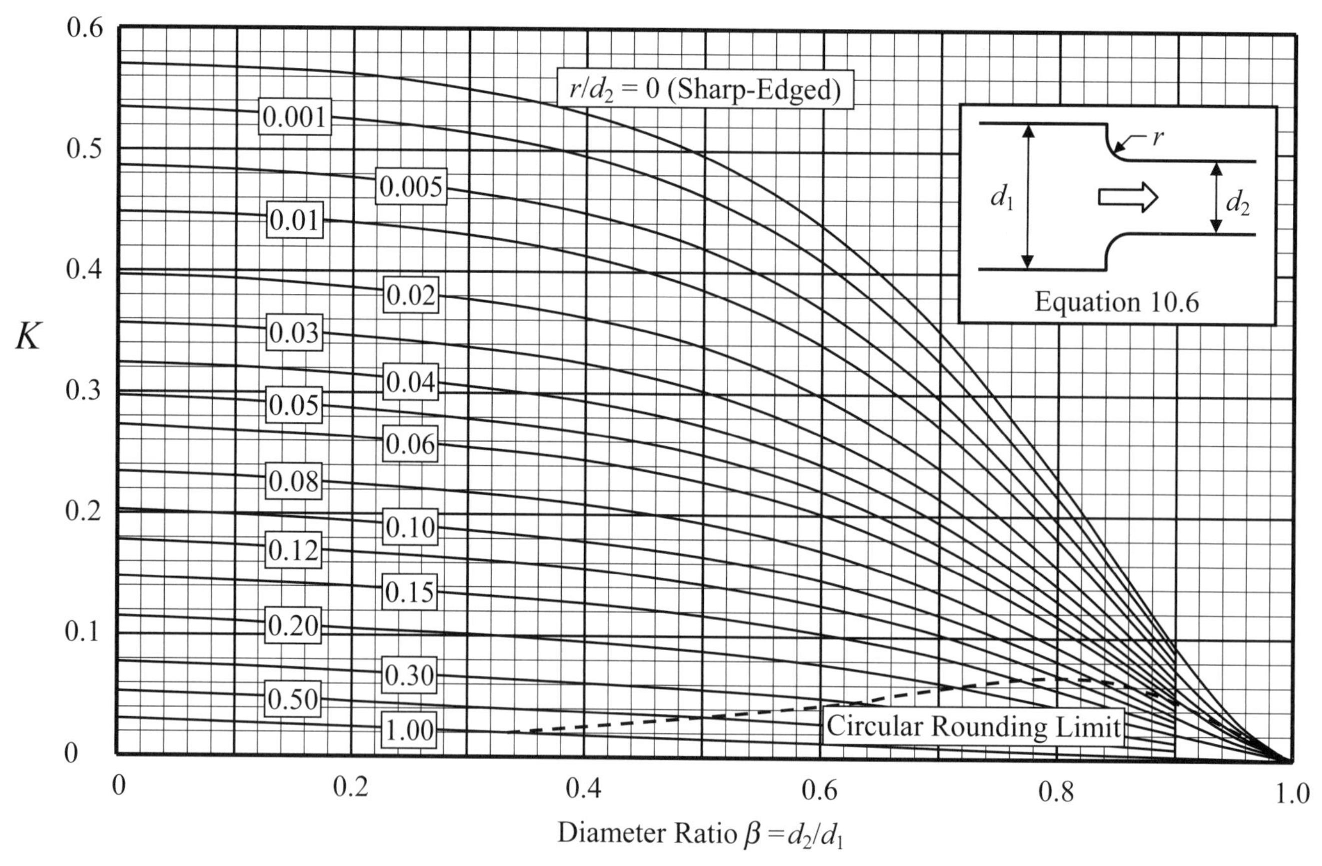

DIAGRAM 10.1. Loss coefficient K_2 of a rounded contraction.

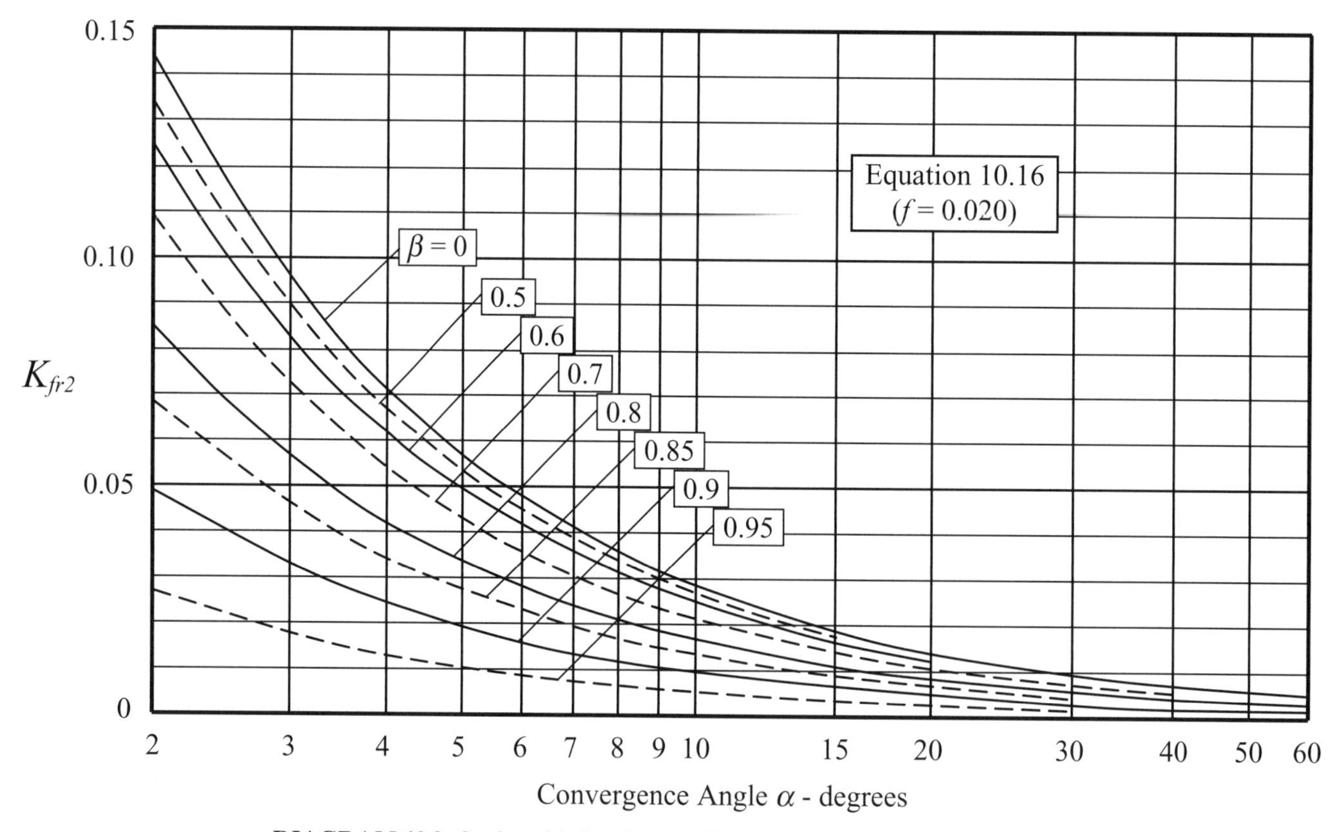

DIAGRAM 10.2. Surface friction loss coefficient K_{fr2} of a conical contraction.

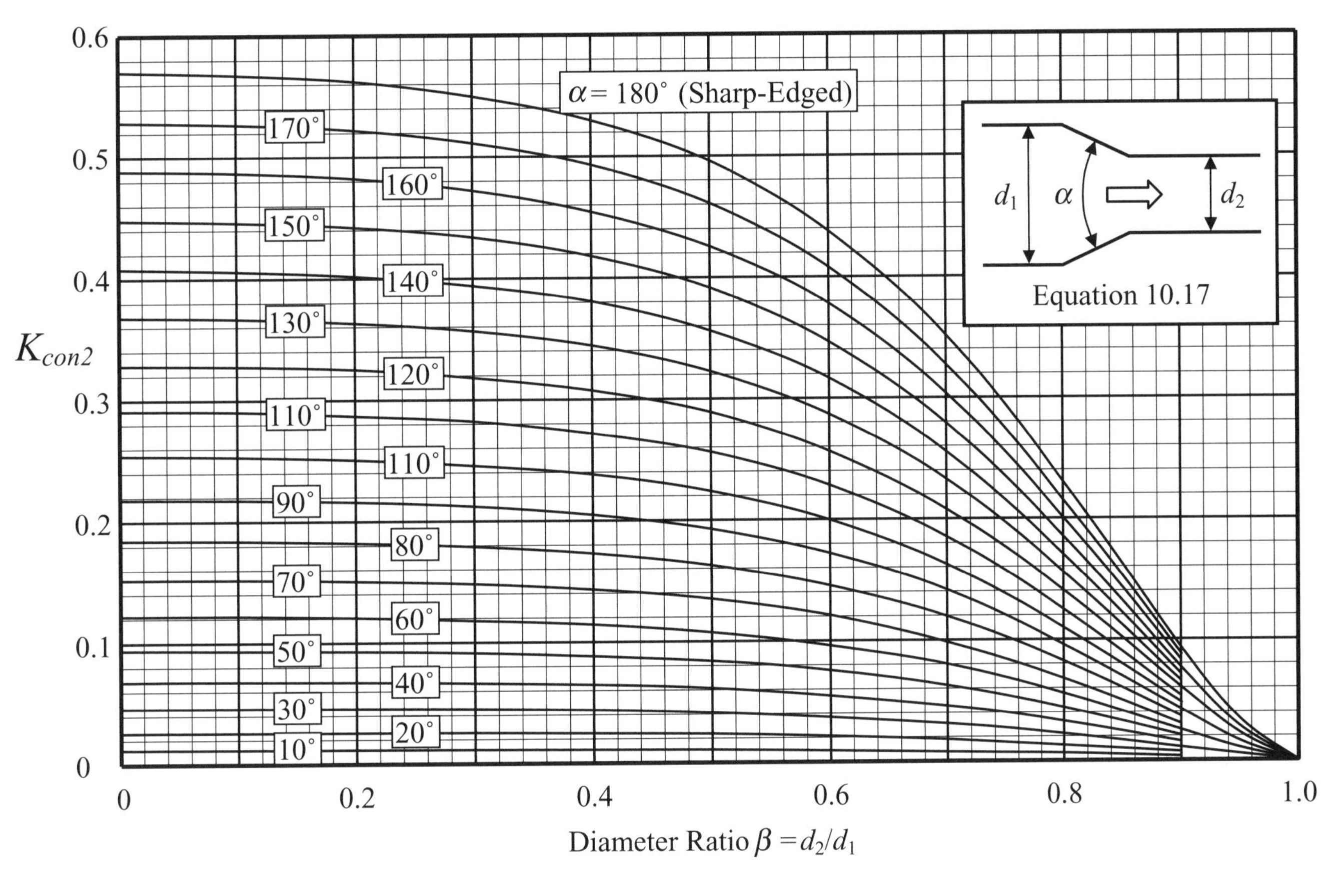

DIAGRAM 10.3. Local loss coefficient K_{con2} of a conical contraction as a function of included angle α.

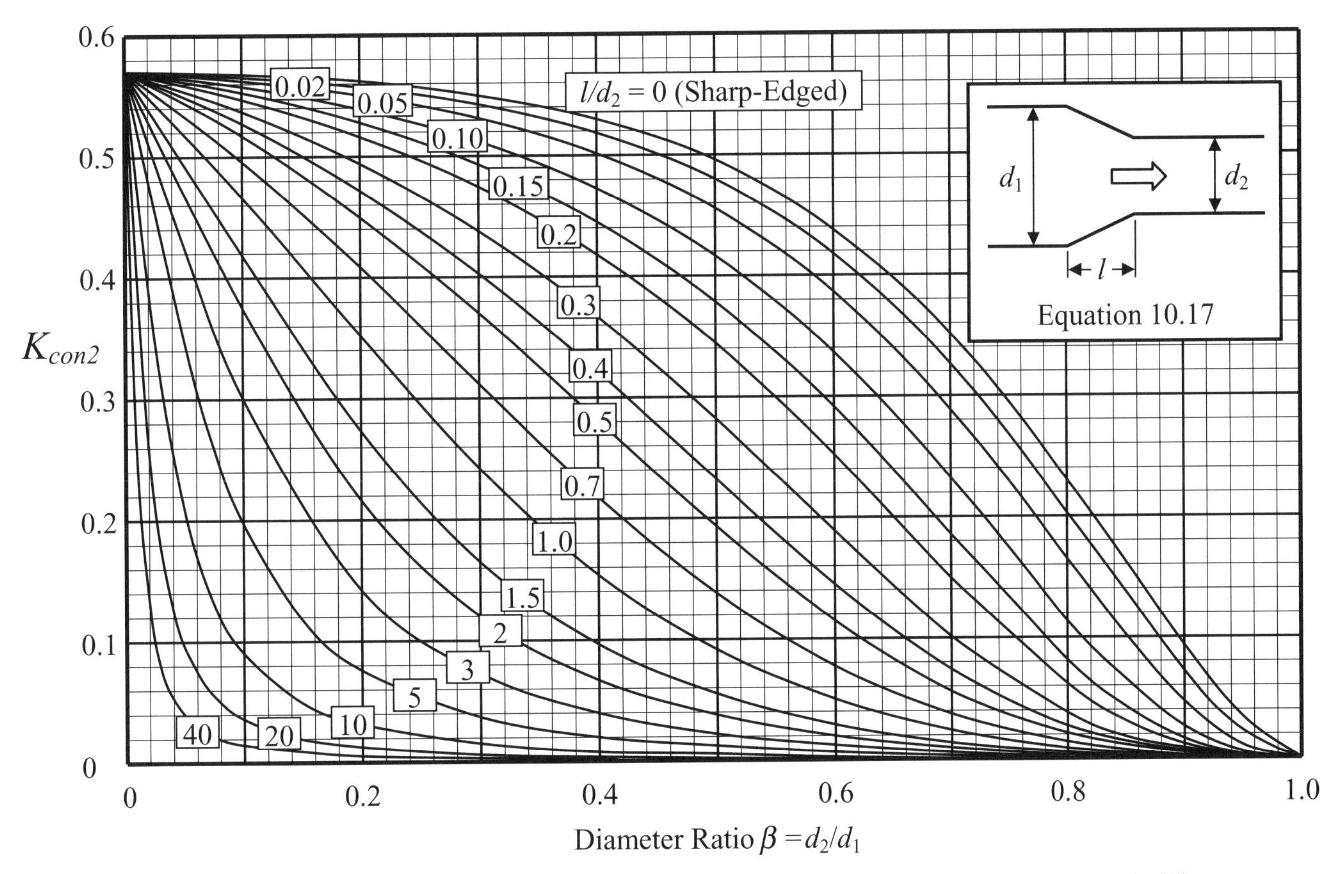

DIAGRAM 10.4. Local loss coefficient K_{con2} of a conical contraction as a function of length ratio l/d_2.

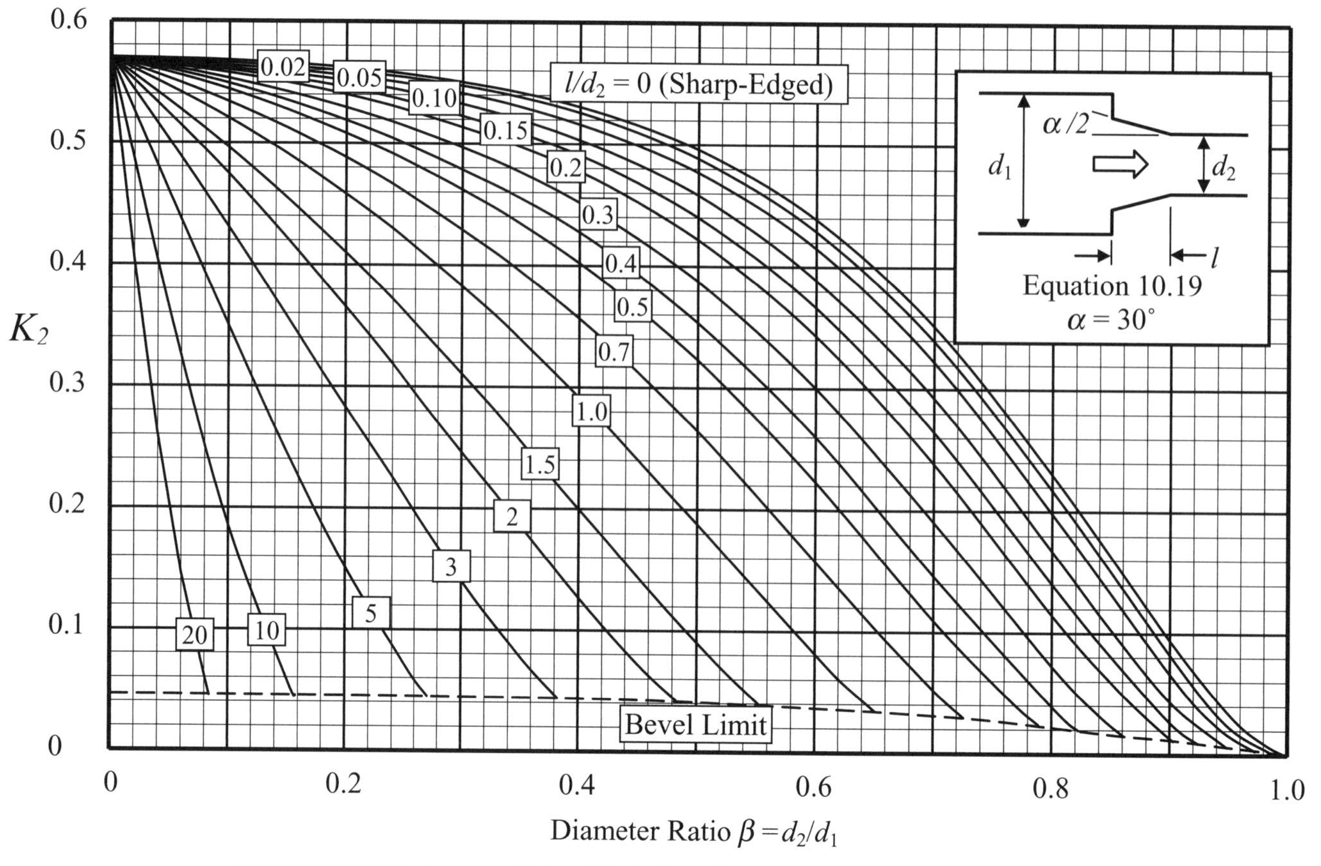

DIAGRAM 10.5. Loss coefficient K_2 of a beveled contraction—30° included angle.

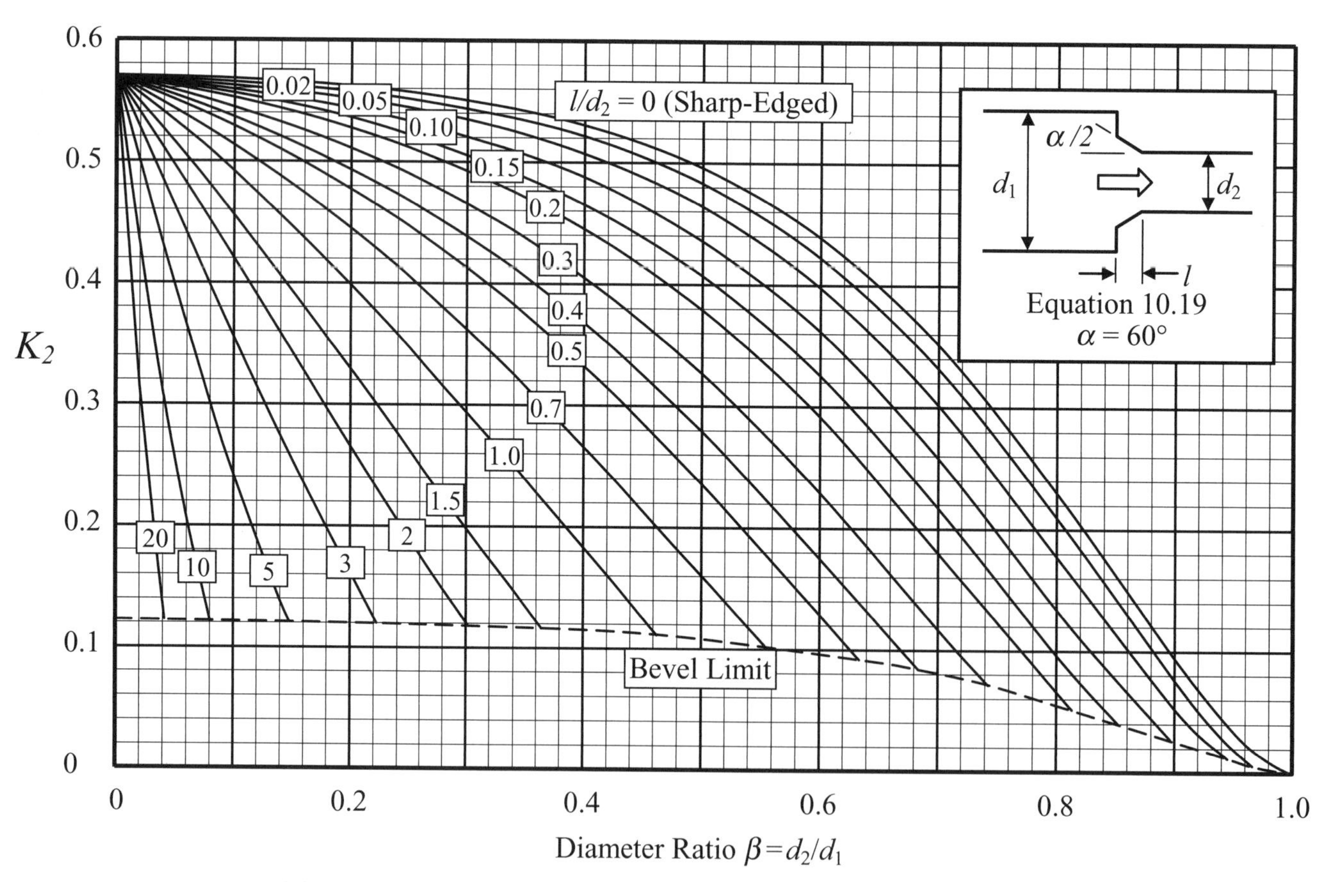

DIAGRAM 10.6. Loss coefficient K_2 of a beveled contraction—60° included angle.

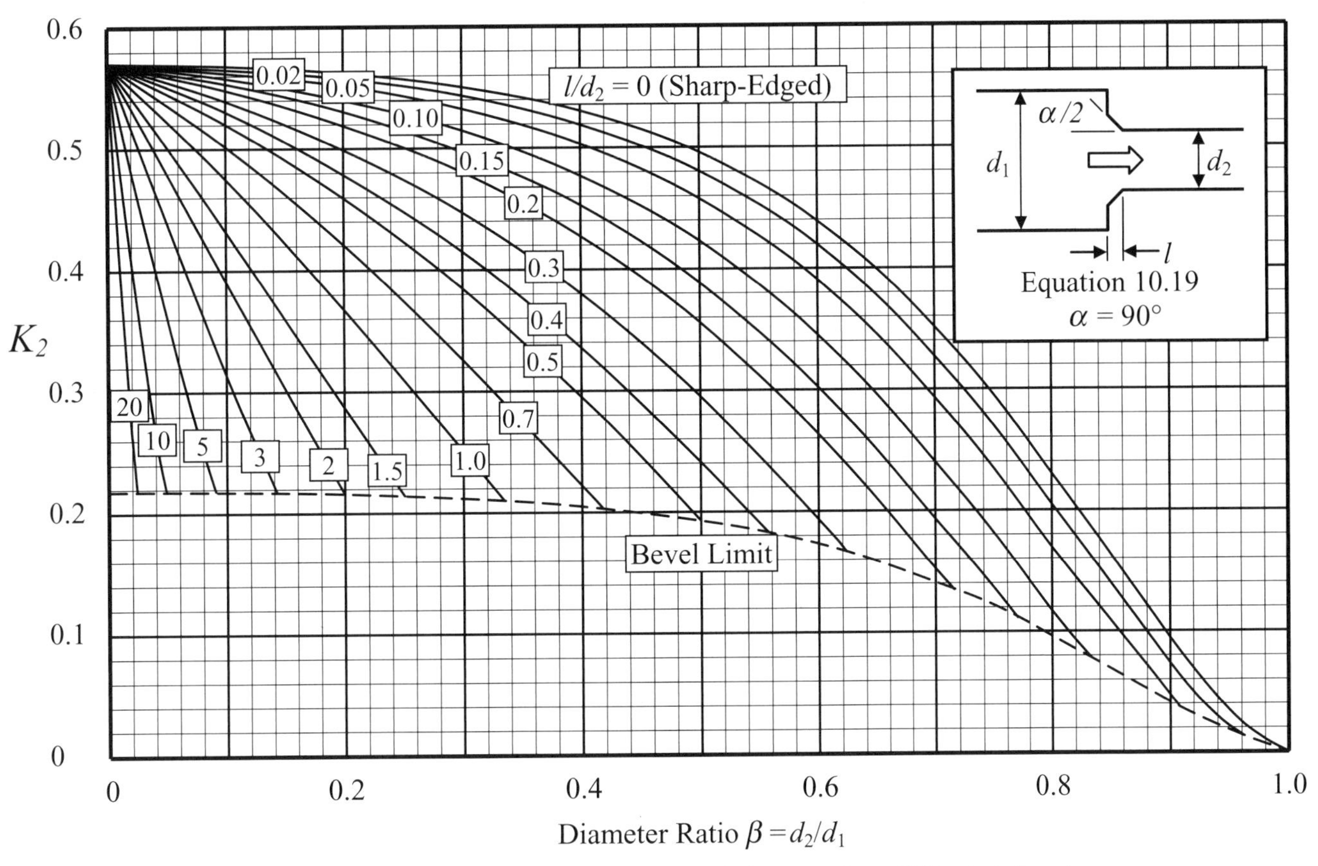

DIAGRAM 10.7. Loss coefficient K_2 of a beveled contraction—90° included angle.

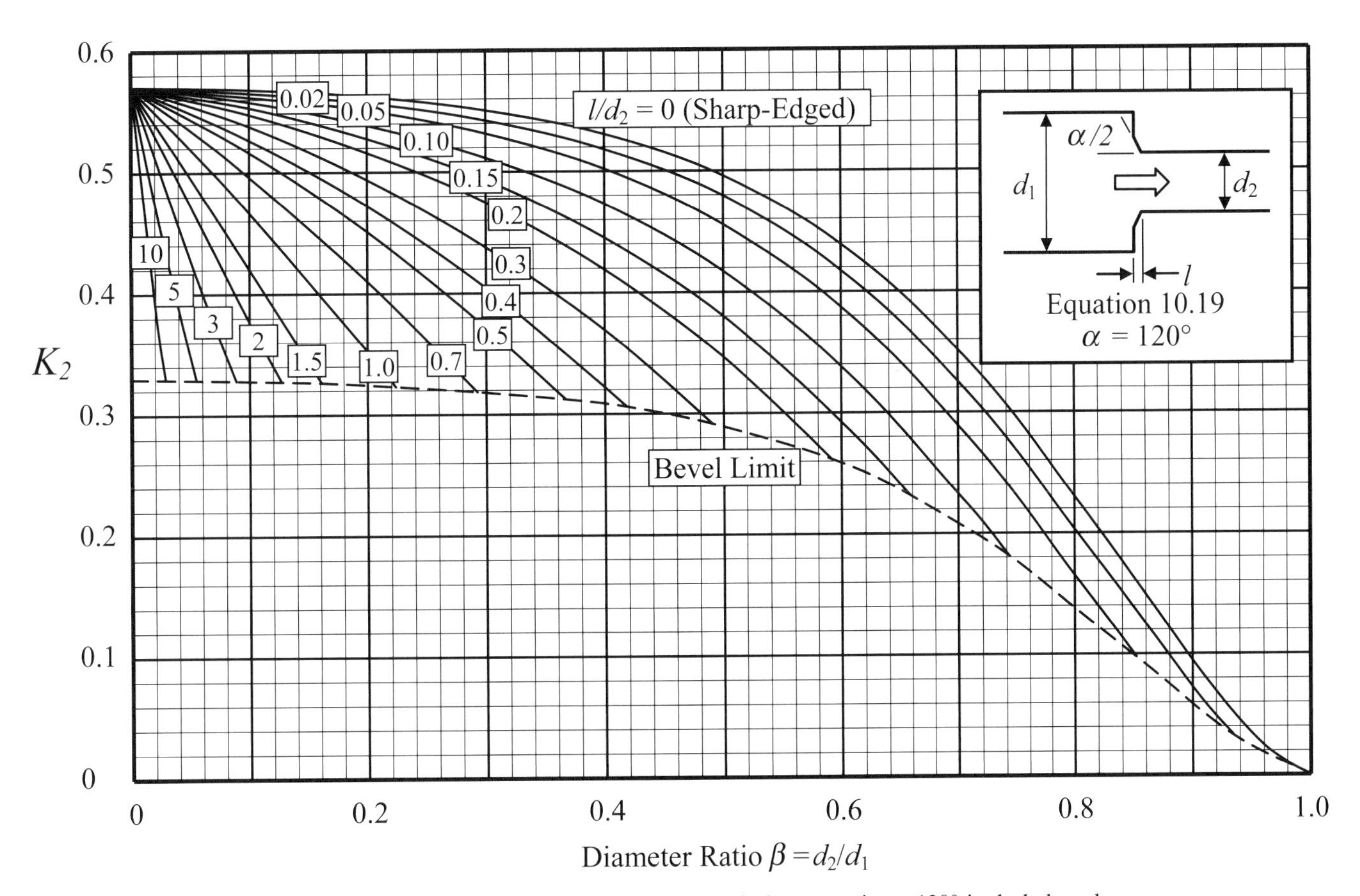

DIAGRAM 10.8. Loss coefficient K_2 of a beveled contraction—120° included angle.

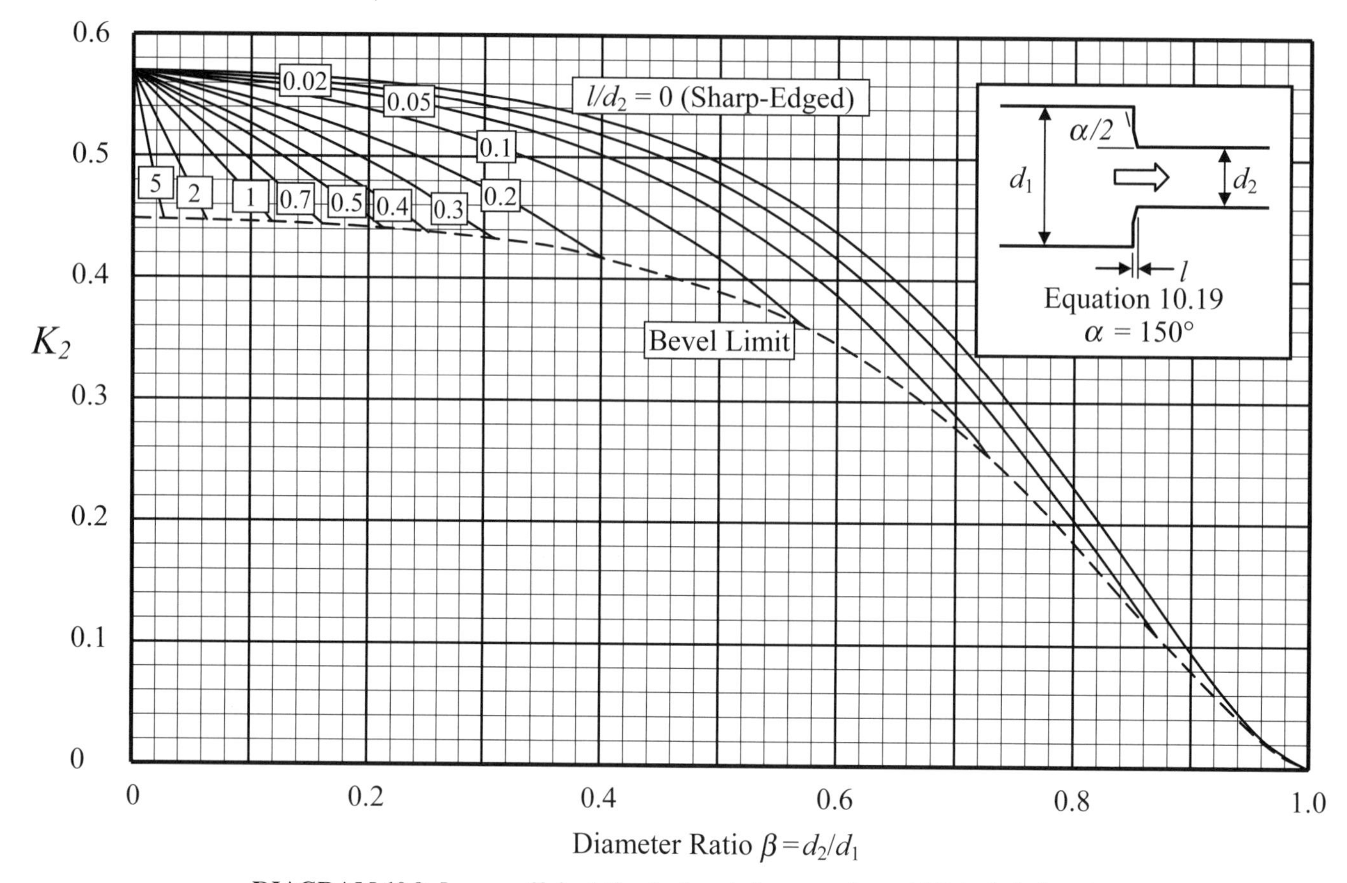

DIAGRAM 10.9. Loss coefficient K_2 of a beveled contraction—150° included angle.

REFERENCES

1. Vennard, J. K., *Elementary Fluid Mechanics*, John Wiley & Sons, Inc, New York, 1961.
2. Weisbach, J., *Mechanics of Engineering*, translated by E. B. Coxe, Van Nostrand Book Co., New York, 1872, p. 821.
3. von Mises, R., Berechnung von Ausfluss-und Uberfall-zahlen, *Z. VOI*, 61, 1917, 477. (Translated as "Calculation of discharge and overfall numbers.")
4. Kirchhoff, G., Zur Theorie freier Flussigkeitsstrahlen, *Crelles Journal*, 70, 1869, 289. (Translated as "On the theory of free fluid jets.")
5. Freeman, J. R., The discharge of water through fire hose and nozzles, *Transactions of the American Society of Civil Engineers*, 21, 1886, 303–482.
6. Benedict, R. P., N. A. Carlucci, and S. D. Swetz, Flow losses in abrupt enlargements and contractions, *Transactions of the American Society of Mechanical Engineers, Journal of Engineering for Power*, 88, 1966, 73–81.
7. Levin, L. and F. Clermont, Étude des pertes de charge singuliéres dans les convergents coniques, *Le Génie Civil*, T. 147, No 10, October 1970.

FURTHER READING

This list includes works that may be helpful to those who wish to pursue further study.

Alvi, S. H., K. Sridharan, and N. S. Lakshmana, Loss characteristics of orifices and nozzles, *Journal of Fluids Engineering, Transactions of the American Society of Mechanical Engineers*, 100, 1978, 299.

Bullen, P. R., D. J. Cheeseman, and L. A. Hussain, The effects of inlet sharpness on the pipe contraction loss coefficient, *International Journal of Heat and Fluid Flow*, 9(4), 1988, 431.

Bullen, P. R., D. J. Cheeseman, L. A. Hussain, and A. E. Ruffell, The determination of pipe contraction pressure loss coefficients for incompressible turbulent flow, *International Journal of Heat and Fluid Flow*, 8(2), 1987, 19.

Durst, F., W. F. Schierholz, and A. M. Wunderlich, Experimental and numerical investigations of plane duct flows with sudden contraction, *Journal of Fluids Engineering, Transactions of the American Society of Mechanical Engineers*, 109, 1987, 376–383.

Teyssandier, R. G., Internal separated flows—Expansions, nozzles, and orifices, PhD dissertation, University of Rhode Island, Kingston, R. I., 1973.

11

EXPANSIONS

Flow through a sudden or abrupt expansion in a piping system (see Fig. 11.1) gives rise to an increase in static pressure at the expense of a drop in kinetic energy. A "potential" core forms in the expanded section. Initially the core has a relatively flat velocity profile. This core spreads out and is separated from the remaining fluid by a surface of separation which disintegrates into powerful eddies in a recirculation or free-mixing stall region. The eddies develop and gradually disappear, and the core expands radially over the section until reattachment to the wall occurs.

Many experimental investigations have been conducted for confined flow in sudden expansions (sometimes referred to as backward facing steps). Test results show that for incompressible, fully developed turbulent flow in circular ducts the reattachment length-to-step height ratio, L/S, ranges from 6 to 9. Many authors assume that complete pressure recovery takes place at the reattachment point. However, beyond the reattachment point, the velocity profile continues to change until a moderately developed turbulent flow profile is achieved at distance ratios L/S on the order of 12–16.

Using a divergent connecting passage, or diffuser to make the transition from a passage of smaller cross section to a passage of larger cross section can substantially reduce expansion losses. The primary purpose of a diffuser is to convert kinetic energy of flow (or dynamic head) into static pressure (or static head) with minimum loss of total pressure. Much data, intimately related to the presence or absence of flow separation, or stall, are available in the literature on the performance and design of straight, two-, and three-dimensional diffusers. Herein, we are simply concerned with loss of total pressure, or the loss coefficient, of three-dimensional diffusers, with or without appreciable stall.

The information presented in this chapter is based on incompressible flow. The information is based on symmetrical inlet conditions between the extremes of uniform velocity and of fully developed turbulent flow at the inlet to the expansion, and assumes a reasonable length of downstream straight pipe. Data on a number of inlet and outlet flow conditions may be found in the literature (see the section "Further Reading" at the end of this chapter).

11.1 SUDDEN EXPANSION

A sudden axisymmetric expansion is shown in Figure 11.2. The energy, momentum, and continuity equations are applied to predict losses through the sudden expansion. While focus is directed to a single circular passage, this treatment is general and applies to both single and multiple passage expansions. The passage may actually be of any cross-sectional shape.

In the constant density fluid case, the continuity relationship for flow rate $\dot{w}$ through the control volume **abcd** is given by:

$$\dot{w} = \rho_w A_1 V_1 = \rho_w A_2 V_2. \tag{11.1}$$

Pipe Flow: A Practical and Comprehensive Guide, First Edition. Donald C. Rennels and Hobart M. Hudson.

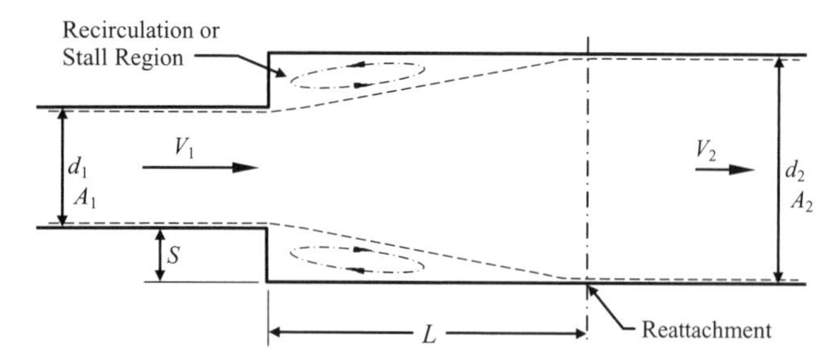

FIGURE 11.1. Sudden expansion.

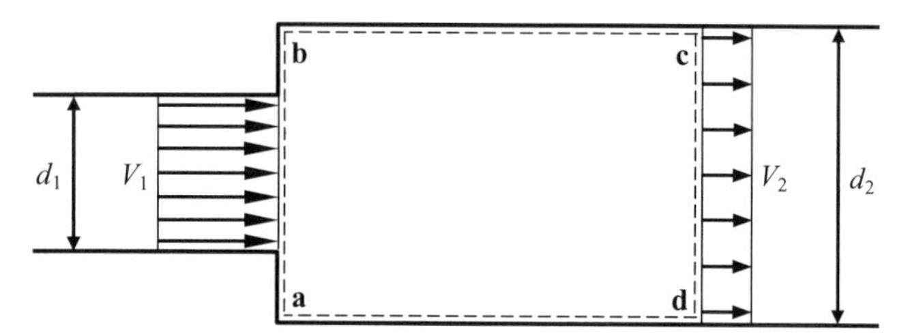

FIGURE 11.2. Sudden expansion.

In practice, the velocity distribution is seldom uniform or flat over the cross section. For the moment, assume uniform velocity profile at the inlet and exit of the control volume. The energy balance is given by:

$$\frac{P_1}{\rho_w}+\frac{V_1^2}{2g}=\frac{P_2}{\rho_w}+\frac{V_2^2}{2g}+H_L.$$

Solving for the head loss H_L gives:

$$H_L=\frac{P_1-P_2}{\rho_w}+\frac{V_1^2}{2g}-\frac{V_2^2}{2g}. \tag{11.2}$$

Nusselt [1] proved experimentally that for subsonic flow the pressure on the downstream face of the enlargement is equal to the static pressure in the stream just prior to expansion. Assuming that the hydrostatic pressures P_1 and P_2 are evenly distributed over the surfaces **ab** and **cd**, respectively, and that the wall friction forces along the control volume are negligible, the momentum balance across the control volume is given by:

$$A_2(P_1-P_2)=\frac{V_2\dot{w}}{g}-\frac{V_1\dot{w}}{g}, \tag{11.3}$$

and substituting $\dot{w}=\rho_w A_2 V_2$ from the continuity equation (Eq. 11.1) into Equation 11.3 gives:

$$\frac{P_1-P_2}{\rho_w}=\frac{V_2^2-V_1V_2}{g}. \tag{11.4}$$

Substitution of Equation 11.4 into Equation 11.2 gives:

$$H_L=\frac{V_1^2-2V_1V_2+V_2^2}{2g}=\frac{(V_1-V_2)^2}{2g}, \tag{11.5}$$

which is a classic formula of early analytical hydraulics and is termed the Borda–Carnot equation, after those who contributed to its original development. Borda [2] was the first to understand the mechanical process and to find a mathematical solution. His formula, in the version of Carnot, is still valid in modern hydrodynamics.*

Substitution of the continuity relationships into Equation 11.5 and letting $H_L=K_1V_1^2/2g$ gives:

$$K_1=(1-A_1/A_2)^2=\left(1-\beta^2\right)^2, \tag{11.6}$$

where the beta ratio β is equal to the ratio of the small diameter to the large diameter, or d_1/d_2. This is the familiar equation for the loss coefficient of a sudden expansion. The engineer generally applies these equations without a correction coefficient (see Section 2.6). The Borda–Carnot equation is plotted in Diagram 11.1. This is an important equation in pipe flow analysis. The utility of the sudden expansion equation is evident throughout Chapters 9 through 14. The Borda–Carnot equation has been experimentally confirmed for incompressible flow many times over the years.

In practice, the velocity profile entering a sudden expansion is not always uniform or follows the power law. This affects the actual losses and can considerably increase them. Several investigators present data to account for the effect of various axisymmetric (or nonuniform) inlet velocity distributions on diffuser loss (see "Further Reading" section).

The Borda–Carnot equation cannot be applied with accuracy to compressible flow where the Mach number at the inlet is greater than about 0.2. Benedict et al. [3] give generalized analytical solutions for incompressible, subsonic, and supercritical flow across an abrupt enlargement. Benedict et al. also present experimental verification of the solution, including tests involving high beta ratios.

11.2 STRAIGHT CONICAL DIFFUSER

A diffuser is a gradually expanding section that is used to make the transition from a smaller flow passage to a larger one as shown in Figure 11.3. The primary purpose

* Borda did not exclusively deal with sudden expansion losses in his paper; rather, he determined the time it takes to fill a submerged vessel with liquid through an orifice in the bottom.

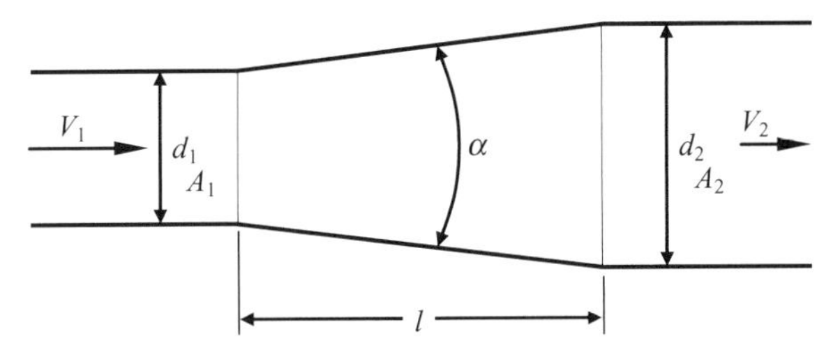

FIGURE 11.3. Straight conical diffuser.

of a diffuser is to recover fluid static pressure with minimal loss of total pressure while reducing the flow velocity. The increase in the cross-sectional area of the diffuser causes a drop in the average flow velocity, and a portion of the kinetic energy of the flow is converted into the potential energy of pressure. An efficient diffuser is one that converts the highest possible percentage of kinetic energy into pressure energy within a given limitation on diffuser length l or divergence (or included) angle α.

The performance level of a diffuser is intimately related to the presence or absence of flow separation (or stall). Regions of stalled flow in a diffuser block the flow, cause low pressure recovery, and may result in severe flow asymmetry, severe unsteadiness, or both. Consequently, much study has centered on the presence or absence of stall, rather than directly on the important design consideration, at least in piping design, of maximum pressure recovery.

The main geometric considerations of conical diffusers with straight walls are the divergence angle α, the beta ratio β, and the length l of the conical section. These quantities are interrelated as follows:

$$l = \frac{d_2 - d_1}{2\tan(\alpha/2)} = \frac{d_1(1/\beta - 1)}{2\tan(\alpha/2)},$$

and

$$\alpha = 2\,\text{atan}\left(\frac{d_2 - d_1}{2 \cdot l}\right) = 2\,\text{atan}\left(\frac{1/\beta - 1}{2l/d_1}\right).$$

Much data on the diffuser has been reported in terms of diffuser efficiency η_d, which is the ratio of the actual static pressure recovery across the diffuser to the ideal pressure recovery. However, here we present diffuser data in terms of the loss coefficient. The relationship between loss coefficient K_1 and efficiency η_d of a diffuser is given by:

$$K_1 = (1 - \eta_d)(1 - \beta^4).$$

At small divergence angles, separation, if present, occurs near the outlet of the diffuser section and usually starts from only one portion of the wall. Separation may alternate from one location to another. At larger divergence angles, the point at which separation occurs progresses toward the inlet of the diffuser section and a major portion of the diffuser is occupied by an extensive region of reverse circulation. At diffuser angles above 40° to 50° the main flow is separated from the diffuser walls over the whole perimeter and the resulting turbulence produces losses greater than for a sudden enlargement. Where, in the design of hydraulic passages, it is necessary for these values of diffuser angle to be exceeded, a sudden enlargement of section will give a more efficient and steady transformation of energy than will a conical diffuser.

The loss coefficient of conical diffusers depends on many parameters besides divergence angle α and beta ratio β. It depends on the boundary layer thickness at the entrance; the shape of the velocity profile at the entrance; the degree of flow turbulence at the entrance; the flow regime; and the length of straight downstream pipe.

A thicker boundary layer at the entrance to the diffuser tends to increase the loss coefficient. Nonuniform velocity profile at the entrance, particularly if it is distorted, can cause earlier onset of flow separation from the wall and greatly increase the loss coefficient. A convex or pointed velocity profile, such as in laminar flow, with maximum velocity at the center and reduced velocities at the walls, aggravates the onset and the extent of flow separation.

Swirl (or tangential rotation of flow) is sometimes present in conical diffusers as a result of rotating machinery, or close-coupled elbows or bends. Swirl has little effect on the performance of separation-free diffusers, but can have a beneficial effect on the performance of diffusers that are moderately or badly separated. The swirl flow apparently helps to spread the core flow to the walls of the diffuser, which yields a more uniform exit velocity profile.

For diffusers discharging into a downstream passage (as is under consideration here), significant pressure recovery continues beyond the diffuser exit. A straight downstream length (or tailpipe length) of two to four pipe diameters is usually sufficient to provide near maximal possible recovery; the longer length is required at higher diffuser angles. Design measures that may improve diffuser performance are the use of stepped diffusers and two-stage diffusers. These multistage diffusers are treated in the next section.

A great deal of data on flow in diverging passages has been amassed in the last 100 years. Much of this data have been on two-dimensional and rectangular diffusers and most data have been on diffusers which discharge into a large plenum (free discharge). Some data are available for conical diffusers which act solely as expansions between constant area circular passages. These data are developed here because it applies to piping system diffuser applications.

In Gibson's classical investigations [4,5], conical diffusers with upstream and downstream pipe sections were tested over a range of angles and area ratios. Gibson developed a head loss equation for values of divergence angle between 7.5° and 35°. He expressed diffuser head loss as a percentage of the Borda–Carnot loss at a sudden enlargement between the same flow areas. However, his equation did not account for surface friction conditions different from his test conditions. Expanding on Gibson's formulations, the author developed equations that separately accounted for local, or expansion, loss, and surface friction loss for values of divergence angle between 0° and 180°. The equations give good agreement with Gibson's test data, particularly for divergence angles between 0° and 20°, which is the range of greatest interest in piping system applications.

Letting K_{fr1} represent the surface friction loss and K_{L1} represent the local or expansion loss, it is fitting to express the loss coefficient of diffusers in the form of:

$$K_1 = K_{fr1} + K_{L1}.$$

The theoretical equation for surface friction loss coefficient that was developed in Section 10.4 for conical contractions can be applied as well to conical diffusers*:

$$K_{fr1} = \frac{f(1-\beta^4)}{8\sin(\alpha/2)}. \tag{11.7}$$

Equation 11.7 was developed for converging flow, but it seems to work just as well for diverging flow. In this case, the friction factor f is the ordinary friction factor based on the relative roughness of the diffuser surface as determined by the hydraulic diameter and Reynolds number at the diffuser inlet.† The equation is plotted as a function of beta ratio β and divergence angle α in Diagram 11.2 for a friction factor f of 0.020. The value of K_{fr1} at friction factors other than 0.020 can be obtained by simple ratio. It is evident that surface friction loss may be generally ignored at divergence angles greater than about 40°.

Using Equation 11.7, surface friction loss was separated from Gibson's data by assigning friction factors at the diffuser inlets ranging from 0.020 to 0.026. A reasonableness check determined that friction factors in this range would be expected.‡

After separating surface friction loss from Gibson's test data, the author developed equations for the local (or expansion) loss portion of diffuser loss. The local loss equations were then recombined with the equation for surface friction loss (Eq. 11.7) to obtain overall equations for conical diffuser loss. For divergence angle from 0° to 20°, expansion loss can be simply expressed as:

$$K_1 = 8.30[\tan(\alpha/2)]^{1.75}(1-\beta^2)^2 + \frac{f(1-\beta^4)}{8\sin(\alpha/2)} \tag{11.8}$$
$$(0° \le \alpha \le 20°)(0 \ge \beta \le 1).$$

The following approximate formula was developed for diffuser loss for divergence angles from 20° to 60° for β less than 0.5:

$$K_1 \approx \left\{1.366\sin\left[\frac{2\pi(\alpha-15°)}{180}\right]^{1/2} - 0.170 - 3.28(0.0625-\beta^4)\sqrt{\frac{\alpha-20°}{40°}}\right\}(1-\beta^2)^2 + \frac{f(1-\beta^4)}{8\sin(\alpha/2)},$$
$$(20° \le \alpha \le 60°)(0 \ge \beta < 0.5). \tag{11.9a}$$

For divergence angles from 20° to 60° for β equal to or greater than 0.5:

$$K_1 \approx \left\{1.366\sin\left[\frac{2\pi(\alpha-15°)}{180}\right]^{1/2} - 0.170\right\}(1-\beta^2)^2 + \frac{f(1-\beta^4)}{8\sin(\alpha/2)}$$
$$(20° \le \alpha \le 60°)(0.5 \ge \beta \le 1). \tag{11.9b}$$

The surface friction term can generally be ignored for divergence angles above 40° or 50°. For divergence

* In the following equations α is generally expressed in radians; the exceptions for using degrees are obvious.

† The magnitude of f may actually vary along the diffuser, but is assumed constant.

‡ Gibson did not report flow rates or Reynolds numbers for his tests. He did, however, note that his test velocities varied from 1.83 to just over 21 ft/s. Assuming 21 ft/s test velocity at the inlet of the narrowest test diffuser diameter of 0.5 in, and ratioing velocity downward by inlet area for the other test diffusers, the author estimates that the Reynolds number at the diffuser inlets ranged from about 20,000 to 70,000. With regard to surface conditions, Gibson's diffusers were "very carefully made of wood, finished off with a coating of shellac varnish." Assuming smooth walls, friction factors at the diffuser inlets ranging from 0.020 to 0.026 were then estimated from the Moody Diagram.

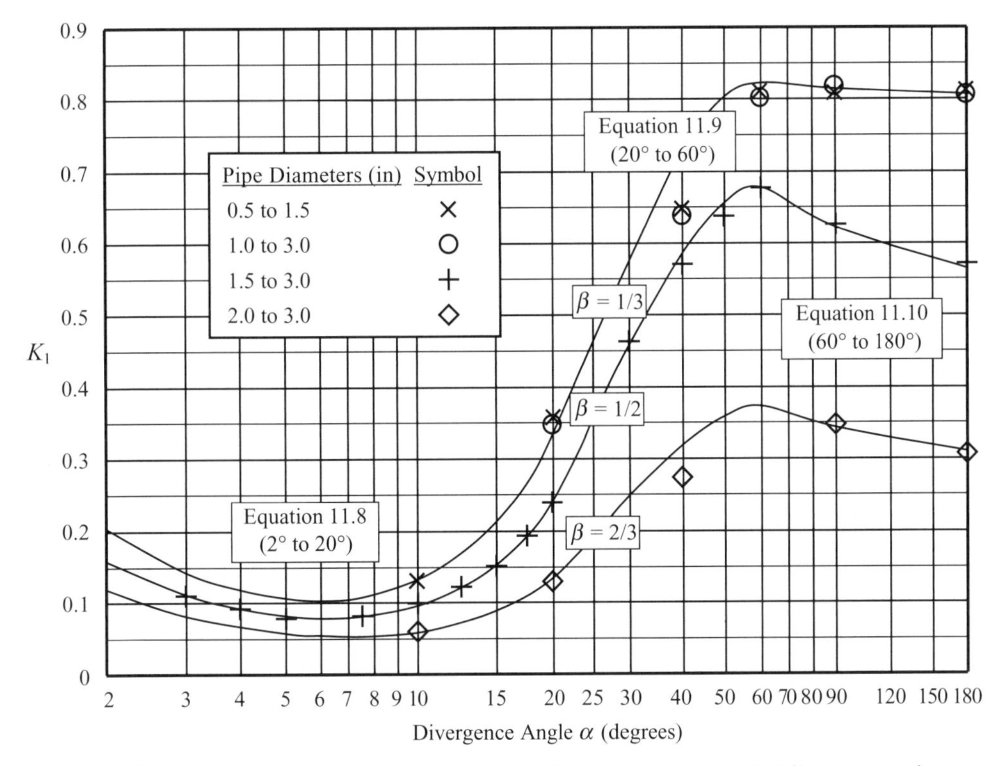

FIGURE 11.4. Comparison of diffuser loss coefficient equations with Gibson's test data.

angles between 60° and 180° for β less than 0.5, the expansion loss is equal to or greater than that of a sudden expansion and can be approximated by:

$$K_1 \approx \left[1.205 - 3.28(0.0625 - \beta^4) - 12.8\beta^6\sqrt{\frac{\alpha - 60^\circ}{120^\circ}}\right](1-\beta^2)^2,$$
$$(60^\circ \le \alpha \le 180^\circ)\,(0 \ge \beta < 0.5). \tag{11.10a}$$

For divergence angles between 60° and 180° for β greater than 0.5, the expansion loss can be approximated by:

$$K_1 \approx \left(1.205 - 0.20\sqrt{\frac{\alpha - 60^\circ}{120^\circ}}\right)(1-\beta^2)^2, \tag{11.10b}$$
$$(60^\circ \le \alpha \le 180^\circ)\,(0.5 \ge \beta \le 1).$$

Equations 11.8 through 11.10 are compared to Gibson's test data in Figure 11.4. Good agreement with test data is evident; especially in the range of from 2° to 20°, the range of greatest interest in pipe flow applications.

The above relationships apply for thin inlet boundary layers, such as would develop within one or two pipe diameters from a nozzle or collector. The available data suggest that the loss coefficients are 5%–10% higher for thick inlet boundary layers such as would develop over long lengths of straight inlet pipe.

Loss coefficients for divergence angles from 0° to 20° are shown in Diagrams 11.3 through 11.7 for friction factors of 0.01, 0.020, 0.030, 0.040, and 0.050, respectively. Loss coefficients for divergence angles from 20° to 180° are shown in Diagram 11.8.

11.3 MULTISTAGE CONICAL DIFFUSERS

When the available length for a diffuser is limited, the energy loss may be reduced by using a multistage diffuser. In a stepped or cropped diffuser, a gradual increase in the cross-sectional area is followed by a sudden expansion as shown in Figure 11.5a. In a two-stage diffuser, point **b** at the exit plane of the stepped diffuser is simply moved backward to form two adjoining conical sections as shown in Figure 11.5b.

11.3.1 Stepped Conical Diffuser

Where no restrictions are placed on the length of a diffuser, a straight wall passage having a divergence angle of from about 4° to 7° will normally give minimum loss of energy between inlet and outlet. The length of such a passage, however, may be impossible or impractical in many cases, and in such cases it becomes important to determine what form of passage will give minimum loss

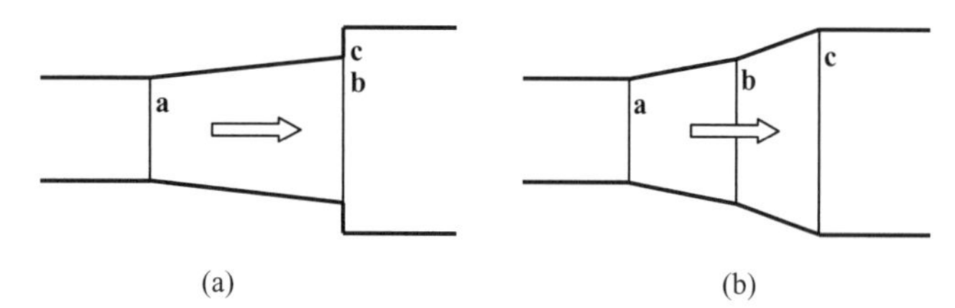

FIGURE 11.5. Multistage conical diffusers. (a) Stepped diffuser. (b) Two-stage diffuser.

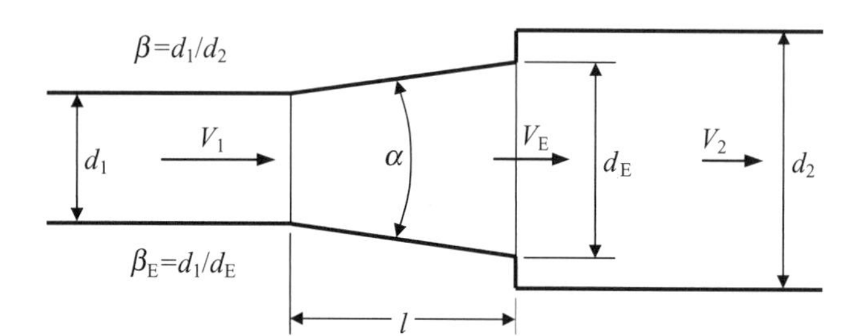

FIGURE 11.6. Stepped conical diffuser.

for a given length and given ratio of enlargement. In this case, a stepped, or truncated, diffuser may significantly reduce the energy loss. In a stepped diffuser a gradual increase in the cross-sectional area is followed by a sudden expansion as shown in Figure 11.6. The sudden expansion loss at the exit step occurs at a relatively low velocity.

The ratio β represents d_1/d_2, the overall diameter ratio as before. The exit step diameter ratio β_E (see Fig. 11.6) is defined as d_1/d_E. The diameter d_E at the exit step is given by:

$$d_E = d_1 + 2l \tan(\alpha/2).$$

A divergence angle α greater than 20° is not anticipated in stepped diffuser design. The loss coefficient of a stepped conical diffuser can be approximately determined by the following equation, which is simply a conical diffuser loss (including local loss and surface friction loss) followed by a sudden expansion:

$$K_1 \approx 8.30[\tan(\alpha/2)]^{1.75}(1-\beta^2)^2 + \frac{f(1-\beta_E^4)}{8\sin(\alpha/2)} + (\beta_E^2 - \beta^2)^2$$
$$(\alpha \le 20°). \tag{11.11}$$

Undoubtedly, the velocity profile at the end of the diffuser section and entering the exit section is not fully developed, thus the exit loss is not fully taken into account. Note, however, that the overall diameter ratio β, rather than the exit step diameter ratio β_E, has been employed in the first term in Equation 11.11. This adjustment slightly increases predicted loss in the diffuser section loss and, in a simple way, tends to make up for underpredicted diffuser exit loss.

For a given length l and a given overall diameter ratio β, the divergence angle α_{opt} that provides minimum loss can be determined by a trial-and-error process. In Table 11.1, values of minimum loss coefficient K_{opt} for optimum stepped diffusers are compared to K_1 for straight conical diffusers of equivalent length l and overall diameter ratio β. Table 11.2 shows the optimum divergence angle α_{opt} that provides the minimum loss coefficient K_{opt}. The dashed boxes in Tables 11.1 and 11.2 indicate the region where α_{opt} becomes greater than the divergence angle α for a straight conical diffuser—an incongruous geometry.

Loss coefficients of various expansion configurations are compared in Table 11.3. The most effective configuration is shown in bold font. It is evident that the optimum stepped diffuser is superior to the other configurations over a wide range of length to diameter and area ratios.

A friction factor of 0.020 was used in constructing Tables 11.1–11.3. At larger length-to-diameter ratios the diffuser loss coefficients are more sensitive to friction factor and the outcome may vary somewhat. This difference may be generally ignored, or it can be accounted for by inserting the appropriate friction factor into the loss coefficient equations.

11.3.2 Two-Stage Conical Diffuser

For a given length l and a given overall diameter ratio β, the two-stage diffuser (Fig. 11.7) can provide a reduction in pressure loss compared to a straight conical diffuser.

The overall diameter ratio β equals d_1/d_2, and the first-stage diameter ratio β_1 equals d_1/d_E as for the stepped diffuser. A first-stage divergence angle α_1 greater than 20° is not anticipated. The second-stage diameter ratio β_2 equals d_E/d_2. Simply treating the two stages as straight conical diffusers in series results in the following tentative equation:

$$K_1 \approx 8.30\tan(\alpha_1/2)^{1.75}(1-\beta_1^2)^2 + \frac{f_1(1-\beta_1^4)}{8\sin(\alpha_1/2)} + K_{2nd}\beta_1^4$$
$$(0° \le \alpha_1 \le 20°), \tag{11.12}$$

where K_{2nd} is taken from Equations 11.8, 11.9, or 11.10, depending on the included angle of the second stage. In the appropriate equation, the second-stage angle α_2 is substituted for α and the second-stage diameter ratio β_2 is substituted for β.

TABLE 11.1. Loss Coefficient K_{opt} for Optimum Stepped Diffusers Compared to Loss Coefficient K_1 for Straight Conical Diffusers of Equal Length and Overall Area Ratio

	l/d_1										
Area Ratio β^2	0.5	1.0	2.0	3.0	4.0	5.0	6.0	8.0	10.0	12.0	15.0
0.00	1.00 *0.838*	1.00 *0.638*	1.00 *0.424*	1.00 *0.321*	1.00 *0.262*	1.00 *0.223*	1.00 *0.197*	1.00 *0.163*	1.00 *0.142*	1.00 *0.129*	1.00 *0.117*
0.05	0.911 *0.744*	0.912 *0.556*	0.913 *0.362*	0.915 *0.270*	0.875 *0.218*	0.785 *0.185*	0.686 *0.163*	0.502 *0.135*	0.364 *0.119*	0.272 *0.109*	0.194 *0.100*
0.10	0.912 *0.655*	0.834 *0.479*	0.837 *0.305*	0.735 *0.225*	0.589 *0.180*	0.456 *0.153*	0.342 *0.134*	0.221 *0.112*	0.160 *0.100*	0.127 *0.093*	0.102 *0.088*
0.15	0.758 *0.571*	0.768 *0.409*	0.712 *0.254*	0.529 *0.185*	0.365 *0.148*	0.254 *0.125*	0.192 *0.111*	0.129 *0.094*	0.102 *0.086*	0.089 *0.082*	0.081 *0.081*
0.20	0.688 *0.493*	0.717 *0.344*	0.562 *0.208*	0.356 *0.151*	0.218 *0.120*	0.156 *0.103*	0.123 *0.093*	0.091 *0.081*	0.080 *0.077*	0.076 *0.076*	–
0.25	0.625 *0.421*	0.667 *0.287*	0.424 *0.169*	0.217 *0.121*	0.142 *0.098*	0.107 *0.085*	0.089 *0.078*	0.074 *0.072*	0.072 *0.072*	–	–
0.30	0.555 *0.354*	0.545 *0.234*	0.286 *0.135*	0.143 *0.097*	0.099 *0.080*	0.079 *0.072*	0.071 *0.068*	0.067 *0.067*	–	–	–
0.35	0.490 *0.293*	0.425 *0.188*	0.175 *0.107*	0.099 *0.078*	0.074 *0.067*	0.064 *0.062*	0.062 *0.062*	–	–	–	–
0.40	0.434 *0.238*	0.318 *0.148*	0.117 *0.083*	0.072 *0.064*	0.059 *0.057*	0.057 *0.057*	–	–	–	–	–
0.45	0.359 *0.189*	0.226 *0.114*	0.080 *0.065*	0.056 *0.053*	0.052 *0.051*	–	–	–	–	–	–
0.50	0.280 *0.146*	0.150 *0.085*	0.066 *0.051*	0.047 *0.046*	–	–	–	–	–	–	–
0.55	0.207 *0.109*	0.089 *0.063*	0.056 *0.042*	0.042 *0.042*	–	–	–	–	–	–	–
0.60	0.144 *0.078*	0.057 *0.045*	0.036 *0.035*	–	–	–	–	–	–	–	–
0.65	0.092 *0.054*	0.037 *0.033*	0.031 *0.031*	–	–	–	–	–	–	–	–
0.70	0.052 *0.035*	0.026 *0.025*	–	–	–	–	–	–	–	–	–
0.75	0.027 *0.022*	0.020 *0.020*	–	–	–	–	–	–	–	–	–

Straight conical diffuser loss coefficients (K_1) are shown in normal font; optimum stepped diffuser loss coefficients (K_{opt}) are shown in *italic* font.

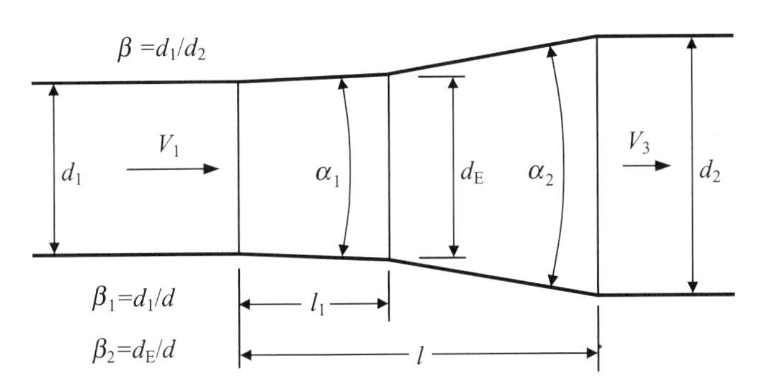

FIGURE 11.7. Two-stage conical diffuser.

For a given overall length l and diameter ratio β, there are two basic variables: first-stage length l_1 and divergence angle α_1. The geometric relationship of the second stage is given by:

$$d_E = d_1 + 2l_1 \tan\left(\frac{\alpha_1}{2}\right)$$

or

TABLE 11.2. Divergence Angle α_{opt} for Optimum Stepped Diffusers Compared to Divergence Angle α for Straight Conical Diffusers of Equal Length and Overall Area Ratio

	l/d_1										
Area Ratio β^2	0.5	1.0	2.0	3.0	4.0	5.0	6.0	8.0	10.0	12.0	15.0
0.00	180	180	180	180	180	180	180	180	180	180	180
	11.1	*13.5*	*12.4*	*11.2*	*11.2*	*9.5*	*8.8*	*7.9*	*7.3*	*6.8*	*6.3*
0.05	147.9	120.1	81.9	60.1	46.9	38.3	32.3	24.5	19.7	16.5	13.2
	11.5	*13.4*	*12.4*	*11.1*	*11.1*	*9.3*	*8.7*	*7.7*	*7.1*	*6.6*	*6.1*
0.10	130.4	94.5	56.8	39.6	30.2	24.4	20.4	15.4	12.3	11.3	8.2
	11.9	*13.5*	*12.3*	*11.9*	*9.9*	*9.1*	*8.4*	*7.3*	*6.8*	*6.4*	*5.9*
0.15	115.4	76.7	43.2	29.5	22.4	18.0	15.0	11.3	9.0	7.5	6.0
	12.3	*13.7*	*12.3*	*11.8*	*9.6*	*8.8*	*8.1*	*7.2*	*6.5*	*6.0*	*5.6*
0.20	102.0	63.4	34.3	23.3	17.6	14.1	11.8	8.8	7.1	5.9	–
	12.7	*13.8*	*12.1*	*11.5*	*9.3*	*8.5*	*7.8*	*6.8*	*6.1*	*5.7*	
0.25	90.0	53.1	28.1	18.9	14.2	11.4	9.5	7.1	5.7	–	–
	13.2	*13.9*	*11.0*	*11.2*	*9.0*	*8.0*	*7.3*	*6.3*	*5.7*		
0.30	79.1	44.9	23.3	15.7	11.8	9.4	7.9	5.9	–	–	–
	13.6	*14.0*	*11.6*	*9.8*	*8.5*	*7.6*	*6.8*	*5.9*			
0.35	69.2	38.1	19.6	13.1	9.9	7.9	6.6	–	–	–	–
	14.0	*14.0*	*11.3*	*9.3*	*8.0*	*7.0*	*6.3*				
0.40	60.3	32.4	16.5	11.1	8.3	6.6	–	–	–	–	–
	14.4	*13.9*	*11.8*	*8.7*	*7.4*	*6.4*					
0.45	52.3	27.6	14.0	9.3	7.0	–	–	–	–	–	–
	14.7	*13.7*	*11.2*	*8.1*	*6.7*						
0.50	45.0	23.4	11.8	7.9	–	–	–	–	–	–	–
	15.0	*13.3*	*9.5*	*7.3*							
0.55	38.4	19.8	11.0	6.6	–	–	–	–	–	–	–
	15.1	*12.7*	*8.7*	*6.5*							
0.60	32.4	16.6	8.3	–	–	–	–	–	–	–	–
	15.1	*12.0*	*7.7*								
0.65	27.0	13.7	6.9		–	–	–	–	–	–	–
	14.8	*11.0*	*6.8*								
0.70	22.1	11.1	–	–	–	–	–	–	–	–	–
	14.2	*9.7*									
0.75	17.6	8.8	–	–	–	–	–	–	–	–	–
	13.0	*8.3*									

Straight conical diffuser included angles (α) are shown in normal font; optimum stepped diffuser included angles (α_{opt}) are shown in *italic* font.

$$\alpha_2 = 2\,\mathrm{atan}\left[\frac{d_2 - d_E}{2(l - l_1)}\right].$$

There are little or no data available in the open literature on the performance of two-stage diffusers. However, the author has utilized Equation 11.12 in a flow model that accurately predicts the performance of jet pumps used as part of the coolant recirculation system of boiling water reactors. There is no effort made here to compare the performance of two-stage diffusers with the other types of diffusers.

11.4 CURVED WALL DIFFUSER

It would appear that a trumpet-shaped passage may well give minimum loss for a given length and given

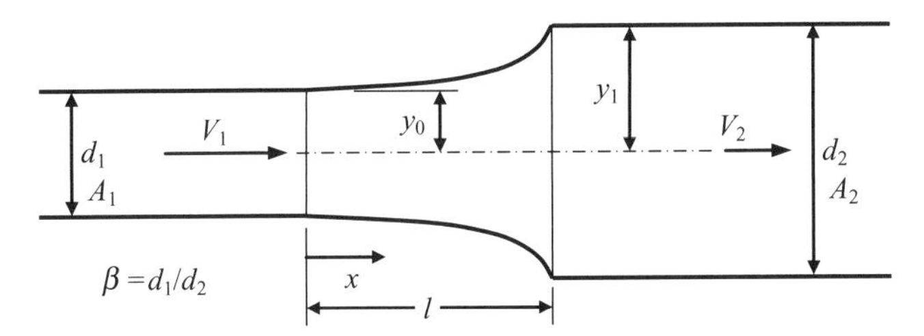

FIGURE 11.8. Curved wall diffuser.

ratio of enlargement (see Fig. 11.8). A diffuser in which the pressure gradient remains constant along the passage (dp/dx = constant) may be the best choice.

Idel'chik [6] presents an equation for dp/dx = constant for the boundary wall of a curved wall diffuser of a circular (or square) cross section,

$$y = \frac{y_1}{\sqrt[4]{1+\left((y_1/y_0)^4-1\right)x/l}},$$

as well as for the diverging wall of a diffuser with a plane cross section,

$$y = \frac{y_1}{\sqrt{1+\left((y_1/y_0)^2-1\right)x/l}}.$$

Based on Idel'chik's experiments, an approximate formula for the loss coefficient of curved wall diffusers, within the limits $0.1 < \beta^2 < 0.9$, is given as:

$$K_1 \approx \varphi_0\left(1.43-1.3\beta^2\right)\left(1-\beta^2\right)^2,$$

where φ_0 is a coefficient that depends on the relative length of the curved wall diffuser as shown in Table 11.4.* A curve fit of φ_0 for circular or square cross-sections gives:

$$\varphi_0 = 1.01 - 0.624\frac{l}{d_1} + 0.30\left(\frac{l}{d_1}\right)^2 - 0.074\left(\frac{l}{d_1}\right)^3 + 0.0070\left(\frac{l}{d_1}\right)^4.$$

The effectiveness of curved wall diffusers is also compared with the other expansion configurations in Table 11.3. Based on the loss coefficients equations developed in this chapter, the most effective configurations for given area and length ratios are shown in bold font.

The curved wall diffuser appears to be generally more efficient than the sudden expansion and the straight conical diffuser. However, except at high area ratios, it does not appear to be as effective as the stepped diffuser. There is considerable uncertainty associated with the calculated loss coefficients in Table 11.3. Future tests and evaluations could change the results. Even then, the curved wall diffuser may not be a viable choice because the improvement may be slight and not worth the extra effort involved in designing and fabricating the curved wall.

11.5 PIPE REDUCER: EXPANDING

Standard butt-weld pipe fittings, ANSI† reducers, are used to join pipe sections of different diameters (Fig. 11.9). Typically, the fittings are generously rounded at the intersection of the conical and cylindrical surfaces. In the case of contracting reducers (see Section 10.7), rounding greatly reduces energy loss through the fitting. However, rounding has little effect in decreasing energy loss when flow through the fitting is expanding (see Fig. 11.9). For a large area expansion, the cone angle may exceed 50° or 60° and the resulting loss may exceed that of a sudden expansion. For smaller area expansions, the conical diffuser section performs more efficiently to transform kinetic energy into pressure energy.

Industry standards define the overall length l of butt-welding reducers. However, there are no standards regarding the dimensions of the straight and conical sections, or the rounding of the intersections. Characteristically, the extended intersection points of the cylindrical inlet and outlet sections with the conical section appear to be about 20% of the length so that the conical section is about 60% of the length as shown in Figure 11.10. Thus the divergence angle can be estimated as:

$$\alpha \approx 2\tan^{-1}\left(\frac{d_1-d_2}{1.20l}\right).$$

Accounting for friction loss in the "straight" sections as well as in the "conical" section, surface friction loss in the reducer can be approximated as:

$$K_f \approx f_1\frac{0.20l}{d_1} + \frac{f_1(1-\beta)}{8\sin(\alpha/2)} + f_2\frac{0.20l}{d_2}\beta^4. \qquad (11.13)$$

Loss coefficients for butt-weld reducers were calculated by substituting Equation 11.13 for the friction loss term

* Friction loss is not separately accounted for as is the case in other diffuser configurations. Idel'chik states, "The frictional losses in very wide-angled diffusers are quite small. It is not necessary to separate these losses from the total losses with curved diffusers which correspond to wide-angle straight diffusers."

† American National Standards Institute.

TABLE 11.3. Comparative Effectiveness of Diffuser Configurations

		l/d_1								
Loss Coefficient K_1	Area Ratio β^2	0.5	1	2	3	4	6	8	10	12
Sudden expansion	0.05	0.903	0.903	0.903	0.903	0.903	0.903	0.903	0.903	0.903
Straight conical		0.933	0.960	1.010	1.082	1.021	0.784	0.561	0.364	0.272
Stepped (optimum)		**0.744**	**0.556**	**0.362**	**0.270**	**0.218**	**0.163**	**0.135**	**0.119**	**0.109**
Curved wall		–	–	–	–	–	–	–	–	–
Sudden expansion	0.15	0.722	0.722	0.722	0.722	0.722	0.722	0.722	0.722	0.706
Straight conical		0.772	0.817	0.784	0.575	0.388	0.192	0.129	0.102	0.089
Stepped (optimum)		**0.571**	**0.409**	**0.254**	**0.185**	**0.148**	**0.111**	**0.094**	**0.086**	**0.082**
Curved wall		0.679	0.547	0.423	0.357	0.330	–	–	–	–
Sudden expansion	0.25	0.563	0.563	0.563	0.563	0.563	0.563	0.563	0.563	0.563
Straight conical		0.622	0.667	0.424	0.217	0.142	0.089	0.074	**0.072**	**0.074**
Stepped (optimum)		**0.421**	**0.287**	**0.169**	**0.121**	**0.098**	**0.078**	**0.072**	**0.072**	–
Curved wall		0.473	0.381	0.294	0.249	0.230	–	–	–	–
Sudden expansion	0.35	0.423	0.423	0.423	0.423	0.423	0.423	0.423	0.423	0.423
Straight conical		0.486	0.425	0.175	0.099	0.074	**0.062**	**0.065**	**0.073**	**0.083**
Stepped (optimum)		**0.293**	**0.188**	**0.106**	**0.078**	**0.067**	**0.062**	–	–	–
Curved wall		0.313	0.253	0.195	0.165	0.152	–	–	–	–
Sudden expansion	0.45	0.303	0.303	0.303	0.303	0.303	0.303	0.303	0.303	0.303
Straight conical		0.359	0.226	0.080	0.056	0.052	**0.058**	**0.071**	**0.085**	**0.100**
Stepped (optimum)		**0.189**	**0.114**	**0.065**	**0.053**	**0.051**	–	–	–	–
Curved wall		0.194	0.157	0.121	0.102	0.095	–	–	–	–
Sudden expansion	0.55	0.202	0.202	0.202	0.202	0.202	0.202	0.202	0.202	0.202
Straight conical		0.207	0.089	0.044	**0.042**	**0.047**	**0.064**	**0.082**	**0.102**	**0.121**
Stepped (optimum)		**0.109**	**0.063**	**0.042**	**0.042**	–	–	–	–	–
Curved wall		0.110	0.089	0.069	0.058	0.054	–	–	–	–
Sudden expansion	0.65	0.122	0.122	0.122	0.122	0.122	0.122	0.122	0.122	**0.122**
Straight conical		0.092	0.037	**0.031**	**0.040**	**0.050**	**0.073**	**0.097**	**0.121**	0.144
Stepped (optimum)		**0.054**	**0.033**	**0.031**	–	–	–	–	–	–
Curved wall		0.055	0.044	0.034	0.029	0.027	–	–	–	–
Sudden expansion	0.75	0.063	0.063	0.063	0.063	0.063	**0.063**	**0.063**	**0.063**	**0.063**
Straight conical		**0.022**	0.020	0.030	0.043	0.057	0.085	0.113	0.142	0.170
Stepped (optimum)		0.023	0.024	–	–	–	–	–	–	–
Curved wall		**0.022**	**0.017**	**0.017**	**0.011**	**0.011**	–	–	–	–
Sudden expansion	0.85	0.022	0.022	0.022	0.022	0.022	**0.022**	**0.022**	**0.022**	**0.022**
Straight conical		0.011	0.017	0.033	0.049	0.066	0.098	0.131	0.164	0.197
Stepped (optimum)		0.011	–	–	–	–	–	–	–	–
Curved wall		**0.006**	**0.005**	**0.004**	**0.003**	**0.003**	–	–	–	–

Note: The most efficient configurations are shown in **bold** font.

TABLE 11.4. Coefficient ϕ_0 as a Function of Relative Length of a Curved Wall Diffuser

l/d_1	0	0.5	1.0	1.5	2.0	2.5	3.0	3.5	4.0	4.5	5.0	6.0
Circular or square cross section												
φ_0	1.01	0.75	0.62	0.53	0.47	0.43	0.40	0.38	0.37	–	–	–
Plane cross section												
φ_0	1.02	0.83	0.72	0.64	0.57	0.52	0.48	0.45	0.43	0.41	0.39	0.37

(the last term) in Equations 11.8–11.10, as appropriate. Surface friction factors f_1 and f_2 at d_1 and d_2, assuming a surface roughness of 0.00015 ft, were assigned using von Kármán's equation for fully turbulent flow in a rough pipe (Eq. 3.5). The calculated results are shown in Table 11.5. The loss coefficient values in Table 11.5 are for concentric reducers. There is some question, but consider adding 15% to the concentric reducer loss coefficient values for eccentric reducers.

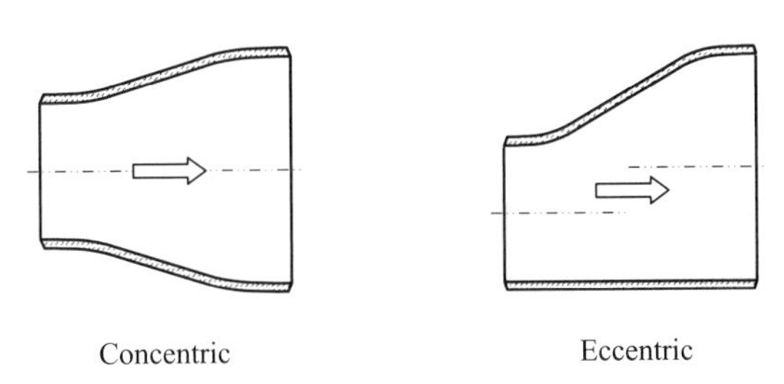

FIGURE 11.9. Butt-weld pipe reducer—expanding.

This method of accounting for butt-weld reducer losses should be sufficient for most engineering purposes. If more definite information regarding internal geometry and surface friction is available, a more accurate loss coefficient value can be calculated.

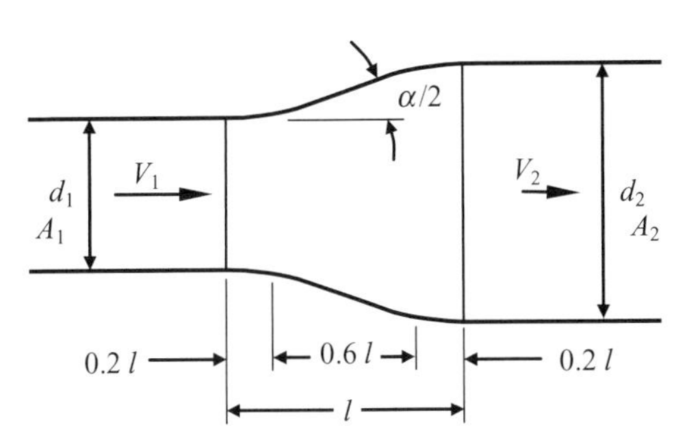

FIGURE 11.10. Concentric butt-weld pipe reducer—expanding.

TABLE 11.5. Loss Coefficient K_1 for Concentric Butt-Weld Reducers—Expanding

Nominal Size (in)	Length l (in)	K_1
3/4 × 1/2	1.5	0.07
3/4 × 3/8		0.21
1 × 3/4	2	0.05
1 × 1/2		0.20
1-1/4 × 1	2	0.07
1-1/4 × 3/4		0.29
1-1/4 × 1/2		0.59
1-1/2 × 1-1/4	2.5	0.03
1-1/2 × 1		0.17
1-1/2 × 3/4		0.45
1-1/2 × 1/2		0.66
2 × 1-1/2	3	0.06
2 × 1-1/4		0.16
2 × 1		0.49
2 × 3/4		0.67
2-1/2 × 2	3.5	0.03
2-1/2 × 1-1/2		0.19
2-1/2 × 1-1/4		0.38
2-1/2 × 1		0.64
3 × 2-1/2	3.5	0.05
3 × 2		0.22
3 × 1-1/2		0.54
3 × 1-1/4		0.67
3-1/2 × 3	4	0.03
3-1/2 × 2-1/2		0.19
3-1/2 × 2		0.41
3-1/2 × 1-1/2		0.67
3-1/2 × 1-1/4		0.76
4 × 3-1/2	4	0.02
4 × 3		0.10
4 × 2-1/2		0.38

Nominal Size (in)	Length l (in)	K_1
4 × 2	4	0.61
4 × 1-1/2		0.77
5 × 4	5	0.06
5 × 3-1/2		0.2
5 × 3		0.39
5 × 2-1/2		0.65
5 × 2		0.76
6 × 5	5.5	0.04
6 × 4		0.30
6 × 3-1/2		0.47
6 × 3		0.65
6 × 2-1/2		0.77
8 × 6	6	0.15
8 × 5		0.40
8 × 4		0.67
8 × 3-1/2		0.73
10 × 8	7	0.10
10 × 6		0.47
10 × 5		0.67
10 × 4		0.76
12 × 10	8	0.05
12 × 8		0.34
12 × 6		0.66
12 × 5		0.74
14 × 12	13	0.01
14 × 10		0.10
14 × 8		0.40
14 × 6		0.69
16 × 14	14	0.02
16 × 12		0.07
16 × 10		0.28

Nominal Size (in)	Length l (in)	K_1
16 × 8	14	0.58
18 × 16	15	0.02
18 × 14		0.10
18 × 12		0.21
18 × 10		0.45
20 × 18	20	0.02
20 × 16		0.05
20 × 14		0.19
20 × 12		0.31
22 × 20	20	0.01
22 × 18		0.05
22 × 16		0.17
22 × 14		0.34
24 × 22	20	0.01
24 × 20		0.04
24 × 18		0.15
24 × 16		0.30
26 × 24	24	0.01
26 × 22		0.03
26 × 20		0.10
26 × 18		0.23
28 × 26	24	0.01
28 × 24		0.03
28 × 22		0.09
28 × 20		0.21
30 × 28	24	0.01
30 × 26		0.02
30 × 24		0.08
30 × 22		0.18

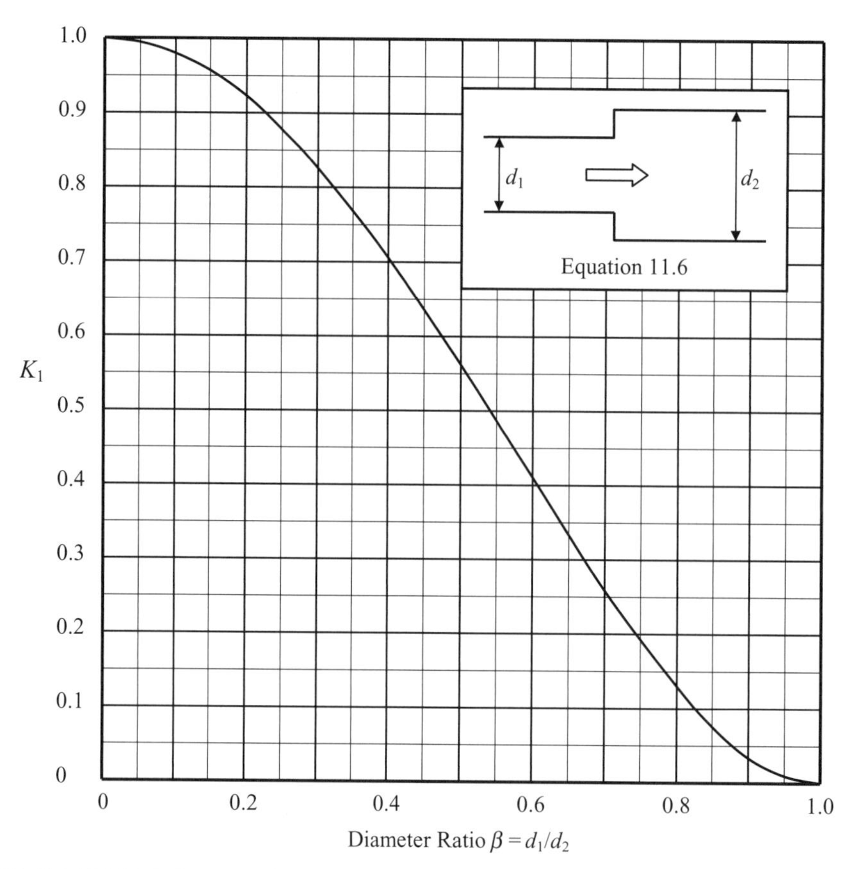

DIAGRAM 11.1. Loss coefficient K_1 of a sudden expansion (Borda–Carnot equation).

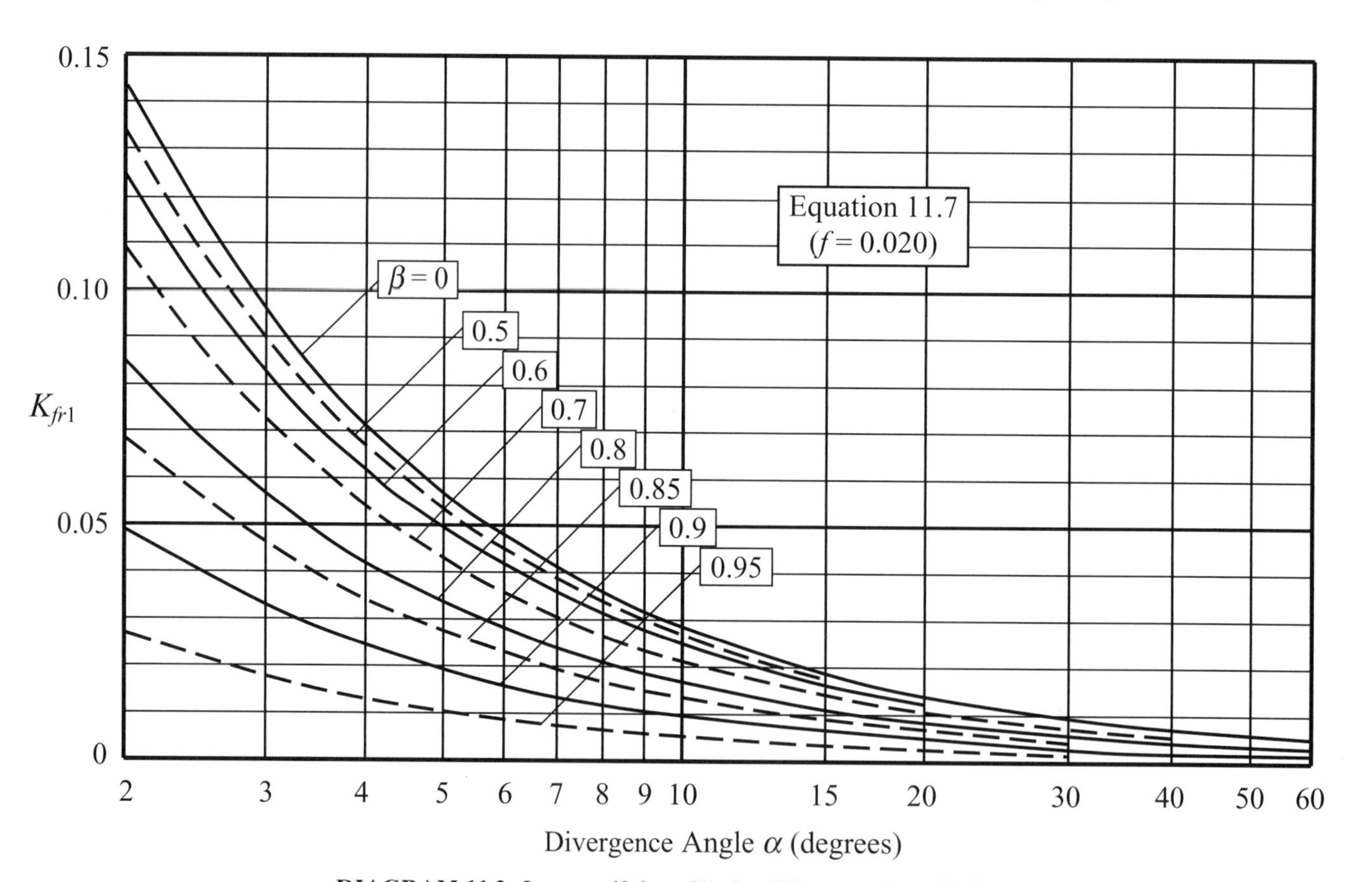

DIAGRAM 11.2. Loss coefficient K_{fr1} for diffuser surface friction.

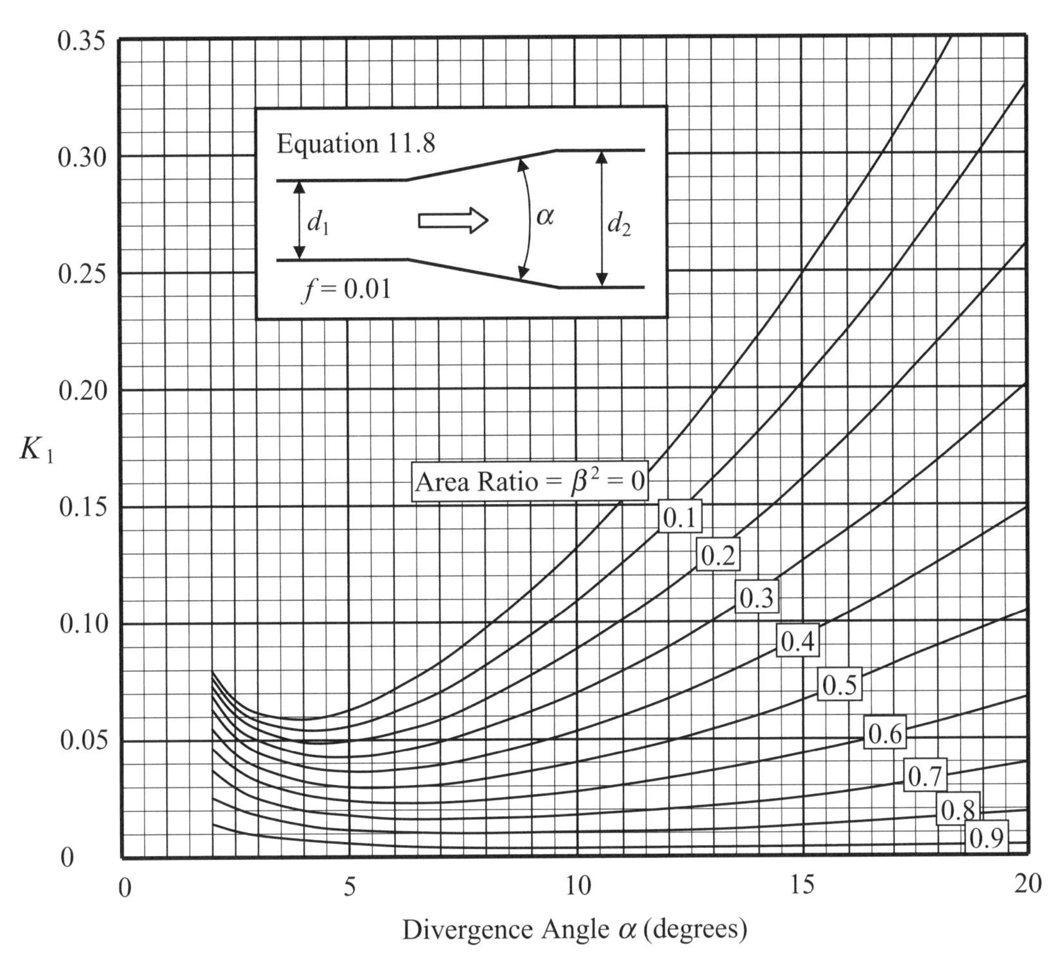

DIAGRAM 11.3. Loss coefficient of a conical diffuser—$\alpha = 2°$ to $20°$ ($f = 0.01$).

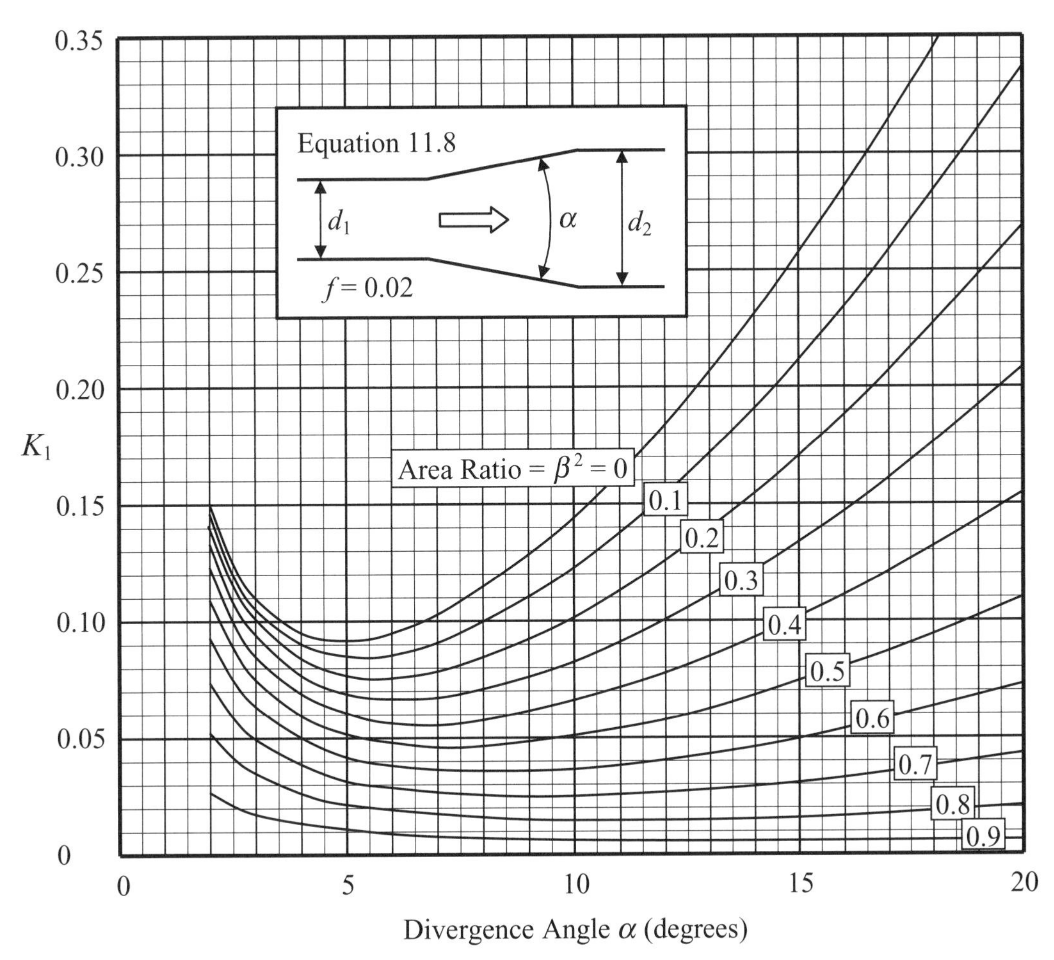

DIAGRAM 11.4. Loss coefficient of a conical diffuser—$\alpha = 2°$ to $20°$ ($f = 0.02$).

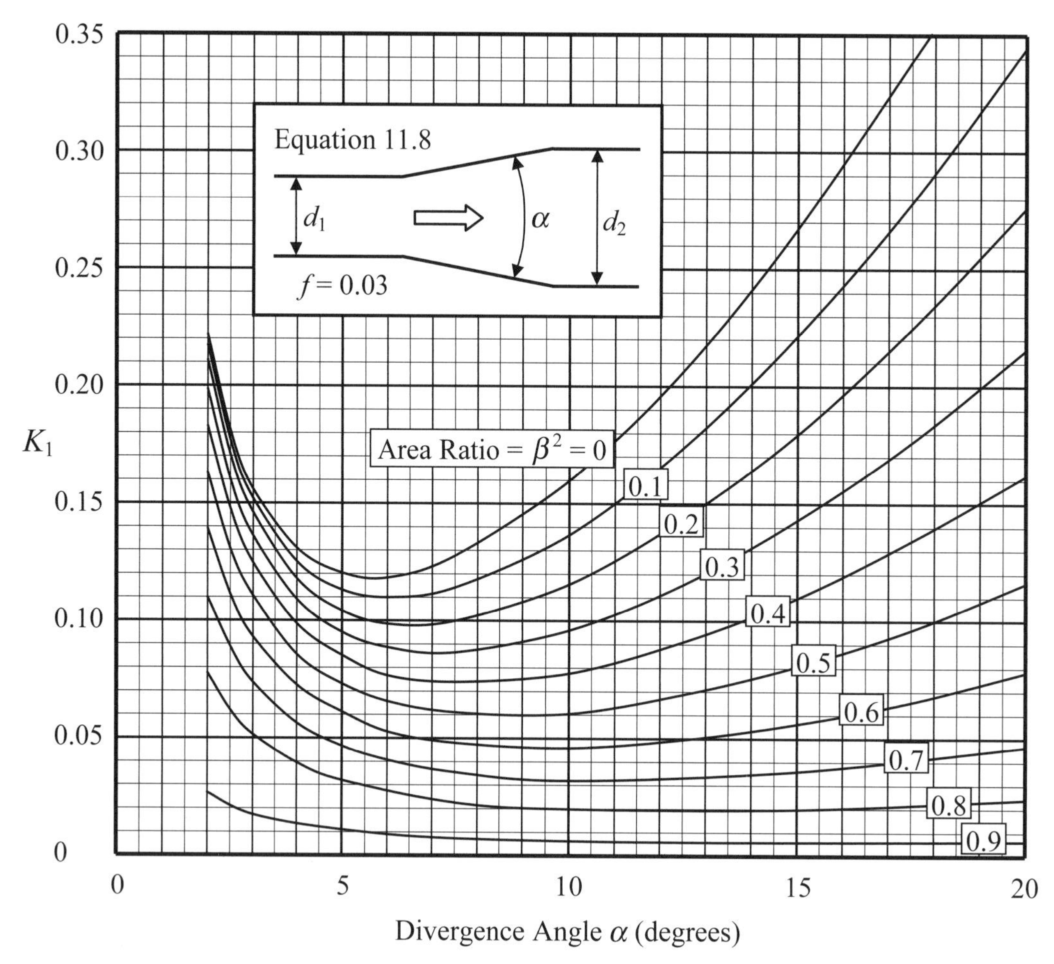

DIAGRAM 11.5. Loss coefficient of a conical diffuser—$\alpha = 2°$ to $20°$ ($f = 0.03$).

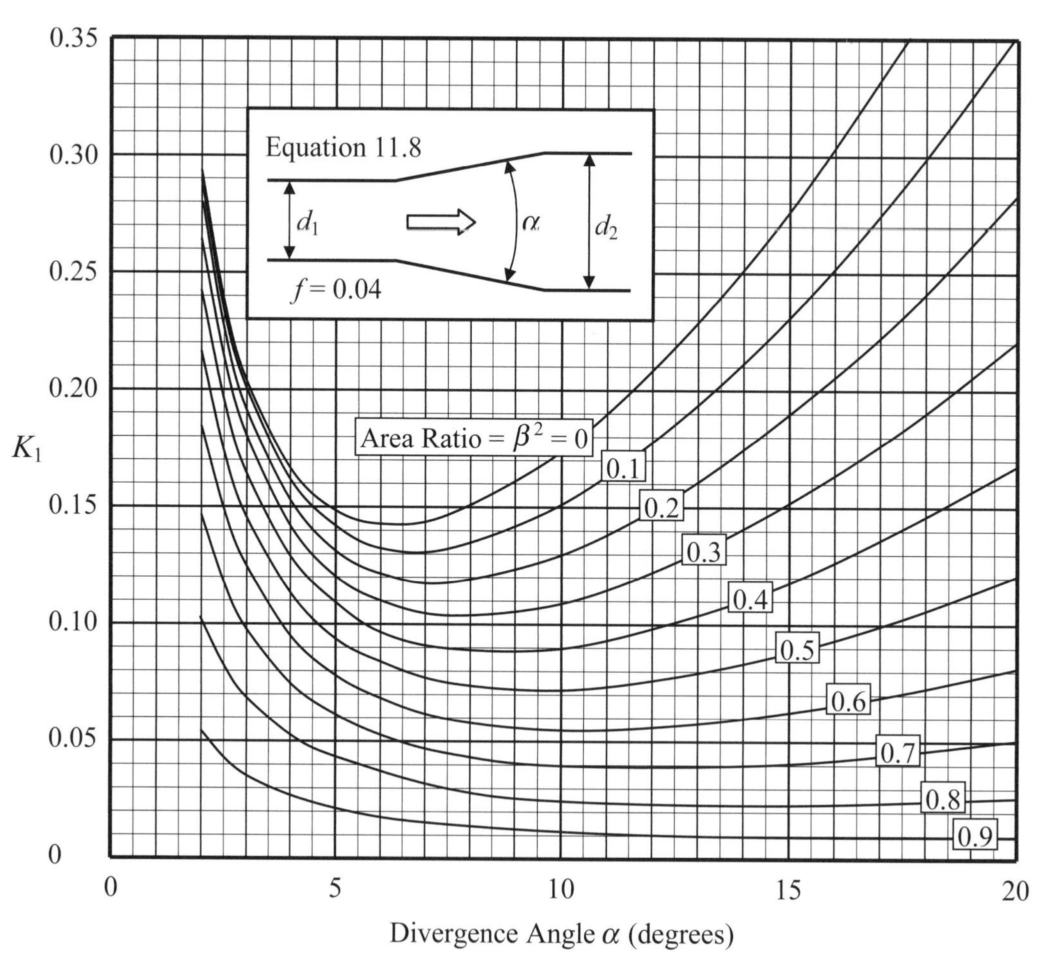

DIAGRAM 11.6. Loss coefficient of a conical diffuser—$\alpha = 2°$ to $20°$ ($f = 0.04$).

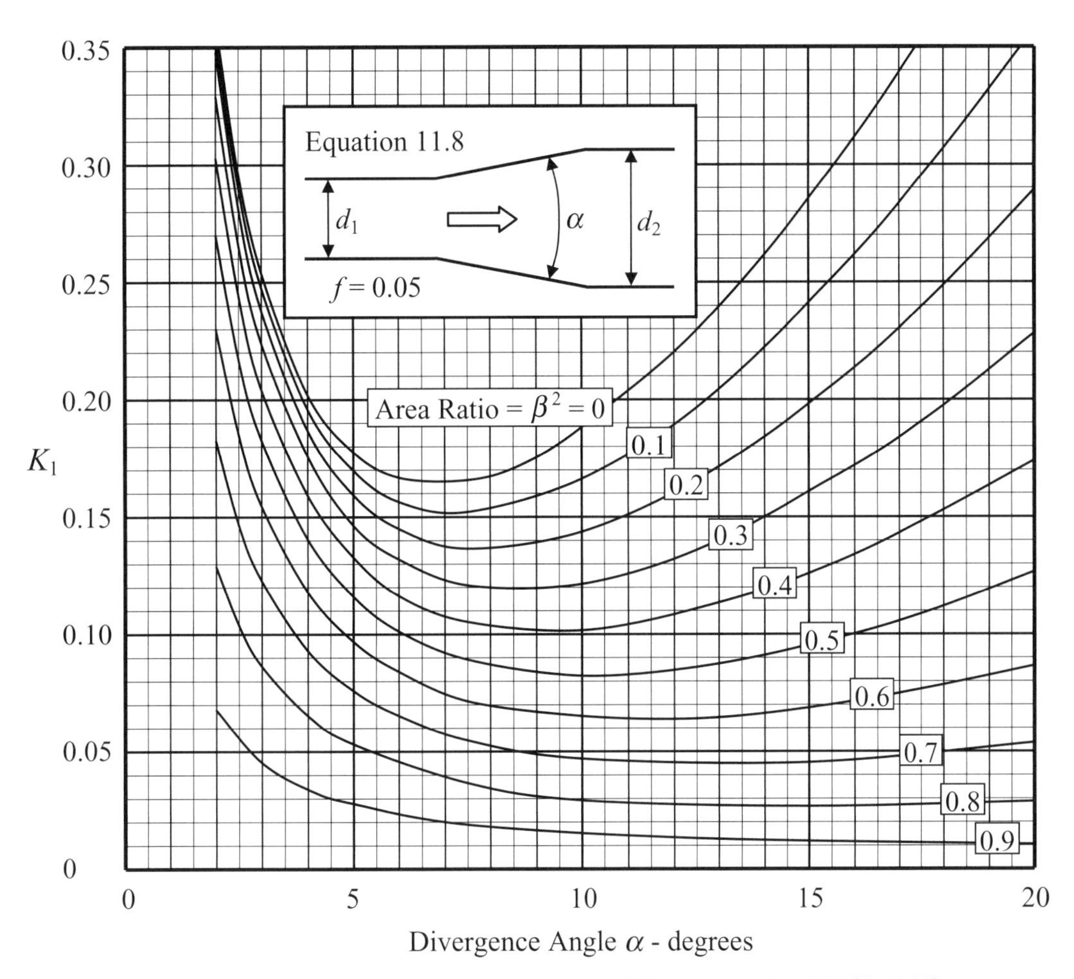

DIAGRAM 11.7. Loss coefficient of a conical diffuser—$\alpha = 2°$ to $20°$ ($f = 0.05$).

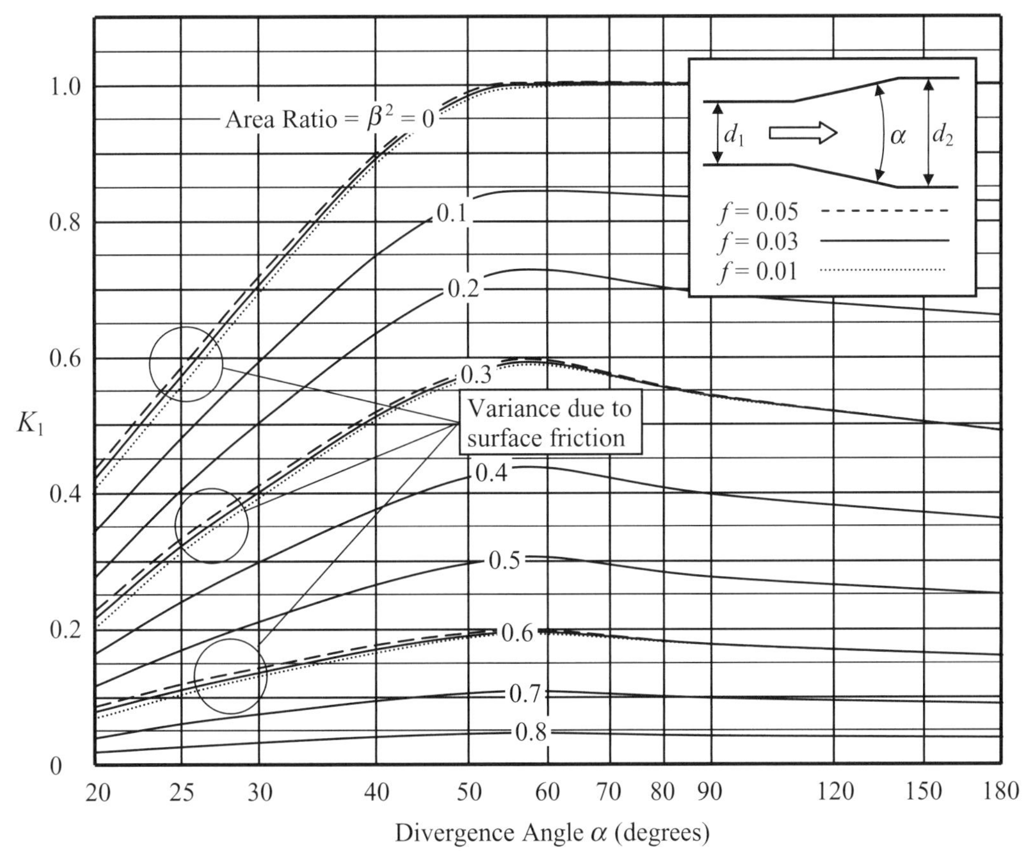

DIAGRAM 11.8. Loss coefficient of a conical diffuser—20° to 180°.

REFERENCES

1. Nusselt, W., The pressure in the annulus of pipes with a sudden increase in cross section for high velocity air flow, *Forschung auf dem Gebiete des Ingenieurwesens*, 11(5), 1940, 250–255.
2. Borda, J. C., Memoire sur l'ecoulement des Fluides par les Orifices des Vases, *Memoire de l'academie Royale des Sciences*, 1766.
3. Benedict, R. P., N. A. Carlucci, and S. D. Swetz, Flow losses in abrupt enlargements and contractions, *Journal of Basic Engineering, Transactions of the American Society of Mechanical Engineers*, 88(1), 1966, 73–81.
4. Gibson, A. H., On the flow of water through pipes and passages having converging or diverging boundaries, *Transactions of the Royal Society, London, Series A*, 83(A563), 1910, 366–378.
5. Gibson, A. H., On the resistance to flow of water through pipes or passages having divergent boundaries, *Proceedings of the Royal Society of Edinburgh*, 48(5), Part 1, 1911–1913, 97–116.
6. Idel'chik, I. E., Aerodynamics of the flow and pressure head losses in diffusers, *Prom. Aerodin.*, No. 3, 1947, pp. 132–209.

FURTHER READING

This list includes books and papers that may be helpful to those who wish to pursue further study.

Archer, W. H., Experimental determination of loss of head due to sudden enlargement in circular pipes, *Transactions of the American Society of Mechanical Engineers*, 76, 1913, 999.

Schutt, H. C., Losses of pressure head due to sudden enlargement of a flow cross-section, *Transactions of the American Society of Mechanical Engineers*, 51, 1929, 83–87.

Peters, H., Conversion of energy in cross-sectional divergences under different conditions of inflow, *Ingenieur-Archiv*, II, 1931, 92–107.

Patterson, G. N., Modern diffuser design, *Aircraft Engineering*, 10, 1938, 267.

Kays, W. M., Loss coefficients for abrupt changes in flow cross section with low Reynolds number in single and multiple-tube systems, *Transactions of the American Society of Mechanical Engineers*, Paper No. 50-S-7, 72, 1950, 1067–1074.

Robertson, J. M. and D. Ross, Effects of entrance conditions on diffuser flow, *Proceedings of the American Society of Civil Engineers*, 78, Separate No. 141, 1952, 1–24.

Squire, H. B., Experiments on Conical Diffusers, *Aeronautical Research Council*, London, R. & M. No. 2751, 1953.

Hall, W. B. and E. M. Orme, Flow of a Compressible Fluid Through a Sudden Enlargement in a Pipe, *The Institution of Mechanical Engineers*, 1955.

Winternitz, F. A. L. and W. J. Ramsay, Effects of inlet boundary layer on pressure recovery, energy conversion and losses in conical diffusers, *Journal of the Royal Aeronautical Society*, 61, 1957, 116.

Rippl, E., Experimental investigations concerning the efficiency of slim conical diffusers and their behavior with regards to flow separation, *Monthly Technical Review*, 2(3), 1958, 64–70.

Kline, S. J., D. E. Abbott, and R. W. Fox, Optimum design of straight-walled diffusers, *Journal of Basic Engineering, Transactions of the American Society of Mechanical Engineers*, Paper No. 58-A-137, 81, 1959, 321–328.

Abbott, D. E. and S. J. Kline, Experimental investigations of subsonic turbulent flow over single and double facing steps, *Journal of Basic Engineering, Transactions of the American Society of Mechanical Engineers*, 84, 1962, 73.

Fox, R. W. and S. J. Kline, Flow regimes in curved subsonic diffusers, *Journal of Basic Engineering, Transactions of the American Society of Mechanical Engineers*, 84, 1962, 303–316.

Lipstein, N. J., Low velocity sudden expansion pipe flow, *American Society of Heating, Refrigerating and Air-Conditioning Engineers Journal*, 5, 1962, 43–47.

Chaturvedi, M. C., Flow characteristics of axisymmetric expansions, *Journal of the Hydraulics Division, Proceedings of the American Society of Civil Engineers*, 89(HY3), 1963, 61–92.

Cockrell, D. J. and E. Markland, A Review of Incompressible Diffuser Flow, *Aircraft Engineering*, October 1963.

McDonald, A. T. and R. W. Fox, An experimental investigation of incompressible flow in conical diffusers, *International Journal of Mechanical Sciences*, 8(2), 1966, 1.

Carlson, J. J., J. P. Johnston, and S. J. Sagi, Effects of wall shape on flow regimes and performance in straight, two-dimensional diffusers, *Journal of Basic Engineering, Transactions of the American Society of Mechanical Engineers*, 89, 1967, 73–81.

Reneau, L. R., J. P. Johnston, and S. J. Kline, Performance and design of straight, two-dimensional diffusers, *Journal of Basic Engineering, Transactions of the American Society of Mechanical Engineers*, 89, 1967, 141–150.

Sagi, C. J. and J. P. Johnston, The design and performance of two-dimensional, curved diffusers, *Journal of Basic Engineering, Transactions of the American Society of Mechanical Engineers*, 89, 1967, 151–160.

Sovran, G. and E. D. Klomp, *Experimentally Determined Optimum Geometries for Rectilinear Diffusers with Rectangular, Conical or Annular Cross-Section*, (reprinted from *Fluid Mechanics of Internal Flow*) Elsevier Publishing Co., Amsterdam, 1967, pp. 270–319.

Masuda, S., I. Ariga, and I. Waranabe, On the behavior of uniform shear flow in diffusers and its effects on diffuser performance, *Journal of Engineering for Power, Transactions of the American Society of Mechanical Engineers*, 93, 1971, 377–385.

McDonald, A. T., R. W. Fox, and R. V. Van Dewoestine, Effects of swirling inlet flow on pressure recovery in conical diffusers, *American Institute of Aeronautics and Astronautics Journal*, 9(10), 1971, 2014–2018.

Runchal, A. K., Mass transfer investigation in turbulent flow downstream of a sudden enlargement of a circular pipe for very high Schmidt numbers, *International Journal of Heat and Mass Transfer*, 14, 1971, 781–791.

Back, L. H. and E. J. Roschke, Shear-layer flow regimes and wave instabilities and reattachment lengths downstream of an abrupt circular channel expansion, *Journal of Applied Mechanics, Transactions of the American Society of Mechanical Engineers*, 94E, 1972, 677–881.

Benedict, R. P., A. R. Gleed, and R. D. Schulte, Air and water studies on a diffuser-modified flow nozzle, *Journal of Fluids Engineering, Transactions of the American Society of Mechanical Engineers*, 95, 1973, 169–179.

Smith, C. R. and S. J. Kline, An experimental investigation of the transitory stall regime in two-dimensional diffusers, *Journal of Fluids Engineering, Transactions of the American Society of Mechanical Engineers*, 95, 1973, 1–5.

Teyssandier, R. G., Internal separated flows—Expansions, nozzles, and orifices, PhD dissertation, University of Rhode Island, Kingston, R. I., 1973.

Teyssandier, R. G. and M. P. Wilson, An analysis of flow through sudden enlargements in pipes, *Journal of Fluid Mechanics*, 64, Part 1, 1974, 85–95.

Benedict, R. P., J. S. Wyler, J. A. Dudek, and A. R. Gleed, Generalized flow across an abrupt enlargement, *Journal of Engineering for Power, Transactions of the American Society of Mechanical Engineers*, 98, 1976, 327–334.

Moon, L. F. and G. Rudinger, Velocity distribution in an abruptly expanding circular duct, *Journal of Fluids Engineering, Transactions of the American Society of Mechanical Engineers*, 99, 1977, 226.

Lohmann, R. P., S. J. Markowski, and E. T. Brookman, Swirling flow through annular diffusers with conical walls, *Journal of Fluids Engineering, Transactions of the American Society of Mechanical Engineers*, 101, 1979, 224–229.

Webb, A. I. C., Head loss of a sudden expansion, *International Journal of Mechanical Engineering Education, Institution of Mechanical Engineers*, University of Manchester Institute of Science and Technology, 8(4), 1980, 173–176.

Dekam, E. I. and J. R. Calvert, Pressure losses in sudden transitions between square and rectangular ducts of the same cross-sectional area, *International Journal of Heat and Fluid Flow*, 9(1), 1988, 2–7.

Hallett, W. L. H., A simple model for the critical swirl in a swirling sudden expansion flow, *Journal of Fluids Engineering, Transactions of the American Society of Mechanical Engineers*, 110, 1988, 155–160.

Stieglmeir, M., C. Tropea, N. Weiser, and W. Nitsche, Experimental investigation of the flow through axisymmetric expansions, *Journal of Fluids Engineering, Transactions of the American Society of Mechanical Engineers*, 111, 1989, 464–471.

Ramamurthy, A. S., R. Balchandar, and H. S. Govina Ram, Some characteristics of flow past backward facing steps, including cavitation effects, *Journal of Fluids Engineering, Transactions of the American Society of Mechanical Engineers*, 113, 1991, 278–284.

Kwong, A. H. M. and A. P. Dowling, Unsteady flow in diffusers, *Journal of Fluids Engineering, Transactions of the American Society of Mechanical Engineers*, 116, 1994, 842–847.

Papadopoulos, G. and M. V. Ötügen, A modified Borda-Carnot relation for the prediction of maximum recovery pressure in planar sudden expansion flows, *Journal of Fluids Engineering, Transactions of the American Society of Mechanical Engineers*, 120, 1998, 400.

12

EXITS

A special case of a sudden expansion occurs when a pipe discharges into a large volume or reservoir. The classic Borda–Carnot equation for a sudden expansion was presented in Chapter 11:

$$H_L = \frac{(V_1 - V_2)^2}{2g}. \qquad (11.5, \text{repeated})$$

Here the velocity V_2 downstream of the expansion goes to zero, and when the head loss is computed from Equation 10.5, it is found to be one velocity head:

$$H_L = \frac{V^2}{2g}.$$

This is the case whether the pipe exit is submerged or open as illustrated in Figure 12.1.*

Many exit configurations are not as simple as a straight pipe. In some cases, exit loss consists of local and friction losses.

12.1 DISCHARGE FROM A STRAIGHT PIPE

The relation between loss coefficient K and head loss H_L is:

$$H_L = K\frac{V^2}{2g}. \qquad (3.7, \text{repeated})$$

When the head loss is given as in Equation 10.7 above, the loss coefficient becomes unity. Equation 3.7 is written as "conventional" head loss, which ignores the kinetic energy correction factor ϕ. If the kinetic energy correction factor were included, Equation 3.7 would become

$$H_L = K\phi\frac{V^2}{2g}.$$

In the Borda–Carnot equation K is taken to be unity for conventional head loss. If the kinetic energy correction factor is taken into account, we may elect to absorb ϕ into K, thus making $K = \phi$. Therefore, in the case of discharge from a straight pipe, the loss coefficient K is simply the kinetic energy correction factor ϕ of the flow stream in the exit stretch. This is so whether the pipe projects into the reservoir, or is sharp-edged or rounded at the exit as shown in Figure 12.2.

In general, a value of 1.0, quite suitable for most engineering purposes, is assigned as the value of the kinetic energy correction factor ϕ (see Section 2.7). In the case of uniform distribution of velocity, ϕ *is* equal to unity. However, in the case of fully developed flow following a long stretch of pipe, the value of ϕ for circular (or square) pipe is 2.0 for laminar flow, and ranges from about 1.04 to 1.10 for turbulent flow.† Nonetheless, in the real world, fully developed flow may also exist at the

* EGL and HGL are the energy grade line and hydraulic grade line, respectively (see Section 2.9).

† From Figure 2.6 we find that the kinetic energy correction factor ϕ ranges from 1.04 to 1.10 for values of friction factor f between 0.01 and 0.03.

Pipe Flow: A Practical and Comprehensive Guide, First Edition. Donald C. Rennels and Hobart M. Hudson.

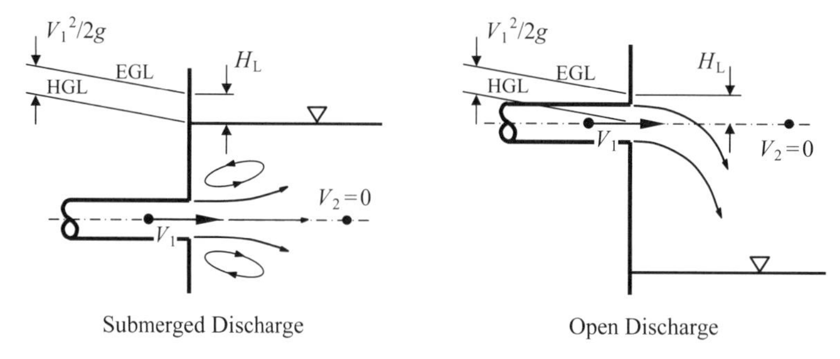

FIGURE 12.1. Pipe exit. EGL, energy grade line; HGL, hydraulic grade line.

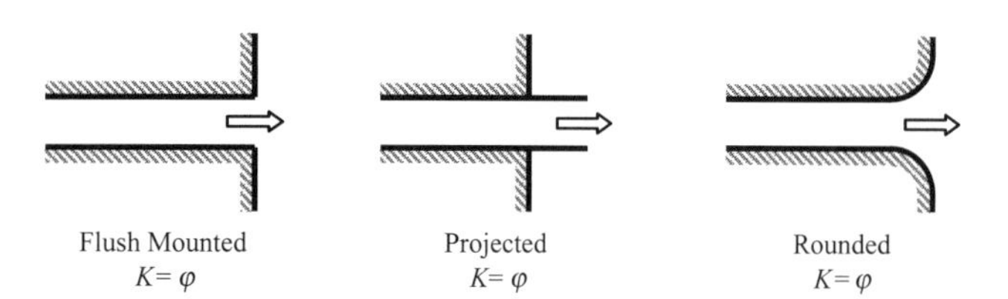

FIGURE 12.2. Straight pipe exit.

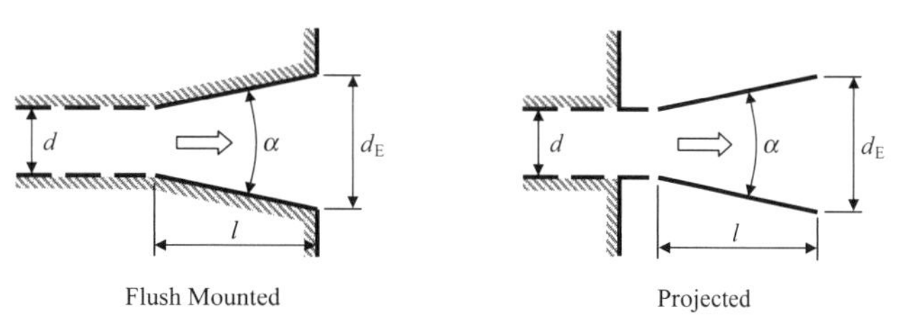

FIGURE 12.3. Discharge from a conical diffuser.

upstream end of the piping system under analysis so that the initial and the exit velocity heads are the same. In this case, assuming a value of 1.0 for ϕ_E at the discharge, along with the generally assigned value of 1.0 for ϕ at the upstream end, is quite adequate.

12.2 DISCHARGE FROM A CONICAL DIFFUSER

Discharge from a conical diffuser into a reservoir is shown in Figure 12.3. The diffuser may be flush mounted to the wall of the reservoir or may be projected into the reservoir.

The loss coefficient of a conical diffuser discharging into a reservoir can be approximately determined by the following equation:

$$K \approx K' + \beta_E^4,$$

where K' is taken as appropriate from Equations 11.8 through 11.10 with β replaced by β_E. The diameter ratio β_E is defined as:

$$\beta_E = \frac{d}{d_E} = \frac{d}{d + 2l\tan(\alpha/2)}.$$

Divergence angles less than 20° provide optimal design, so the following equation adapted from Equation 11.8 is most useful:

$$K \cong 8.30\tan(\alpha/2)(1-\beta_E^2)^2 + \frac{f_d(1-\beta_E^4)}{8\sin(\alpha/2)} + \beta_E^4 \quad (0° \geq \alpha \leq 20°). \tag{12.1}$$

The loss coefficient of discharge from a conical diffuser into a reservoir is shown in Diagrams 12.1–12.3 for friction factors f equal to 0.01, 0.03, and 0.05, respectively.

12.3 DISCHARGE FROM AN ORIFICE

Loss coefficient equations developed in Chapter 13 for various orifice configurations in a transition section can be modified to represent discharge from an orifice into a reservoir by recognizing that d_2 is in effect equal to infinity so that d_o/d_2 goes to zero. The orifice may be flush mounted or projected into the reservoir.

Note that the orifice loss coefficients (K_os) presented in this section are based on the velocity (or flow area) of the orifice restriction. When summing the loss coefficients in a piping stretch they must be transformed to the "standardized" area used in the ΔP formula; usually the pipe flow area (see Section 3.2.3):

$$K = \frac{A^2}{A_o^2}K_o = \frac{d^4}{d_o^4}K_o.$$

12.3.1 Sharp-Edged Orifice

A sharp-edged orifice discharging into a reservoir is shown in Figure 12.4. For this orifice configuration, Equation 13.5 can be transformed into:

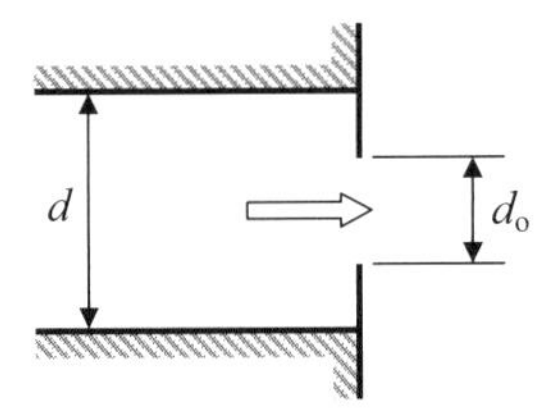

FIGURE 12.4. Discharge from a sharp-edged orifice.

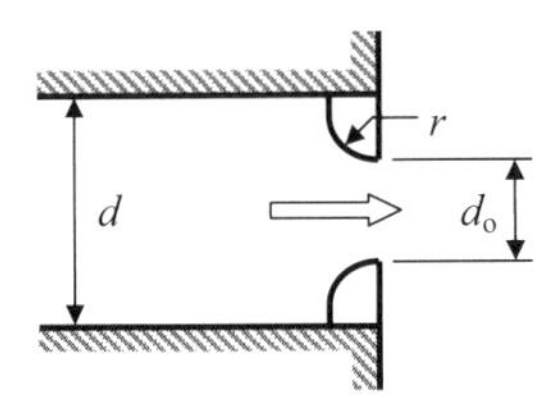

FIGURE 12.5. Discharge from a round-edged orifice.

$$K_o = 0.0696(1-\beta^5)\lambda^2 + \lambda^2, \qquad (12.2)$$

where the diameter ratio $\beta = d_0/d$ and where the jet velocity ratio λ is given by:

$$\lambda = 1 + 0.622(1 - 0.215\beta^2 - 0.785\beta^5). \qquad (12.3)$$

Equation 12.2, the loss coefficient for pipe discharge from a sharp-edged orifice into a reservoir, is depicted as the uppermost curve in Diagrams 12.4–12.6.

12.3.2 Round-Edged Orifice

A round-edged orifice discharging into a reservoir is shown in Figure 12.5. For this case Equation 13.8 can be transformed into:

$$K_o = 0.0696\left(1 - 0.569\frac{r}{d_o}\right)\left(1 - \sqrt{\frac{r}{d_o}}\beta\right)(1-\beta^5)\lambda^2 + \lambda^2, \qquad (12.4)$$

where the jet contraction ratio λ is:

$$\lambda = 1 + 0.622\left(1 - 0.30\sqrt{\frac{r}{d_o}} - 0.70\frac{r}{d_o}\right)^4 (1 - 0.215\beta^2 - 0.785\beta^5) \quad (r/d_o \le 1).$$

Loss coefficients for pipe discharge through a round-edged orifice into a reservoir are shown in Diagram 12.4. The dashed line in Diagram 12.4 represents the boundary where full rounding cannot be achieved by simple circular rounding because of geometry limitations (see Section 13.2).

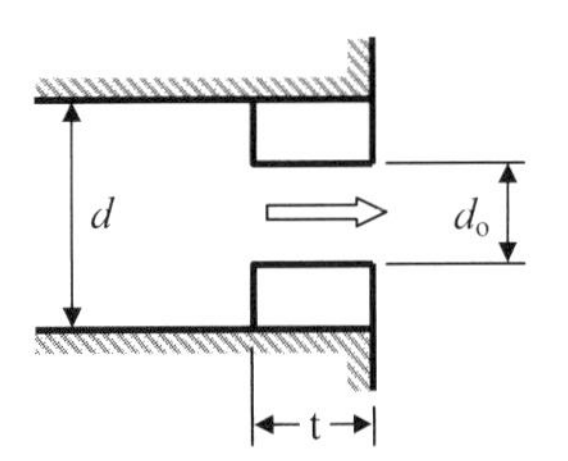

FIGURE 12.6. Discharge from a thick-edged orifice.

12.3.3 Thick-Edged Orifice

A thick-edged orifice discharging into a reservoir is shown in Figure 12.6. For this configuration, Equation 13.16 can be transformed by letting d_o/d_2 go to zero. Thus the loss coefficient for discharge from a bevel-edged orifice where thickness t is equal to or less than $1.6d_o$ becomes:

$$K_o = 0.0696(1-\beta^5)\lambda^2 + C_{th}\lambda^2 + (1 - C_{th})[(\lambda-1)^2 + 1] \quad (t/d_o \le 1.4), \qquad (12.5)$$

where the jet contraction coefficient λ is given by:

$$\lambda = 1 + 0.622(1 - 0.215\beta^2 - 0.785\beta^5), \qquad (12.3, \text{repeated})$$

and where C_{th} is given by:

$$C_{th} = \left\{1 - 0.50\left[\left(\frac{t}{1.4d_o}\right)^{2.5} + \left(\frac{t}{1.4d_o}\right)^3\right]\right\}^{4.5}. \qquad (13.14, \text{repeated})$$

For thickness t equal to or greater than $1.4d_o$, surface friction loss becomes significant and the loss coefficient can be determined from the following equation:

$$K_o = 0.0696(1-\beta^5)\lambda^2 + [\lambda - 1]^2 + 1 + f_o\left(\frac{t}{d_o} - 1.4\right) \quad (t/d_o \le 1.4).$$

Loss coefficients for pipe discharge through a thick-edged orifice into a reservoir for thickness t equal to or less than $1.4d_o$ are shown in Diagram 12.5.

12.3.4 Bevel-Edged Orifice

A bevel-edged orifice discharging into a reservoir is shown in Figure 12.7. Equation 13.12 (bevel-edged orifice in a transition section) can be transformed in this case by letting d_o/d_2 go to zero. Thus the loss coefficient for discharge from a bevel-edged orifice is*:

* In the following equations α is generally expressed in radians; the modifications for using degrees are obvious.

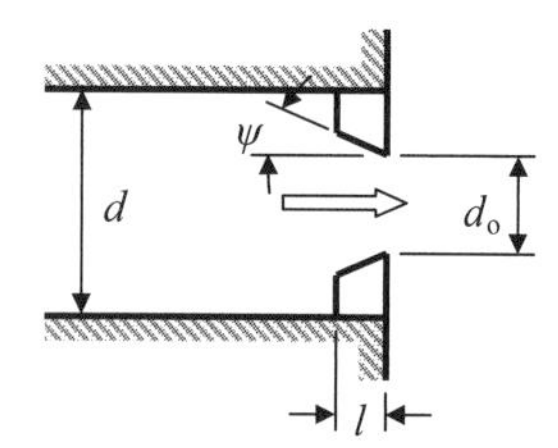

FIGURE 12.7. Discharge from a bevel-edged orifice.

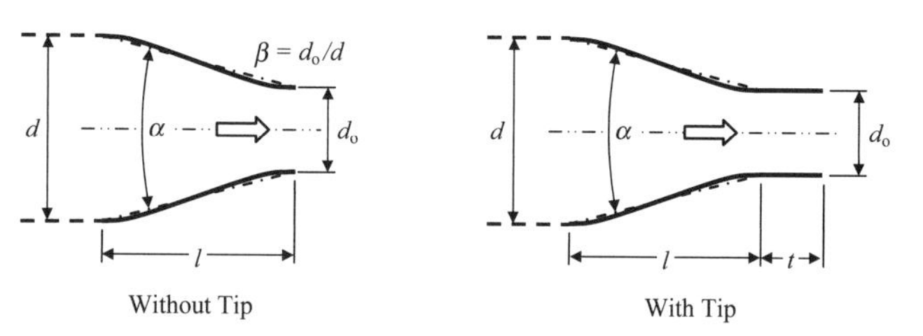

FIGURE 12.8. Discharge from a smooth nozzle.

$$K_2 \approx 0.0696\left(1 - C_b \frac{l}{d_o}\right)\left(1 - 0.42\sqrt{\frac{1}{d_o}}\beta^2\right)(1-\beta^5)\lambda^2 + \lambda^2, \quad (12.6)$$

where the diameter ratio $\beta = d_2/d_1$, where the jet contraction coefficient λ is given by:

$$\lambda = 1 + 0.622\left[1 - C_b \left(\frac{l}{d_o}\right)^{\frac{1-\sqrt[4]{l/d_o}}{2}}\right](1 - 0.215\beta^2 - 0.785\beta^5), \quad (13.10, \text{repeated})$$

and where C_b is the ratio of bevel length l to the length of the maximum bevel possible for given diameter ratio β and included angle α:

$$C_b = \left(1 - \frac{\psi}{90}\right)\left(\frac{\psi}{90}\right)^{\frac{1}{2+l/d_o}} \quad (13.11, \text{repeated})$$

The loss coefficient for pipe discharge from a 45° bevel-edged orifice (a 90° included angle) as a function of length to diameter ratio l/d_o can be obtained from Diagram 12.6. The radial distance available between the upstream pipe wall and the orifice face may limit the actual extent of beveling as shown by the dashed line.

Keep in mind that substantial rounding or chamfering may be applied to the edges of manufactured items. If such is the case for a bevel-edged orifice, the loss coefficient may best be determined by treating it as a round-edged orifice, or as somewhere between a round-edged orifice and a bevel-edged orifice.

12.4 DISCHARGE FROM A SMOOTH NOZZLE

Discharge from a smooth nozzle into a reservoir is shown in Figure 12.8. The nozzle may be flush-mounted to the wall of the reservoir or it may be projected into the reservoir.* Equation 10.16 for surface friction loss in a conical contraction can be transformed into a smooth nozzle discharging into a reservoir, with or without a tip of length t:

$$K_o \approx \frac{f_n(1-\beta^4)}{8\sin(\alpha/2)} + \frac{f_t t}{d_o} + 1,$$

where f_n is the friction factor in the nozzle and f_t is the friction factor at the tip (if there) based on the relative roughness of the surfaces as determined by the hydraulic diameter and Reynolds number at the outlet. The effective included angle α can be determined as:

$$\alpha = 2\,\text{atan}\left(\frac{d - d_o}{2l}\right) = 2\,\text{atan}\left(\frac{1/\beta - 1}{2l/d_o}\right).$$

* Of course the nozzle may be attached to the end of a hose.

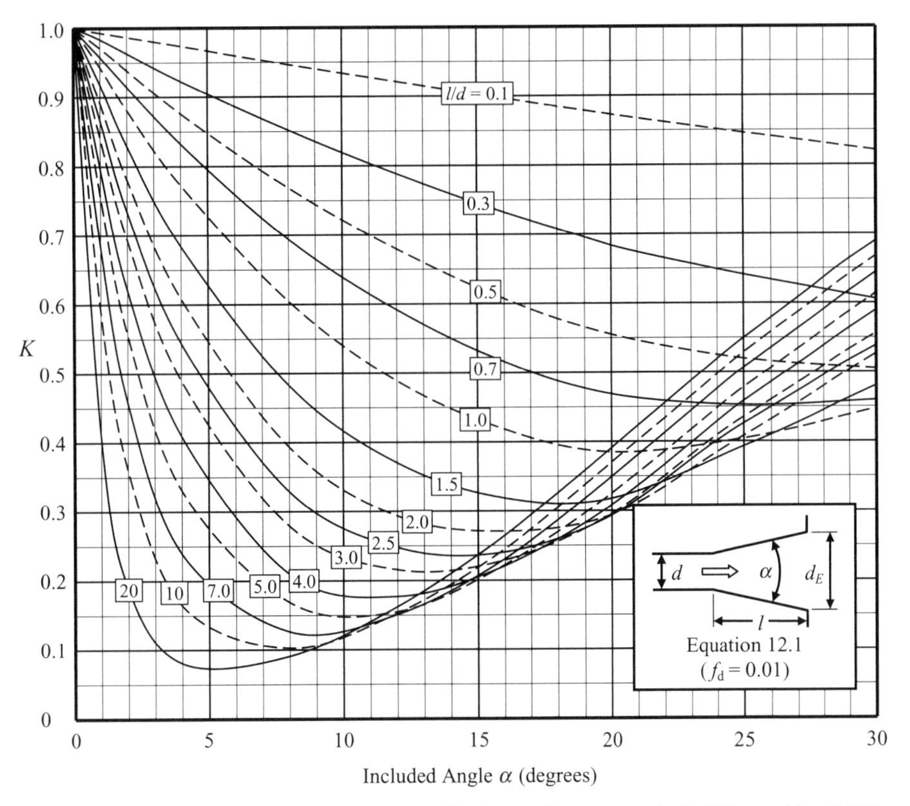

DIAGRAM 12.1. Loss coefficient K for discharge from a conical diffuser ($f_d = 0.01$).

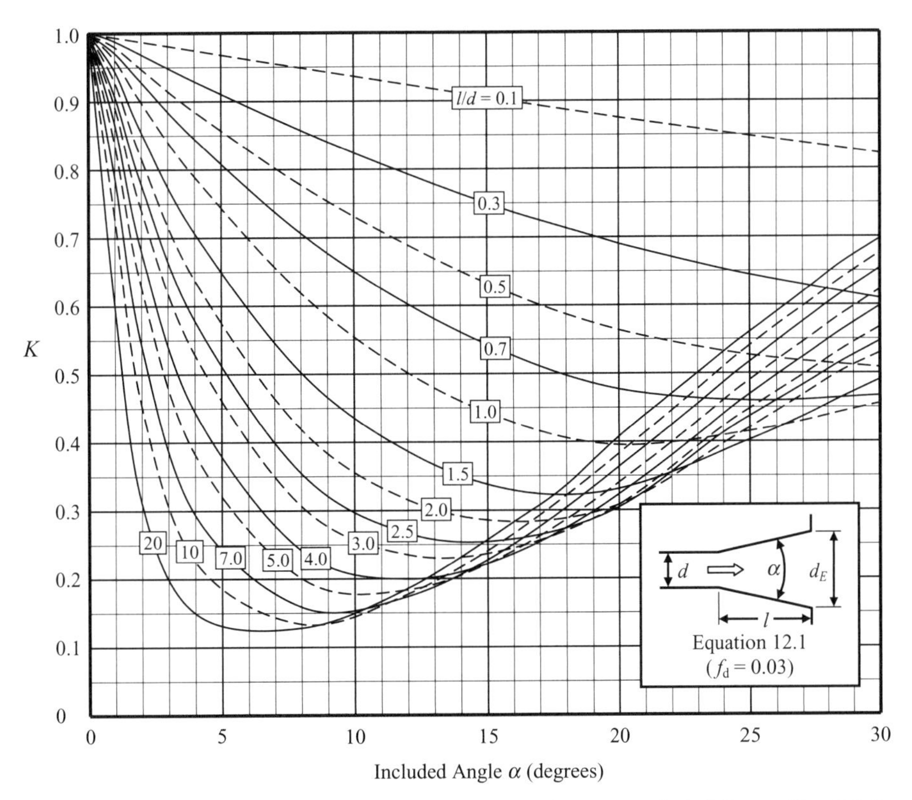

DIAGRAM 12.2. Loss coefficient K for discharge from a conical diffuser ($f_d = 0.03$).

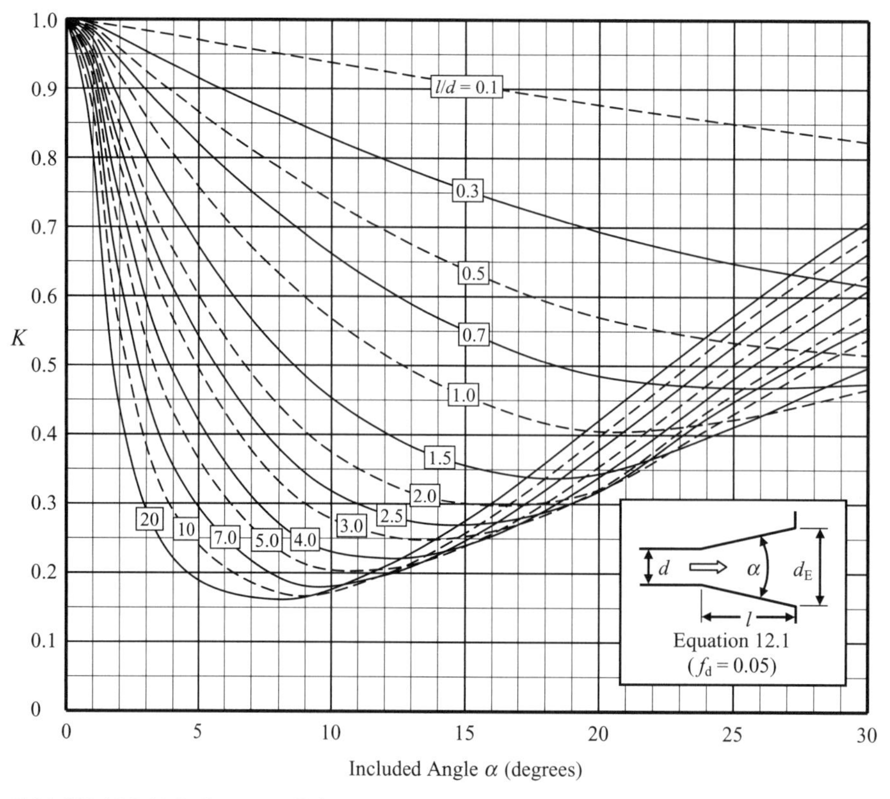

DIAGRAM 12.3. Loss coefficient K for discharge from a conical diffuser ($f_d = 0.05$).

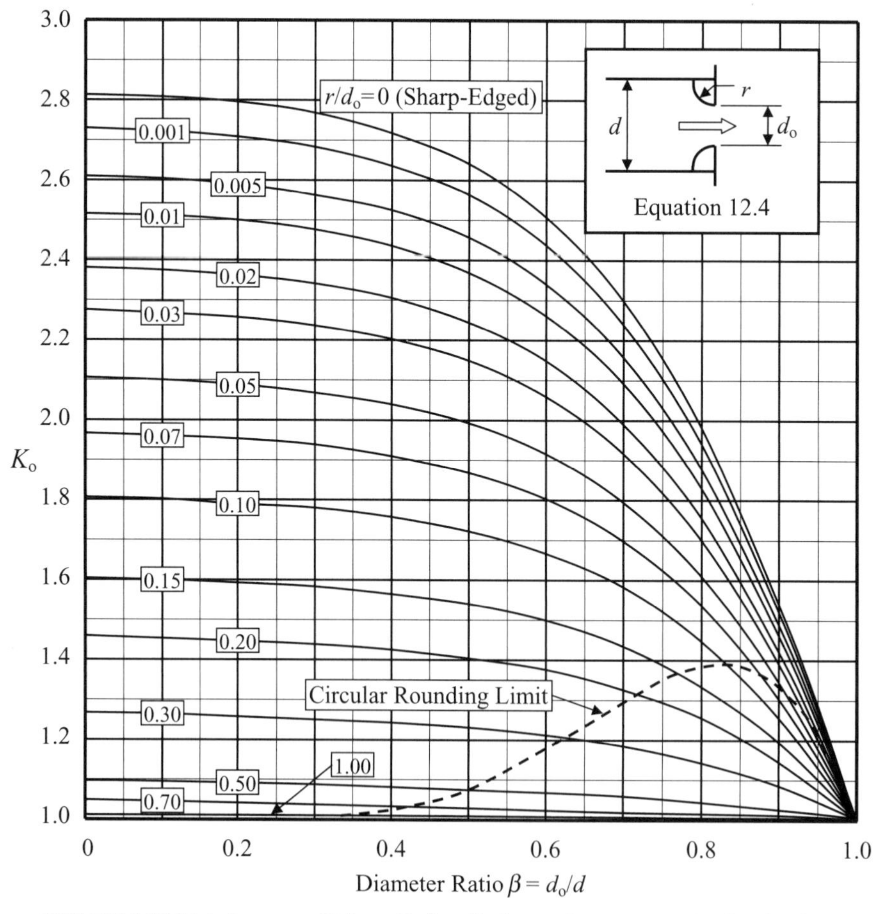

DIAGRAM 12.4. Loss coefficient K_o for discharge from a round-edged orifice.

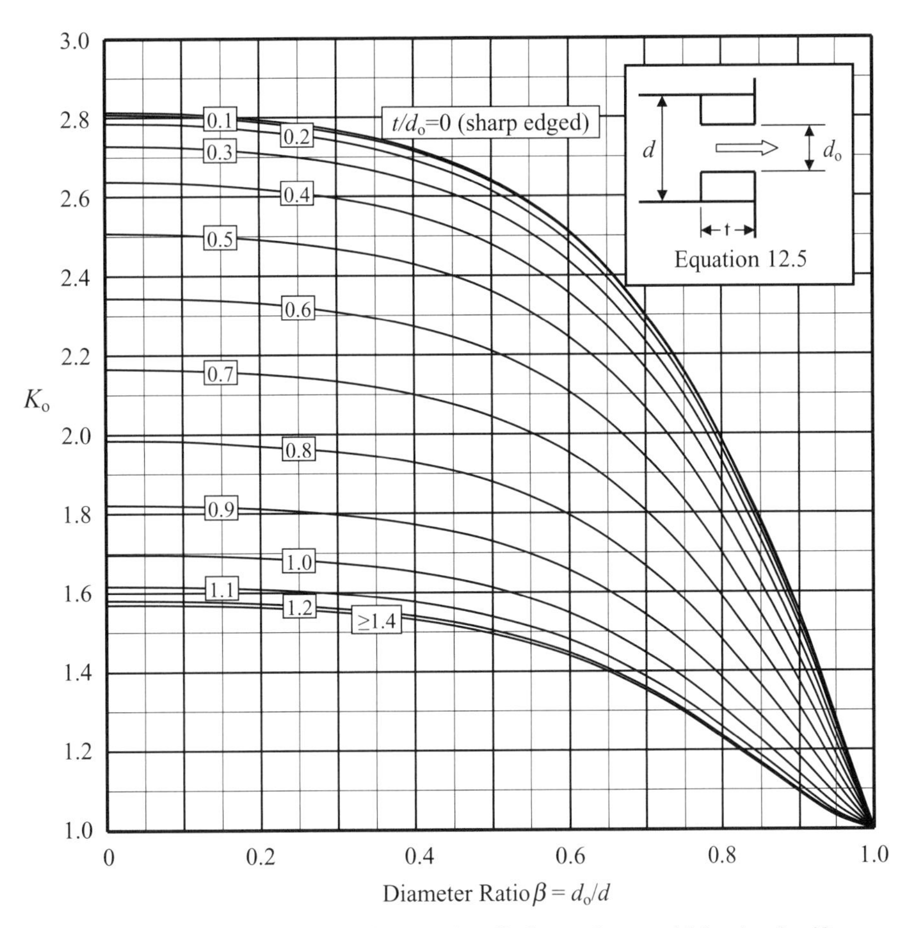

DIAGRAM 12.5. Loss coefficient K_o for discharge from a thick-edged orifice.

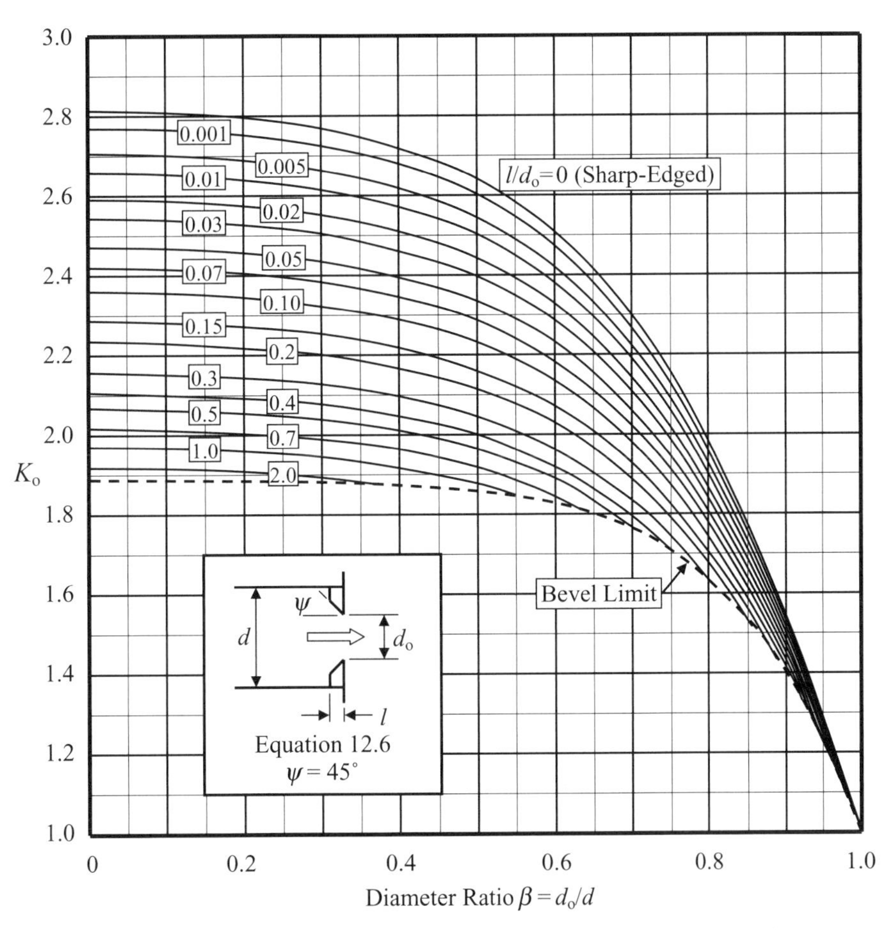

DIAGRAM 12.6. Loss coefficient K_o for discharge from a 45° bevel-edged orifice (90° included angle).

13

ORIFICES

Orifices are widely installed in piping systems and hydraulic machinery to produce a regular and reproducible loss of pressure. Orifices are used to measure flow, to limit flow, or, in branching systems, to balance or otherwise distribute flow. Single-hole orifices are commonly used. Cylindrical tube orifices and multiholed orifices have been tested in an attempt to find improved metering characteristics without much success. Orifices to limit flow are sometimes installed in series to avoid cavitation in low pressure applications.

13.1 GENERALIZED FLOW MODEL

Information on orifices derives largely from thin-plate or sharp-edged orifices that are used extensively for flow measurement. Some information is available on round-edged, bevel-edged, and thick-edged orifices. The essential geometrical similarities between all these types of orifices indicate that they may be considered as members of a single family of constrictions. These constrictions consist of a contraction of the flow area followed by a sudden expansion.

The available pressure drop information has not been treated uniformly in the literature. The common geometrical properties have not been used as the basis for a consistent assessment of the data. In some cases the pressure drop characteristics have been expressed as a discharge coefficient. In other cases, the data have simply been presented as a plot of pressure drop versus flow rate or a plot of pressure drop as a percentage of flow measurement differential pressure.

Here an understanding of the broad physical features of the flow leads to a generalized model of the flow characteristics. The available experimental data are evaluated to develop loss coefficients for the various types of orifices. The data were derived basically from symmetrical circular holes in circular plates, but apply quite well for square holes, square ducts, and small departures from symmetry.

Flow through a sharp-edged orifice is illustrated in Figure 13.1. The flow accelerates as it approaches the orifice. The outer filaments adjacent to the wall achieve a high radially inward velocity of about the same order of magnitude as the axial velocity. The flow separates at the edge of the orifice. The high radially inward velocity causes the jet to contract and form a vena contracta or minimum jet cross section immediately downstream of the orifice. At this point, the separated jet begins to entrain some of the fluid from the recirculation vortex formed between the jet and the pipe wall. The jet decelerates and expands toward the wall until it reattaches and fills the entire pipe. Rounding or beveling the inlet edge of the orifice reduces or prevents the formation of the vena contracta. The total loss through the orifice is thus reduced.

The conventional and preferred use of orifices is to locate the center of the orifice on the centerline of the pipe. Eccentric and segmental orifices are suitable when the fluid carries a considerable amount of sediment or material in suspension. The eccentric orifice opening is installed tangent to the bottom surface of the pipe to allow passage of sediment or material. On the other hand, the segmental orifice plate is installed at the top

Pipe Flow: A Practical and Comprehensive Guide, First Edition. Donald C. Rennels and Hobart M. Hudson.

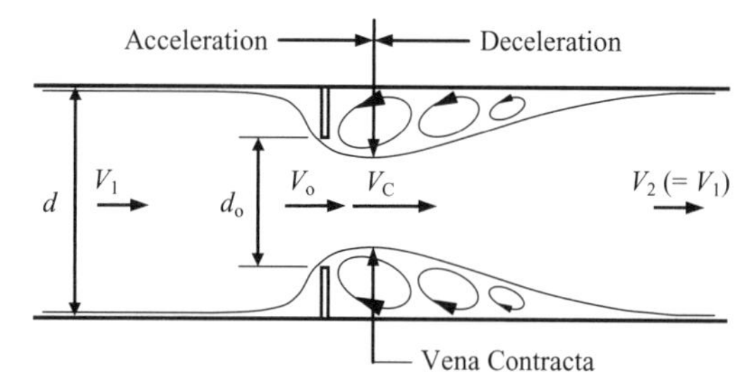

FIGURE 13.1. Orifice flow.

of the pipe, leaving the bottom open. Eccentric and segmental orifices ostensibly have higher discharge coefficients, and presumably lower loss coefficients, than centered orifices but the difference is small and may be ignored.

Orifice losses are sensitive to upstream and downstream conditions. The American Society of Mechanical Engineers (ASME) Fluid Meters Report [1] gives recommended minimum lengths of straight pipe preceding and following orifices to limit flow measurement errors due to interference of less than 0.5%. Head loss is no doubt similarly affected and the ASME recommendations may be used as a guide.

Taking the total head loss of an orifice as the sum of the losses in the acceleration and deceleration regions and treating them as a gradual contraction and sudden enlargement, respectively, gives:

$$H_o = K_o \frac{V_o^2}{2g} = k_{acc} \frac{V_C^2}{2g} + \frac{(V_C - V_2)^2}{2g}.$$

Rearrangement gives:

$$K_o = k_{acc} \frac{V_2^2}{V_o^2} + \left(\frac{V_C}{V_o} - \frac{V_2}{V_o} \right)^2.$$

The velocity ratio V_C/V_o can be defined as the jet velocity coefficient λ (see Section 10.2), and the equation becomes:

$$K_o = k_{acc}\lambda^2 + \left(\lambda - \frac{V_2}{V_o} \right)^2.$$

By use of the continuity equation $\dot{w} = \rho A V$, the equation becomes:

$$K_o = k_{acc}\lambda^2 + \left(\lambda - \frac{A_o}{A_2} \right)^2. \tag{13.1}$$

Equation 13.1 is the universal case, where the upstream and downstream pipe sizes are *not* the same. In the case of an orifice in a straight pipe, the downstream flow area A_2 is equal to the upstream flow area A_1, and the loss coefficient equation becomes:

$$K_o = k_{acc}\lambda^2 + \left(\lambda - \beta^2\right)^2, \tag{13.2}$$

where $\beta = d_o/d$, the ratio of the orifice diameter to the diameter of the pipe The loss coefficients of various orifice configurations can be expressed by employing appropriate expressions for k_{acc} and λ based on available test data.

Undoubtedly, the universal velocity profile exists at the vena contracta as well as in the fully developed flow regions in the upstream and downstream pipes. Nonetheless, the assumption is made that the velocity profile is uniform in the pipe and at the vena contracta. The justification of this simplification for the various orifice configurations is demonstrated in the following sections through successfully developing formulations that accurately match available test data.

Note that the orifice loss coefficients (K_os) developed in this chapter are based on the area (or diameter) of the orifice restriction. When summing the loss coefficients in a piping stretch they must be modified by the ratio of the "standardized" area (used in the ΔP formula) to the orifice area (see Section 3.2.3).

13.2 SHARP-EDGED ORIFICE

The problem of flow through a sharp-edged or thin-plate orifice has long been of interest due to its practical use in flow measurement. The ASME Fluid Meters Report [1] assigns the loss for sharp-edged orifices and flow nozzles as a percentage of the measured pressure differential across the orifice meter. The objective here is to accurately express the overall loss coefficient on the essential geometry of the orifice independent of any flow measurement function.

13.2.1 In a Straight Pipe

Sharp-edged orifices in a straight pipe are shown in Figure 13.2. According to the ASME Fluid Meters Report, the edge width t of the cylindrical surface of the orifice itself should be $d_o/8$ or between $0.01d$ and $0.02d$, whichever is smaller. If the thickness of the orifice plate exceeds the minimum, perhaps for structural reasons, the outlet corner of the orifice should then be beveled at an angle of about 45° to the face of the plate sufficiently to provide the minimum face width. Face widths in excess of the minimum can be evaluated as a thick-edged orifice (see Section 13.4).

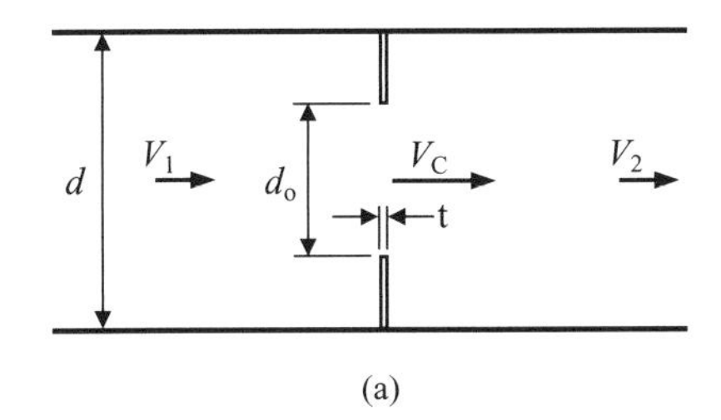

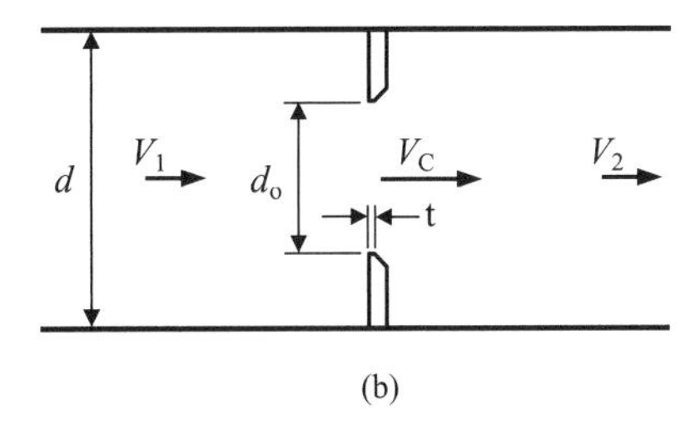

(a) (b)

FIGURE 13.2. Sharp-edged orifice in a straight pipe. (a) Thin plate; (b) thick plate (with beveled outlet).

The upstream or inlet edge of the orifice must be square, sharp, and free from any rounding or beveling. It must be free from burrs, nicks, or wire edges. Early on, it was found that even very slight rounding (or chamfering) of the inlet edge had a significant effect on orifices. It was found that it was not practical to attempt to give any values for the discharge coefficients* of orifices with slightly rounded edges. It was very difficult, if not impossible, to measure the amount of this rounding without destroying the orifice plate, and the amount of rounding had a very definite effect on the value of the coefficient.

Because the effect of rounding (or beveling) was not easily determined, "sharp edged" was frequently defined as one whose inlet edge would not appreciably reflect a beam of light when viewed without magnification. In recent years more sophisticated methods of measuring edge sharpness have been used and a sharp edge has been defined as one of radius $r \le 0.0004d_o$. One method that seems quite suitable for obtaining accurate measurements is to obtain an edge impression by pressing a soft metal disk against the inlet edge of the orifice plate. Several impressions are taken at equally spaced points on the orifice plate. The edge impressions are magnified and projected onto a viewing screen where templates are used to measure the radii of the projected images. Even so, an investigation by Crockett and Upp [2] indicates that the coefficient of discharge C_d may deviate from the ASME Fluid Meters values by as much as 1% or 2% when the edge radius r is equal to or less than $0.0004d_o$. Thus, rounding, however slight, has a significant effect on orifice discharge coefficients, and can be expected to have a similar effect on orifice loss coefficients.

Based on the flow model developed in Section 13.1, the following expression was derived for the loss coefficient of a sharp-edged orifice in a straight pipe:

$$K_o = 0.0696\left(1-\beta^5\right)\lambda^2 + \left(\lambda-\beta^2\right)^2, \qquad (13.3)$$

where the diameter ratio $\beta = d_o/d$, and where the jet velocity ratio λ is given by:

$$\lambda = 1 + 0.622(1 - 0.215\beta^2 - 0.785\beta^5). \qquad (13.4)$$

Equation 13.3 is compared to sharp-edged orifice test data in Figure 13.3. The equation closely matches and tends to bound the test data. Loss coefficients for sharp-edged orifices in a straight pipe are shown as the upper curve in Diagrams 13.2 through 13.8.

13.2.2 In a Transition Section

Sharp-edged orifices in a transition section, where the upstream and downstream pipe sizes are *not* the same, are illustrated in Figure 13.4.

For the universal, or transition case, the loss coefficient equation becomes:

$$K_o = 0.0696\left(1-\beta^5\right)\lambda^2 + \left(\lambda - \left(\frac{d_o}{d_2}\right)^2\right)^2, \qquad (13.5)$$

where, as before, the diameter ratio $\beta = d_o/d_1$ and the jet velocity ratio λ is given by:

$$\lambda = 1 + 0.622(1 - 0.215\beta^2 - 0.785\beta^5). \qquad (13.4, repeated)$$

Sharp-edged orifice loss coefficients in a transition section are shown in Diagram 13.1.

13.2.3 In a Wall

A sharp-edged orifice in a wall between infinite flow areas is illustrated in Figure 13.5.

In this passage from one large volume to another, the effective diameters, d_1 and d_2 are in effect equal to infinity; thus $\beta = 0$ and Equation 13.4 reduces to $\lambda = 1.622$. Also $d_o/d_2 = 0$, so that substitution into Equation 13.5 yields the following result for a sharp-edged orifice in a wall:

$$K_o = 2.81.$$

* The coefficient of discharge is defined by the equation: C_d = actual rate of flow/theoretical rate of flow.

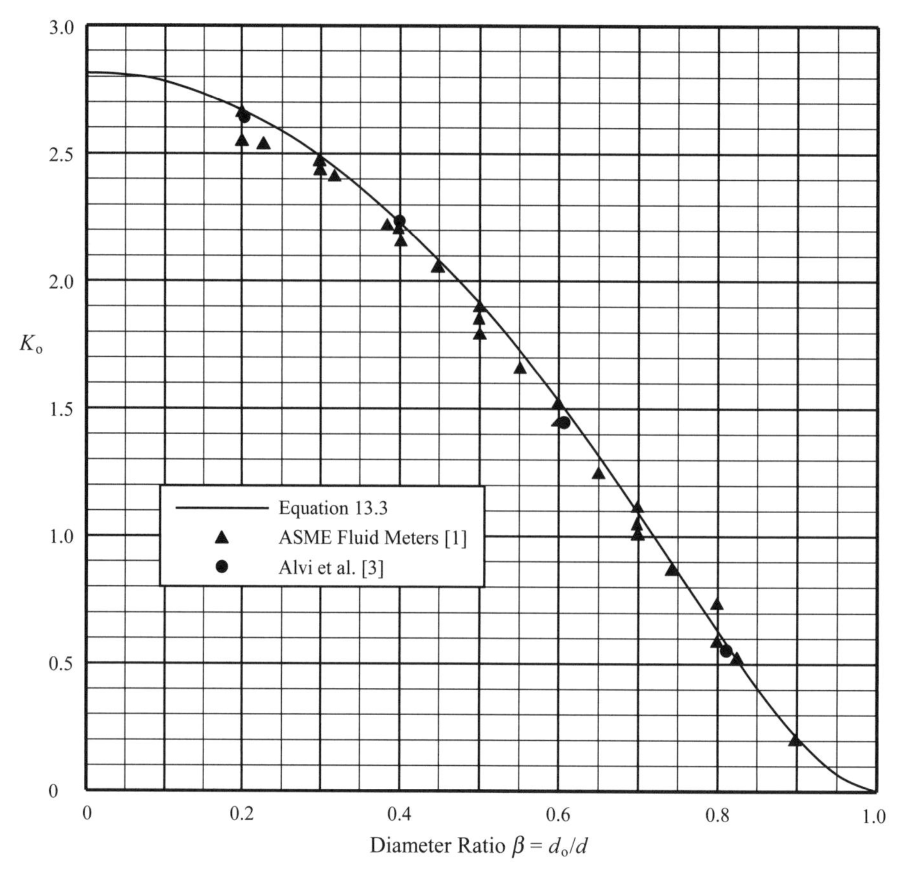

FIGURE 13.3. Sharp-edged orifice in a straight pipe—comparison to test data.

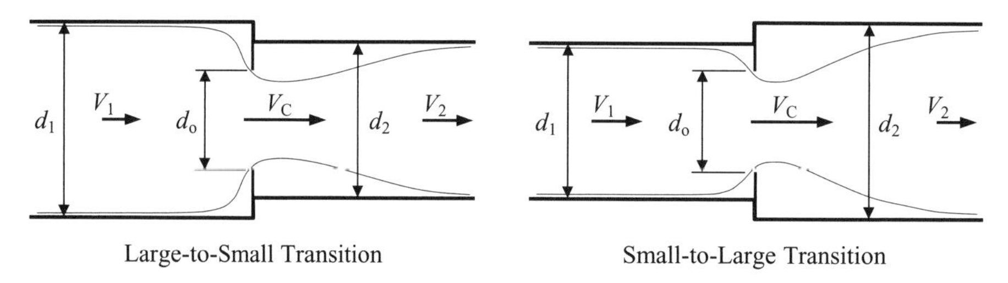

FIGURE 13.4. Sharp-edged orifice in a transition section.

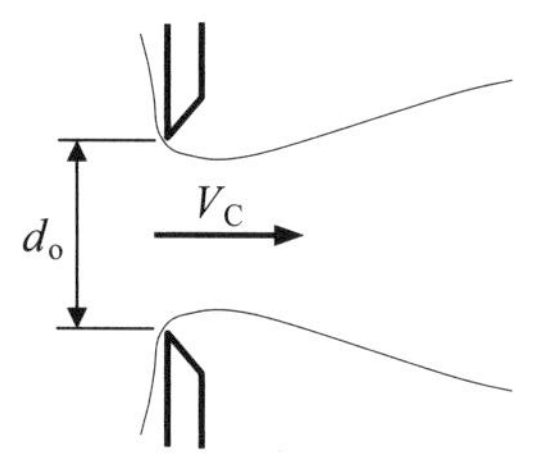

FIGURE 13.5. Sharp-edged orifice in a wall.

13.3 ROUND-EDGED ORIFICE

Rounding of the leading edge of an orifice can considerably diminish or eliminate the vena contracta and thus substantially reduce the head loss. For a circular edge, the rounding radius r is simply the radius of the quarter circle. In cases where the amount of rounding is limited by the radial distance available between the pipe wall and the orifice face, rounding may take the form of an ellipse or other curved shape. In the case of such noncircular edges, the rounding radius r can be expressed as:

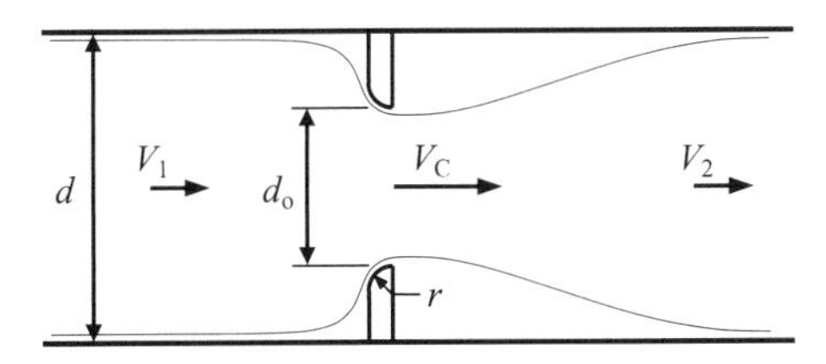

FIGURE 13.6. Round-edged orifice in a straight pipe.

$$r = \sqrt[3]{r_1^2 r_2}, \qquad (10.5, \text{repeated})$$

where r_1 and r_2 are the semimajor (longitudinal) and semiminor (radial) axes, respectively.

13.3.1 In a Straight Pipe

The following expression was developed for the loss coefficient of a round-edged orifice in a straight pipe (see Fig. 13.6) when the rounding ratio r/d_o is equal to or less than 1:

$$K_o = 0.0696\left(1 - 0.569\frac{r}{d_o}\right)\left(1 - \sqrt{\frac{r}{d_o}}\beta\right)(1-\beta^5)\lambda^2 + (\lambda - \beta^2)^2 \quad (r/d_o \le 1), \qquad (13.6)$$

where the diameter ratio $\beta = d_o/d$ and where the jet contraction coefficient λ is given by:

$$\lambda = 1 + 0.622\left(1 - 0.30\sqrt{\frac{r}{d_o}} - 0.70\frac{r}{d_o}\right)^4 (1 - 0.215\beta^2 - 0.785\beta^5). \qquad (13.7)$$

In the case of a generously rounded orifice where r/d_o is equal to or greater than 1, the jet contraction ratio λ equals 1 and the loss coefficient becomes:

$$K_o = 0.030(1-\beta)(1-\beta^5) + (1-\beta^2)^2 \quad (r/d_o > 1).$$

The above expressions, in the basic form set up by Equation 13.2, were derived by empirically curve fitting to available test data from Alvi et al. [3], as shown in Figure 13.7.* Equation 13.6 matches the test data quite well. Beyond $r/d_o = 0.20$, at r/d_o approaching 1.0, Equation 13.6 closely matches data for ASME flow nozzles that, despite a short cylindrical throat, are, in effect, rounded orifices.† It should be noted that most sources indicate little or no reduction in head loss beyond $r/d_o = 0.20$. The comparisons in Figure 13.7 belie that notion.

Loss coefficients of round-edged orifices in a straight pipe can be determined from Diagram 13.2. The curve for $r/d_o = 0.0004$, the recommended limiting value of edge sharpness for metering orifices, shown as a dashed line, is evidence that the loss coefficient is extremely sensitive to even very slight rounding of the inlet edge. This may well account for the scatter in data for professed sharp-edged orifices, as well as the scatter in data for sharp-edged contractions and entrances. This may be the reason why the maximum value of a sharp-edged entrance has been reported as 0.50 or lower by many sources.

Because of this sensitivity, "sharp-edged" orifices are frequently *not* sharp-edged. One anecdote told by a fluid mechanics class instructor was that their lab experiments with sharp-edged orifices were giving strange, inconsistent data. Upon investigation it was found that the machinist making the orifices for the experiments was doing what any good machinist would do—touching the inlet and outlet edges with a file to break the sharp edge! Wherever "sharp-edged" orifices are used, it is recommended that the true inlet edge radius be determined accurately, if possible, and the loss coefficient de-rated to a likely finite value if necessary.

The lower dashed line in Diagram 13.2 defines the boundary where simple circular rounding is limited by the radial distance available between the pipe wall and the orifice face. Below this line, rounding must take the form of an ellipse or other curved shape in accordance with Equation 10.5 in order to obtain a further reduction in loss. The diameter ratio β_{limit} at which circular rounding is limited by geometry is given by:

$$\beta_{limit} = \frac{1}{1 + 2\frac{r}{d_o}}.$$

13.3.2 In a Transition Section

This is the case where the upstream and downstream pipe sizes are *not* the same. Large-to-small and small-to-large transitions are shown in Figure 13.8. In this case, the loss coefficient for a round-edged orifice becomes:

$$K_o = 0.0696\left(1 - 0.569\frac{r}{d_o}\right)\left(1 - \sqrt{\frac{r}{d_o}}\beta\right)(1-\beta^5)\lambda^2 + \left(\lambda - \left(\frac{d_o}{d_2}\right)^2\right)^2 \quad (r/d_o \le 1), \qquad (13.8)$$

* The test data are for Reynolds number at 10,000—above this the loss coefficient would be substantially constant.

† Note that Equation 10.5 was successfully applied to determine the effective radius R of the elliptical shape of the ASME flow nozzles in Chapter 14.

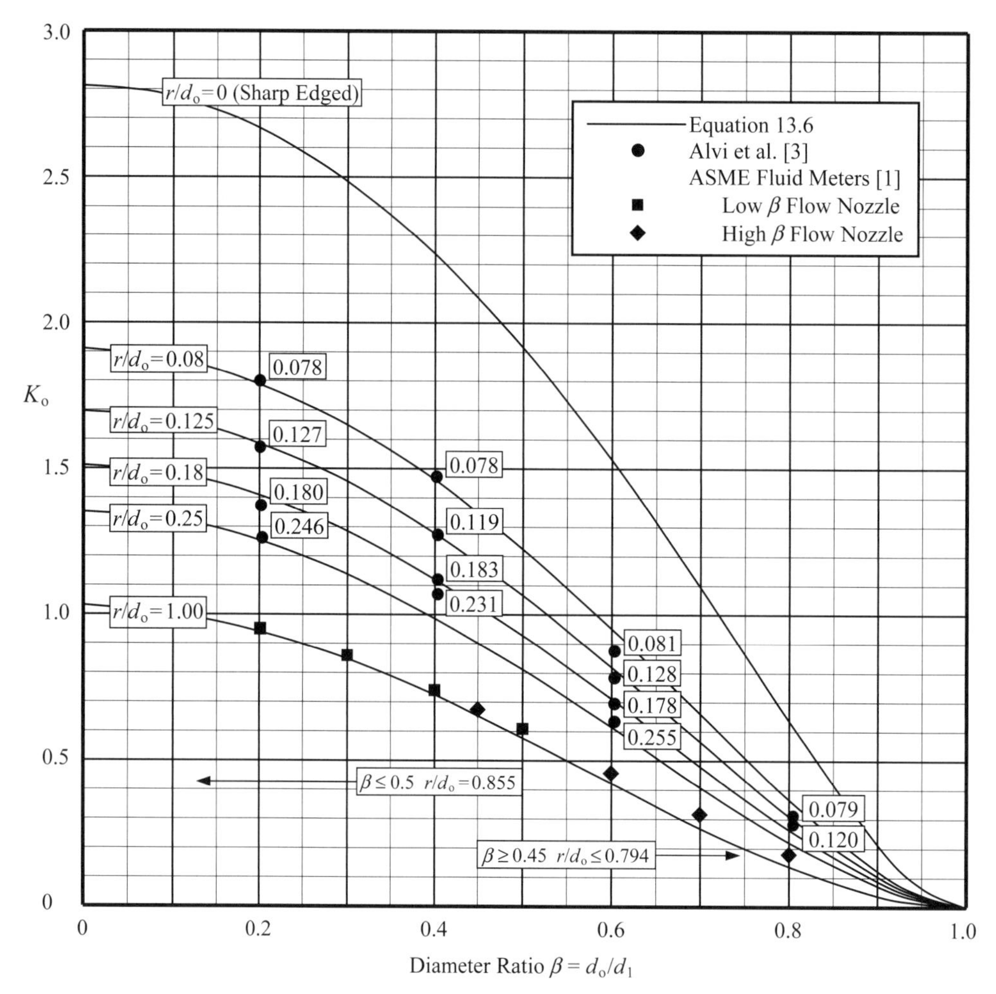

FIGURE 13.7. Round-edged orifice in a straight pipe—comparison to data.

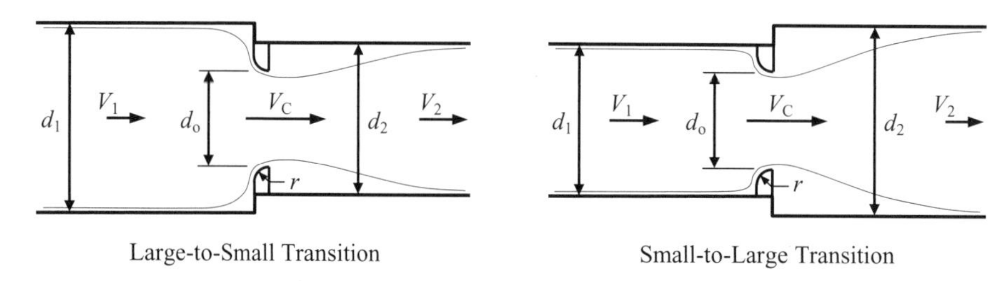

FIGURE 13.8. Round-edged orifice in a transition section.

where the diameter ratio $\beta = d_o/d_1$, and where the jet contraction coefficient λ is given by:

$$\lambda = 1 + 0.622\left(1 - 0.30\sqrt{\frac{r}{d_o}} - 0.70\frac{r}{d_o}\right)^4 \left(1 - 0.215\beta^2 - 0.785\beta^5\right). \quad (13.7, repeated)$$

For the case of a generously rounded orifice where r/d_o is equal to or greater than 1, the jet contraction ratio $\lambda = 1$ and the loss coefficient becomes:

$$K_o = 0.030(1-\beta)(1-\beta^5) + \left[1 - \left(\frac{d_o}{d_2}\right)^2\right]^2 \quad (r/d_o \ge 1).$$

As was the case for a rounded orifice in a straight pipe, the radial distance available between the upstream pipe wall and the orifice face may limit the actual amount of rounding.

13.3.3 In a Wall

A round-edged orifice in a wall between infinite flow areas is shown in Figure 13.9. In this passage from one large volume to another, the diameters d_1 and d_2 are effectively equal to infinity so that Equation 13.7 reduces to:

$$K_o = \left[0.0696\left(1-0.569\frac{r}{d_o}\right)+1\right]\lambda^2,$$

where the jet contraction coefficient λ is given by:

$$\lambda = 1+0.622\left(1-0.30\sqrt{\frac{r}{d_o}}-0.70\frac{r}{d_o}\right).$$

For values of the loss coefficient as a function of rounding ratio r/d_o, see Table 13.1.

13.4 BEVEL-EDGED ORIFICE

Beveling (or chamfering) the inlet edge of an orifice reduces the head loss. The important parameters are the nondimensional bevel length-to-orifice diameter ratio, l/d_o, and the bevel angle α.

13.4.1 In a Straight Pipe

A bevel-edged orifice in a straight pipe is shown in Figure 13.10. The following approximate equation was developed for a contraction with a bevel of length-to-orifice diameter ratio, l/d_o, less than or equal to 1:

$$K_o = 0.0696\left(1-C_b\frac{l}{d_o}\right)\left(1-0.42\sqrt{\frac{l}{d_o}}\beta^2\right)(1-\beta^5)\lambda^2 + (\lambda-\beta^2)^2, \quad (13.9)$$

where the diameter ratio $\beta = d_o/d$, where the jet contraction coefficient λ is given by:

$$\lambda = 1+0.622\left[1-C_b\left(\frac{l}{d_o}\right)^{\frac{1-\sqrt[4]{l/d_o}}{2}}\right](1-0.215\beta^2-0.785\beta^5), \quad (13.10)$$

and where C_b, a function of bevel angle ψ in degrees, and bevel length to diameter ratio l/d_o, is given by:

$$C_b = \left(1-\frac{\psi}{90}\right)\left(\frac{\psi}{90}\right)^{\frac{1}{2+l/d_o}}. \quad (13.11)$$

The above expressions are related to similar expressions developed for beveled contractions (see Section 9.3) and beveled entrances (see Section 8.3).

Loss coefficients for bevel angles ψ of 5°, 15°, 30°, 45°, 60°, and 75° can be approximately determined from Diagrams 13.3 through 13.8. The lower dashed line in each figure defines the boundary where beveling is limited by the radial distance available between the pipe wall and the orifice face. The diameter ratio β_{limit} at which beveling is limited is given by:

$$\beta_{limit} = \frac{1}{1+2\frac{l}{d_o}\tan(\psi)}.$$

13.4.2 In a Transition Section

As shown in Figure 13.11, this is the case where the upstream and downstream pipe sizes are *not* the same. For this case, the loss coefficient for bevel-edged orifices becomes:

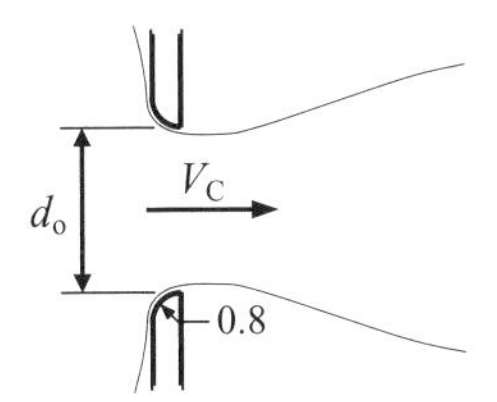

FIGURE 13.9. Round-edged orifice in a wall.

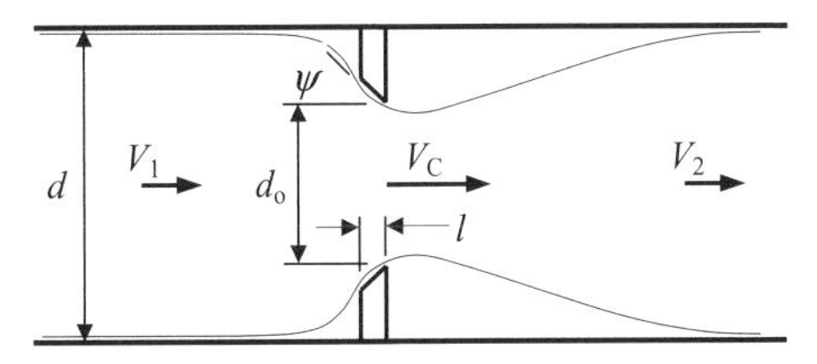

FIGURE 13.10. Bevel-edged orifice in a straight pipe.

TABLE 13.1. Loss Coefficient K_o for a Round-Edged Orifice in a Wall

r/d_o	0	0.0004	0.001	0.005	0.01	0.02	0.03	0.04	0.05	0.06
K_o	2.81	2.76	2.73	2.61	2.52	2.38	2.28	2.19	2.11	2.04
r/d_o	0.08	0.10	0.12	0.15	0.20	0.25	0.30	0.40	0.50	1.00
K_o	1.91	1.81	1.72	1.61	1.46	1.35	1.27	1.16	1.10	1.03

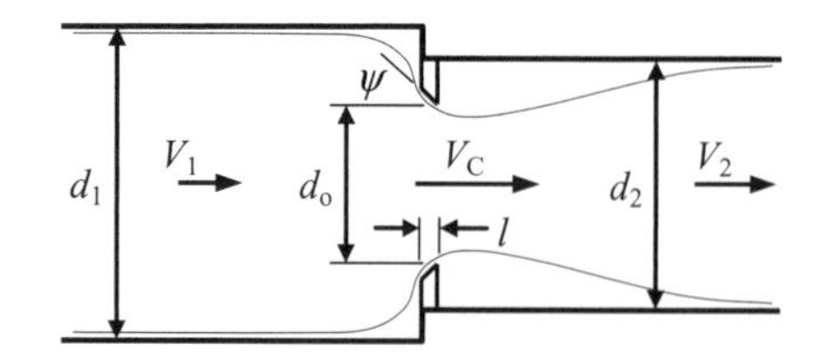

Large-to-Small Transition

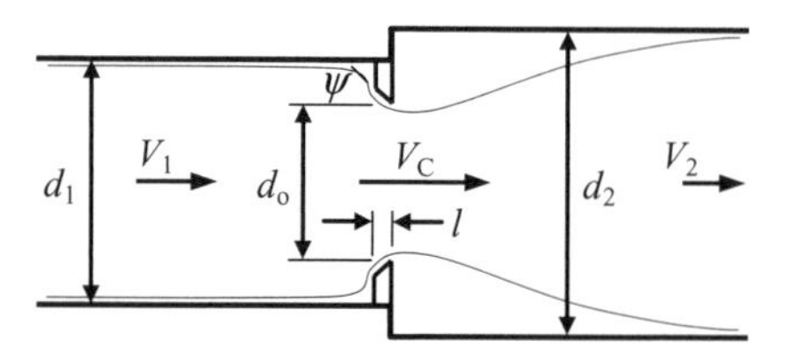

Small-to-Large Transition

FIGURE 13.11. Bevel-edged orifice in a transition section.

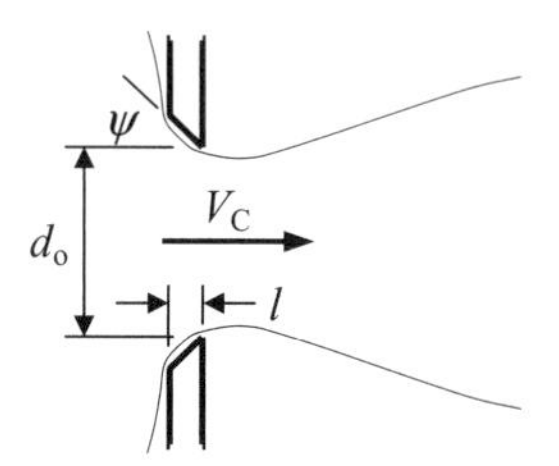

FIGURE 13.12. Bevel-edged orifice in a wall.

$$K_o \approx 0.0696\left(1-C_b\frac{l}{d_o}\right)\left(1-0.42\sqrt{\frac{l}{d_o}}\beta^2\right)\left(1-\beta^5\right)\lambda^2+\left(\lambda-\left(\frac{d_o}{d_2}\right)^2\right)^2, \qquad (13.12)$$

where the diameter ratio $\beta = d_o/d_1$, where the jet contraction coefficient λ is given by:

$$\lambda = 1+0.622\left[1-C_b\left(\frac{l}{d_o}\right)^{\frac{1-\sqrt[4]{l/d_o}}{2}}\right]\left(1-0.215\beta^2-0.785\beta^5\right), \qquad (13.10, \text{repeated})$$

and where C_b is given by:

$$C_b = \left(1-\frac{\psi}{90}\right)\left(\frac{\psi}{90}\right)^{\frac{1}{2+l/d_o}}. \qquad (13.11, \text{repeated})$$

Again, the radial distance available between the upstream pipe wall and the orifice face may limit the actual extent of beveling.

13.4.3 In a Wall

A bevel-edged orifice in a wall between infinite flow areas is shown in Figure 13.12. In this passage from one large volume to another, the diameters d_1 and d_2 are effectively infinite so that Equation 13.9 reduces to:

$$K_o \approx 0.0696\left(1-C_b\frac{l}{d_o}\right)\lambda^2+\lambda^2,$$

where the diameter ratio $\beta = d_o/d_1$, where the jet contraction coefficient λ is given by:

$$\lambda = 1+0.622\left[1-C_b\left(\frac{l}{d_o}\right)^{\frac{1+\sqrt[4]{l/d_o}}{2}}\right],$$

and where C_b is given by:

$$C_b = \left(1-\frac{\psi}{90}\right)\left(\frac{\psi}{90}\right)^{\frac{1}{2+l/d_o}}. \qquad (13.11, \text{repeated})$$

13.5 THICK-EDGED ORIFICE

The important parameter of the thick-edged (or square-edged) orifice is the nondimensional orifice thickness to diameter ratio, t/d_o. For a vanishingly thin thickness ($t/d_o \to 0$), the orifice acts as a sharp-edged orifice. For a wide orifice thickness ($t/d_o \geq 1.4$), the thick-edged orifice acts simply as a sudden contraction followed by a sudden expansion. The performance of the orifice between these two extremes is investigated below.

13.5.1 In a Straight Pipe

A thick-edged orifice in a straight pipe is shown in Figure 13.13. The upstream edge is sharp—free from any rounding or chamfering. The flow breaks away from the surface of the orifice to form a discrete jet which contracts to a minimum flow region at the vena contracta. Downstream from the vena contracta the flow expands to finally rejoin the duct wall within about six duct diameters from the constriction. For a thin orifice, illustrated in Figure 13.13a, the flow fully separates from the orifice surface throughout its journey. For a thick orifice, illustrated in Figure 13.13b, the flow attaches to the orifice surface at a distance of about $0.8d_o$ from the orifice entrance and eventually separates at the downstream face of the orifice. From there, the flow undergoes a sudden enlargement before finally rejoining the duct wall.

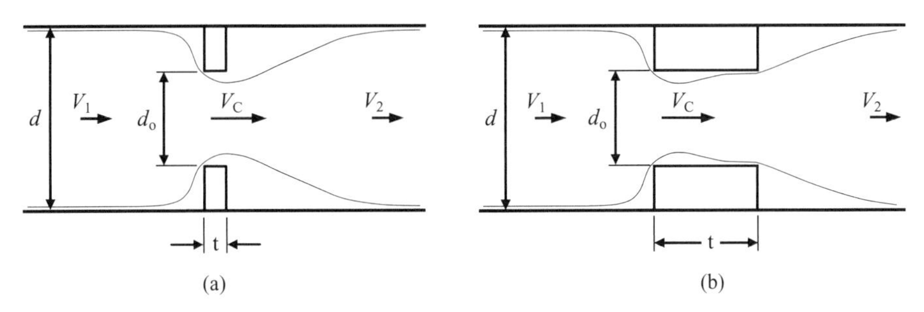

FIGURE 13.13. Thick-edged orifice in a straight pipe. (a) Separated flow; (b) attached flow.

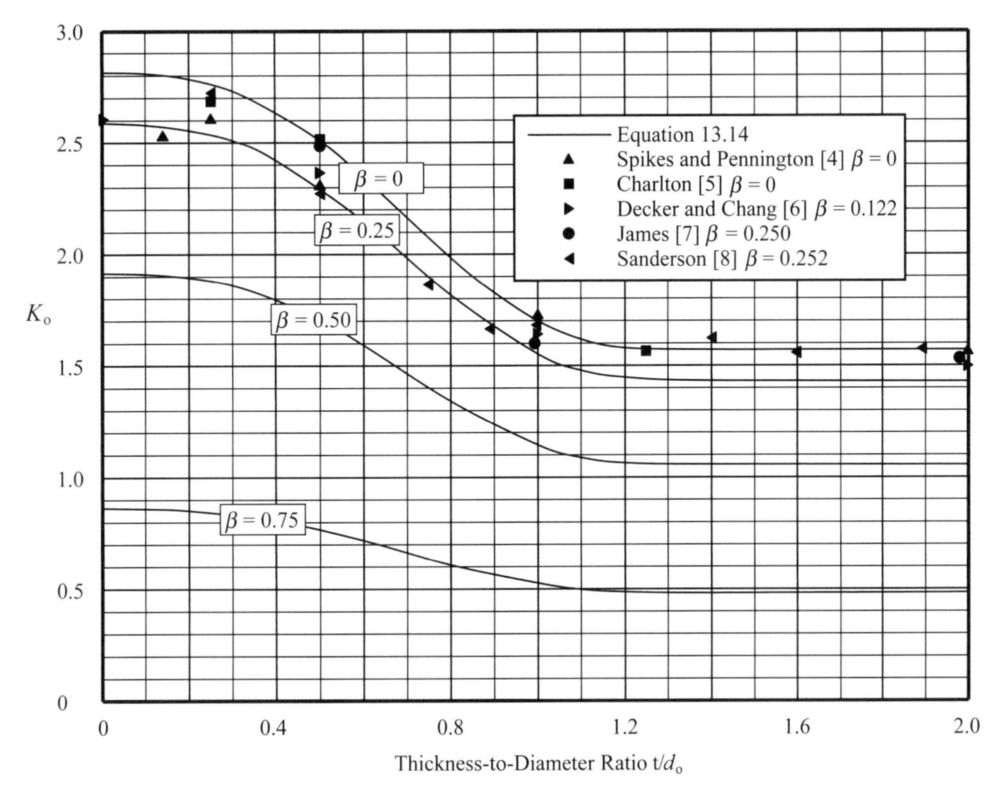

FIGURE 13.14. Thick-edged orifice—comparison to test data.

The loss coefficient of local resistance for thickness t equal to or less than $1.4d_o$ can be determined from the following equation:

$$K_o = 0.0696(1-\beta^5)\lambda^2 + C_{th}(\lambda-\beta^2)^2 + (1-C_{th})\left[(\lambda-1)^2+(1-\beta^2)^2\right] \quad (t/d_o \le 1.4), \tag{13.13}$$

where the jet contraction coefficient λ is given by:

$$\lambda = 1+0.622(1-0.215\beta^2-0.785\beta^5), \quad \text{(13.4, repeated)}$$

and where C_{th} is given by:

$$C_{th} = \left[1-0.50\left(\frac{t}{1.4d_o}\right)^{2.5}-0.50\left(\frac{t}{1.4d_o}\right)^3\right]^{4.5}. \tag{13.14}$$

For thickness t greater than $1.4d_o$, the orifice acts as a sudden contraction followed by a sudden expansion and surface friction loss becomes significant. For this case, the loss coefficient can be determined from the following equation, where f_o is the friction factor of the cylindrical surface of the orifice:

$$K_o = 0.0696(1-\beta^5)\lambda^2 + (\lambda-1)^2 + (1-\beta^2)^2 + f_o\left(\frac{t}{d_o}-1.4\right) \quad (t/d_o > 1.4). \tag{13.15}$$

Equations 13.13 and 13.15 (without friction loss) are presented in Figure 13.14. Note that the results at the left (at t/d_o equals zero) represent sharp-edged orifice performance that has already been demonstrated, and

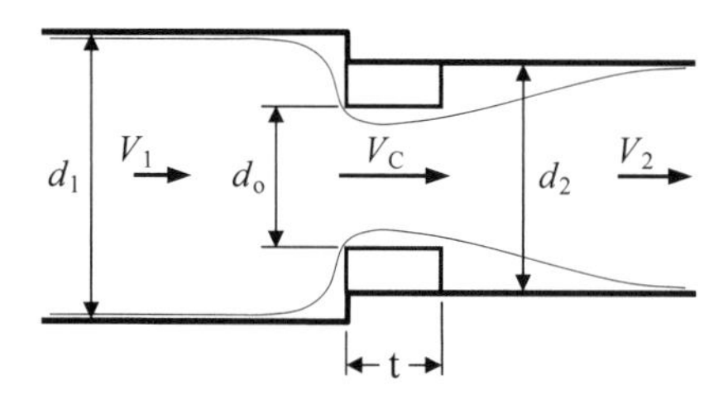

Large-to-Small Transition

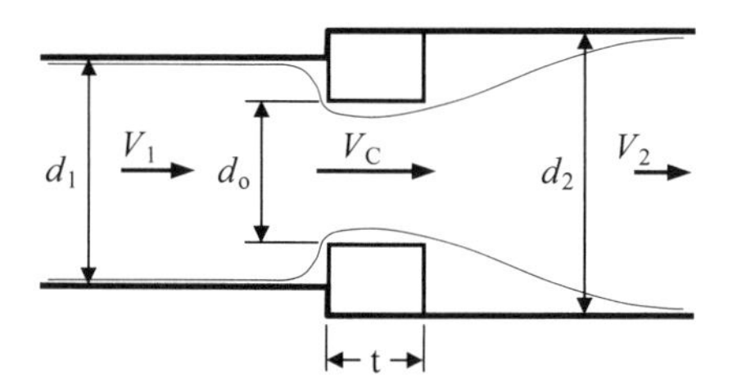

Small-to-Large Transition

FIGURE 13.15. Thick-edged orifice in a transition section.

that the results at the right (at t/d_o greater than 1.4) simply correspond to a sudden contraction followed by a sudden expansion. The comparison with test data from various sources shows that Equation 13.13 effectively represents the transition between these two regions in the range of $0 \leq \beta \geq 0.25$. Adequacy of the equation at higher beta ratios is assured because the generalized flow model developed in Section 13.1 was employed in deriving Equation 13.13.

Loss coefficients of thick-edged orifices in a straight pipe can be determined from Diagram 13.9. The ASME Fluid Meters Report specifies that the face width t of the cylindrical surface of a sharp edged orifice should be $d_o/8$ ($0.125d_o$) or between $0.01d_1$ and $0.02d_1$, whichever is smaller. From Diagram 13.8 we can see that there is only a very modest departure from sharp-edged values until the thickness t far exceeds $0.1d_o$.

13.5.2 In a Transition Section

As shown in Figure 13.15, this is the case where the upstream and downstream pipe sizes are *not* the same. For this case, the loss coefficient for thick-edged orifices where thickness t is equal to or less than $1.4d_o$ becomes:

$$K_o = 0.0696(1-\beta^5)\lambda^2 + C_{th}\left(\lambda - \left(\frac{d_o}{d_2}\right)^2\right)^2 +$$
$$(1-C_{th})\left[(\lambda-1)^2 + \left(1-\left(\frac{d_o}{d_2}\right)^2\right)^2\right] \quad (\text{t}/d_o \leq 1.4), \tag{13.16}$$

where the jet contraction coefficient λ is given by:

$$\lambda = 1 + 0.622(1 - 0.215\beta^2 - 0.785\beta^5), \quad \text{(13.4, repeated)}$$

and where C_{th} is given by:

$$C_{th} = \left[1 - 0.50\left(\frac{\text{t}}{1.4d_o}\right)^{2.5} - 0.50\left(\frac{\text{t}}{1.4d_o}\right)^3\right]^{4.5}. \quad \text{(13.14, repeated)}$$

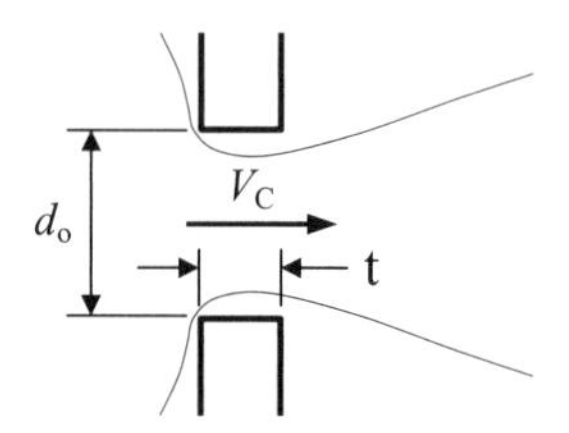

FIGURE 13.16. Thick-edged orifice in a wall.

For thickness t equal to or greater than $1.4d_o$, surface friction loss becomes significant and the loss coefficient can be determined from the following equation:

$$K_o = 0.0696(1-\beta^5)\lambda^2 + (\lambda-1)^2 + \left[1-\left(\frac{d_o}{d_2}\right)^2\right]^2 +$$
$$f_o\left(\frac{\text{t}}{d_o} - 1.4\right) \quad (\text{t}/d_o \geq 1.4).$$

13.5.3 In a Wall

A thick-edged orifice in a wall is illustrated in Figure 13.16. The effective diameters d_1 and d_2 are infinite so that Equation 13.16 for thickness t equal to or less than $1.4d_o$ reduces to:

$$K_o = 0.0696(1-\beta^5)\lambda^2 + C_{th}\lambda^2 + (1-C_{th})\left[(\lambda-1)^2 + 1\right]$$
$$(\text{t}/d_o \leq 1.4),$$

where the diameter ratio $\beta = d_o/d_1$, and where the jet contraction coefficient λ is given by:

$$\lambda = 1 + 0.622(1 - 0.215\beta^2 - 0.785\beta^5), \quad \text{(13.4, repeated)}$$

and where C_{th} is given by:

$$C_{th} = \left[1 - 0.50\left(\frac{\text{t}}{1.4d_o}\right)^{2.5} - 0.50\left(\frac{\text{t}}{1.4d_o}\right)^3\right]^{4.5}. \quad \text{(13.14, repeated)}$$

TABLE 13.2. Loss Coefficient K_o for a Thick-Edged Orifice in a Wall

t/d_o	0	0.1	0.2	0.3	0.4	0.5	0.6	0.7
K_o	2.81	2.81	2.78	2.71	2.62	2.50	2.35	2.19
t/d_o	0.8	0.9	1.0	1.1	1.2	1.3	≥1.4	
K_o	2.02	1.87	1.75	1.66	1.60	1.58	1.57 + f(f_o,)	

For thickness t greater than $1.4d_o$, the loss coefficient can be determined from the following equation, where f_o is the friction factor in the cylindrical surface of the orifice:

$$K_o = 1.57 + f_o\left(\frac{t}{d_o} - 1.4\right) \quad (t/d_o \geq 1.4).$$

Loss coefficients of thick-edged orifices in a wall as a function of t/d_o are shown in Table 13.2.

13.6 MULTIHOLE ORIFICES

The important nondimensional parameter for multihole orifices is the porosity ϕ, the ratio of the total cross-sectional area of the orifice holes to the total cross-sectional area of the duct. The relationship between porosity and diameter ratio β is given by:

$$\phi = \left(\frac{d_o}{d_1}\right)^2 = \beta^2 \quad \text{or} \quad \beta = \sqrt{\phi}.$$

More often than not, the geometry of the orifice holes is the same or similar. In that case, simply substitute $\sqrt{\phi}$ for β in the applicable loss coefficient equation.

Perforated plate may be treated in this manner. Typically, the perforations are punched into the plate. The punch produces slightly rounded or beveled edges on the side of the plate that the punch enters, and produces sharp, outward projecting edges on the side of the plate that the punch exits.* Thus the pressure drop will depend on the direction of flow through the perforated plate. If flow is to enter through the sharp, outward projecting side of the plate, treat it as a thick-edged orifice. If flow is to enter through the rounded or beveled side of the plate, treat it as a rounded or beveled orifice. It may be difficult, if not impossible; to accurately measure the amount of rounding or beveling so judgment may be necessary.

If the geometry of the holes is greatly dissimilar, consider treating them as parallel paths (see Section 5.2).

13.7 NONCIRCULAR ORIFICES

The orifice loss coefficient equations in this chapter were primarily derived from data on symmetrical circular holes in circular passages. However, they apply quite well for square holes and passages, and for small departures from symmetry. They can be applied to other odd flow shapes and to larger departures from symmetry with reasonable accuracy when specific data are unavailable. Substitute $\sqrt{\phi}$ for β in the various orifice loss coefficient equations as described in Section 13.6 above.

* You can easily determine the entry and exit sides by running your fingers over the surface.

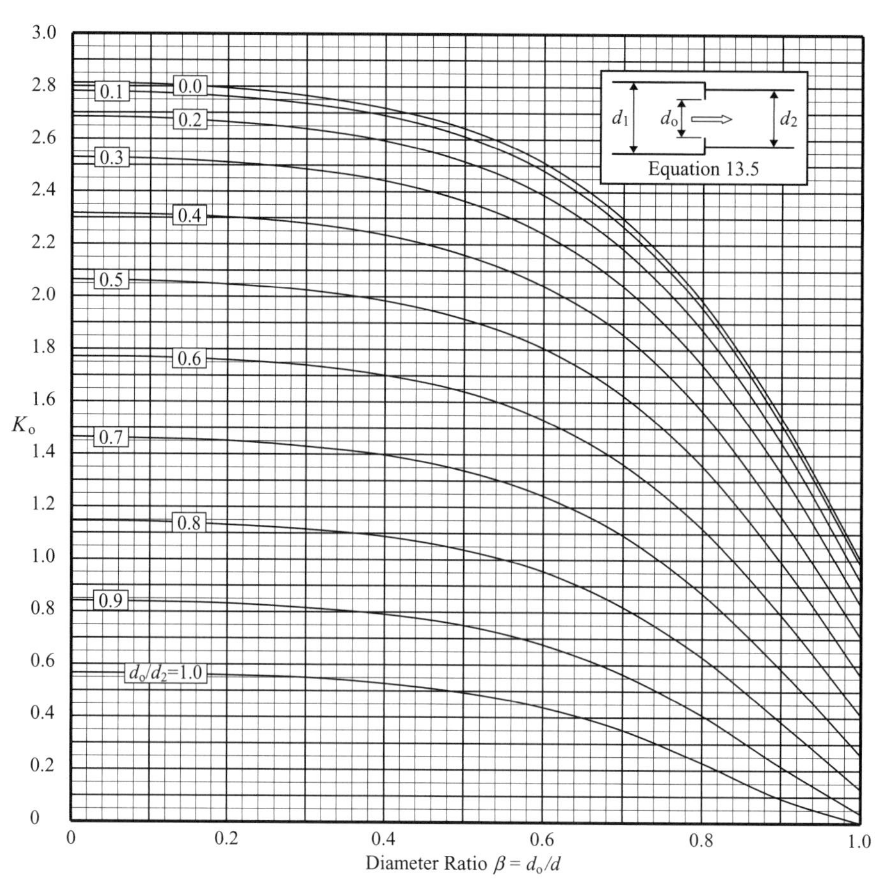

DIAGRAM 13.1. Loss coefficient K_o for sharp-edged orifice in a transition section.

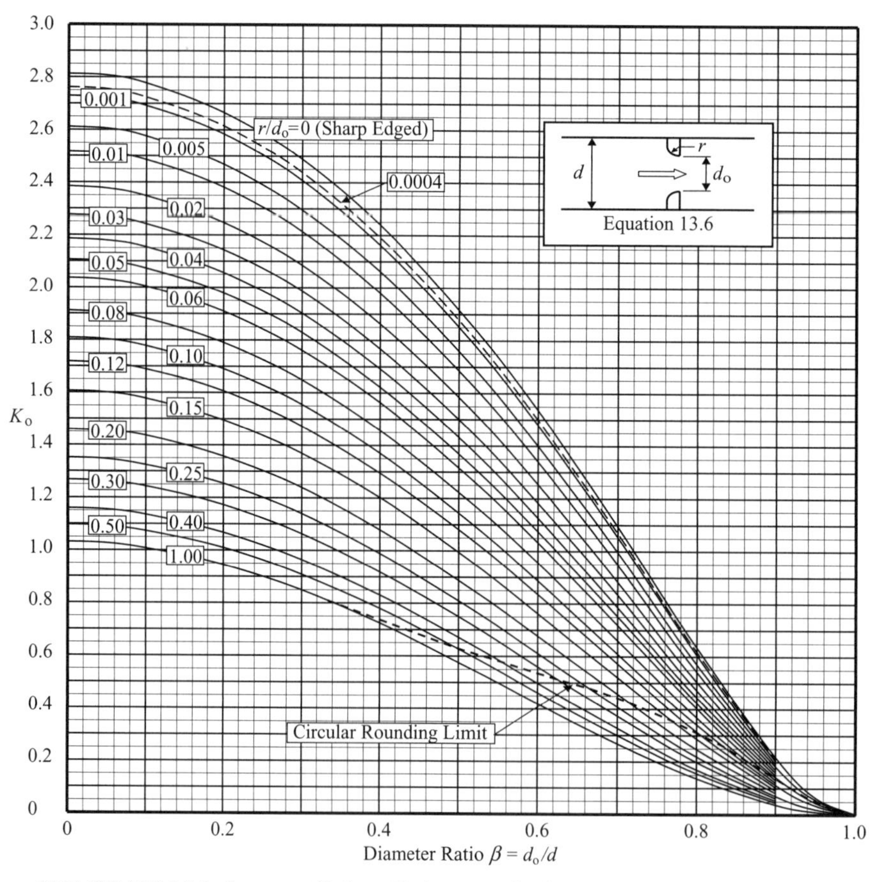

DIAGRAM 13.2. Loss coefficient K_o for round-edged orifice in a straight pipe.

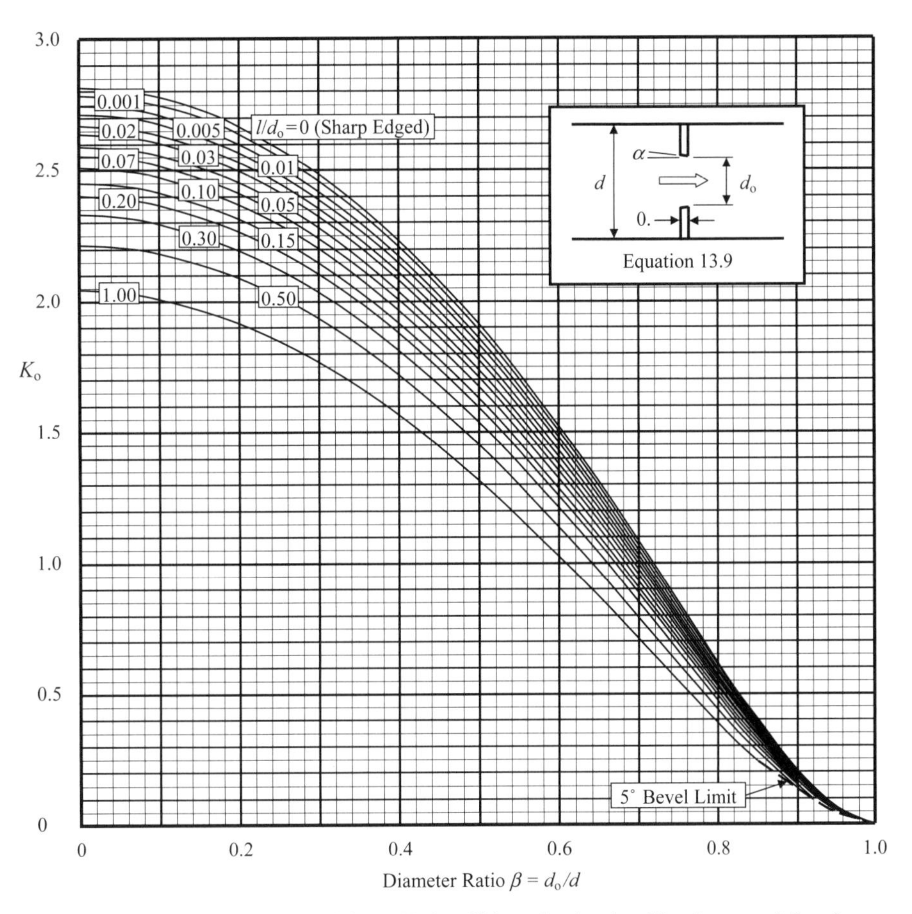

DIAGRAM 13.3. Loss coefficient K_o for 5° bevel-edged orifice in a straight pipe.

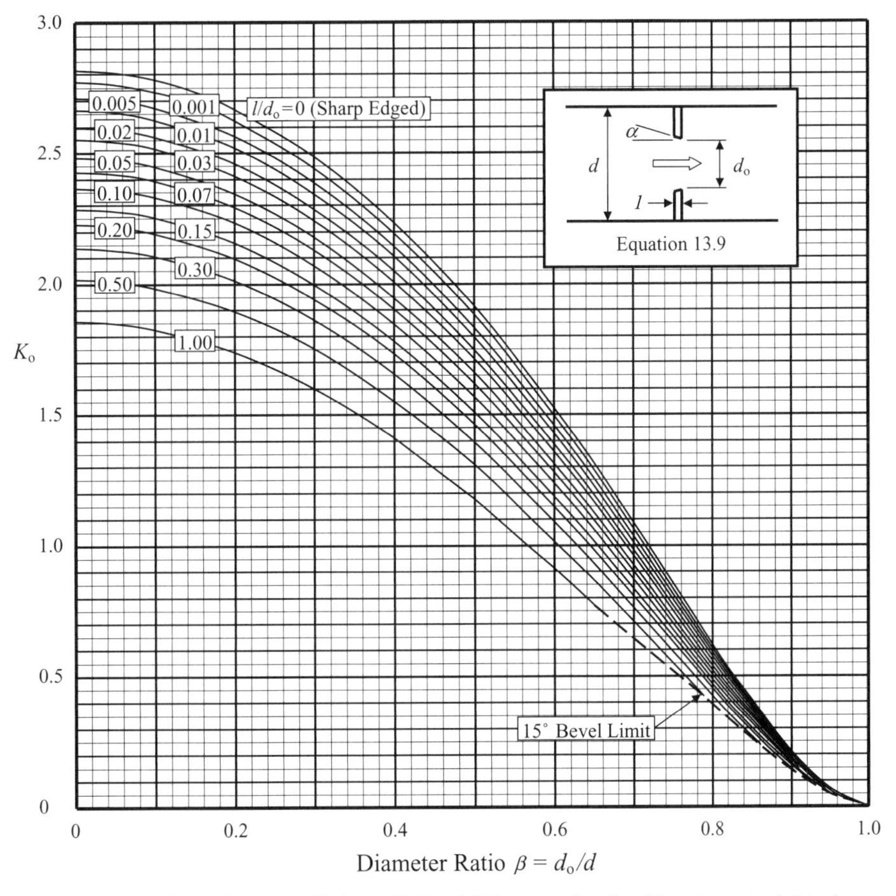

DIAGRAM 13.4. Loss coefficient K_o for 15° bevel-edged orifice in a straight pipe.

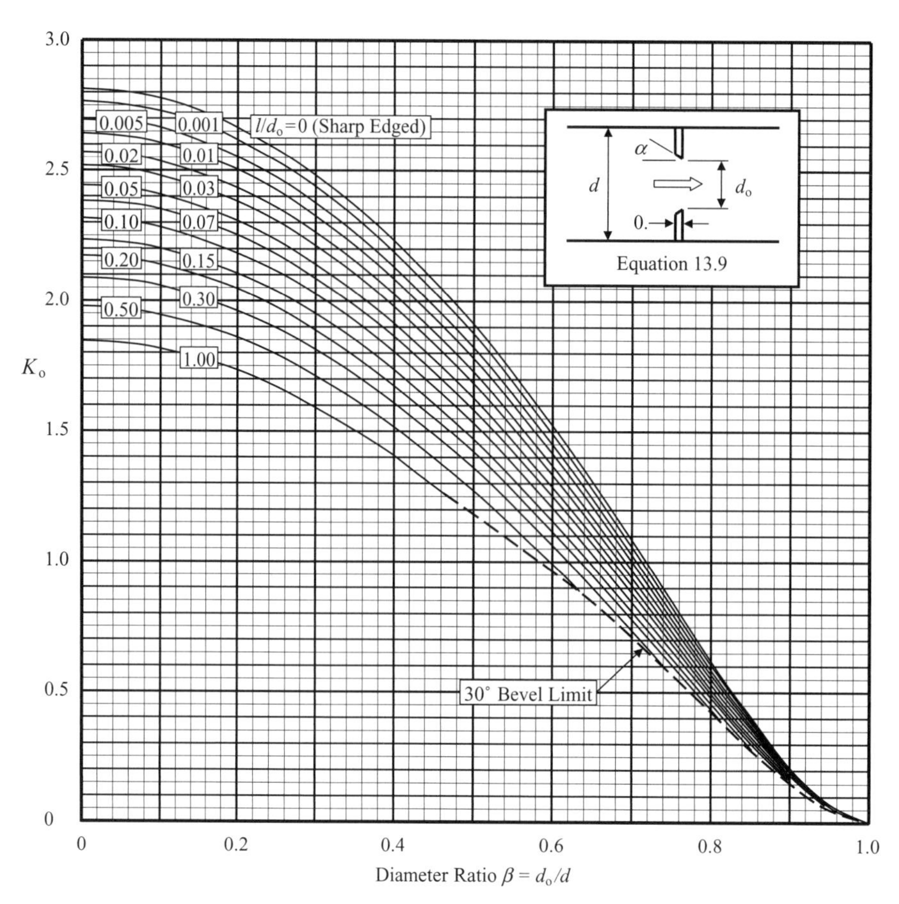

DIAGRAM 13.5. Loss coefficient K_o for 30° bevel-edged orifice in a straight pipe.

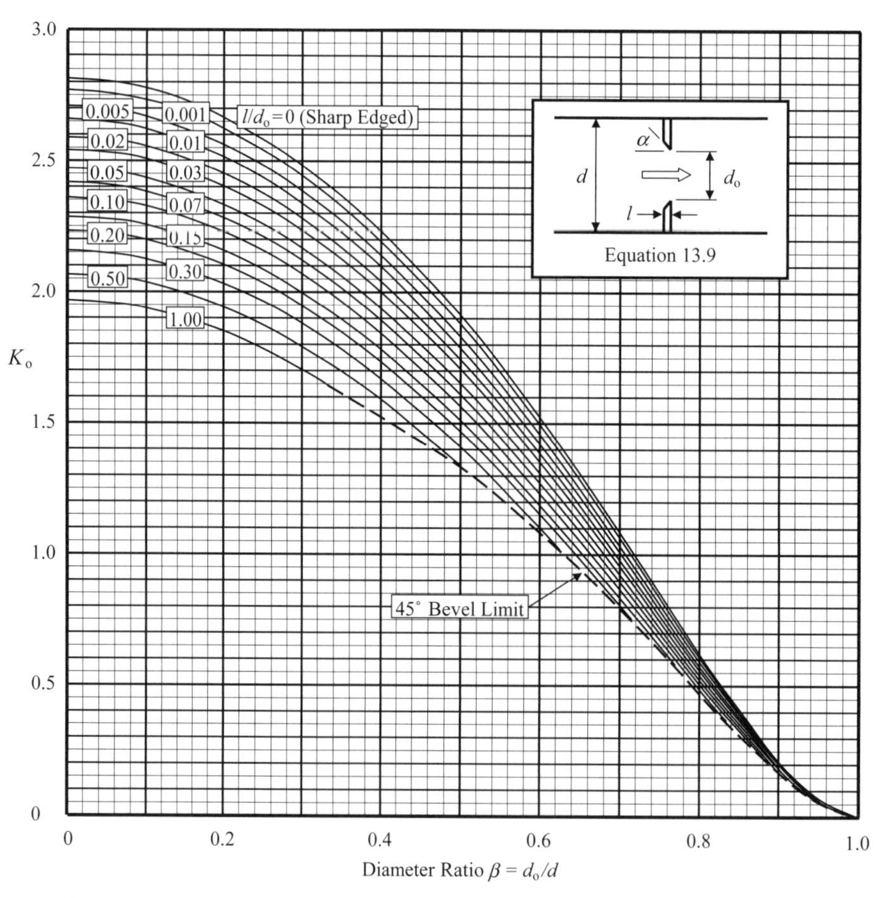

DIAGRAM 13.6. Loss coefficient K_o for 45° bevel-edged orifice in a straight pipe.

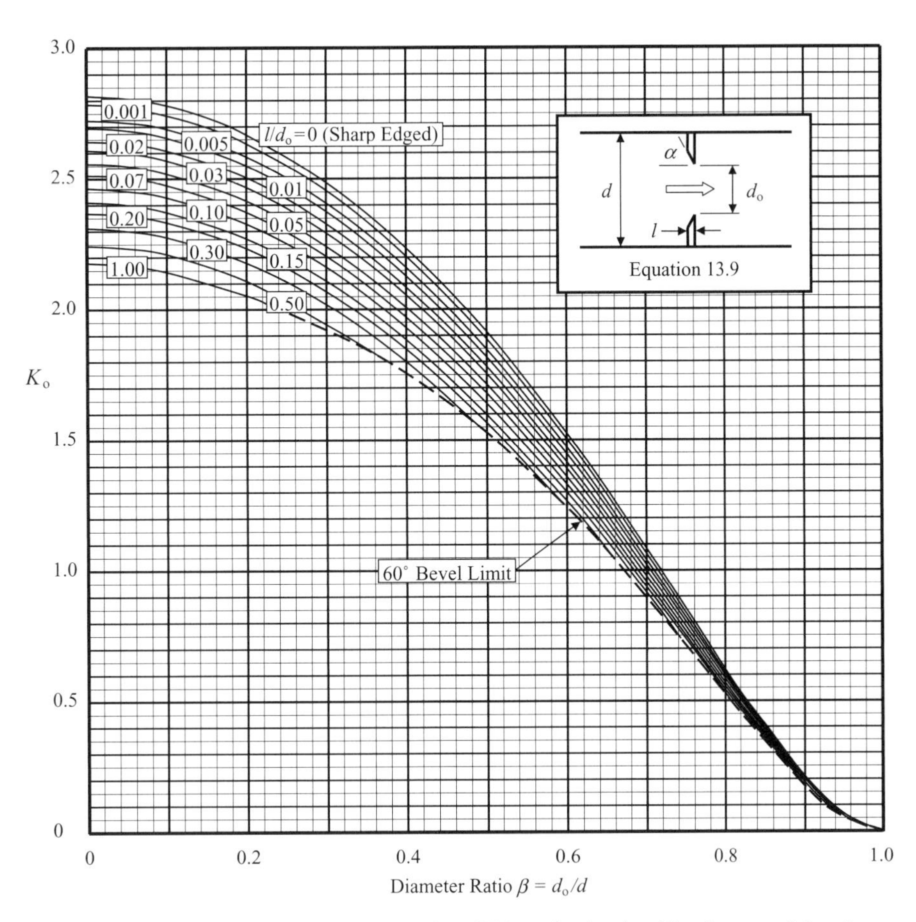

DIAGRAM 13.7. Loss coefficient K_o for 60° bevel-edged orifice in a straight pipe.

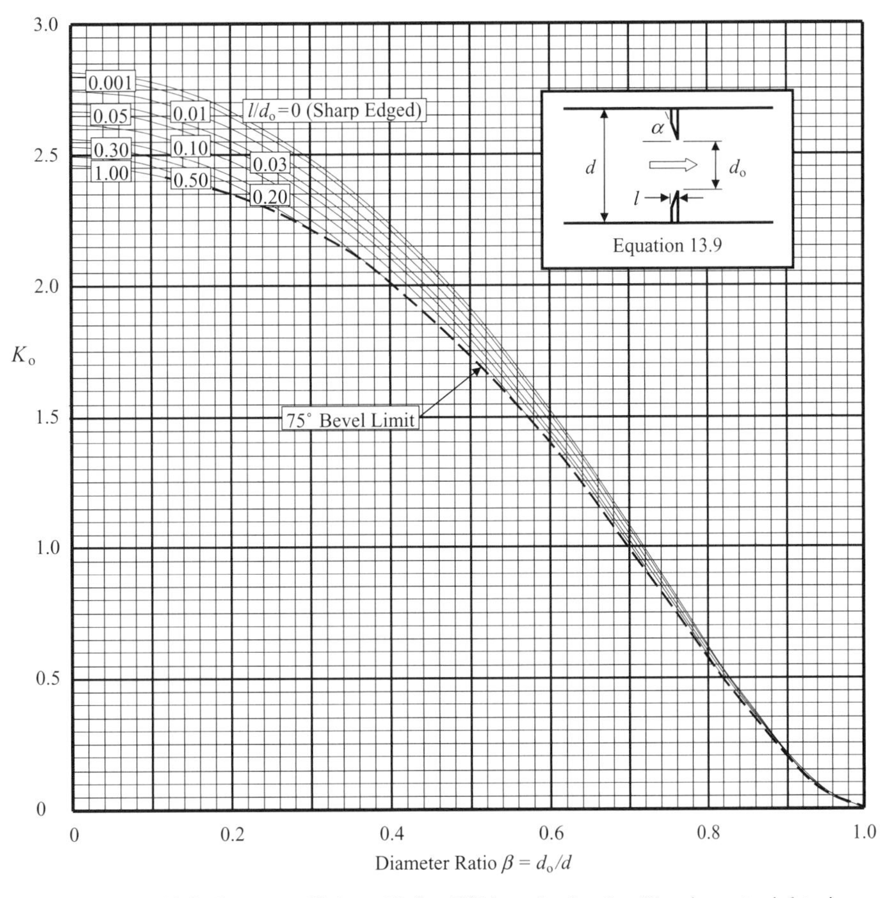

DIAGRAM 13.8. Loss coefficient K_o for 75° bevel-edged orifice in a straight pipe.

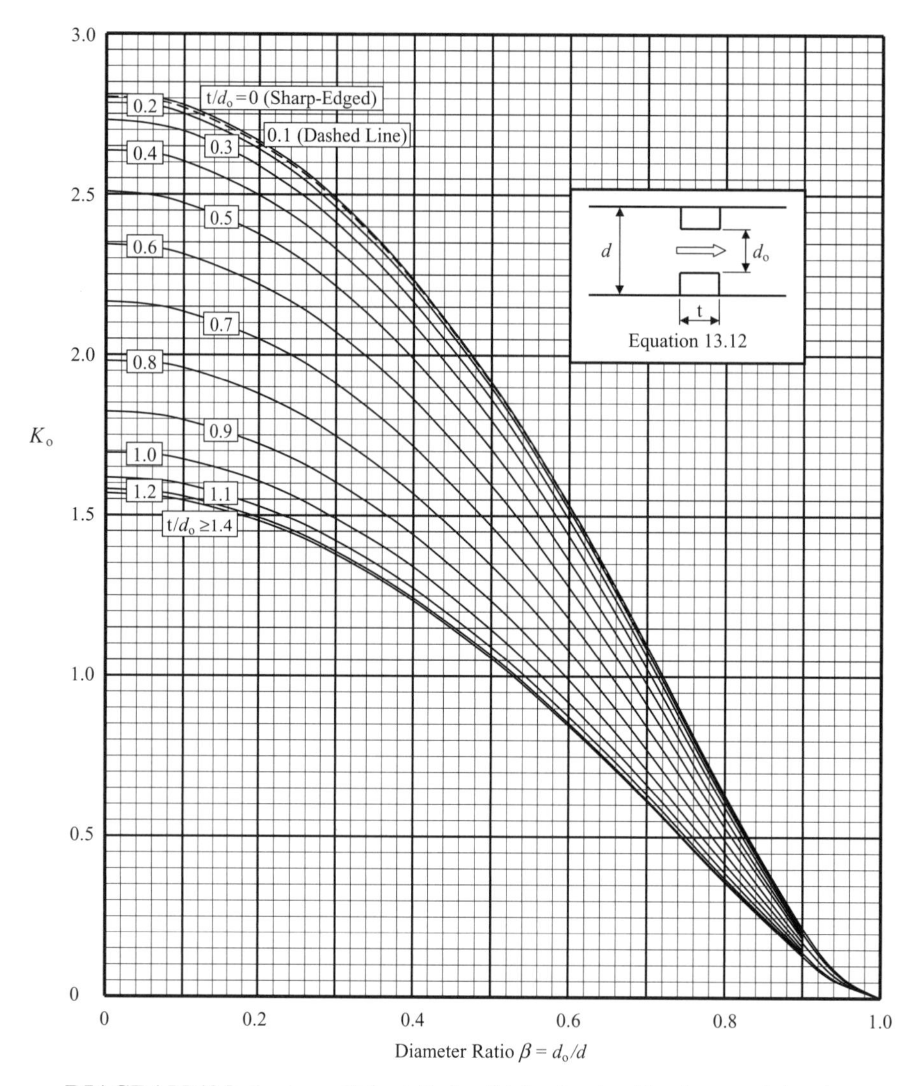

DIAGRAM 13.9. Loss coefficient K_o for thick-edged orifice in a straight pipe.

REFERENCES

1 Bean, H. S., ed., *Fluid Meters, Their Theory and Application*, 6th ed., Report of ASME Committee on Fluid Meters, 1971.

2 Crockett, K. A. and E. L. Upp, The measurement and effects of edge sharpness on the flow coefficients of standard orifices, *Journal of Fluids Engineering, Transactions of the American Society of Mechanical Engineers*, 95, 1973, 271–275.

3 Alvi, S. H., K. Sridharan, and N. S. Lakshmana Rao, Loss characteristics of orifices and nozzles, *Journal of Fluids Engineering, Transactions of the American Society of Mechanical Engineer*, 100, 1978, 299–307.

4 Spikes, R. H. and G. A. Pennington, Discharge coefficient of small submerged orifices, *Proceedings of the Institution of Mechanical Engineers*, 173, 1959, 661.

5 Charleton, J. A., Pneumatic Breakwaters, *British Hydrodynamics Research Association*, RR 684, February 1961.

6 Decker, B. E. I. and Y. F. Chang, An investigation of steady compressible flow through thick orifices, *Proceeding of the Institution of Mechanical Engineers*, 180, Part 3J, 1965–1966, 312–323.

7 James, A. J., Flow through a long orifice, B. Sc. Thesis, Nottingham University, 1961.

8 Sanderson, E. W., Flow through long orifices, B. Sc. Thesis, Nottingham University, 1962.

FURTHER READING

This list includes books and papers that may be helpful to those who wish to pursue further study.

Stuart, M. C. and D. R. Yarnell, Fluid flow through two orifices in series, *Transactions of the American Society of Mechanical Engineers*, 58, 1936, 479–484.

Medaugh, F. W. and G. D. Johnson, Investigation of the discharge and coefficients of small circular orifices, *Transactions of the American Society of Civil Engineers*, 10(9), 1940, 422.

Stuart, M. C. and D. R. Yarnell, Fluid flow through two orifices in series(II, *Transactions of the American Society of Mechanical Engineers*, 66, 1944, 387–397.

Rouse, H. and A. Abul-Fetouh, Characteristics of the irotational flow through axially-symmetric orifices, *Journal of Applied Mechanics, Transactions of the American Society of Mechanical Engineers*, 17(4), 1950, 421–426.

Landstra, J. A., Quarter-circle orifices, *Transactions of the Institution of Chemical Engineers*, 38, 1960, 26–32.

Bogema, M. and P. L. Monkmeyer, The quadrant edge orifice(a fluid meter for low Reynolds number, *Journal of Basic Engineering, Transactions of the American Society of Mechanical Engineers*, 82, 1960, 729–734.

Bogema, M., B. Spring, and M. V. Ramamoorthy, Quadrant edge orifice performance-effect of upstream velocity distribution, *Journal of Basic Engineering, Transactions of the American Society of Mechanical Engineers*, 84, 1962, 415–418.

Leutheusser, H. J., Flow nozzles with zero beta ratio, *Journal of Basic Engineering, Transactions of the American Society of Mechanical Engineers*, 86, 1964, 538–542.

Lenkei, A., Close Clearance Orifices, *Product Engineering*, April 26, 1965, pp. 57–61.

Ramamoorthy, M. V. and K. Seetharamiah, Quadrant-edge orifice and performance at very high Reynolds numbers, *Journal of Basic Engineering, Transactions of the American Society of Mechanical Engineers*, 88, 1966, 9–13.

Ghazi, H. S., On nonuniform flow characteristics at the vena contracta, *Journal of Basic Engineering, Transactions of the American Society of Mechanical Engineers*, 92(70-FE-34), 1970, 1–6.

Miller, D. S., *Internal Guide to Losses in Pipe and Duct Systems*, The British Hydromechanics Research Association, 1971.

Teyssandier, R. G., Internal separated flows—Expansions, nozzles, and orifices, PhD dissertation, University of Rhode Island, Kingston, R. I., 1973.

Miller, R. W. and O. Kneisel, A comparison between orifice and flow nozzle laboratory data and published coefficients, *Journal of Fluids Engineering, Transactions of the American Society of Mechanical Engineers*, 96, 1974, 139–149.

Head, V. P., Improved expansion factors for nozzles, orifices, and variable-area meters, *Journal of Fluids Engineering, Transactions of the American Society of Mechanical Engineers*, 96, 1974, 150–157.

Wilson, M. P. Jr. and R. G. Teyssandier, The paradox of the vena contracta, *Journal of Fluids Engineering, Transactions of the American Society of Mechanical Engineers*, 97, 1975, pp. 366–371.

Nigro, F. E. B., A. B. Strong, and S. A. Alpay, A numerical study of the laminar viscous incompressible flow through a pipe orifice, *Journal of Fluids Engineering, Transactions of the American Society of Mechanical Engineers*, 100, 1978, 467–472.

Sparrow, E. M., J. W. Ramsey, and S. C. Lau, Flow and pressure characteristics downstream of a segmental blockage in a turbulent pipe flow, *Journal of Fluids Engineering, Transactions of the American Society of Mechanical Engineers*, 101, 1979, 200–207.

Lienhard, J. H. V and J. H. IV Lienhard, Velocity coefficients for free jets from sharp-edged orifices, *Journal of Fluids Engineering, Transactions of the American Society of Mechanical Engineering*, 106, 1984, 13.

Grose, R. D., Orifice contraction coefficient for invisid incompressible flow, *Journal of Fluids Engineering, Transactions of the American Society of Mechanical Engineers*, 107, 1985, 36–43.

Bullen, P. R., D. J. Cheeseman, L. A. Hussain, and A. E. Ruffell, The Determination of pipe contraction pressure loss coefficients for incompressible turbulent flow, *International Journal of Heat and Fluid Flow*, 8(2), 1987, 111–118.

Bullen, P. R., D. J. Cheeseman, and L. A. Hussain, The effects of inlet sharpness on the pipe contraction loss coefficient, *International Journal of Heat and Fluid Flow*, 9(4), 1988, 431–433.

Andrews, K. A. and R. H. Sabersky, Flow through an orifice from a transverse stream, *Journal of Fluids Engineering, Transactions of the American Society of Mechanical Engineers*, 112, 1990, 524–526.

Faramarzi, J. and E. Logan, Reattachment length behind a single roughness element in turbulent pipe flow, *Journal of Fluids Engineering, Transactions of the American Society of Mechanical Engineers*, 113, 1991, 712–714.

Brundrett, E., Prediction of pressure drop for incompressible flow through screens, *Journal of Fluids Engineering, Transactions of the American Society of Mechanical Engineers*, 115, 1993, 239–242.

Brower, W. B., Jr., E. Eisler, E. J. Filkorn, J. Gonenc, C. Plati, and J. Stagnitti, On the compressible flow through an orifice, *Journal of Fluids Engineering, Transactions of the American Society of Mechanical Engineers*, 115, 1993, 660–664.

Agarwal, N. K., Mean separation and reattachment in turbulent pipe flow due to an orifice plate, *Journal of Fluids Engineering, Transactions of the American Society of Mechanical Engineers*, 116, 1994, 373–376.

14

FLOW METERS

A constriction that produces an accelerated flow and a resulting drop in static pressure is an excellent meter in which this pressure difference can be measured and related to the mass or volume rate of flow. The distinctive feature of this group of meters is that there is a marked pressure difference or pressure drop associated with the flow of a fluid through the device. In this pressure differential group of flow measuring devices are the flow nozzle, the Venturi tube, and the nozzle/Venturi. The loss coefficient of another member of this group, the sharp-edged orifice, was dealt with in Section 12.1. Other members are the elbow (centrifugal) meter and the pipe section (frictional resistance) meter.

This chapter deals with overall head loss through the first three pressure differential devices mentioned above. The flow measuring characteristics of all these devices, as well as many other types of flow measuring devices, are extensively dealt with in References [1,2].

14.1 FLOW NOZZLE

When the radial distance available between the pipe wall and the nozzle face is limited, rounding may take the form of an ellipse or other curved shape. Such is the case for flow nozzles. The recommended form of the flow nozzle is the "long radius" or elliptical inlet nozzle, in which the curvature of the inlet to the nozzle throat is the quadrant of an ellipse as shown in Figure 14.1. In the case of such noncircular inlets, the rounding radius r is can be expressed as:

$$r = \sqrt[3]{r_1^2 r_2}, \qquad \text{(10.5, repeated)}$$

where r_1 and r_2 are the semimajor (longitudinal) and semiminor (radial) axes, respectively.*

The flow nozzle is basically a well-rounded orifice except for the addition of a length of cylindrical throat section. Thus the loss coefficient equation developed in Section 13.3.1 for a rounded orifice, supplemented with a surface friction loss term, can be applied to a flow nozzle. When the rounding ratio r/d_T is less than 1:

$$K_T = 0.0696\left(1-0.569\frac{r}{d_T}\right)\left(1-\sqrt{\frac{r}{d_T}}\beta\right)(1-\beta^5)\lambda^2 + (\lambda-\beta^2)^2 + f_T\frac{l_T}{d_T} \quad (r/d_T < 1), \qquad (14.1)$$

where the diameter ratio $\beta = d_T/d$, and where the jet contraction coefficient λ is given by:

$$\lambda = 1 + 0.622\left(1-0.30\sqrt{\frac{r}{d_T}}-0.70\frac{r}{d_T}\right)^4 (1-0.215\beta^2-0.785\beta^5) \quad (r/d_T < 1). \qquad \text{(13.7, repeated)}$$

The results of Equation 14.1 are compared in Figure 14.2 to data from the American Society of Mechanical Engineers (ASME) Fluid Meters [1] for low β and high β flow nozzles. The relative nozzle radii ratio r/d_T noted

* There is no analytical basis for this expression; however, it works quite well.

Pipe Flow: A Practical and Comprehensive Guide, First Edition. Donald C. Rennels and Hobart M. Hudson.

in the figures are proportions specified in the handbook. Equation 14.1 compares well with the data [1].* This comparison also validates the loss coefficient expression developed in Section 13.3 for a rounded orifice that is quite similar to a flow nozzle. The loss coefficient of a flow nozzle built to ASME Fluid Meters specifications can be found in Diagram 14.1.

14.2 VENTURI TUBE

The classical or Herschel Venturi tube is usually made of cast iron or cast steel in the smaller sizes. In the larger sizes, Venturi tubes are generally made with rough-cast inlet cones—the throat and diffuser sections are usually made of a smoother surface material. As shown in Figure 14.3, the inlet section consists of a short cylindrical Venturi tube joined by an easy curvature to a converging inlet cone having an included angle α_I of 21°. The inlet cone is joined by another smooth curve to a short cylindrical section called the throat. The exit from this throat section leads by another easy curve into the diverging outlet cone or diffuser. The recommended included angle α_D of the outlet cone is 7–8°; however, it may be as large as 15°.

The inlet section of the Venturi tube has a very smooth generatrix so its losses are mainly due to surface friction. Accordingly, the overall pressure loss is made up of friction losses in the inlet cone, throat, and diffuser sections, as well as an expansion loss in the diffuser section:

* The ASME Fluid Meters data, presented in terms of percent of differential pressure, were converted to loss coefficient by the relationship $K = \%\Delta P(1 - \beta^4)/(100C^2)$ assuming a calibration coefficient C of 0.997.

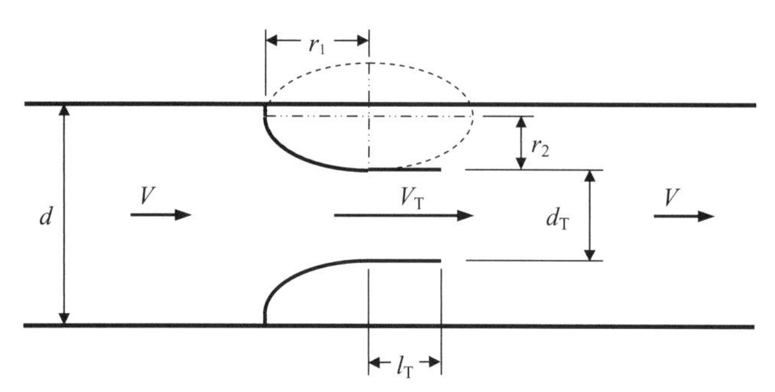

FIGURE 14.1. Flow nozzle.

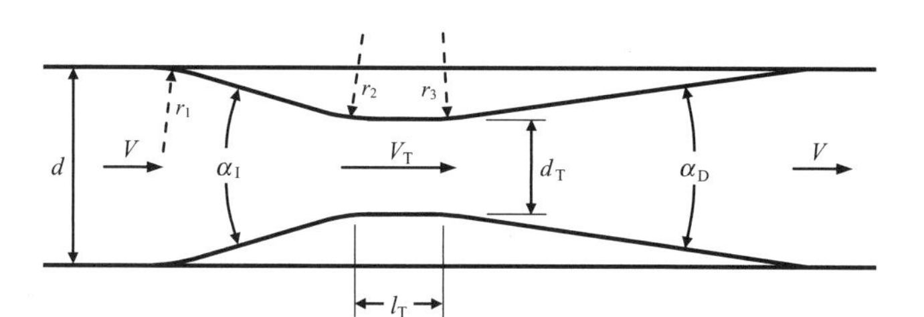

FIGURE 14.3. Venturi tube.

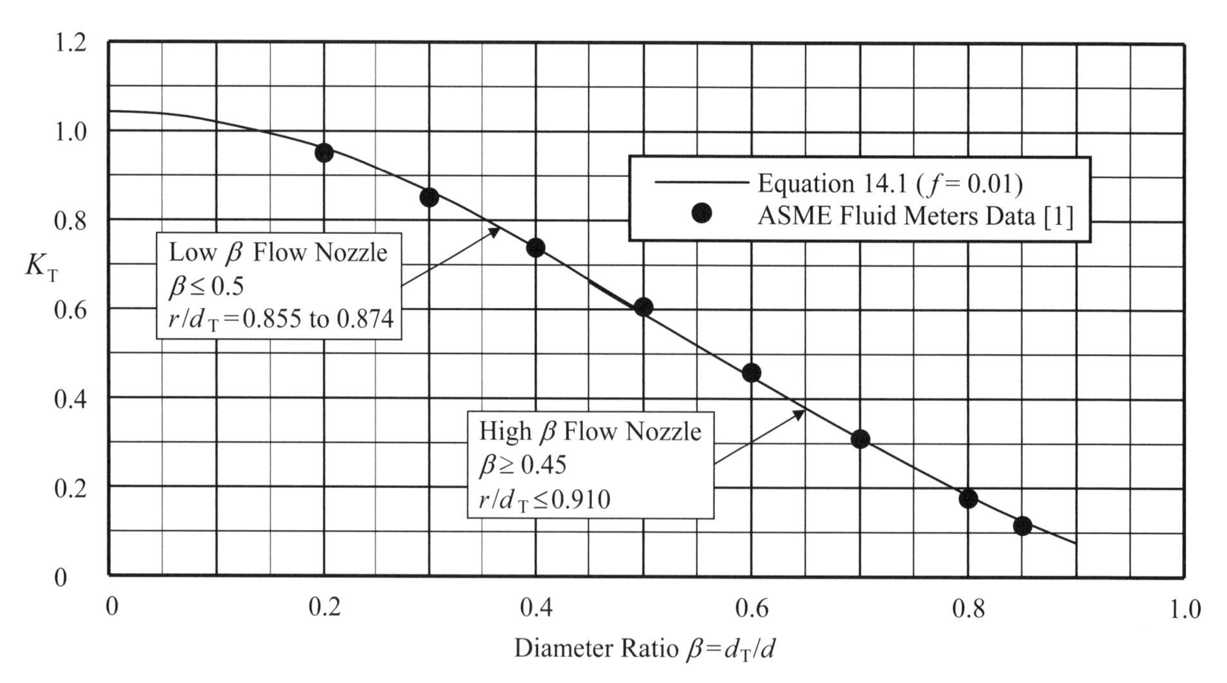

FIGURE 14.2. Loss coefficients for ASME long radius flow nozzles.

$$K_T = \frac{f_I(1-\beta^4)}{8\sin(\alpha_I/2)} + f_T\frac{l_T}{d_T} + 8.30[\tan(\alpha_D/2)]^{1.75}(1-\beta^2)^2 + \frac{f_D(1-\beta^4)}{8\sin(\alpha_D/2)}, \quad (14.2)$$

where f_I, f_T, and f_D are friction factors in the inlet cone, the throat, and the diffuser sections, respectively. The friction factors are based on the surface roughness of the inlet, throat, and discharge sections, and consistent with the Reynolds number and diameter of the throat section.

Calculated results using Equation 14.2 are compared to ASME Fluid Meters data in Figure 14.4.* The friction factors were chosen considering that the surface of the inlet section is usually much rougher than the throat and diffuser surfaces. The calculated results generally match the ASME Fluid Meters data.

The downstream end of the outlet cone may be truncated as much as 30% or 40% of its normal length with only a small effect on pressure loss. In fact, truncating the outlet cone of a low β Venturi tube may actually decrease pressure loss. In this case the equation for loss coefficient becomes†:

$$K_T = \frac{f_I(1-\beta^4)}{8\sin(\alpha_I/2)} + f_T\frac{l_T}{d_T} + 8.30[\tan(\alpha_D/2)]^{1.75}(1-\beta_E^2)^2 + \frac{f_D(1-\beta_E^4)}{8\sin(\alpha_D/2)} + (\beta_E^2-\beta^2)^2,$$

where the diffuser angle α_D is equal to or less than 20°.

14.3 NOZZLE/VENTURI

In a nozzle/Venturi, the conical inlet section of the classical Venturi tube is replaced with a rounded inlet section. In addition, the downstream end of the diverging outlet cone is usually truncated as shown in Figure 14.5. The overall pressure loss is made up of an entrance loss, friction losses in the throat and diffuser, an expansion loss in the diffuser, and a sudden expansion loss at the exit.‡

* See previous footnote.

† Refer to Section 11.3.1 for the nomenclature of a stepped conical diffuser.

‡ Gibson's tests, which were used to quantify diffuser performance in Chapter 11, were made with long straight lengths of pipe upstream and downstream. Substitution of a short nozzle for the upstream pipe length will alter the inlet velocity distribution from the standard turbulent flow profile to a practically uniform one. The effect is that the actual head loss in the nozzle/Venturi could be somewhat less, say by 5%–10%, than predicted herein.

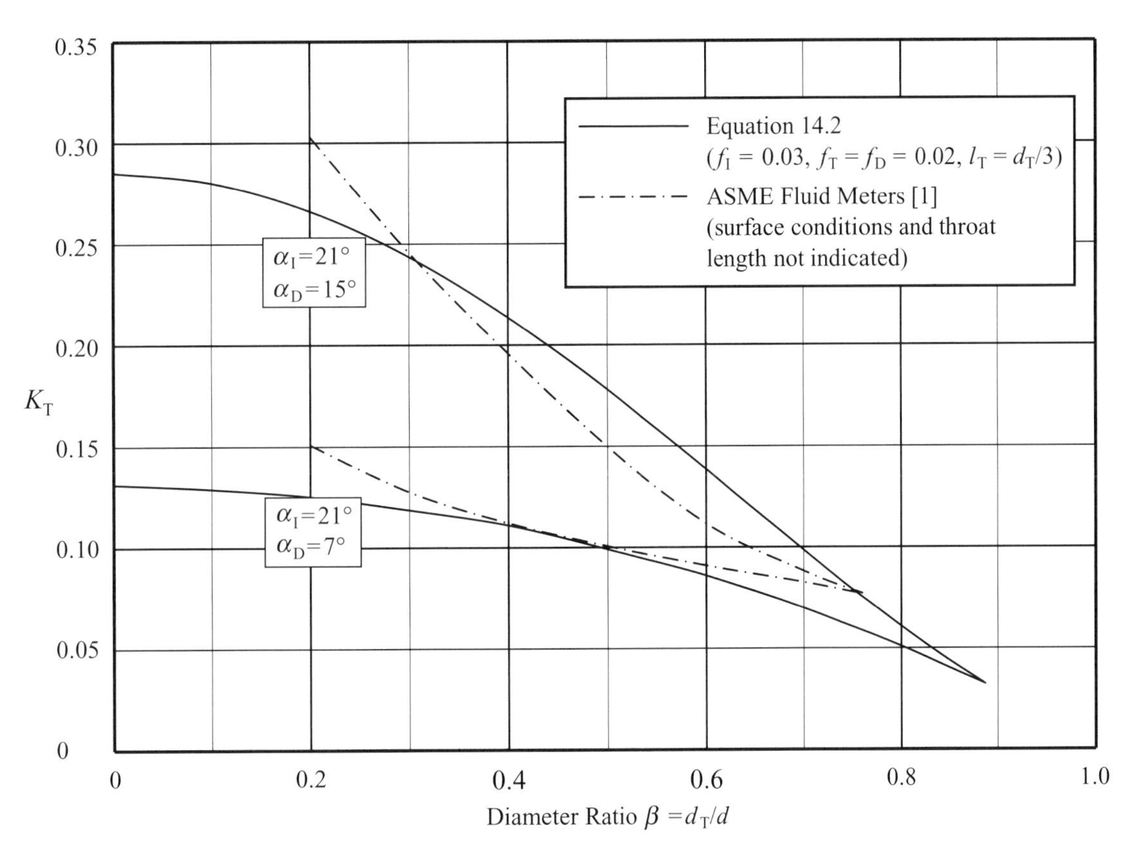

FIGURE 14.4. Comparison of Equation 14.2 to ASME Fluid Meters data.

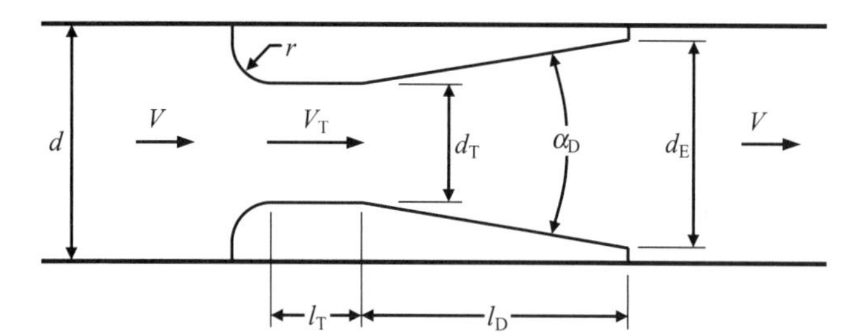

FIGURE 14.5. Nozzle/Venturi.

The overall diameter ratio β equals d_T/d, and the exit step diameter ratio β_E equals d_T/d_E for the stepped diffuser. For diffuser included angle α_D less than 20°, the loss coefficient of a stepped conical diffuser can be approximately determined by the following equation adapted from Equation 10.6 for a rounded contraction and from Equation 11.11 for a stepped diffuser:

$$K_T \approx 0.0696\left(1-0.569\frac{r}{d_T}\right)\left(1-\sqrt{\frac{r}{d_T}}\beta\right)(1-\beta^5)\lambda^2 + (\lambda-1)^2 + f_T\frac{l_T}{d_T} + 8.30(\tan(\alpha_D/2))^{1.75}(1-\beta_E^2)^2 + \frac{f_D(1-\beta_E^4)}{8\sin(\alpha_D/2)} + (\beta_E^2-\beta^2)^2$$
$$(\alpha \le 20°) \quad (0 \ge \beta_E \le 1), \tag{14.3}$$

where the jet contraction coefficient λ is given by:

$$\lambda = 1+0.622\left(1-0.30\sqrt{\frac{r}{d_T}}-0.70\frac{r}{d_T}\right)^4 (1-0.215\beta^2-0.785\beta^5). \tag{13.7, repeated}$$

The rounding contour may be the arc of a circle, or it may take the form of an ellipse, lemniscate, or other smoothly curved shape. For a circular inlet, the rounding radius r is simply the radius of the circle. In the case of an elliptical inlet contour, the rounding radius can be expressed as:

$$r = \sqrt[3]{r_1^2 r_2}, \tag{10.5, repeated}$$

where r_1 and r_2 are the semimajor (longitudinal) and semiminor (radial) axes, respectively.

When the nozzle/Venturi is not stepped, β_E equals β in Equations 14.3 and the last term, the sudden expansion term, vanishes.

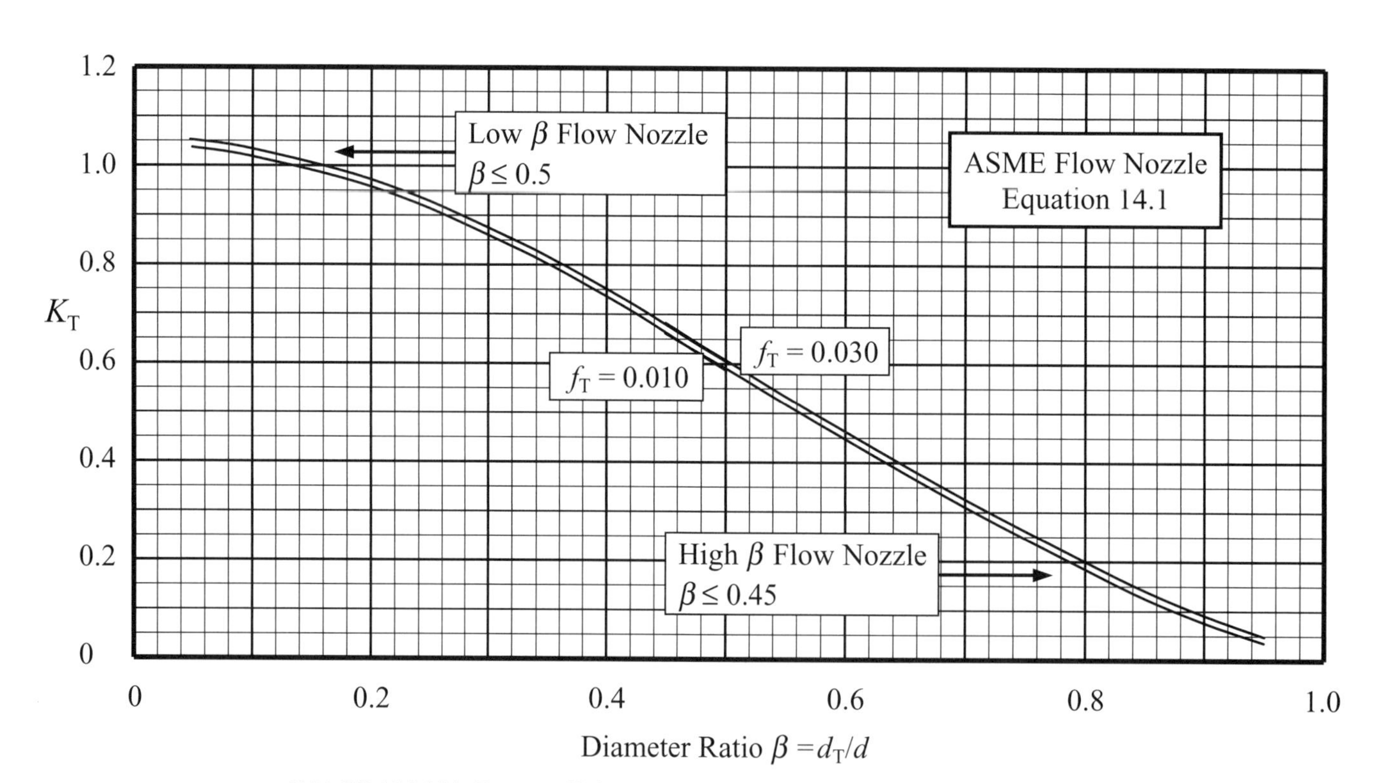

DIAGRAM 14.1. Loss coefficient of flow nozzle built to ASME specifications.

REFERENCES

1. Bean, H. S., ed., *Fluid Meters, Their Theory and Application*, 6th ed., Report of ASME Committee on Fluid Meters, 1971.
2. Spink, L. K., *Principles and Practices of Flow Meter Engineering*, 9th ed., The Foxboro Company, Foxboro, MA, 1978.

FURTHER READING

This list includes books and papers that may be helpful to those who wish to pursue further study.

Warren, J., A study of head loss in Venturi-meter diffuser sections, *Transactions of the American Society of Mechanical Engineers*, Paper No. 50-A-65, 73, 1951, 1–4.

Methods for the Measurement of Fluid Flow, *British Standard 1042; Part 1,* 1964.

Leutheusser, H. J., Flow nozzles with zero beta ratio, *Journal of Basic Engineering, Transactions of the American Society of Mechanical Engineers*, 86, 1964, 538–542.

Replogle, J. A., L. E. Myers, and K. J. Brust, Evaluation of pipe elbow as flow meters, *Journal of the Irrigation and Drainage Division, Proceedings of the American Society of Civil Engineers*, 92(IR3), 1966, 17–34.

Arnberg, B. T., C. L. Britton, and W. F. Seidl, Discharge coefficient correlations for circular-arc Venturi flowmeters at critical (sonic) flow, *Transactions of the American Society of Mechanical Engineers*, Paper No. 73-WA/FM-8, 96, 1974, 1–13.

Miller, R. W. and O. Kneisel, A comparison between orifice and flow nozzle laboratory data and published coefficients, *Journal of Fluids Engineering, Transactions of the American Society of Mechanical Engineers*, 96, 1974, 139–149.

Head, V. P., Improved expansion factors for nozzles, orifices, and variable-area meters, *Journal of Fluids Engineering, Transactions of the American Society of Mechanical Engineers*, 96, 1974, 150–157.

Benedict, R. P., Loss coefficients of fluid meters, *Journal of Fluids Engineering, Transactions of the American Society of Mechanical Engineers*, Paper No. 76-WA/FM-2, 99, 1977, 245–248.

Benedict, R. P. and J. S. Wyler, Analytical and experimental studies of ASME flow nozzles, *Journal of Fluids Engineering, Transactions of the American Society of Mechanical Engineers*, 100, 1978, 265–275.

Alvi, S. H., K. Sridharan, and N. S. Lakshmana Rao, Loss characteristics of orifices and nozzles, *Journal of Fluids Engineering, Transactions of the American Society of Mechanical Engineers*, 100, 1978, 299–307.

Buzzard, W., *Flowmeter Orifice Sizing*, Handbook No. 10B9000, Fischer and Porter Company, 1978.

Kopp, J., Flowmeter selection(Part 1, in *Oil, Gas & Petrochem Equipment*, The Petroleum Publishing Company, 1979.

Kopp, J., Flowmeter selection(Part 2, in *Oil, Gas & Petrochem Equipment*, The Petroleum Publishing Company, 1979.

ASME Fluid Meters Research Committee, The ISO-ASME Orifice Coefficient Equation, *Mechanical Engineering,* July 1981, pp. 44–45.

Miller, R. W., *Flow Measurement Engineering Handbook*, 3rd ed., McGraw-Hill, 1996.

Roberson, J. A. and C. T. Crowe, *Engineering Fluid Mechanics*, 6th ed., John Wiley & Sons, 1997 (Chapter 13 Flow Measurements).

Baker, R. C., *Flow Measurement Handbook: Industrial Design, Operating Principles, Performance, and Applications*, Cambridge University Press, 2000.

Furness, R., Don't Install That Flowmeter (Until You Read This Article on Installation Claims, Tips(and Reality), *Flow Control,* March 2002, pp. 31–47.

Falcone, G., *Multiphase Flow Metering*, Elsevier, 2009.

15

BENDS

This chapter is primarily concerned with the flow of an incompressible turbulent fluid in pipe bends and elbows. Deflection angle α and bend radius ratio r/d (the ratio of the bend centerline radius to the inside diameter of the pipe) are important geometric parameters. Surface roughness of the wall of the bend, as well as of the connecting pipe, is also important. Hence, friction factor is important.

The work presented here is for bends where there is no change of flow area between the inlet and outlet. It is assumed that roughness of the connecting pipe is similar to roughness of the bend. The work is for circular passages, but can be reasonably applied to square ducts or to rectangular ducts of low aspect ratio.

15.1 ELBOWS AND PIPE BENDS

The pressure loss in pipe bends may be thought of as made up of three components. One component is the pressure loss due to ordinary *surface friction* that corresponds to fully developed flow in a straight pipe having the same length as the centerline of the bend. A second component is due to a twin-eddy *secondary flow* superimposed on the main or primary flow due to the combined action of centrifugal force and frictional resistance of the pipe walls. A third component is due to *separation* of the main flow from the inner and outer radius of the bend and subsequent expansion of the contracted stream.* For bends of small radius of curvature, flow separation and secondary flow dominate. For bends of large radius of curvature, ordinary surface friction and secondary flow prevail. Flow separation and secondary flow are illustrated in Figure 15.1.†

As noted above, the bend radius ratio r/d is defined as the ratio of the centerline radius of the bend to the inside diameter of the pipe. In this context, a bend of radius ratio 0.5 represents a bend with a sharp (zero radius) inner corner and an outer bend radius of 1.0. At the extreme is a miter bend in which two pipes are joined together in a sharp angle without any rounding at the plane of intersection. The rounding of the corner at the inner wall, or simply beveling the corner, greatly attenuates the separation and reduces the pressure loss. At the opposite extreme, bend losses, excluding friction losses, are at a minimum when the bend radius ratio is at a maximum.

A bend must always be considered with relation to the straight pipes, or tangents, connected to its ends. This brings about experimental difficulties. Whatever the velocity distribution may be at the upstream end, the downstream length must be sufficient for the gradual adjustment of the distribution until it regains a normal velocity profile. An example of the measured pressure distribution along a bend is shown in Figure 15.2. In this bend of circular cross section, a marked increase in pressure along the outer wall is accompanied by a corresponding decrease in pressure along the inner wall. The bend loss is found by measuring the pressure difference between static pressure taps located just before

* At higher bend radius ratios, the flow stream may not actually separate from the walls. However, contraction of flow stream and subsequent redevelopment of the velocity profile contributes to pressure loss.

† The flow phenomena are shown separately for clarity. In reality, flow separation and secondary flow occur at the same time in bends of small bend radius ratio.

Pipe Flow: A Practical and Comprehensive Guide, First Edition. Donald C. Rennels and Hobart M. Hudson.

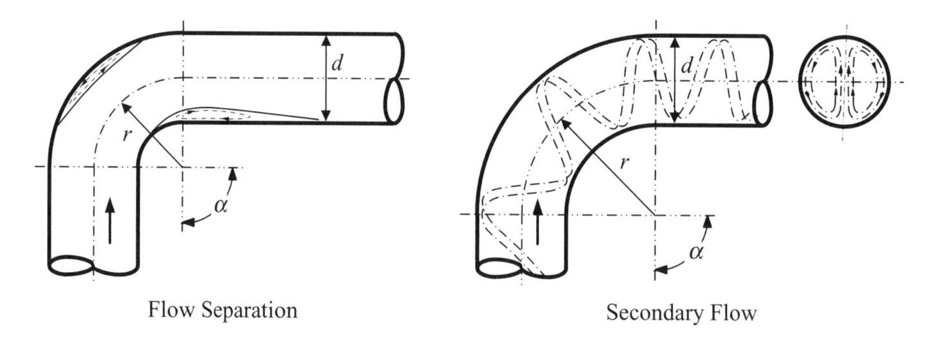

FIGURE 15.1. Curved pipe flow.

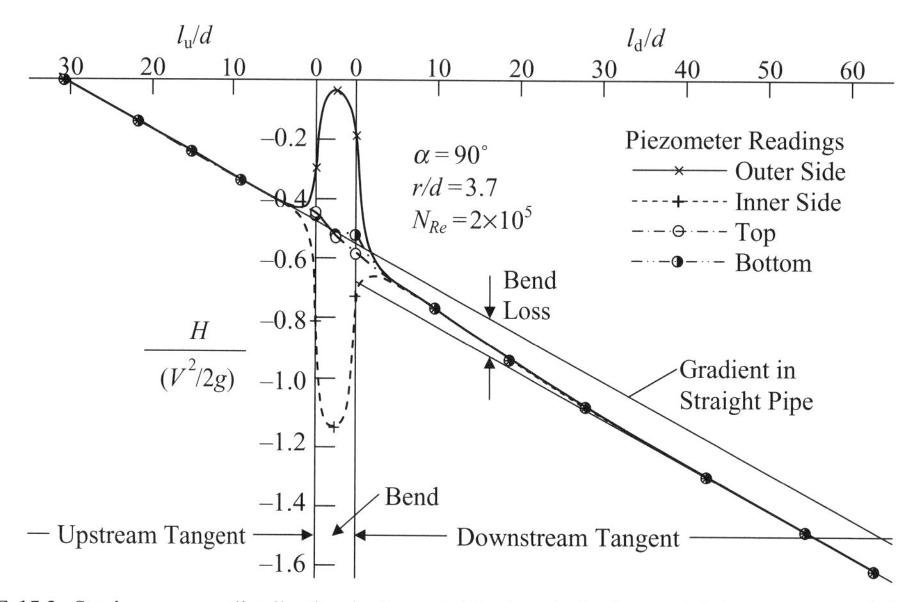

FIGURE 15.2. Static pressure distribution in the neighborhood of a bend with long tangents (after Ito [1]).

the bend and taps located 40 diameters or so downstream of the bend and then subtracting the ordinary friction loss for developed flow in straight pipe between the two taps. Because the friction loss over this distance may be many times the bend loss, particularly for small deflection angles, the pressure loss is often the difference between two large values. Very careful and accurate measurements are required under these conditions to obtain accurate results.

Because of its considerable importance in the design and analysis of fluid machinery and piping systems, a vast amount of experimental and theoretical data on flow through bends has been reported over the past century. However, a review of the literature shows wide variations in loss coefficients quoted by the various investigators. Because actual details of their test conditions are often lacking, it is not possible to correct their results to provide meaningful data. In any case, all investigators report that bend loss is a strong function of friction factor. Many investigators go so far as to characterize bend loss as a direct function of friction factor.

The Dean number, a dimensionless number giving the ratio of the viscous force acting on a fluid in a curved pipe to the centrifugal force, has frequently been employed in the study of flow in curved pipes and channels.* Nonetheless, the authors did not employ the Dean number to aid (or hamper) their formulation of bend loss coefficient.

Experiments on curved pipe carried out at Munich from 1927 to 1932 [2–5] were the first to carefully specify their test conditions. Later, H. Ito [1,6–9] of Tohoku University, Japan, extended the investigation of circular section bends to cover a wider range than used in the

* The Dean number is equal to the Reynolds number times the square root of the ratio of the inside diameter d of the pipe to twice the radius of curvature of the bend.

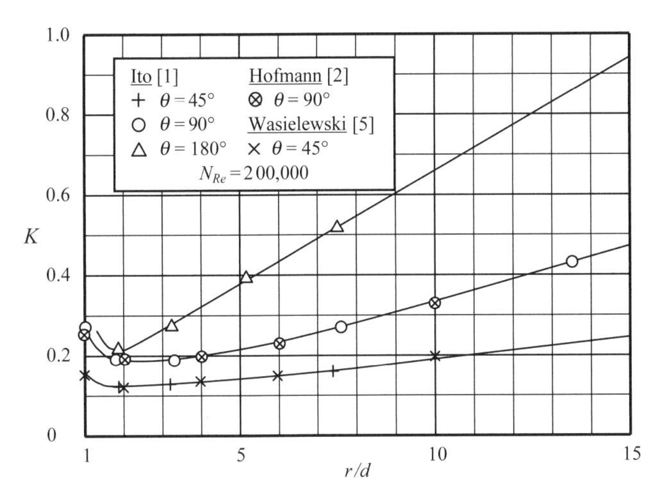

FIGURE 15.3. Loss coefficients for smooth pipe bends (after Ito [1]).

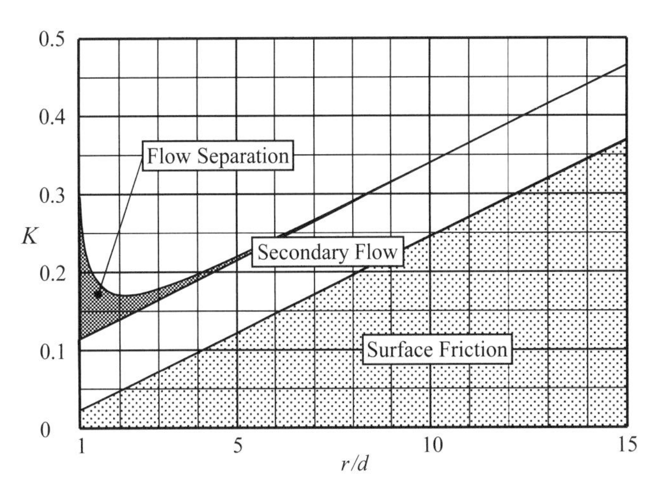

FIGURE 15.4. The distinct effects of surface friction, secondary flow, and flow separation.

Munich tests. His reported data, which included the work of the Munich investigators, provided the bulk of information used to develop a loss coefficient formula applicable to a wide range of elbow and pipe bend configurations.

A portion of Ito's work depicting variation of total head loss with bend radius ratio r/d and deflection angle α is shown in Figure 15.3. The solid lines in Figure 15.3 represent the results of multipart formulas developed by Ito for bends with smooth walls. Several writers have reproduced this diagram to illustrate an interesting, but perhaps misleading, feature of head loss in bends, that is, that a minimum of head loss occurs at certain low values of r/d.*

Investigators agree that the loss coefficient of elbows and curved pipe is practically a direct function of friction factor. The friction factor in smooth-walled tests can be fairly accurately predicted as a function of Reynolds number. Accordingly, most experimental and theoretical data have been for smooth-walled bends. There are very little useful data on rough-walled bends. The experimenters often used artificially roughened pipe and actual surface roughness was usually not reported.

For curved pipe, the effects of surface friction, secondary flow, and flow separation can be rationally divided into three distinct effects as illustrated in Figure 15.4. The lower region represents ordinary surface friction loss as in a straight stretch of pipe equal to the centerline distance of the bend. The mid- and upper regions can be attributed to secondary flow loss and flow separation loss, respectively.

* See Section 15.5, "Bend Economy," to explore this issue.

Employing smooth-walled test data, the author developed an empirical equation for circular bends for r/d equal to or greater than 0.5, and for bend angle α from zero to $\pi/2$ (180°):

$$K = f\alpha\frac{r}{d} + (0.10 + 2.4f)\sin(\alpha/2) + \frac{6.6f\left(\sqrt{\sin(\alpha/2)} + \sin(\alpha/2)\right)}{(r/d)^{\frac{4\alpha}{\pi}}}. \quad (15.1)$$

The first term in Equation 15.1 represents surface friction loss, the second term represents secondary flow, and the third term represents flow separation. Note that Equation 15.1 essentially encompasses the entire range of elbow and pipe bend configurations.

The results of Equation 15.1 are compared with test data reported by Ito [1] and Miller [10] as shown in Figures 15.5–15.7 for smooth-walled pipe bends at Reynolds numbers of 20,000, 200,000 ,and 1,000,000, respectively. The results compare very well with Ito's reported data and reasonably well with Miller's data. In extending the formulation to cover radius ratios less than 1, it was taken into account that the loss at radius ratio 0.5 approaches that of a miter bend (see Section 15.3).

From Equation 15.1, loss coefficient values K_T for welded elbows and returns, and for fabricated pipe bends, are presented in Tables 15.5 through 15.12 located at the end of this chapter. Loss coefficients for pipe schedules other than provided in the tables can be interpolated. Note that loss coefficient K_T provided in the tables is for clean commercial steel pipe and pipe fittings in the zone of complete

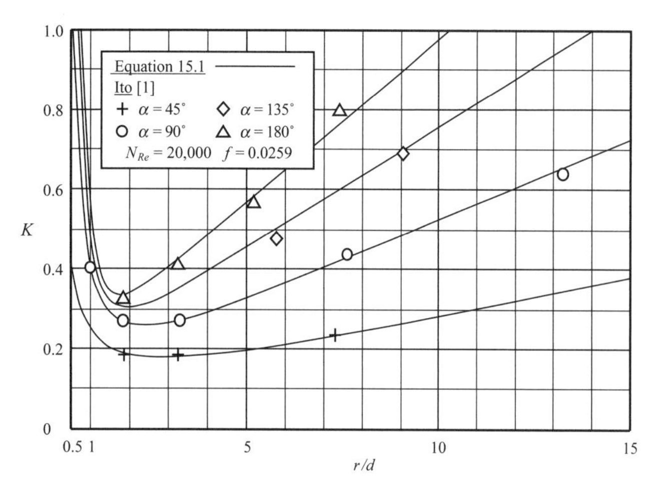

FIGURE 15.5. Comparison of Equation 15.1 to test data at Reynolds number = 20,000.

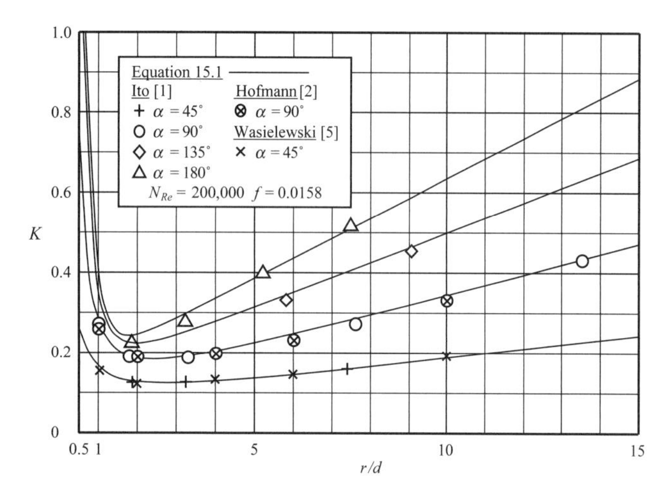

FIGURE 15.6. Comparison of Equation 15.1 to test data at Reynolds number = 200,000.

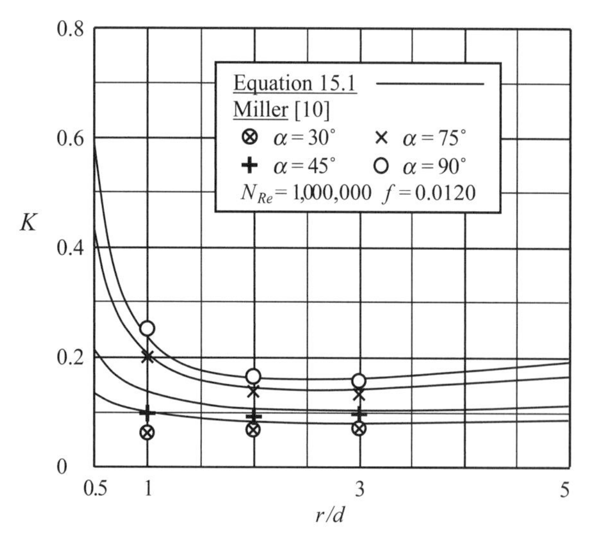

FIGURE 15.7. Comparison of Equation 15.1 to test data at Reynolds number = 1,000,000.

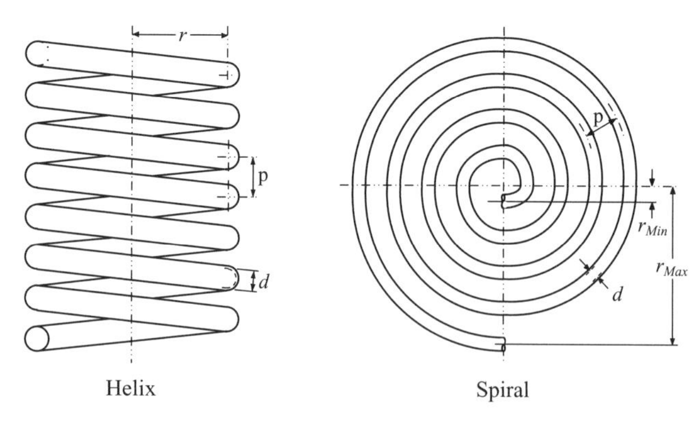

FIGURE 15.8. Constant pitch coils.

turbulence. To obtain loss coefficient K in the case of turbulent flow in the transition zone, or for other than clean steel pipe, simply adjust the tabulated values as follows:

$$K = \frac{f}{f_T} K_T, \tag{15.2}$$

where f is friction factor at the flow condition of interest, and f_T and K_T are obtained from Tables 15.5 through 15.12. This method is reasonable because, as evident in Equation 15.1, the loss coefficient for curved bends is very nearly a direct function of friction factor. If greater precision is desired, calculate the loss coefficient directly using Equation 15.1.

15.2 COILS

Coils of pipe or tubing can be classified as those with constant curvature, *helices*, and those with variable curvature, *spirals*. A helix is a three-dimensional coil that runs along the surface of a cylinder. A spiral is typically a planar curve (that is flat), like the grooves of a phonograph record or a DVD. Coils provide for a relatively large amount of surface area within a confined space, as in a heat exchanger. Furthermore, heat transfer coefficients in coils are higher than in straight pipes.

The curvature of a coil is defined as the ratio of the radius r of the circle into which the tubing is bent to the inside diameter d of the pipe. The distance between the central lines of two consecutive turns is the pitch p. In general, a helix has a constant pitch. An Archimedean spiral has a constant pitch; other spirals do not. Constant pitch coils are illustrated in Figure 15.8.

A number of equations for calculating pressure drop in coils have been reported in the literature. The equations are for smooth tubes only; they are not applicable to rough pipe. The equations either predict a dimensionless Fanning friction factor f_c for coils, or

TABLE 15.1. Comparison of Equation 15.3 with Other Formulations for Smooth Helical Coils (N = 5)

					Loss Coefficient K			
N_{Re}	d (in)	f[a]	p (in)	r (in)	Ito [1]	Kubair and Varrier [11]	Mori and Nakayama [12]	Equation 15.3
20,000	2	0.0259	4	10	5.5	5.5	5.4	5.7
				16	8.3	8.7	8.3	8.1
				24	11.9	13.0	12.0	11.4
	4	0.0259	12	16	4.5	4.4	4.4	4.9
				24	6.4	6.5	6.4	6.5
				36	9.2	9.8	9.2	9.0
	8	0.0259	16	36	5.0	4.9	4.9	5.3
				48	6.4	6.5	6.3	6.5
				72	9.2	9.8	9.2	9.0
200,000	1	0.0158	4	10	3.7	(Formula out of range)	3.4	3.9
				16	5.4		5.1	5.4
				24	7.7		7.6	7.3
	2	0.0158	12	16	3.1		2.7	3.4
				24	4.3		4.0	4.4
				36	6.0		5.7	5.8
	4	0.0157	16	36	3.4		3.0	3.6
				48	4.3		4.0	4.3
				72	6.0		5.7	5.8

[a] Friction factor was calculated using Equation 8.3, assuming a surface roughness e = 0.000060 in for smooth pipe or tubing.

predict the ratio of f_c to the friction factor for straight tubes f_s. Typically, they are limited to specific coil geometry and ranges of Reynolds number bounded by the authors test data.

15.2.1 Constant Pitch Helix

Equation 15.1 was adapted to the geometry of a constant pitch helix as follows:

$$K = \mathrm{N}\left[f\frac{\sqrt{(2\pi r)^2 + \mathrm{p}^2}}{d} + 0.20 + 4.8f\right], \qquad (15.3)$$

where N is the number of coils and f is the friction factor for straight pipe.* The formulation is based on peak secondary flow as for two 180° bends per coil. The flow separation term was ignored. Equation 15.3 compares well to equations developed by Ito [1], Kubair and Varrier [11], and Mori and Nakayama [12] for smooth-walled helical coils as shown in Table 15.1.

Equation 15.2 is applicable to rough pipes as well as to smooth pipes, whereas the other formulations are limited to smooth wall pipe. Also, the other formulations have Reynolds number and geometry limitations.

15.2.2 Constant Pitch Spiral

Similarly as for a constant pitch helix, Equation 15.1 was adapted to the geometry of a constant pitch spiral:

$$K = f\frac{\pi\left(r_{Max}^2 - r_{Min}^2\right)}{\mathrm{p}d} + \mathrm{N}(0.20 + 4.8f) + \frac{13.2f}{(r_{Min}/d)^2} \qquad (15.4a)$$

or

$$K = \frac{r_{Max} - r_{Min}}{\mathrm{p}}\left[f\pi\left(\frac{r_{Max} + r_{Min}}{d}\right) + 0.20 + 4.8f\right] + \frac{13.2f}{(r_{Min}/d)^2}, \qquad (15.4b)$$

where N is the number of coils, f is the friction factor for straight pipe, and r_{Max} and r_{Min} are maximum and minimum radii.† The last term, flow separation, is small and is negligible in most cases.

Limited data are available for constant pitch spirals. Equation 15.4 is compared to a formulation by Kubair and Kuloor [13] in Table 15.2. The Reynolds number

* Note that $\sqrt{(2\pi r)^2 + \mathrm{p}^2}$ is the centerline length of one 360° helical coil.

† Note that $\pi\left(r_{Max}^2 - r_{Min}^2\right)/\mathrm{p}$ is the approximate centerline length of a constant pitch spiral. Also, note that $r_{Max} - r_{Min}/\mathrm{p}$ is equal to N.

TABLE 15.2. Comparison of Equation 15.4 with Kubair and Kuloor's Formulation

							K at N_{Re} = 10,000	
Spiral No.	d (cm)	f[a]	r_{Max} (cm)	r_{Min} (cm)	p (cm)	N[b]	Kubair and Kuloor [13]	Equation 15.4
I	1.260	0.0312	29.5	6.0	6.5	3.62	11.5	11.4
II	0.642	0.0314	23.5	6.0	3.5	5.0	26.9	25.1
III	0.957	0.0313	26.0	5.0	4.0	5.25	18.4	18.8

[a] Friction factor was calculated using Equation 8.3, assuming an absolute roughness e of 0.000060 in for smooth pipe or tubing.
[b] Kubair and Kuloor reported the number of coils as 3.5, 4.5 and 5.0 respectively for the three spirals. The tabulated numbers are calculated values based on their reported geometry features.

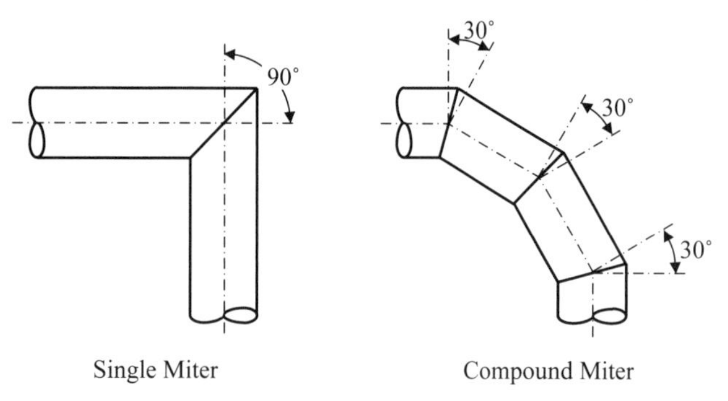

FIGURE 15.9. Ninety-degree miter bends.

was assumed to be 10,000, which is within the range of Kubair and Kuloor's experiments on three spirals of different geometry on which their formulation was based. Equation 15.4 compares well with Kubair and Kuloor's formulation. Whereas their formulation is applicable to smooth pipes only, Equation 15.4 can be applied to rough as well as smooth pipes.

15.3 MITER BENDS

In a miter bend, two passages are joined together in a sharp angle without any rounding at the plane of intersection. Fittings made in this manner from several miter bends placed one after another were once frequently used in place of smoothly curved bends in welded or riveted pipelines. They are still employed today in the construction of large-size conduits such as for wind tunnels and penstocks. Internal flow passages in hydraulic machinery often take the form of a miter bend. A 90° single miter bend is illustrated in Figure 15.9 along with a 90° multijoint or compound miter bend. The compound miter bend can be constructed of any number of segments and, of course, both types of bends can be constructed to practically any overall angle. Nonetheless, the following study is restricted to single miter bends.

TABLE 15.3. Comparison of Equation 15.5 with Single Bend Loss Coefficient Data

Bend Angle α	Shubart [4]	Crane [14][a]	Haidar [15][b]	Equation 15.5
150°	–	–	2.70	2.71
120°	–	–	2.00	2.03
90°	1.20	60 f_T	1.20	1.20
75°	–	40 f_T	–	0.83
60°	0.54	25 f_T	0.52	0.53
45°	0.29	15 f_T	–	0.30
30°	0.14	8 f_T	0.14	0.15
15°	0.06	4 f_T	–	0.06
0°	–	2 f_T (sic)	–	0

[a] Crane alone presents K as a function of friction factor. Assuming a friction factor of 0.020, the Crane values agree quite well with the other sources.
[b] Compressible flow test data was extrapolated to a Mach number of zero.

From various data sources, the following empirical equation was developed for single miter bends for bend angle α from 0° to 150°:

$$K = 0.42\sin(\alpha/2) + 2.56\sin^3(\alpha/2). \qquad (15.5)$$

As shown in Table 15.3, the results of Equation 15.5 compare favorably with single bend loss coefficient data from various sources.

It would seem that pressure loss through a miter bend should be a strong function of friction factor but this has not been noted by most sources. Loss coefficients of single miter bends as a function of bend angle α can be determined from Diagram 15.1. For data on compound miter bends of a number of constructions, refer to Kirchbach [3], Schubart [4], and Idel'chik [16].

15.4 COUPLED BENDS

Basic loss coefficients are for isolated bends having sufficiently long inlet and outlet lengths to ensure that developed flow exists at the inlet to the bend and redevelops again downstream of the bend. When two bends are closely spaced (or coupled), the flow from the first bend into the second bend is not fully developed and the combined loss coefficient is no longer merely the sum of the two bends; it can be greater or less. The combined loss coefficient is a function of the spacer length l (distance between exit of the first bend and entrance to the second), the order of the bends (if they are dissimilar), and the orientation (twist) of the bends (see Fig. 15.10). A spacer length of four or five pipe diameters is usually sufficient to isolate coupled bends.

It turns out that interactions between closely spaced (or coupled) bends are ignored more often than not. The combined loss coefficient of many bend configurations is less than the sum of the two bends, so that ignoring their interaction leads to an overestimation of pressure loss rather than an underestimation. In a piping arrangement with several coupled bends in various configurations, the plusses and minuses may tend to even out.

If precision is required, data for a number of coupled bend arrangements are available in the literature—see Miller [10], Corp and Hartwell [17], and Murikami, et al. [18–20].

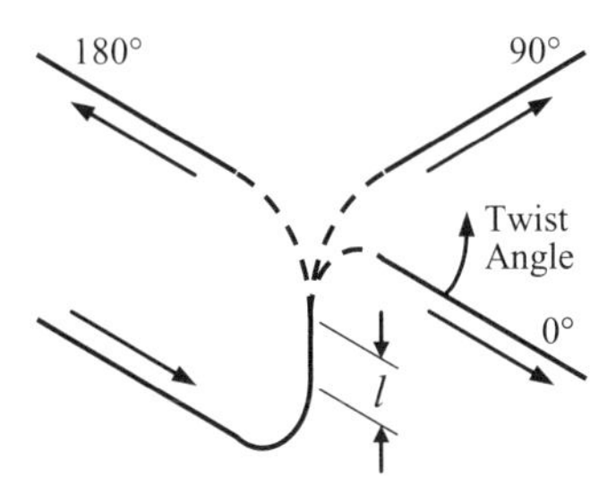

FIGURE 15.10. Configuration of two 90° coupled bends.

15.5 BEND ECONOMY

From Figure 15.3 we see that a minimum loss of head in bends occurs at a radius ratio r/d of about 2 or 3. This simple observation can be misleading. In many piping applications, using bends of larger radius ratio can result in decreased pressure loss. Substituting 45° elbows for 90° elbows can also decrease pressure loss. The piping configurations shown in Figure 15.11 are used to demonstrate this subject.

Several possible configurations of 6-in schedule 40 pipe transporting fluid from point 1 to point 2 are illustrated in Figure 15.11. Assuming fully turbulent flow, data from Tables 15.6 and 15.10 were used to calculate the total loss coefficient for each configuration as shown in Table 15.4.

As demonstrated in the example configurations featured in Figure 15.11, it is possible to significantly decrease piping system losses using pipe bends and 45° elbows. Of course, the piping designer has many other considerations—location of piping system components, proximity to nearby equipment, walkways, and so on—that may influence the piping layout. Where these considerations allow, fabrication costs, as well as pressure loss, can be appreciably reduced.

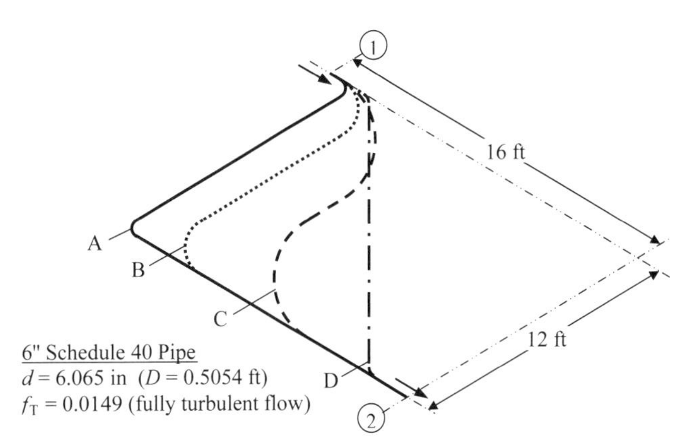

FIGURE 15.11. Example piping configurations. A, two 90° long radius (LR) elbows + 25 ft straight pipe. B, two 90° 5D pipe bends + 18 ft straight pipe. C, two 90° 10D pipe bends + 8 ft straight pipe. D, two 45° LR elbows + 16 ft straight pipe.

TABLE 15.4. Demonstration of Possible Reduction in Pressure Loss for Various Piping Configurations

Configuration	Calculation $K_{Total} = 2 \times K_T + f_T \times L \div D$	Total Loss Coefficient K_{Total}	Reduction
A	$2 \times 0.195 + 0.0149 \times 25 \div 0.5054$	1.13	–
B	$2 \times 0.216 + 0.0149 \times 18 \div 0.5054$	0.96	15%
C	$2 \times 0.328 + 0.0149 \times 8 \div 0.5054$	0.89	21%
D	$2 \times 0.132 + 0.0149 \times 16 \div 0.5054$	0.74	35%

TABLE 15.5. Loss Coefficient K_T for Welded Elbows and Returns in Zone of Complete Turbulence—Clean Commercial Steel Pipe Fittings—Schedule 10

Nominal Pipe Size d_{Nom} (in)	Outside Diameter d_{OD} (in)	Inside Diameter d (in)	Friction Factor[a] f_T	45° Elbow Long Radius[b]	45° Elbow 3R[c]	90° Elbow Short Radius[d]	90° Elbow Long Radius[b]	90° Elbow 3R[c]	180° Return Short Radius[d]	180° Return Long Radius[b]
1	1.315	1.097	0.0222	0.190	–	0.414	0.278	–	0.642	0.333
1-1/4	1.660	1.442	0.0207	0.184	–	0.416	0.274	–	0.691	0.330
1-1/2	1.900	1.682	0.0200	0.176	–	0.389	0.260	–	0.620	0.314
2	2.375	2.157	0.0188	0.165	0.141	0.353	0.243	0.210	0.535	0.293
2-1/2	2.875	2.635	0.0179	0.158	0.136	0.331	0.232	0.204	0.488	0.281
3	3.500	3.260	0.0171	0.154	0.132	0.330	0.228	0.197	0.505	0.277
3-1/2	4.000	3.760	0.0165	0.150	0.129	0.318	0.222	0.193	0.479	0.270
4	4.500	4.260	0.0161	0.146	0.126	0.308	0.216	0.190	0.459	0.264
5	5.563	5.295	0.0153	0.141	0.122	0.295	0.209	0.185	0.437	0.255
6	6.625	6.357	0.0148	0.137	0.119	0.287	0.204	0.180	0.424	0.249
8	8.625	8.329	0.0139	0.131	0.115	0.270	0.195	0.175	0.392	0.239
10	10.750	10.420	0.0133	0.127	0.111	0.261	0.189	0.170	0.379	0.233
12	12.750	12.390	0.0129	0.121	0.109	0.238	0.179	0.169	0.324	0.223
14	15.000	13.624	0.0127	0.119	0.108	0.233	0.177	0.167	0.316	0.221
16	16.000	15.624	0.0123	0.117	0.106	0.230	0.174	0.165	0.313	0.218
18	18.000	17.624	0.0120	0.115	0.104	0.227	0.172	0.162	0.310	0.215
20	20.000	19.564	0.0118	0.114	0.103	0.224	0.170	0.160	0.307	0.213
24	24.000	23.500	0.0114	0.112	0.101	0.220	0.167	0.157	0.302	0.209
30	30.000	29.376	0.0109	0.109	0.098	0.214	0.163	0.154	0.294	0.205
36	36.000	35.376	0.0106	0.106	0.096	0.210	0.160	0.151	0.289	0.202

[a] Friction factor for fully turbulent flow defined by Equation 8.2 using absolute roughness $\varepsilon = 0.00015$ ft for new/clean steel pipe.
[b] Long radius is defined as bend radius r equals $1.5 \cdot d_{Nom}$.
[c] 3R is defined as bend radius r equals 3.0 d_{Nom}.
[d] Short radius is defined as bend radius r equals $1.0 \cdot d_{Nom}$.

TABLE 15.6. Loss Coefficient K_T for Welded Elbows and Returns in Zone of Complete Turbulence—Clean Commercial Steel Pipe Fittings—Schedule 40

Nominal Pipe Size d_{Nom} (in)	Outside Diameter d_{OD} (in)	Inside Diameter d (in)	Friction Factor[a] f_T	45° Elbow Long Radius[b]	45° Elbow 3R[c]	90° Elbow Short Radius[d]	90° Elbow Long Radius[b]	90° Elbow 3R[c]	180° Return Short Radius[d]	180° Return Long Radius[b]
1	1.315	1.049	0.0225	0.188	–	0.395	0.272	–	0.581	0.326
1-1/4	1.660	1.380	0.0210	0.182	–	0.397	0.267	–	0.621	0.321
1-1/2	1.900	1.610	0.0202	0.174	–	0.372	0.255	–	0.561	0.307
2	2.375	2.067	0.0190	0.164	0.142	0.339	0.238	0.212	0.489	0.289
2-1/2	2.875	2.469	0.0182	0.156	0.138	0.312	0.226	0.209	0.430	0.276
3	3.500	3.068	0.0173	0.152	0.133	0.312	0.222	0.200	0.445	0.271
3-1/2	4.000	3.548	0.0168	0.148	0.130	0.301	0.216	0.197	0.426	0.264
4	4.500	4.020	0.0163	0.144	0.127	0.292	0.211	0.193	0.409	0.259
5	5.563	5.047	0.0155	0.139	0.123	0.282	0.205	0.187	0.398	0.251
6	6.625	6.065	0.0149	0.136	0.120	0.275	0.200	0.184	0.387	0.246
8	8.625	7.981	0.0141	0.130	0.115	0.260	0.191	0.177	0.362	0.237
10	10.750	10.020	0.0134	0.126	0.112	0.252	0.186	0.172	0.353	0.231
12	12.750	11.938	0.0130	0.120	0.110	0.231	0.177	0.171	0.306	0.222
14	15.000	13.124	0.0127	0.119	0.108	0.228	0.175	0.169	0.303	0.220
16	16.000	15.000	0.0124	0.116	0.106	0.224	0.172	0.166	0.298	0.217
18	18.000	16.876	0.0121	0.115	0.105	0.220	0.170	0.164	0.293	0.214
20	20.000	18.812	0.0119	0.113	0.104	0.218	0.168	0.162	0.29	0.212
24	24.000	22.624	0.0115	0.111	0.101	0.213	0.165	0.159	0.286	0.208
32	32.000	30.624	0.0109	0.107	0.098	0.208	0.161	0.154	0.282	0.20
36	36.000	34.500	0.0106	0.106	0.097	0.206	0.159	0.152	0.279	0.201

[a] Friction factor for fully turbulent flow defined by Equation 8.2 using absolute roughness $\varepsilon = 0.00015$ ft for new/clean steel pipe.
[b] Long radius is defined as bend radius r equals $1.5 \cdot d_{Nom}$.
[c] 3R is defined as bend radius r equals 3.0 d_{Nom}.
[d] Short radius is defined as bend radius r equals $1.0 \cdot d_{Nom}$.

TABLE 15.7. Loss Coefficient K_T for Welded Elbows and Returns in Zone of Complete—Turbulence Clean Commercial Steel Pipe Fittings—Schedule 120

Nominal Pipe Size d_{Nom} (in)	Outside Diameter d_{OD} (in)	Inside Diameter d (in)	Friction Factor[a] f_T	45° Elbow		90° Elbow			180° Return	
				Long Radius[b]	3R[c]	Short Radius[d]	Long Radius[b]	3R[c]	Short Radius[d]	Long Radius[b]
1	1.315	0.957	0.0230	0.185	–	0.363	0.262	–	0.486	0.319
1-1/4	1.660	1.278	0.0214	0.179	–	0.368	0.258	–	0.525	0.311
1-1/2	1.900	1.500	0.0205	0.172	–	0.348	0.247	–	0.485	0.300
2	2.375	1.939	0.0193	0.162	0.144	0.320	0.218	0.218	0.434	0.284
2-1/2	2.875	2.323	0.0185	0.154	0.140	0.296	0.221	0.214	0.388	0.274
3	3.500	2.900	0.0175	0.150	0.134	0.296	0.218	0.205	0.401	0.267
3-1/2	4.000	3.364	0.0170	0.146	0.131	0.287	0.212	0.200	0.387	0.261
4	4.500	3.826	0.0165	0.143	0.129	0.280	0.208	0.197	0.376	0.257
5	5.563	4.813	0.0157	0.138	0.124	0.271	0.201	0.190	0.366	0.249
6	6.625	5.761	0.0151	0.134	0.121	0.263	0.196	0.186	0.354	0.243
8	8.625	7.625	0.0142	0.128	0.116	0.250	0.188	0.180	0.336	0.235
10	10.750	9.562	0.0136	0.125	0.113	0.243	0.183	0.175	0.327	0.229
12	12.750	11.374	0.0131	0.121	0.110	0.235	0.179	0.171	0.314	0.224
14	15.000	12.500	0.0129	0.118	0.109	0.220	0.173	0.172	0.284	0.220
16	16.000	14.232	0.0125	0.115	0.108	0.215	0.170	0.170	0.278	0.217
18	18.000	16.124	0.0122	0.114	0.106	0.213	0.168	0.167	0.276	0.214
20	20.000	17.938	0.0120	0.112	0.104	0.211	0.166	0.165	0.273	0.212
22	22.000	19.750	0.0118	0.111	0.103	0.208	0.165	0.163	0.270	0.210
24	24.000	21.562	0.0116	0.110	0.102	0.206	0.163	0.162	0.268	0.208

[a] Friction factor for fully turbulent flow defined by Equation 8.2 using absolute roughness $\varepsilon = 0.00015$ ft for new/clean steel pipe.
[b] Long radius is defined as bend radius r equals $1.5 \cdot d_{Nom}$.
[c] 3R is defined as bend radius r equals 3.0 d_{Nom}.
[d] Short radius is defined as bend radius r equals $1.0 \cdot d_{Nom}$.

TABLE 15.8. Loss Coefficient K_T for Welded Elbows and Returns in Zone of Complete Turbulence—Clean Commercial Steel Pipe Fittings—Schedule 160

Nominal Pipe Size d_{Nom} (in)	Outside Diameter d_{OD} (in)	Inside Diameter d (in)	Friction Factor[a] f_T	45° Elbow		90° Elbow			180° Return	
				Long Radius[b]	3R[c]	Short Radius[d]	Long Radius[b]	3R[c]	Short Radius[d]	Long Radius[b]
1	1.315	0.815	0.0240	0.181	–	0.321	0.254	–	0.390	0.324
1-1/4	1.660	1.160	0.0219	0.176	–	0.338	0.249	–	0.441	0.306
1-1/2	1.900	1.338	0.0211	0.169	–	0.315	0.239	–	0.402	0.297
2	2.375	1.687	0.0199	0.159	0.155	0.287	0.227	0.232	0.355	0.286
2-1/2	2.875	2.125	0.0189	0.152	0.143	0.277	0.217	0.223	0.345	0.275
3	3.500	2.624	0.0179	0.148	0.138	0.274	0.212	0.213	0.346	0.267
4	4.500	3.438	0.0169	0.141	0.132	0.257	0.202	0.206	0.324	0.257
5	5.563	4.313	0.0160	0.136	0.127	0.249	0.196	0.199	0.314	0.249
6	6.625	5.187	0.0154	0.132	0.124	0.242	0.191	0.194	0.307	0.243
8	8.625	6.813	0.0145	0.126	0.119	0.230	0.183	0.188	0.289	0.235
10	10.750	8.500	0.0139	0.122	0.116	0.222	0.178	0.183	0.280	0.229
12	12.750	10.126	0.0134	0.119	0.113	0.216	0.174	0.179	0.272	0.225
14	15.000	11.188	0.0131	0.116	0.112	0.205	0.170	0.180	0.254	0.223
16	16.000	12.812	0.0128	0.114	0.110	0.201	0.167	0.177	0.250	0.220
18	18.000	15.438	0.0125	0.112	0.109	0.199	0.165	0.174	0.247	0.217
20	20.000	16.062	0.0122	0.111	0.107	0.196	0.163	0.172	0.244	0.214
22	22.000	17.750	0.0120	0.110	0.106	0.194	0.162	0.170	0.243	0.212
24	24.000	19.312	0.0118	0.108	0.105	0.192	0.160	0.169	0.240	0.210

[a] Friction factor for fully turbulent flow defined by Equation 8.2 using absolute roughness $\varepsilon = 0.00015$ ft for new/clean steel pipe.
[b] Long radius is defined as bend radius r equals $1.5 \cdot d_{Nom}$.
[c] 3R is defined as bend radius r equals 3.0 d_{Nom}.
[d] Short radius is defined as bend radius r equals $1.0 \cdot d_{Nom}$.

TABLE 15.9. Loss Coefficient K_T for Fabricated Pipe Bends in Zone of Complete Turbulence—Clean Commercial Steel Pipe—Schedule 10

Nominal Pipe Size d_{Nom} (in)	Outside Diameter d_{OD} (in)	Inside Diameter d (in)	Friction Factor[a] f_T	5D Bend[b]						10D Bend[c]					
				15°	30°	45°	60°	75°	90°	15°	30°	45°	60°	75°	90°
1	1.315	1.097	0.0222	0.090	0.134	0.170	0.206	0.242	0.278	0.108	0.172	0.234	0.298	0.364	0.429
1-1/4	1.660	1.442	0.0207	0.084	0.125	0.159	0.192	0.232	0.258	0.099	0.158	0.214	0.272	0.332	0.391
1-1/2/	1.900	1.682	0.0200	0.082	0.122	0.156	0.189	0.222	0.255	0.097	0.155	0.211	0.269	0.328	0.387
2	2.375	2.157	0.0188	0.078	0.117	0.151	0.183	0.216	0.248	0.094	0.150	0.206	0.263	0.321	0.379
2-1/2	2.875	2.635	0.0179	0.076	0.114	0.146	0.178	0.211	0.243	0.091	0.146	0.201	0.257	0.314	0.370
3	3.500	3.260	0.0171	0.072	0.109	0.140	0.171	0.201	0.231	0.086	0.138	0.190	0.242	0.295	0.348
3-1/2	4.000	3.760	0.0165	0.071	0.106	0.137	0.167	0.197	0.227	0.084	0.136	0.186	0.238	0.290	0.342
4	4.500	4.260	0.0161	0.069	0.104	0.135	0.165	0.194	0.224	0.082	0.133	0.183	0.234	0.286	0.337
5	5.563	5.295	0.0153	0.066	0.101	0.131	0.160	0.189	0.217	0.079	0.129	0.177	0.226	0.276	0.326
6	6.625	6.357	0.0148	0.064	0.103	0.136	0.163	0.187	0.212	0.077	0.125	0.171	0.219	0.268	0.316
8	8.625	8.329	0.0139	0.062	0.094	0.123	0.151	0.178	0.206	0.074	0.120	0.166	0.212	0.260	0.306
10	10.750	10.420	0.0133	0.060	0.091	0.119	0.146	0.173	0.200	0.071	0.116	0.160	0.205	0.251	0.296
12	12.750	12.390	0.0129	0.059	0.090	0.119	0.146	0.174	0.202	0.071	0.117	0.162	0.209	0.256	0.302
14	15.000	13.624	0.0126	0.058	0.090	0.118	0.145	0.173	0.200	0.070	0.116	0.161	0.207	0.254	0.300
16	16.000	15.624	0.0123	0.057	0.086	0.115	0.142	0.169	0.196	0.069	0.113	0.157	0.202	0.248	0.293
18	18.000	17.624	0.0120	0.056	0.086	0.113	0.140	0.167	0.193	0.067	0.111	0.154	0.198	0.243	0.287
20	20.000	19.564	0.0118	0.056	0.085	0.112	0.138	0.164	0.190	0.066	0.109	0.152	0.195	0.239	0.282
24	24.000	23.500	0.0114	0.053	0.083	0.109	0.135	0.161	0.186	0.064	0.106	0.147	0.190	0.232	0.274
30	30.000	29.376	0.0109	0.052	0.081	0.106	0.131	0.157	0.181	0.062	0.103	0.143	0.184	0.225	0.266
36	36.000	35.376	0.0106	0.050	0.079	0.104	0.129	0.153	0.177	0.060	0.100	0.139	0.179	0.219	0.25F

[a] Friction factor for fully turbulent flow defined by Equation 8.2 using absolute roughness $\varepsilon = 0.00015$ ft for new/clean steel pipe.
[b] 5D is defined as bend radius r equals $5 \cdot d_{Nom}$.
[c] 10D is defined as bend radius r equals $10 \cdot d_{Nom}$.

TABLE 15.10. Loss Coefficient K_T for Fabricated Pipe Bends in Zone of Complete Turbulence—Clean Commercial Steel Pipe—Schedule 40

Nominal Pipe Size d_{Nom} (in)	Outside Diameter d_{OD} (in)	Inside Diameter d (in)	Friction Factor[a] f_T	5D Bend[b]						10D Bend[c]					
				15°	30°	45°	60°	75°	90°	15°	30°	45°	60°	75°	90°
1	1.315	1.049	0.0225	0.092	0.136	0.174	0.212	0.249	0.287	0.111	0.177	0.243	0.310	0.379	0.448
1-1/4	1.660	1.380	0.0210	0.086	0.127	0.163	0.197	0.231	0.266	0.102	0.163	0.222	0.283	0.345	0.407
1-1/2/	1.900	1.610	0.0202	0.083	0.124	0.159	0.193	0.228	0.262	0.100	0.160	0.219	0.279	0.341	0.403
2	2.375	2.067	0.0190	0.079	0.119	0.154	0.187	0.221	0.255	0.096	0.155	0.213	0.272	0.333	0.394
2-1/2	2.875	2.469	0.0182	0.077	0.117	0.151	0.185	0.219	0.254	0.094	0.153	0.212	0.271	0.332	0.393
3	3.500	3.068	0.0173	0.074	0.111	0.144	0.176	0.200	0.240	0.089	0.144	0.199	0.255	0.311	0.368
3-1/2	4.000	3.548	0.0168	0.072	0.109	0.141	0.172	0.204	0.236	0.087	0.141	0.195	0.249	0.305	0.360
4	4.500	4.020	0.0163	0.070	0.107	0.138	0.170	0.201	0.232	0.085	0.139	0.191	0.245	0.300	0.355
5	5.563	5.047	0.0155	0.068	0.103	0.133	0.164	0.194	0.224	0.082	0.133	0.183	0.235	0.288	0.340
6	6.625	6.065	0.0149	0.065	0.100	0.130	0.159	0.189	0.218	0.079	0.129	0.178	0.228	0.279	0.329
8	8.625	7.981	0.0141	0.063	0.096	0.125	0.154	0.182	0.211	0.076	0.124	0.171	0.220	0.269	0.317
10	10.750	10.020	0.0134	0.060	0.093	0.121	0.149	0.177	0.204	0.073	0.119	0.165	0.211	0.259	0.305
12	12.750	11.938	0.0130	0.059	0.092	0.121	0.149	0.178	0.206	0.072	0.120	0.167	0.215	0.263	0.311
14	15.000	13.124	0.0127	0.058	0.090	0.119	0.147	0.176	0.204	0.071	0.118	0.164	0.212	0.259	0.307
16	16.000	15.000	0.0124	0.057	0.089	0.117	0.145	0.173	0.200	0.070	0.116	0.161	0.208	0.254	0.301
18	18.000	16.876	0.0121	0.056	0.087	0.115	0.143	0.170	0.197	0.069	0.114	0.158	0.204	0.250	0.296
20	20.000	18.812	0.0119	0.055	0.086	0.114	0.141	0.168	0.194	0.067	0.112	0.156	0.201	0.246	0.290
24	24.000	22.624	0.0115	0.054	0.084	0.111	0.137	0.164	0.190	0.065	0.109	0.152	0.195	0.239	0.282
32	32.000	30.624	0.0109	0.052	0.081	0.106	0.132	0.157	0.182	0.062	0.103	0.144	0.186	0.227	0.268
36	36.000	34.500	0.0106	0.051	0.079	0.105	0.130	0.155	0.180	0.061	0.102	0.142	0.183	0.223	0.264

[a] Friction factor for fully turbulent flow defined by Equation 8.2 using absolute roughness $\varepsilon = 0.00015$ ft for new/clean steel pipe.
[b] 5D is defined as bend radius r equals $5 \cdot d_{Nom}$.
[c] 10D is defined as bend radius r equals $10 \cdot d_{Nom}$.

TABLE 15.11. Loss Coefficient K_T for Fabricated Pipe Bends in Zone of Complete Turbulence—Clean Commercial Steel Pipe—Schedule 80

Nominal Pipe Size d_{Nom} (in)	Outside Diameter d_{OD} (in)	Inside Diameter d (in)	Friction Factor[a] f_T	5D Bend[b]						10D Bend[c]					
				15°	30°	45°	60°	75°	90°	15°	30°	45°	60°	75°	90°
1	1.315	0.957	0.0230	0.095	0.142	0.183	0.224	0.265	0.307	0.117	0.191	0.263	0.338	0.414	0.490
1-1/4	1.660	1.278	0.0214	0.088	0.131	0.169	0.206	0.243	0.280	0.107	0.172	0.237	0.303	0.370	0.438
1-1/2/	1.900	1.500	0.0205	0.085	0.128	0.165	0.201	0.238	0.275	0.104	0.169	0.232	0.297	0.364	0.430
2	2.375	1.939	0.0193	0.081	0.123	0.159	0.194	0.231	0.267	0.100	0.163	0.224	0.288	0.353	0.418
2-1/2	2.875	2.323	0.0185	0.079	0.120	0.156	0.192	0.228	0.265	0.098	0.161	0.223	0.286	0.351	0.416
3	3.500	2.900	0.0175	0.075	0.114	0.148	0.182	0.216	0.250	0.092	0.150	0.208	0.267	0.327	0.387
3-1/2	4.000	3.364	0.0170	0.073	0.111	0.145	0.178	0.211	0.244	0.090	0.147	0.203	0.261	0.320	0.378
4	4.500	3.826	0.0165	0.072	0.109	0.142	0.174	0.207	0.240	0.088	0.144	0.199	0.256	0.313	0.371
5	5.563	4.813	0.0157	0.069	0.105	0.136	0.168	0.199	0.231	0.084	0.137	0.190	0.245	0.300	0.354
6	6.625	5.761	0.0151	0.067	0.102	0.133	0.163	0.194	0.225	0.081	0.133	0.185	0.238	0.291	0.344
8	8.625	7.625	0.0142	0.064	0.098	0.128	0.157	0.187	0.217	0.078	0.128	0.177	0.228	0.279	0.330
10	10.750	9.562	0.0136	0.061	0.094	0.124	0.152	0.181	0.210	0.075	0.123	0.171	0.219	0.269	0.318
12	12.750	11.374	0.0131	0.060	0.092	0.121	0.149	0.178	0.206	0.073	0.120	0.167	0.215	0.263	0.311
14	15.000	12.500	0.0129	0.059	0.092	0.122	0.151	0.181	0.210	0.073	0.122	0.171	0.220	0.270	0.320
16	16.000	14.232	0.0125	0.058	0.091	0.120	0.149	0.178	0.207	0.072	0.120	0.168	0.217	0.266	0.314
18	18.000	16.124	0.0122	0.057	0.089	0.118	0.146	0.174	0.203	0.070	0.117	0.164	0.212	0.259	0.307
20	20.000	17.938	0.0120	0.056	0.088	0.116	0.144	0.172	0.200	0.069	0.115	0.161	0.208	0.255	0.302
22	22.000	19.750	0.0118	0.055	0.086	0.115	0.142	0.170	0.198	0.068	0.114	0.159	0.205	0.252	0.298
24	24.000	21.562	0.0116	0.055	0.085	0.113	0.141	0.168	0.195	0.067	0.112	0.157	0.203	0.248	0.294

[a] Friction factor for fully turbulent flow defined by Equation 8.2 using absolute roughness $\varepsilon = 0.00015$ ft for new/clean steel pipe.
[b] 5D is defined as bend radius r equals $5 \cdot d_{Nom}$.
[c] 10D is defined as bend radius r equals $10 \cdot d_{Nom}$.

TABLE 15.12. Loss Coefficient K_T for Fabricated Pipe Bends in Zone of Complete Turbulence—Clean Commercial Steel Pipe—Schedule 160

Nominal Pipe Size d_{Nom} (in)	Outside Diameter d_{OD} (in)	Inside Diameter d (in)	Friction Factor[a] f_T	5D Bend[b]						10D Bend[c]					
				15°	30°	45°	60°	75°	90°	15°	30°	45°	60°	75°	90°
1	1.315	0.815	0.0240	0.102	0.154	0.202	0.250	0.300	0.350	0.132	0.218	0.305	0.394	0.485	0.576
1-1/4	1.660	1.160	0.0219	0.091	0.137	0.178	0.218	0.260	0.301	0.114	0.186	0.257	0.331	0.406	0.481
1-1/2/	1.900	1.338	0.0211	0.089	0.135	0.176	0.216	0.258	0.299	0.112	0.184	0.256	0.330	0.405	0.480
2	2.375	1.687	0.0199	0.086	0.131	0.172	0.213	0.254	0.296	0.110	0.181	0.253	0.327	0.402	0.477
2-1/2	2.875	2.125	0.0189	0.082	0.125	0.164	0.203	0.243	0.283	0.104	0.172	0.240	0.311	0.382	0.453
3	3.500	2.624	0.0179	0.078	0.119	0.156	0.193	0.230	0.268	0.098	0.162	0.226	0.292	0.358	0.425
4	4.500	3.438	0.0169	0.074	0.114	0.150	0.186	0.222	0.259	0.094	0.156	0.217	0.281	0.345	0.409
5	5.563	4.313	0.0160	0.071	0.110	0.144	0.179	0.214	0.249	0.090	0.149	0.208	0.269	0.330	0.391
6	6.625	5.187	0.0154	0.069	0.106	0.140	0.174	0.208	0.242	0.087	0.144	0.201	0.260	0.319	0.378
8	8.625	6.813	0.0145	0.066	0.102	0.135	0.168	0.201	0.234	0.083	0.139	0.194	0.251	0.308	0.365
10	10.750	8.500	0.0139	0.064	0.099	0.131	0.163	0.195	0.227	0.080	0.134	0.187	0.242	0.297	0.352
12	12.750	10.126	0.0134	0.062	0.097	0.128	0.159	0.191	0.222	0.078	0.130	0.183	0.236	0.290	0.344
14	15.000	11.188	0.0131	0.062	0.097	0.129	0.161	0.193	0.226	0.079	0.132	0.186	0.242	0.297	0.352
16	16.000	12.812	0.0128	0.061	0.095	0.126	0.168	0.190	0.221	0.077	0.129	0.182	0.236	0.290	0.344
18	18.000	15.438	0.0125	0.059	0.093	0.124	0.155	0.187	0.218	0.075	0.127	0.179	0.232	0.285	0.337
20	20.000	16.062	0.0122	0.058	0.092	0.122	0.153	0.184	0.214	0.074	0.125	0.176	0.228	0.280	0.332
22	22.000	17.750	0.0120	0.058	0.090	0.121	0.151	0.181	0.211	0.073	0.123	0.173	0.224	0.275	0.326
24	24.000	19.312	0.0118	0.057	0.089	0.119	0.149	0.179	0.209	0.072	0.121	0.171	0.221	0.272	0.322

[a] Friction factor for fully turbulent flow defined by Equation 8.2 using absolute roughness $\varepsilon = 0.00015$ ft for new/clean steel pipe.
[b] 5D is defined as bend radius r equals $5 \cdot d_{Nom}$.
[c] 10D is defined as bend radius r equals $10 \cdot d_{Nom}$.

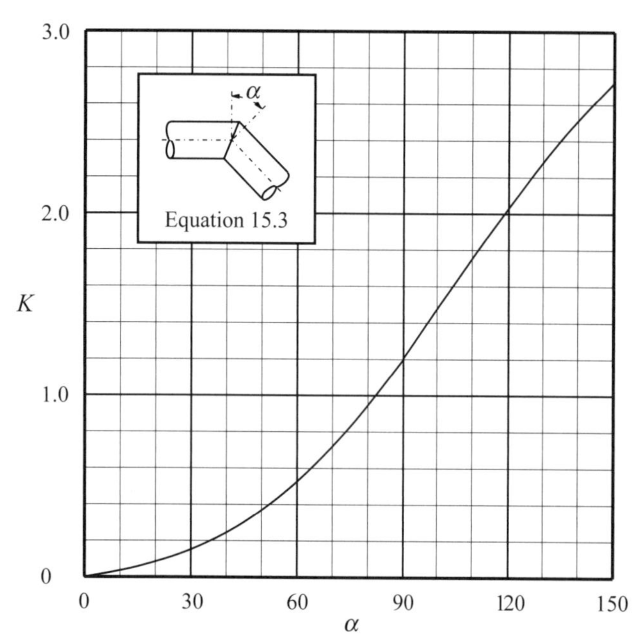

DIAGRAM 15.1. Loss coefficient K for a single miter bend.

REFERENCES

1. Ito, H., Pressure losses in smooth pipe bends, *Journal of Basic Engineering, Transactions of the American Society of Mechanical Engineers*, 95, March 1960, 131–143.
2. Hoffmann, A., Loss in 90-degree pipe bends of constant circular cross-section, Transactions of the Munich Hydraulic Institute, Munich Technical University, Bulletin No. 3, 1929. (Translation published by American Society of Mechanical Engineers, 1935, pp. 29–41.)
3. Kirchbach, H., Loss of energy in miter bends, Transactions of the Hydraulic Institute, Munich Technical University, Bulletin No. 3, 1929. (Translation published by American Society of Mechanical Engineers, 1935, pp. 43–64.)
4. Schubart, W., Energy loss in smooth- and rough-surfaced bends and curves in pipe lines, Transactions of the Hydraulic Institute, Munich Technical University, Bulletin No. 3, 1929. (Translation published by the American Society of Mechanical Engineers, 1935, pp. 81–99.)
5. Wasielewski, R., Losses in smooth pipe bends of circular cross section with less than 90 degrees change, *Mitteilungen Hydraulische Instituts, Technischen Hochschule Munchen (in German)*, 5, 1932, 53–67.
6. Ito, H., On the pressure for turbulent flow in smooth pipe bends, *R. Institute of High Speed Mechanics*, 6(54), 1955, 54–102.
7. Ito, H., Friction factors for turbulent flow in curved pipes, *Journal of Basic Engineering, Transactions of the American Society of Mechanical Engineers*, 94, June 1959, 123–134.
8. Ito, H., Laminar flow in curved pipe, *Japan Society of Mechanical Engineers International Journal*, 12(262), 1969, 653–563.
9. Ito, H., Flow in curved pipe, *Japan Society of Mechanical Engineers International Journal*, 30(262), 1987, 543–552.
10. Miller, D. S., *Internal Flow, a Guide to Losses in Pipe and Duct Systems*, The British Hydromechanics Research Association, 1971.
11. Kubair, V. and C. B. S. Varrier, Pressure drop for liquid flow in helical coils, *Transactions Indian Institute Chemical Engineers*, 14, 1961/62, 93–97.
12. Mori, Y. and W. Nakayama, Study on forced convective heat transfer in curved pipes (2nd Report, Turbulent Region), *International Journal Heat Mass Transfer*, 10, 1967, 37–59.
13. Kubair, V. and N. R. Kuloor, Comparison of performance of helical & spiral coil heat exchangers, *Indian Journal of Technology*, 4, January 1966, 1–3.
14. Flow of Fluids Through Valves, Fittings and Pipe, Technical Paper No. 410, Crane Company, Chicago, 1985, p. A-26.
15. Haidar, N., Prediction of compressible flow pressure losses in 30–150 deg sharp-cornered bends, *Journal of Fluids Engineering, Transactions of the American Society of Mechanical Engineers*, 117, December 1995, 589–592.
16. Idel'chik, I. E., *Handbook of Hydraulic Resistance—Coefficients of Local Resistance and of Friction*, Moskva-Leningrad, 1960. (Translated from Russian; Published for the U.S. Atomic Energy Commission and the National Science Foundation, Washington, D.C. by the Israel Program for Scientific Translations, Jerusalem, 1966.)
17. Corp, C. I. and H. T. Hartwell, Experiments on loss of head in U, S, and Twisted S pipe bends, Bulletin of the University of Wisconsin, Engineering Experiment Station Series No. 66, 1927, pp. 1–181.
18. Murikami, M., Y. Shimuzu, and H. Shiragami, Studies on fluid flow in three-dimensional bend conduits, *Bulletin of Japan Society of Mechanical Engineers*, 12(54), 1969, 1369–1379.
19. Murakami, M. and Y. Shimuzu, Hydraulic losses and flow patterns in pipes with two bends combined, *Bulletin of the Japan Society of Mechanical Engineers*, 20(147), 1977, 1136–1144.
20. Murakami, M. and Y. Shimuzu, Asymmetric swirling flows in composite pipe bends, *Bulletin of the Japan Society of Mechanical Engineers*, 21(157), 1978, 1144–1151.

FURTHER READING

This list includes books and papers that may be helpful to those who wish to pursue further study.

Williams, G. S. et al., Experiments at Detroit, Michigan, on the effect of curvature upon the flow of water in pipes,

Transactions of the American Society of Civil Engineers, Paper No. 911, Presented at the Meeting of September 4th, 1901.

Eustice, J., Flow of water in curved pipes, *Proceedings Royal Society of London, Series A*, 84, 1910, 107–118.

Eustice, J., Experiments on stream-line motion in curved pipes, *Proceedings Royal Society of London, Series A*, 85, 1911, 119–131.

White, C. M., Streamline flow through curved pipes, *Proceedings of the Royal Society, London*, A123, 1929, 645–663.

Keulegan, G. H. and K. H. Beij, Pressure loss for fluid flow in curved pipes, *Journal of Research of the National Bureau of Standards*, 18, January 1937, 89–115.

Beij, K. H., Pressure loss for fluid flow in 90° pipe bends, *Journal of Research of the National Bureau of Standards*, 21, July 1938, 1–17.

Freeman, J. R., *Experiments Upon the Flow of Water in Pipes and Pipe Fittings Made At Nashua, New Hampshire, June 28 to October 22, 1892*, The American Society of Mechanical Engineers, 1941.

Vazsonyi, A. and D. Branches, Pressure loss in elbows and duct branches, *Transactions of the American Society of Mechanical Engineers*, 1966, April 1944, 177–182.

Kubair, V. and N. R. Kuloor, Flow of Newtonian fluids in Archimedean spiral tube coils: Correlation of the laminar, transition & turbulent flows, *Indian Journal of Technology*, 4, January 1966, 3–8.

Zanker, K. J. and T. E. Brock, A review of the literature on bend flow through closed conduit bends, *The British Hydromechanics Research Laboratory*, TN 901, July 1967.

Srinivasan, P. S., S. S. Nandapurkar, and F. A. Holland, Pressure drop and heat transfer in coils, *The Chemical Engineer*, (218), May 1968, 113–119.

Sprenger, H., Pressure head loss in 90° bends for tubes or ducting of rectangular cross-section, *Schweizerische Bauzeitung*, 87(13), March 27, 1969, 233–231 (in German).

Srinivasan, P. S., S. S. Nandapurkar, and F. A. Holland, Friction factor for coils, *Transactions of the Institution of Chemical Engineers*, 48, 1970, 156–161.

Sankaraiah, M. and Y. V. N. Rao, Analysis of steady laminar flow of an incompressible Newtonian fluid through curved pipes of small curvature, *Journal of Fluids Engineering, Transactions of the American Society of Mechanical Engineers*, 95, March 1973, 75–80.

Smith, F. T., Steady motion within a curved pipe, *Proceedings of the Royal Society, London*, A347, 1976, 345–370.

Singh, R. P. and P. Mishra, Friction factor for Newtonian and non-Newtonian fluid flow in curved pipes, *Journal of Chemical Engineering of Japan*, 13(4), 1980, 275–280.

Taylor, A. M. K. P., J. H. Whitelaw, and M. Yianneskis, Curved ducts with strong secondary motion: Velocity measurements of developing laminar and turbulent flow, *Journal of Fluids Engineering, Transactions of the American Society of Mechanical Engineers*, 104, September 1982, 350–359.

Takami, T. and K. Susou, Flow through curved pipes with elliptic cross section, *Bulletin of the Japan Society of Mechanical Engineers*, 27(228), June 1984, 1176–1181.

Daskopoulos, P. and A. M. Lenhoff, Flow in curved ducts: Bifurcation structure for stationary ducts, *Journal of Fluid Mechanics*, 203, 1989, 125–148.

Belaidi, A., M. W. Johnson, and J. A. C. Humphrey, Flow instability in a curved duct of rectangular cross section, *Journal of Fluids Engineering, Transactions of the American Society of Mechanical Engineers*, 114, December 1992, 585–592.

Shimuzu, Y., Y. Futaki, and S. S. Martin, Secondary flow and hydraulic losses within sinuous conduits of rectangular cross section, *Journal of Fluids Engineering, Transactions of the American Society of Mechanical Engineers*, 114, December 1992, 593–600.

Hamakiotes, C. C. and S. A. Berger, Periodic flow through curved tubes: The effect of the frequency parameter, *Journal of Fluid Mechanics*, 210, 1997, 353–370.

Graf, E. and S. Neti, Two-phase pressure drop in right angle bends, *Journal of Fluids Engineering, Transactions of the American Society of Mechanical Engineers*, 122, December 2000, 761–768.

16

TEES

Determining energy losses caused by the division and combination of flow at pipe junctions is of great importance in the design and analysis of piping systems. The energy loss in junctions not only depends on geometric properties (angle of branch with respect to the run, branch-to-run diameter ratio, and curvature of the joining edge) but also upon the direction of flow and the proportion of flow division. In practice, however, constant loss coefficient values have often been used.

A great deal of theoretical and experimental research on junctions has been reported over the years. However, experimental data and published formulas for loss coefficients have provided results that are in considerable disagreement. Most researchers have not taken into account all possible configurations of flow. Except for Gardel [1,2], and Ito and Imai [3], published equations are for sharp-edged pipe junctions only. Experimental results and published equations are, as a rule, presented based on velocity in the common channel, the channel containing the maximum (or combined) flow. Herein, however, loss coefficients are also presented in terms of velocity in the flow paths *entering* and *exiting* the common channel because this format is most useful in engineering calculations.

The lack of quantitative agreement between the results of different investigators is not surprising as the test results depend on the extension of hydraulic grade lines that are difficult to establish accurately. It is also evident that in some cases errors in geometry, in flow measurement, or in pressure measurement account for discrepancies among results. Notwithstanding, the following work presents loss coefficients of reasonable accuracy.

This chapter is concerned with 90° tees that are most frequently used in piping systems. Semiempirical formulas are developed based on conservation of mass, energy, and momentum principles. Coefficients are added to match available data. The formulas account for (1) the ratio of flow rates through the tee, (2) the ratio of the branch (lateral channel) to run (main channel) diameter, and (3) the ratio of the radius r of the branch edge to the diameter of the branch.

The various configurations of flow through a tee are shown in Figure 16.1. The directions of flow under consideration are shown by heavy dashed lines. It can be seen that there are six kinds of flow that differ fundamentally from each other. In all cases, the flow in the common channel is denoted by the subscript $_1$, and the flows entering or exiting the common channel are denoted by the subscripts $_2$ and $_3$. Thus, the flow rate relationship of the various configurations always takes the form of:

$$\dot{w}_1 = \dot{w}_2 + \dot{w}_3.$$

The loss coefficient equations are first developed in terms of velocity in the common channel in which form they are compared to experimental results and published formulas. The equations are then rearranged in terms of the velocity in the flow paths entering or exiting the common channel. Finally, they are

Pipe Flow: A Practical and Comprehensive Guide, First Edition. Donald C. Rennels and Hobart M. Hudson.

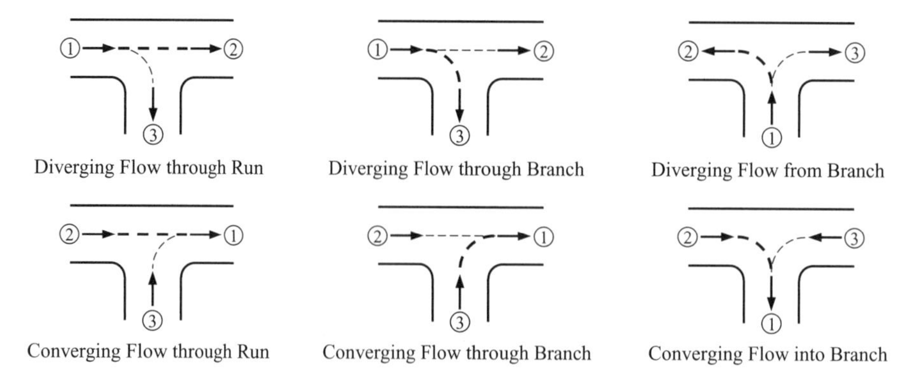

FIGURE 16.1. Flow configurations through tees.

presented in a pressure drop (or energy) equation, in which format they can be readily employed in an engineering calculation.

The Reynolds number is not a significant variable in the performance of junctions if it is above 10^4 in the common flow channel. Surface roughness does not appear to be a significant factor. There are a lack of reliable data concerning the effects of inlet and outlet conditions, but it is thought that loss will not be significantly affected if components are located three or more diameters before or after a tee. The loss coefficient of certain flow paths can actually have a negative value under certain flow conditions, which means that an energy increase has occurred in that flow path. Energy loss in the other flow path more than compensates for the increase so that the net result is an energy loss.

Investigators typically relate the radius of the branch edge to the diameter of the *run* in reported test results and formulations. Herein, the radius ratio is related to the diameter of the branch where it makes more sense, particularly when the diameter of the branch is smaller than that of the run. This allows the use of entrance loss data from Chapter 9 in the case of diverging flow through the branch.

There are no industry standards regarding the degree of rounding of the branch edge of commercial tees. For lack of manufacturers' information, a radius ratio r/d_{branch} of 0.10 is a reasonable and, likely, a conservative assumption. In most cases about 50% of the benefit of rounding is provided by a radius ratio of 0.10, and about 90% of the benefit is provided by a radius ratio of 0.30.

16.1 DIVERGING TEES

Local losses of diverging tees mainly consist of more or less sudden expansion losses in the main channel (run) and of losses due to flow turning into or from the branch. A radius or chamfer at the leading edge of the joining branch has a beneficial effect on branch loss but has little effect on flow through run loss.

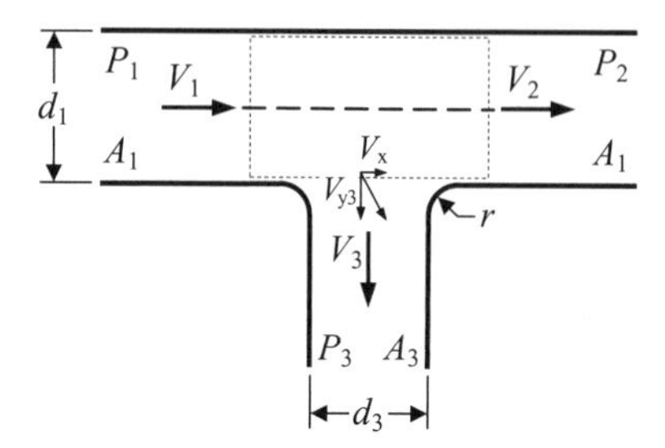

FIGURE 16.2. Flow diverging through run of tee.

16.1.1 Flow through Run

As part of the fluid leaves a run of uniform size through the branch, the velocity in the run decreases suddenly and produces an effect similar to that accompanying a sudden increase in pipe size. The energy equation for flow through run (see Fig. 16.2) can be written as:

$$\frac{P_1}{\rho_w}+\frac{V_1^2}{2g}=\frac{P_2}{\rho_w}+\frac{V_2^2}{2g}+K_{12_1}\frac{V_1^2}{2g},$$

where K_{12_1} is the loss coefficient for flow through run (from point 1 to point 2) in terms of the velocity at point 1.

Rearrangement of the energy equation gives:

$$K_{12_1}=\frac{2g}{\rho_w V_1^2}(P_1-P_2)+1-\frac{V_2^2}{V_1^2}. \tag{16.1}$$

A momentum balance in the x-direction gives:

$$A_1(P_1-P_2)=\frac{V_2\dot{w}_2}{g}-\frac{V_1\dot{w}_1}{g}+\frac{V_x\dot{w}_3}{g}.$$

We expect $V_x < V_1$ and express this relationship as $V_x = C_{xD}V_1$, where the coefficient C_{xD} allows for an uncertainty in axial momentum transported through the branch due to the turning of the flow. By using this relationship and rearranging, the momentum equation becomes:

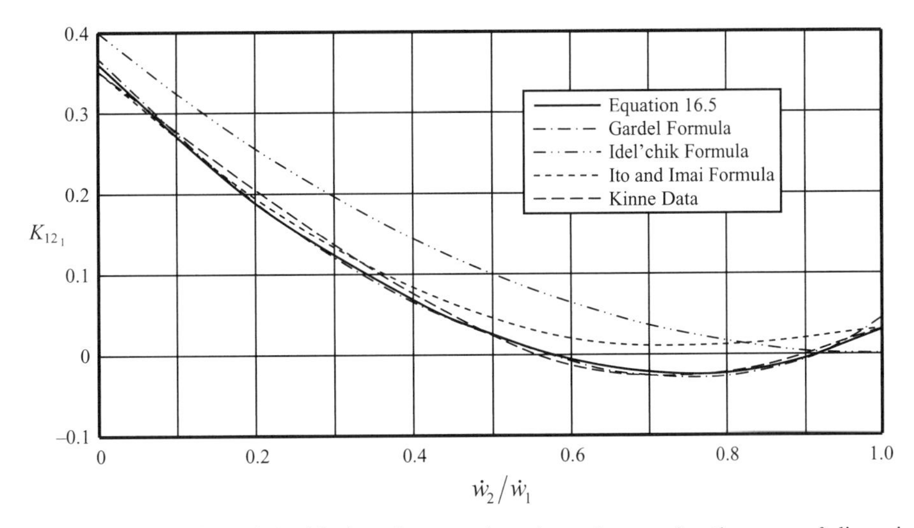

FIGURE 16.3. Comparison of Equation 16.5 with data from various investigators for the case of diverging flow through run of tee.

$$P_1 - P_2 = \frac{1}{gA_1}(V_2\dot{w}_2 - V_1\dot{w}_1 + C_{xD}V_1\dot{w}_3).$$

By use of the continuity equations $\dot{w}_3 = \dot{w}_1 - \dot{w}_2$, $\dot{w}_1 = V_1A_1\rho_w$, and $\dot{w}_2 = V_2A_1\rho_w$, the momentum equation becomes:

$$P_1 - P_2 = \frac{\rho_w V_1^2}{g}\left(\frac{V_2^2}{V_1^2} - 1 + C_{xD} - C_{xD}\frac{V_2}{V_1}\right). \quad (16.2)$$

Substituting Equation 16.2 into Equation 16.1 and letting $V_1 = \dot{w}_1/A_1\rho_w$ and $V_2 = \dot{w}_2/A_1\rho_w$, the loss coefficient equation becomes:

$$K_{12_1} = \frac{\dot{w}_2^2}{\dot{w}_1^2} + (2C_{xD} - 1) - 2C_{xD}\frac{\dot{w}_2}{\dot{w}_1}. \quad (16.3)$$

Experimental results indicate that the loss coefficient is fairly insensitive to the diameter ratio d_3/d_1, and to the radius r of the branch inlet. Correlation with available data indicates that C_{xD} is a function of the flow rate ratio as follows:

$$C_{xD} = 0.68 + 0.19\frac{\dot{w}_2}{\dot{w}_1}. \quad (16.4)$$

As in the case of converging flow through run, a small loss results from the abrupt enlargement and contraction of flow area past the branch when $\dot{w}_2$ approaches $\dot{w}_1$. This loss was not accounted for in the derivation of Equation 16.3. Adding a term to account for this loss, plus substitution of Equation 16.4 into Equation 16.3, gives:

$$K_{12_1} = 0.36 - 0.98\frac{\dot{w}_2}{\dot{w}_1} + 0.62\frac{\dot{w}_2^2}{\dot{w}_1^2} + 0.03\frac{\dot{w}_2^8}{\dot{w}_1^8}. \quad (16.5)$$

Results from Equation 16.5 are compared with published formulas by Gardel [2], Idel'chik [4], and Ito and Imai [3], as well as with test data by Kinne [5], in Figure 16.3. The results compare favorably with Kinne's data and with Ito and Imai's formula; they compare less so with Gardel's and Idel'chik's formulas.

Multiplying Equation 16.5 by $\dot{w}_1^2/\dot{w}_2^2$ produces the loss coefficient for diverging flow through run in terms of the velocity at point 2:

$$K_{12_2} = 0.62 - 0.98\frac{\dot{w}_1}{\dot{w}_2} + 0.36\frac{\dot{w}_1^2}{\dot{w}_2^2} + 0.03\frac{\dot{w}_2^6}{\dot{w}_1^6}. \quad (16.6)$$

Loss coefficient K_{12_2} for diverging flow through run can be obtained from Diagram 16.1. The energy equation in terms of the velocity at point 2 may be written as:

$$P_1 - P_2 = \frac{\dot{w}_2^2}{2g\rho_w A_1^2}\left(K_{12_2} + 1 - \frac{\dot{w}_1^2}{\dot{w}_2^2}\right).$$

By substituting Equation 16.6 into the energy, the pressure drop equation for diverging flow through run becomes:

$$P_1 - P_2 = \frac{\dot{w}_2^2}{2g\rho_w A_1^2}\left(1.62 - 0.98\frac{\dot{w}_1}{\dot{w}_2} - 0.64\frac{\dot{w}_1^2}{\dot{w}_2^2} + 0.03\frac{\dot{w}_2^6}{\dot{w}_1^6}\right). \quad (16.7)$$

16.1.2 Flow through Branch

The change in direction of the flow entering the branch causes a detachment that partly depends on the degree

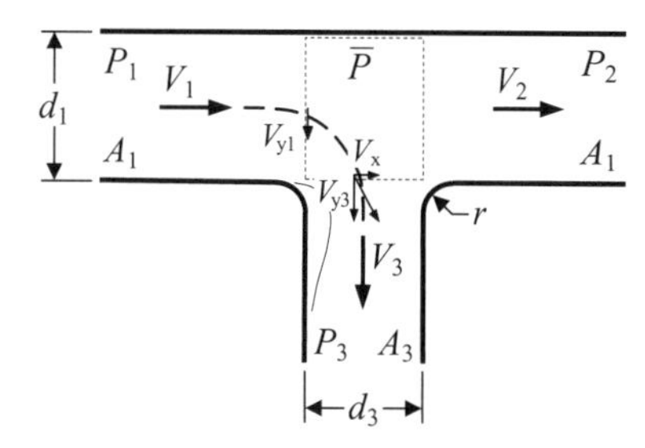

FIGURE 16.4. Diverging flow through branch of tee.

of rounding of the corners. The energy equation for flow through branch (see Fig. 16.4) can be written as:

$$\frac{P_1}{\rho_w}+\frac{V_1^2}{2g}=\frac{P_3}{\rho_w}+\frac{V_3^2}{2g}+K_{13_1}\frac{V_1^2}{2g},$$

where K_{13_1} is the loss coefficient for flow through branch (from point 1 to point 3) in terms of the velocity at point 1. Rearrangement of the energy equation gives:

$$K_{13_1}=\frac{2g}{\rho_w V_1^2}(P_1-P_3)+1-\frac{V_3^2}{V_1^2}. \tag{16.8}$$

A momentum balance in the y-direction can be expressed as:

$$A_3(\bar{P}-P_3)=\frac{V_{y3}\dot{w}_3}{g},$$

where uncertainty in momentum can be expressed as $Vy_3 = C_{y3}V_3$. By applying the continuity equation, $\dot{w}_3 = V_3A_3\rho_w$, and rearranging, the momentum equation becomes:

$$\bar{P}-P_3=C_{y3}\frac{\rho_w V_3^2}{g}.$$

Letting $\bar{P}=P_1-(P_1-P_2)/2$, the momentum equation becomes:

$$P_1-P_3=C_{y3}\frac{\rho_w V_3^2}{g}+\frac{P_1-P_2}{2}. \tag{16.9}$$

Substitution of Equation 16.2 into Equation 16.9 gives:

$$P_1-P_3=C_{y3}\frac{\rho_w V_3^2}{g}+\frac{\rho_w V_1^2}{2g}\left[\frac{V_2^2}{V_1^2}-1+C_{xD}\left(1-\frac{V_2}{V_1}\right)\right]. \tag{16.10}$$

Substitution of Equation 16.10 into Equation 16.8 gives:

$$K_{13_1}=(2C_{y3}-1)\frac{V_3^2}{V_1^2}+\frac{V_2^2}{V_1^2}+C_{xD}\left(1-\frac{V_2}{V_1}\right).$$

A turning/entrance loss into the branch is essentially located downstream of the control volume. At this point an expression, C_{Turn}, is added to account for this loss:

$$K_{13_1}=(2C_{y3}+C_{Turn}-1)\frac{V_3^2}{V_1^2}+\frac{V_2^2}{V_1^2}+C_{xD}\left(1-\frac{V_2}{V_1}\right).$$

Because $V_1=\dot{w}_1/A_1\rho_w$, $V_2=\dot{w}_2/A_1\rho_w$, $V_3=\dot{w}_3/A_1\rho_w$, and $\dot{w}_2=\dot{w}_1-\dot{w}_3$, the loss coefficient equation for flow through branch becomes:

$$K_{13_1}=1-(2-C_{xD})\frac{\dot{w}_3}{\dot{w}_1}+\left\{1+\left[2C_{y3}+C_{Turn}-1\right]\frac{A_1^2}{A_3^2}\right\}\frac{\dot{w}_3^2}{\dot{w}_1^2}, \tag{16.11}$$

C_{xD} is a function of flow rate ratio $\dot{w}_2/\dot{w}_1$ as before:

$$C_{xD}=0.68+0.19\frac{\dot{w}_2}{\dot{w}_1}. \tag{16.4, repeated}$$

Correlation with test data indicates that C_{y3} and C_{Turn} are functions of the diameter ratio and the rounding of the branch inlet. The combined effect is given by:

$$2C_{y3}+C_{Turn}=1+1.12\frac{d_3}{d_1}-1.08\frac{d_3^3}{d_1^3}+K_{9.3}, \tag{16.12}$$

where $K_{9.3}$ is determined as for a rounded entrance for $r/d_3 \le 1.00$ from Chapter 9:

$$K_{9.3}=0.57-1.07(r/d_3)^{1/2}-2.13(r/d_3)+8.24(r/d_3)^{3/2}-8.48(r/d_3)^2+2.90(r/d_3)^{5/2}. \tag{9.3, modified}$$

Substitution of Equations 16.4 and 16.12 into Equation 16.11, and recognizing that d_1^4/d_3^4 equals A_1^2/A_3^2, gives the loss coefficient equation for diverging flow through branch in terms of the velocity at point 1:

$$K_{13_1}=1.00-1.13\frac{\dot{w}_3}{\dot{w}_1}+\left[0.81+\left(1.12\frac{d_3}{d_1}-1.08\frac{d_3^3}{d_1^3}+K_{9.3}\right)\frac{d_1^4}{d_3^4}\right]\frac{\dot{w}_3^2}{\dot{w}_1^2}. \tag{16.13}$$

Equation 16.13 for the case of tees with a diameter ratio equal to unity is compared with data from Kinne [5], Gardel [2], and Ito and Imai [3] in Figure 16.5. There is fair agreement with all three sources. For the case of tees with radius ratio r/d_3 equal to 0.10, and diameter ratios equal to 1.0, 0.583, and 0.349, Equation 16.13 is shown to compare well with data from Kinne [5] in Figure 16.6. Equation 16.13 generally agrees with limited data at other diameter and radius ratios as well (not shown).

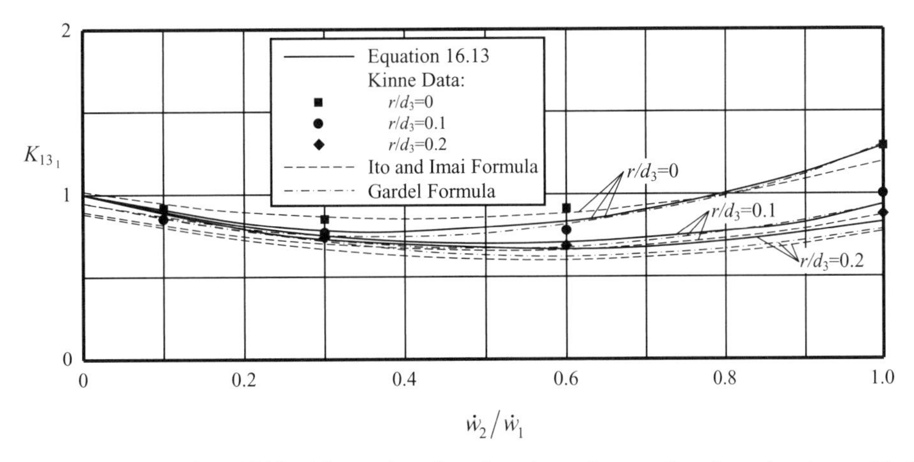

FIGURE 16.5. Comparison of Equation 16.13 with results of various investigators for diverging tees with diameter ratio equal to 1.0.

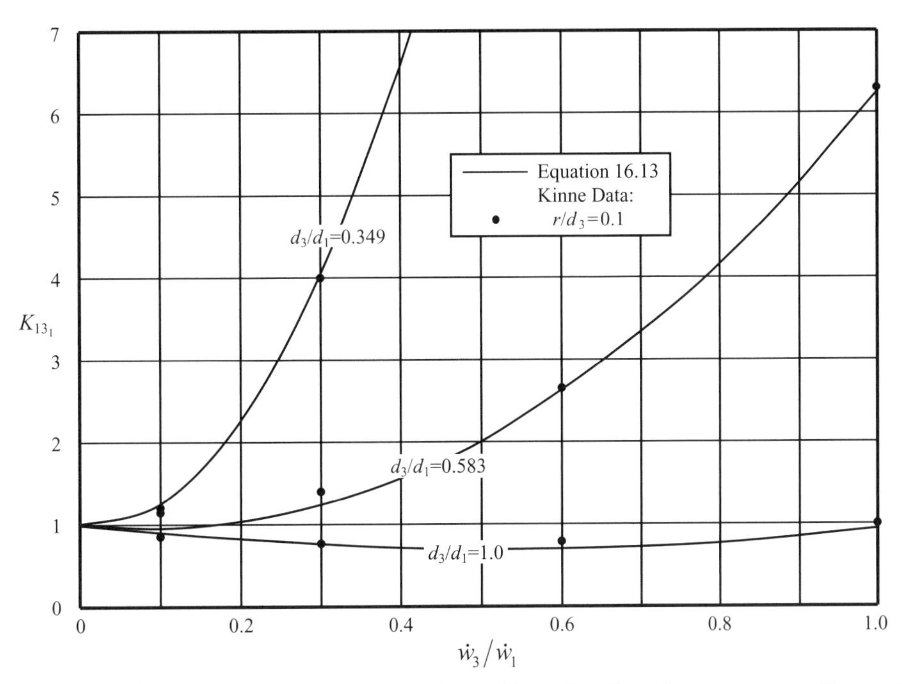

FIGURE 16.6. Comparison of Equation 16.13 with data from Kinne for diverging tees with radius ratio equal to 0.10.

Multiplying Equation 16.13 by $d_3^4\dot{w}_1^2/d_1^4\dot{w}_3^2$ produces the loss coefficient for diverging flow through branch in terms of the velocity at point 3:

$$K_{13_3} = \left(0.81 - 1.13\frac{\dot{w}_1}{\dot{w}_3} + \frac{\dot{w}_1^2}{\dot{w}_3^2}\right)\frac{d_3^4}{d_1^4} + 1.12\frac{d_3}{d_1} - 1.08\frac{d_3^3}{d_1^3} + K_{9.3}. \tag{16.14}$$

Loss coefficient K_{13_3} for diverging flow through branch can be obtained from Diagrams 16.2 through 16.7. The energy equation in terms of the velocity at point 3 may be written as:

$$P_1 - P_3 = \frac{\dot{w}_3^2}{2g\rho_w A_3^2}\left(K_{13_3} + 1 - \frac{\dot{w}_1^2 d_3^4}{\dot{w}_3^2 d_1^4}\right).$$

By substituting Equation 16.14 into the energy equation, the pressure drop equation for diverging flow through branch becomes:

$$P_1 - P_3 = \frac{\dot{w}_3^2}{2g\rho_w A_3^2}\left[\left(0.81 - 1.13\frac{\dot{w}_1}{\dot{w}_3}\right)\frac{d_3^4}{d_1^4} + 1.00 + 1.12\frac{d_3}{d_1} - 1.08\frac{d_3^3}{d_1^3} + K_{9.3}\right]. \tag{16.15}$$

16.1.3 Flow from Branch

The common flow channel is located in the branch in the case of diverging flow from branch of tee as shown in Figure 16.7. Very little test data are available for this configuration. However, Ito and Imai [3] have developed a formula that applies to tees with a run-to-branch diameter ratio of unity that are most frequently used for this purpose. The agreement with their experimental data is satisfactory except where notable peaks in loss appear over the ranges of extremely unequal division of flow.* In terms of the velocity at point 1 (or in the common flow channel), Ito and Imai's formula, transposed to the form and symbols used in this document, is:

$$K_{12_1} = 0.59 + \left(1.18 - 1.84\sqrt{\frac{r}{d}} + 1.16\frac{r}{d}\right)\frac{\dot{w}_2}{\dot{w}_1} - \left(0.68 - 1.04\sqrt{\frac{r}{d}} + 1.16\frac{r}{d}\right)\frac{\dot{w}_2^2}{\dot{w}_1^2} \quad (0.2 \le \dot{w}_2/\dot{w}_1 \le 0.8). \tag{16.16}$$

Multiplying Equation 16.16 by $\dot{w}_1^2/\dot{w}_2^2$ produces the loss coefficient for diverging flow from branch in terms of the velocity at point 2:

$$K_{12_2} = 0.59\frac{\dot{w}_1^2}{\dot{w}_2^2} + \left(1.18 - 1.84\sqrt{\frac{r}{d}} + 1.16\frac{r}{d}\right)\frac{\dot{w}_1}{\dot{w}_2} - 0.68 + 1.04\sqrt{\frac{r}{d}} - 1.16\frac{r}{d} \quad (0.2 \le \dot{w}_2/\dot{w}_1 \le 0.8). \tag{16.17}$$

The energy equation in terms of the velocity at point 2 may be written as:

$$P_1 - P_2 = \frac{\dot{w}_2^2}{2g\rho_w A^2}\left(K_{12_2} - 1 + \frac{\dot{w}_1^2}{\dot{w}_2^2}\right).$$

* In the case of tees with r/d values of 0.1 and 0.2, there was a notable peak of high loss coefficient over each of the two ranges $0 < \dot{w}_2/\dot{w}_1 < 0.2$ and $0.8 < \dot{w}_2/\dot{w}_1 < 1$; the maximum excess in loss coefficient was about 50%. The magnitude of the peak depended on r/d. The peak was not perceptible for a r/d of zero and diminished as r/d approached 0.5.

By substituting Equation 16.17 into the energy equation, the pressure drop equation for diverging flow from branch in terms of the velocity at point 2 is:

$$P_1 - P_2 = \frac{\dot{w}_2^2}{2g\rho_w A^2}\left[1.59\frac{\dot{w}_1^2}{\dot{w}_2^2} + \left(1.18 - 1.84\sqrt{\frac{r}{d}} + 1.16\frac{r}{d}\right)\frac{\dot{w}_1}{\dot{w}_2} - 1.68 + 1.04\sqrt{\frac{r}{d}} - 1.16\frac{r}{d}\right] \quad (0.2 \le \dot{w}_2/\dot{w}_1 \le 0.8). \tag{16.18}$$

Loss coefficient K_{21_2} for diverging flow from branch of tee using Equation 16.17 can be determined from Diagram 16.8. For diverging flow into the other side run, replace subscript $_2$ by subscript $_3$ in Equations 16.17 and 16.18.

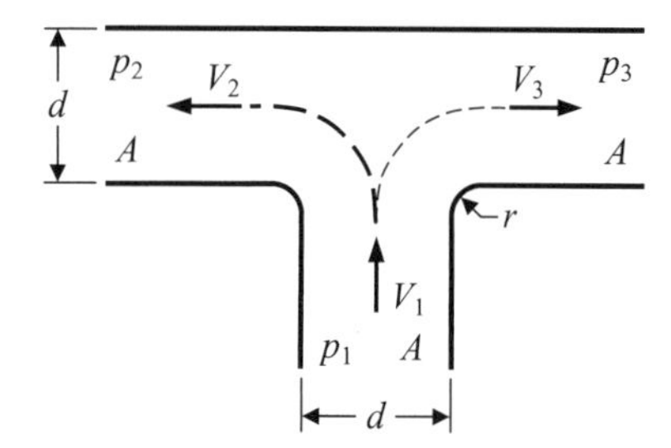

FIGURE 16.7. Diverging flow from branch of tee.

16.2 CONVERGING TEES

Pressure losses of converging streams consist mainly of loss due to turbulent mixing of the two streams with different velocities and loss due to the curving of the stream at its passage from the branch into the run. A radius (or chamfer) at the trailing edge of the branch significantly reduces the loss of both the branch and run flow. The equations developed in this section are valid for rounding ratio r/d up to 0.4; very little reduction in loss is gained at higher ratios.

16.2.1 Flow through Run

The energy equation for flow through run (see Fig. 16.8) can be written as:

$$\frac{P_2}{\rho_w} + \frac{V_2^2}{2g} = \frac{P_1}{\rho_w} + \frac{V_1^2}{2g} + K_{21_1}\frac{V_1^2}{2g},$$

where K_{21_2} is the loss coefficient for flow through run (from point 2 to point 1) in terms of the velocity at point 1. Rearrangement of the energy equation gives:

$$K_{21_1} = \frac{2g}{\rho_w V_1^2}(P_2 - P_1) + \frac{V_2^2}{V_1^2} - 1. \tag{16.19}$$

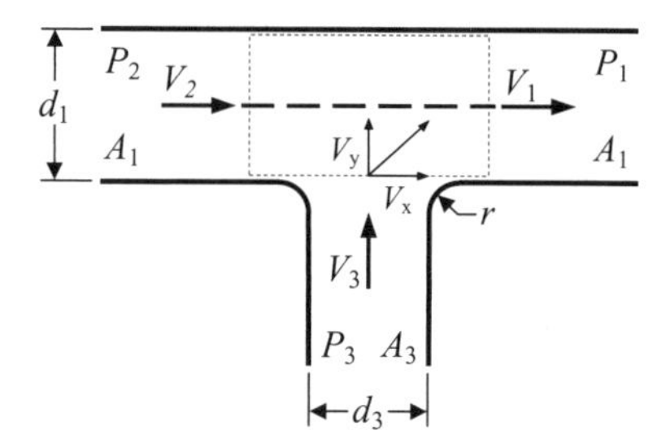

FIGURE 16.8. Converging flow through run of tee.

A momentum balance in the x-direction gives:

$$A_1(P_2 - P)_1 = \frac{V_1\dot{w}_1}{g} - \frac{V_2\dot{w}_2}{g} - \frac{V_x\dot{w}_3}{g} - \frac{C_M V_1\dot{w}_3}{g},$$

where the last term accounts for the uncertainty associated with fluid from the lateral channel, or branch, piercing the flow field that is a violation of the model. The coefficients C_{xC} and C_M are defined in Equations 16.23 and 16.24 below. We expect $V_x < V_2$ and express this as $V_x = C_{xC}V_2$. By using this relationship and rearranging, the momentum equation becomes:

$$P_2 - P_1 = \frac{1}{gA_1}(V_1\dot{w}_1 - V_2\dot{w}_2 - C_{xC}V_2\dot{w}_3 - C_M V_1\dot{w}_3).$$

By use of the continuity equations $\dot{w}_3 = \dot{w}_1 - \dot{w}_2$, $\dot{w}_1 = V_1 A_1 \rho_w$, and $\dot{w}_2 = V_2 A_2 \rho_w$, the momentum equation becomes:

$$P_2 - P_1 = \frac{\rho_w V_1^2}{g}\left[1 - \frac{V_2^2}{V_1^2} - C_{xC}\left(\frac{V_2}{V_1} - \frac{V_2^2}{V_1^2}\right) - C_M\left(1 - \frac{V_2}{V_1}\right)\right]. \quad (16.20)$$

Substitution of Equation 16.20 into Equation 16.19, and letting $V_1 = \dot{w}_1/A_1\rho_w$ and $V_2 = \dot{w}_2/A_1\rho_w$ gives:

$$K'_{21_1} = 1 - \frac{\dot{w}_2^2}{\dot{w}_1^2} - 2C_{xC}\left(\frac{\dot{w}_2}{\dot{w}_1} - \frac{\dot{w}_2^2}{\dot{w}_1^2}\right) - 2C_M\left(1 - \frac{\dot{w}_2}{\dot{w}_1}\right). \quad (16.21)$$

Equation 16.21 produces a value of zero when there is no flow from the branch, that is, when $\dot{w}_2 = \dot{w}_1$. Actually a small loss results from the enlargement and expansion of flow across the branch opening. The second term in the equation was modified to account for this loss:

$$K_{21_1} = 1 - 0.95\frac{\dot{w}_2^2}{\dot{w}_1^2} - 2C_{xC}\left(\frac{\dot{w}_2}{\dot{w}_1} - \frac{\dot{w}_2^2}{\dot{w}_1^2}\right) - 2C_M\left(1 - \frac{\dot{w}_2}{\dot{w}_1}\right). \quad (16.22)$$

Correlation with data from various investigators indicates that the loss coefficient for flow through run in the converging flow case is virtually independent of the diameter ratio d_3/d_1, but is a strong function of the radius of curvature r of the joining edge. The coefficients C_{xC} and C_M in Equation 16.22 are functions of r and may be determined by the following equations for $r/d_3 \le 0.3$:

$$C_M = 0.23 + 1.46(r/d_3) - 2.75(r/d_3)^2 + 1.65(r/d_3)^3 \quad (16.23)$$

and

$$C_{xC} = 0.08 + 0.56(r/d_3) - 1.75(r/d_3)^2 + 1.83(r/d_3)^3. \quad (16.24)$$

Results from Equation 16.22 are compared with data from various sources in Figure 16.9. There is general

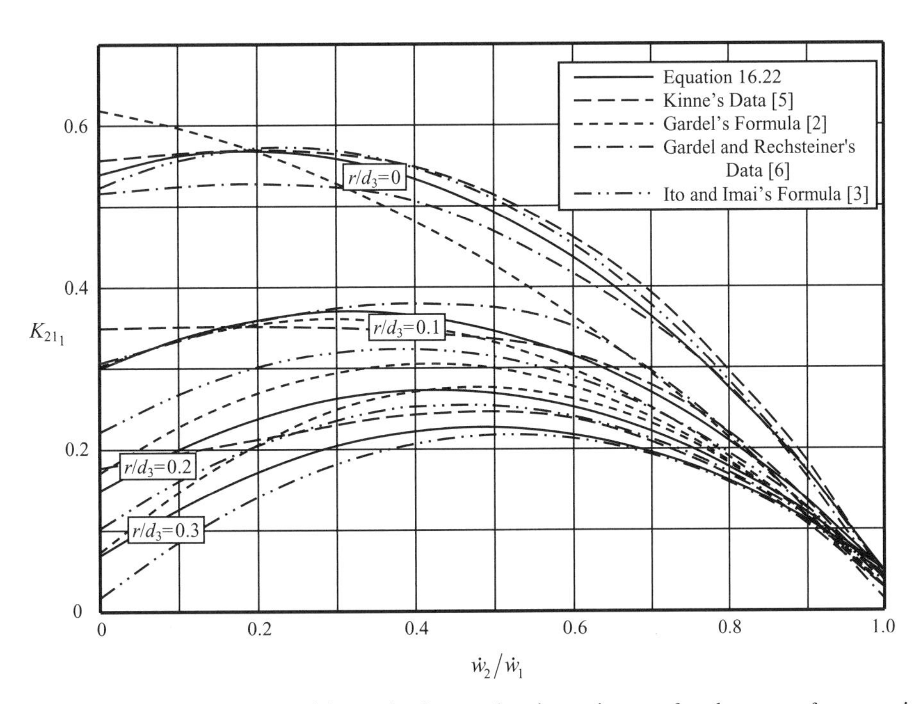

FIGURE 16.9. Comparison of Equation 16.22 with results from other investigators for the case of converging flow through run of tee.

agreement between sources at r/d_3 equal to zero, but there is considerable scatter among the various sources elsewhere. Equation 16.22 was developed to provide more or less average values throughout the range of data.

Multiplying Equation 16.22 by $\dot{w}_1^2/\dot{w}_2^2$ produces the loss coefficient for converging flow through run in terms of the velocity at point 2:

$$K_{21_2} = \frac{\dot{w}_1^2}{\dot{w}_2^2} - 0.95 - 2C_{xC}\left(\frac{\dot{w}_1}{\dot{w}_2} - 1\right) - 2C_M\left(\frac{\dot{w}_1^2}{\dot{w}_2^2} - \frac{\dot{w}_1}{\dot{w}_2}\right). \quad (16.25)$$

Loss coefficient K_{21_2} for converging flow through run from point 2 to point 1 can be obtained from Diagram 16.9. The energy equation in terms of the velocity at point 2 may be written as:

$$P_2 - P_1 = \frac{\dot{w}_2^2}{2g\rho_w A_1^2}\left(K_{21_2} + \frac{\dot{w}_1^2}{\dot{w}_2^2} - 1\right).$$

By substituting Equation 16.25 into the energy equation, the pressure drop equation for converging flow through run becomes:

$$P_2 - P_1 = \frac{\dot{w}_2^2}{2g\rho_w A_1^2}\left[2\frac{\dot{w}_1^2}{\dot{w}_2^2} - 1.95 - 2C_{xC}\left(\frac{\dot{w}_1}{\dot{w}_2} - 1\right) - 2C_M\left(\frac{\dot{w}_1^2}{\dot{w}_2^2} - \frac{\dot{w}_1}{\dot{w}_2}\right)\right]. \quad (16.26)$$

16.2.2 Flow through Branch

The energy equation for converging flow through branch (see Fig. 16.10) can be written as:

$$\frac{P_3}{\rho_w} + \frac{V_3^2}{2g} = \frac{P_1}{\rho_w} + \frac{V_1^2}{2g} + K_{31_1}\frac{V_1^2}{2g},$$

where K_{31_1} is the loss coefficient for flow through branch (from point 3 to point 1) in terms of the velocity at point 1. Rearrangement of the energy equation gives:

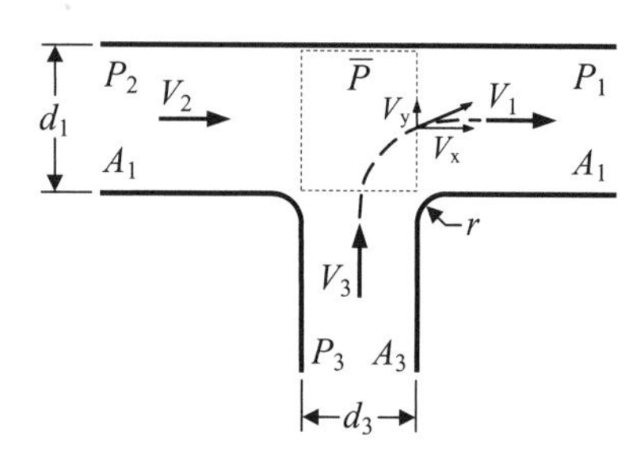

FIGURE 16.10. Converging flow through branch of tee.

$$K_{31_1} = \frac{2g}{\rho_w V_1^2}(P_3 - P_1) + \frac{V_3^2}{V_1^2} - 1. \quad (16.27)$$

A momentum balance in the y-direction gives:

$$A_3(P_3 - \bar{P}) = \frac{V_y\dot{w}_3}{g} - \frac{V_3\dot{w}_3}{g}.$$

We expect $V_y \le V_3$ and express this as $V_y = C_{yC}V_3$. By using this relationship and rearranging, the momentum equation becomes:

$$P_3 - \bar{P} = \frac{1}{gA_3}(C_{yC}V_3\dot{w}_3 - V_3\dot{w}_3).$$

By use of the continuity equation $\dot{w}_3 = V_3A_3\rho_w$, the momentum equation becomes:

$$P_3 - \bar{P} = \frac{\rho_w V_3^2}{g}(C_{yC} - 1).$$

Because most of the loss through the run takes place downstream of the branch the pressure $\bar{P}$ is in effect equal to P_2. Therefore:

$$P_3 - P_1 = \frac{\rho_w V_3^2}{g}(C_{yC} - 1) + P_2 - P_1. \quad (16.28)$$

Substituting Equation 16.28 into Equation 16.27, and using the continuity equations $V_1 = \dot{w}_1/A_1\rho_w$, $V_2 = \dot{w}_2/A_1\rho_w$, $V_3 = \dot{w}_3/A_3\rho_w$ and $\dot{w}_2 = \dot{w}_1 - \dot{w}_3$, gives:

$$K_{31_1} = -1 + 2(2 - C_{xC} - C_M)\frac{\dot{w}_3}{\dot{w}_1} + \left[(2C_{yC} - 1)\frac{d_1^4}{d_3^4} + 2(C_{xC} - 1)\right]\frac{\dot{w}_3^2}{\dot{w}_1^2}, \quad (16.29)$$

where C_M and C_{xC} are determined by Equations 16.23 and 16.24:

$$C_M = 0.23 + 1.46(r/d_3) - 2.75(r/d_3)^2 + 1.65(r/d_3)^3, \quad \text{(16.23, repeated)}$$

and

$$C_{xC} = 0.08 + 0.56(r/d_3) - 1.75(r/d_3)^2 + 1.83(r/d_3)^3. \quad \text{(16.24, repeated)}$$

Correlation with data from various sources indicates that C_{yC} is a function of the diameter ratio and of the rounding of the branch inlet. C_{yC} may be determined by the following equation for $r/d_3 \le 0.3$:

$$C_{yC} = 1 - 0.25\left(\frac{d_3}{d_1}\right)^{1.3} - \left(0.11\frac{r}{d_3} - 0.65\frac{r^2}{d_3^2} + 0.83\frac{r^3}{d_3^3}\right)\frac{d_3^2}{d_1^2}.$$

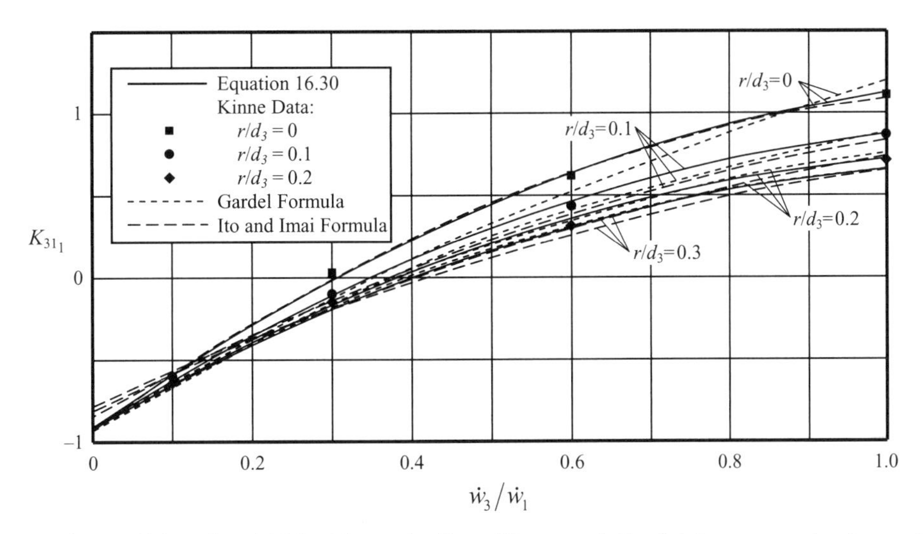

FIGURE 16.11. Comparison of Equation 16.30 with results from Kinne and Gardel for converging tees with a diameter ratio $d_3/d_1 = 1.0$.

The first term in Equation 16.29 was modified to further match experimental results as follows:

$$K_{31_1} = -0.92 + 2(2 - C_{xC} - C_M)\frac{\dot{w}_3}{\dot{w}_1} + \left[(2C_{yC} - 1)\frac{d_l^4}{d_3^4} + 2(C_{xC} - 1)\right]\frac{\dot{w}_3^2}{\dot{w}_1^2}. \quad (16.30)$$

Results from Equation 16.30 for the case of tees with a diameter ratio equal to unity are compared with data from various sources in Figure 16.11. There is general agreement with Kinne's data [5], with Gardel's formula [2], and with Ito and Imai's formula [3].

Results from Equation 16.30 for tees with diameter ratios equal to 1.0, 0.583, and 0.349 are compared with results from Kinne [5] and Gardel [2] in Figure 16.12. There is excellent agreement with their results for the case of radius ratio r/d_3 equal to 0.10. Equation 16.30 generally agrees with Gardel's formula at other diameter and radius ratios as well (not shown).

Multiplying Equation 16.30 by $d_3^4\dot{w}_1^2/d_1^4\dot{w}_3^2$ produces the loss coefficient for converging flow through branch in terms of the velocity at point 3:

$$K_{31_3} = 2C_{yC} - 1 + \frac{d_3^4}{d_1^4}\left[2(C_{xC} - 1) + 2(2 - C_{xC} - C_M)\frac{\dot{w}_1}{\dot{w}_3} - 0.92\frac{\dot{w}_1^2}{\dot{w}_3^2}\right]. \quad (16.31)$$

Loss coefficient K_{31_1} for converging flow through branch from point 3 to point 1 can be determined from Diagrams 16.10 through 16.14. The energy equation in terms of the velocity at point 3 may be written as:

$$P_3 - P_1 = \frac{\dot{w}_3^2}{2g\rho_w A_3^2}\left(K_{31_3} - 1 + \frac{\dot{w}_1^2 d_3^4}{\dot{w}_3^2 d_1^4}\right).$$

By substituting Equation 16.31 into the energy equation, the pressure drop equation for flow through branch in terms of the velocity at point 3 is:

$$P_3 - P_1 = \frac{\dot{w}_3^2}{2g\rho_w A_3^2}\left\{2(C_{yC} - 1) + \frac{d_3^4}{d_1^4}\left[2(C_{xC} - 1) + 2(2 - C_{xC} - C_M)\frac{\dot{w}_1}{\dot{w}_3} + 0.08\frac{\dot{w}_1^2}{\dot{w}_3^2}\right]\right\}. \quad (16.32)$$

16.2.3 Flow into Branch

Very little data are available for this configuration where the combined flow channel is located in the branch (see Fig. 16.13). Ito and Imai [3] have developed a formula that applies to tees with a branch-to-run diameter ratio of unity that are most frequently used for this purpose. The results of their formula agree quite well with their experimental data.

In terms of the velocity at point 1 (in the common flow channel), Ito and Imai's formula, transposed to the form and symbols use in this document, is:

$$K_{21_1} = 0.81 - 1.16\sqrt{\frac{r}{d}} + 0.50\frac{r}{d} - \left(0.95 - 1.65\frac{r}{d}\right)\frac{\dot{w}_2}{\dot{w}_1} + \left(1.34 - 1.69\frac{r}{d}\right)\frac{\dot{w}_2^2}{\dot{w}_1^2}. \quad (16.33)$$

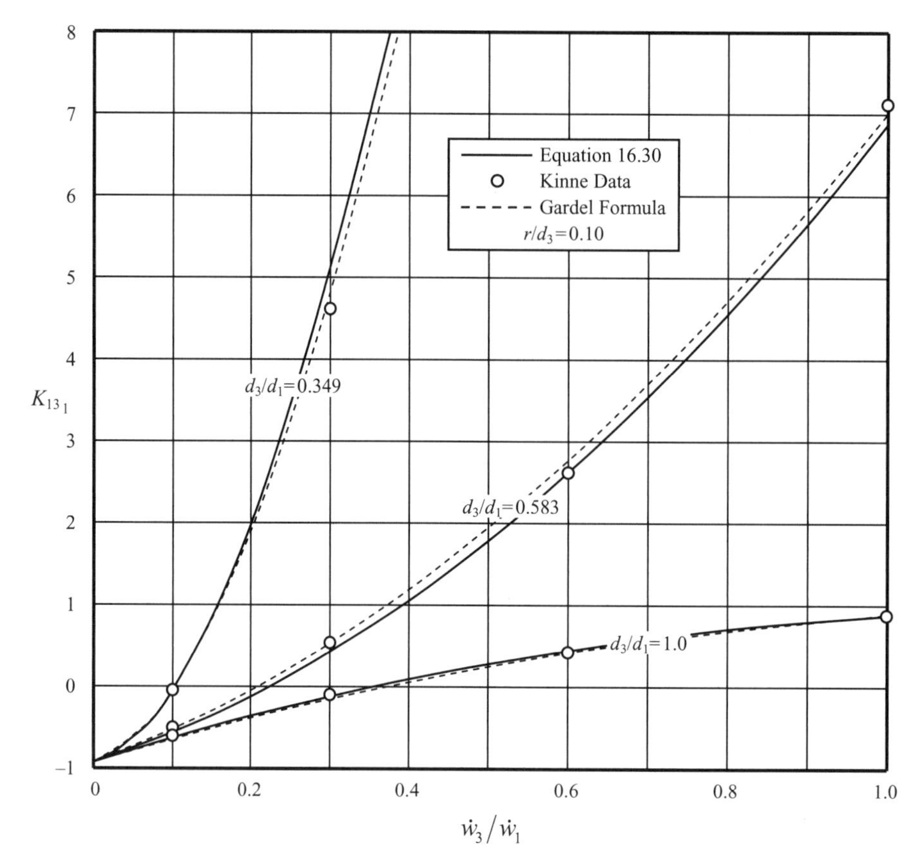

FIGURE 16.12. Comparison of Equation 16.30 with results of Kinne and Gardel for converging tees with radius ratio r/d_3 equal to 0.10.

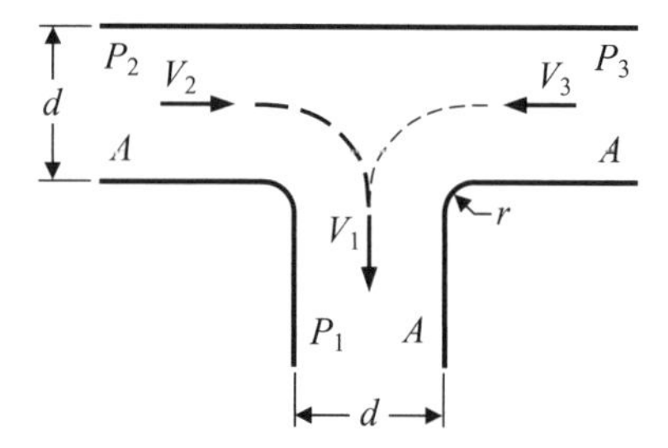

FIGURE 16.13. Converging flow into branch of tee.

Multiplying Equation 16.33 by $\dot{w}_1^2/\dot{w}_2^2$ produces the loss coefficient for converging flow into branch in terms of the velocity at point 2:

$$K_{21_2} = \left(0.81 - 1.16\sqrt{\frac{r}{d}} + 0.50\frac{r}{d}\right)\frac{\dot{w}_2^2}{\dot{w}_1^2} - \left(0.95 - 1.65\frac{r}{d}\right)\frac{\dot{w}_2}{\dot{w}_1} + 1.34 - 1.69\frac{r}{d}. \tag{16.34}$$

The energy equation in terms of the velocity at point 2 may be written as:

$$P_2 - P_1 = \frac{\dot{w}_2^2}{2g\rho_w A_2^2}\left(K_{21_2} - 1 + \frac{\dot{w}_1^2 d_2^4}{\dot{w}_2^2 d_1^4}\right).$$

By substituting Equation 16.34 into the energy equation, the pressure drop equation for converging flow into branch in terms of the velocity at point 2 is:

$$P_2 - P_1 = \frac{\dot{w}_2^2}{2g\rho_w A_2^2}\left[\left(1.81 - 1.16\sqrt{\frac{r}{d}} + 0.50\frac{r}{d}\right)\frac{\dot{w}_1^2}{\dot{w}_2} - \left(0.95 - 1.65\frac{r}{d}\right)\frac{\dot{w}_1}{\dot{w}_2} + 0.34 - 1.69\frac{r}{d}\right]. \tag{16.35}$$

Loss coefficient K_{21_2} for converging flow into the branch employing Equation 16.34 can be determined from Diagram 16.15. For converging flow from the other side run, replace subscript $_2$ by subscript $_3$ in Equations 16.34 and 16.35.

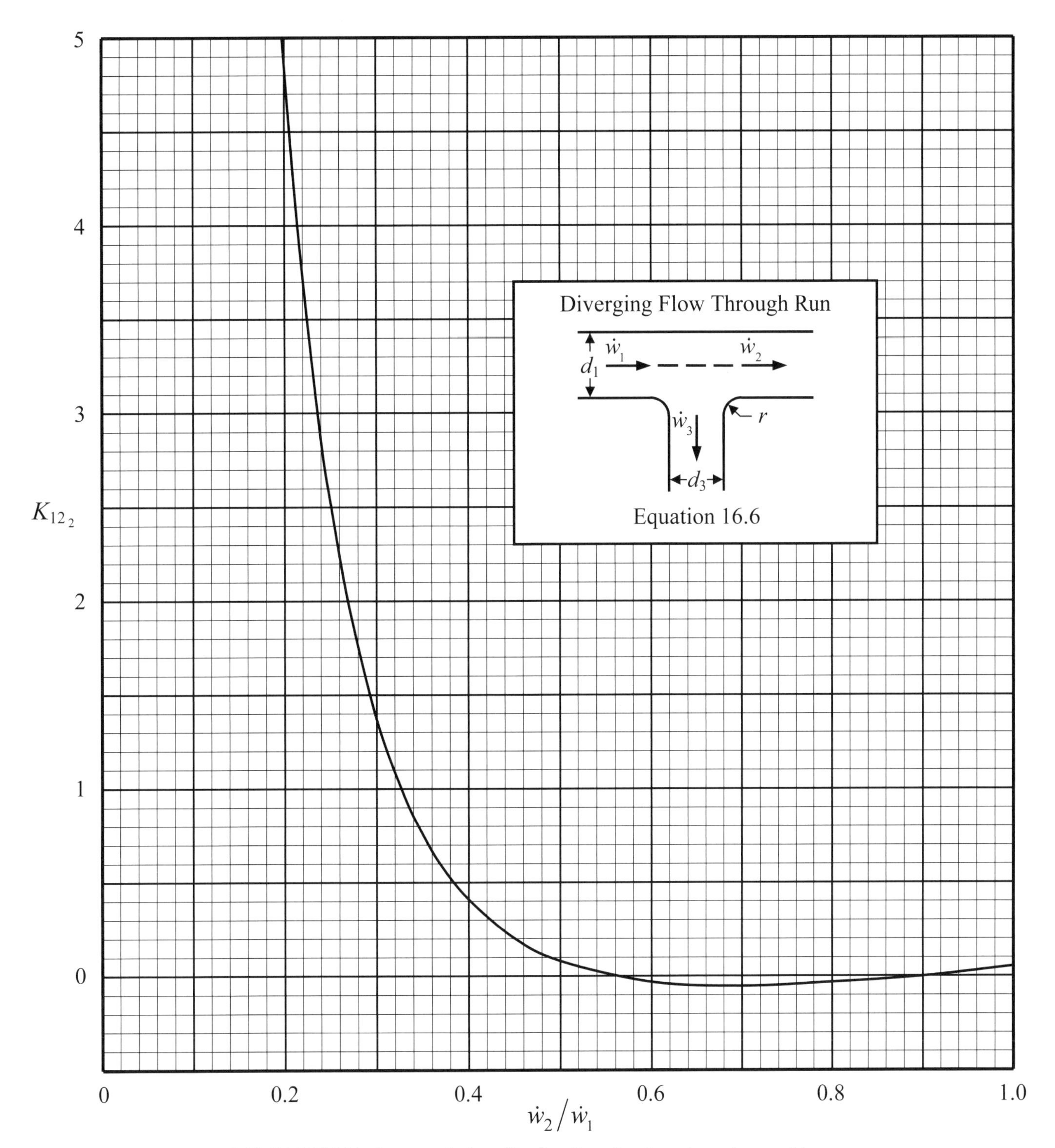

DIAGRAM 16.1. Loss coefficient K_{12_2} for diverging flow through run of tee.

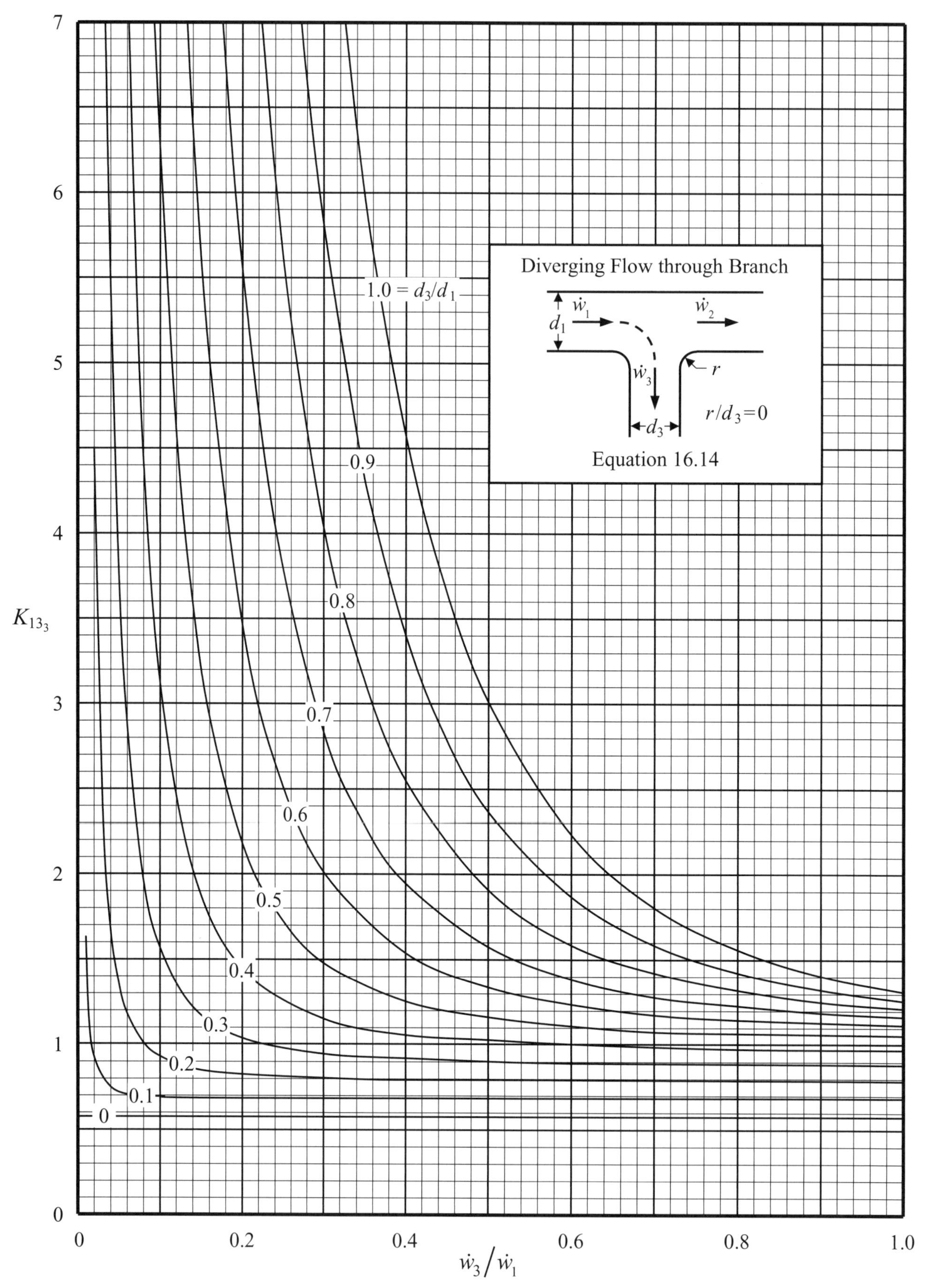

DIAGRAM 16.2. Loss coefficient K_{13_3} for diverging flow through branch of tee ($r/d_3 = 0$).

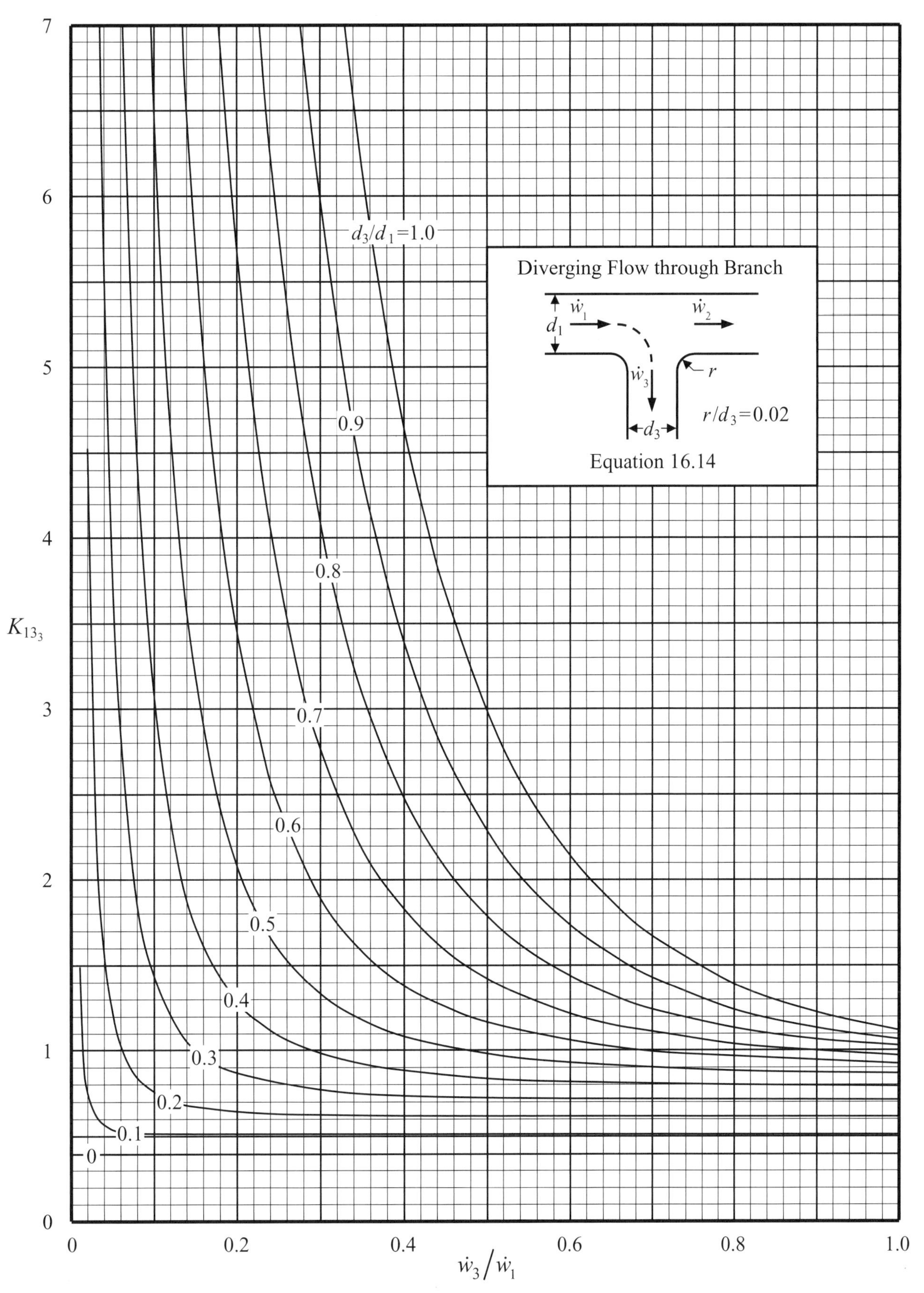

DIAGRAM 16.3. Loss coefficient K_{13_3} for diverging flow through branch of tee—$r/d_3 = 0.02$.

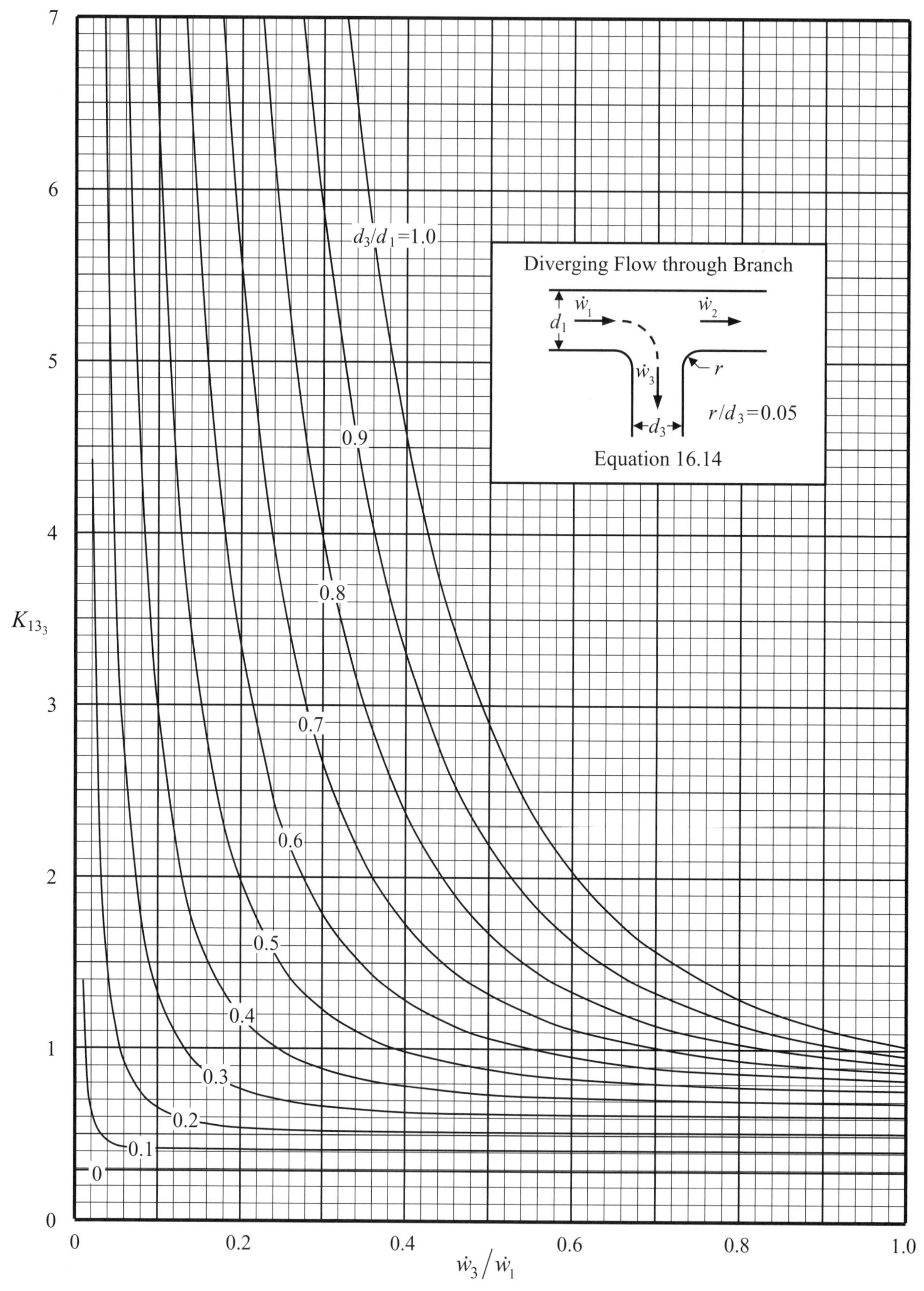

DIAGRAM 16.4. Loss coefficient K_{13_3} for diverging flow through branch of tee—$r/d_3 = 0.05$.

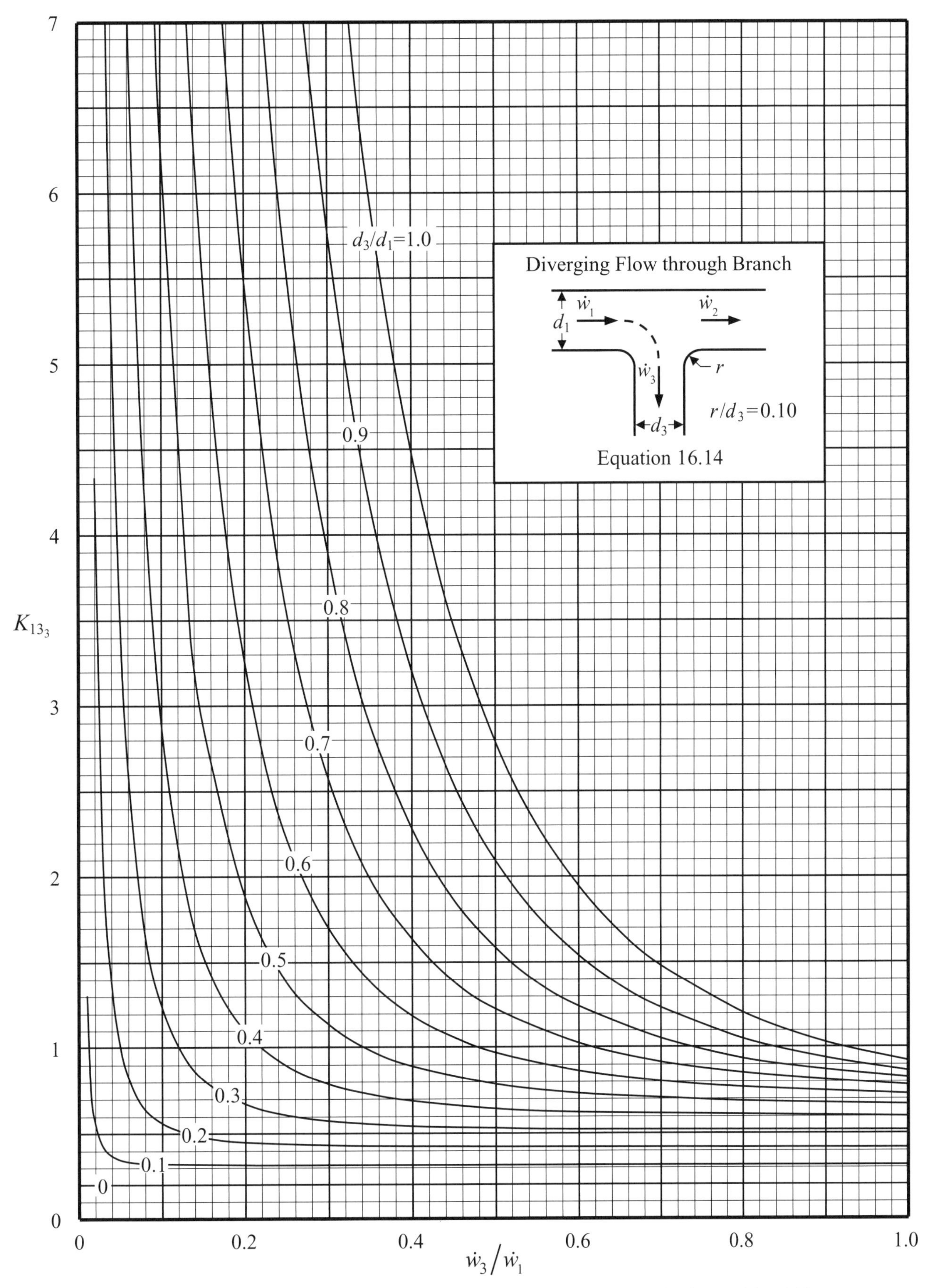

DIAGRAM 16.5. Loss coefficient K_{13_3} for diverging flow through branch of tee—$r/d_3 = 0.10$.

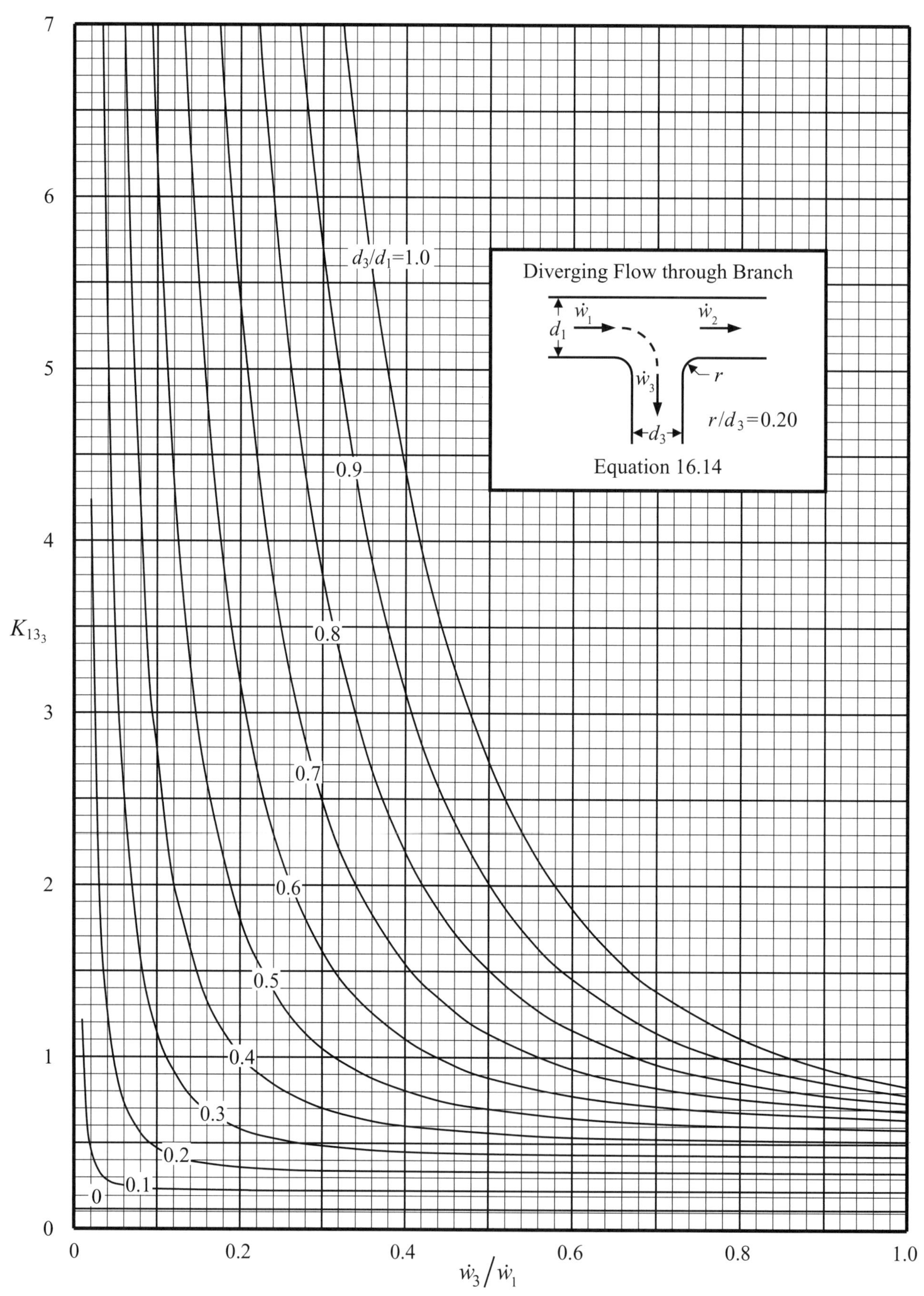

DIAGRAM 16.6. Loss coefficient K_{13_3} for diverging flow through branch of tee—$r/d_3 = 0.20$.

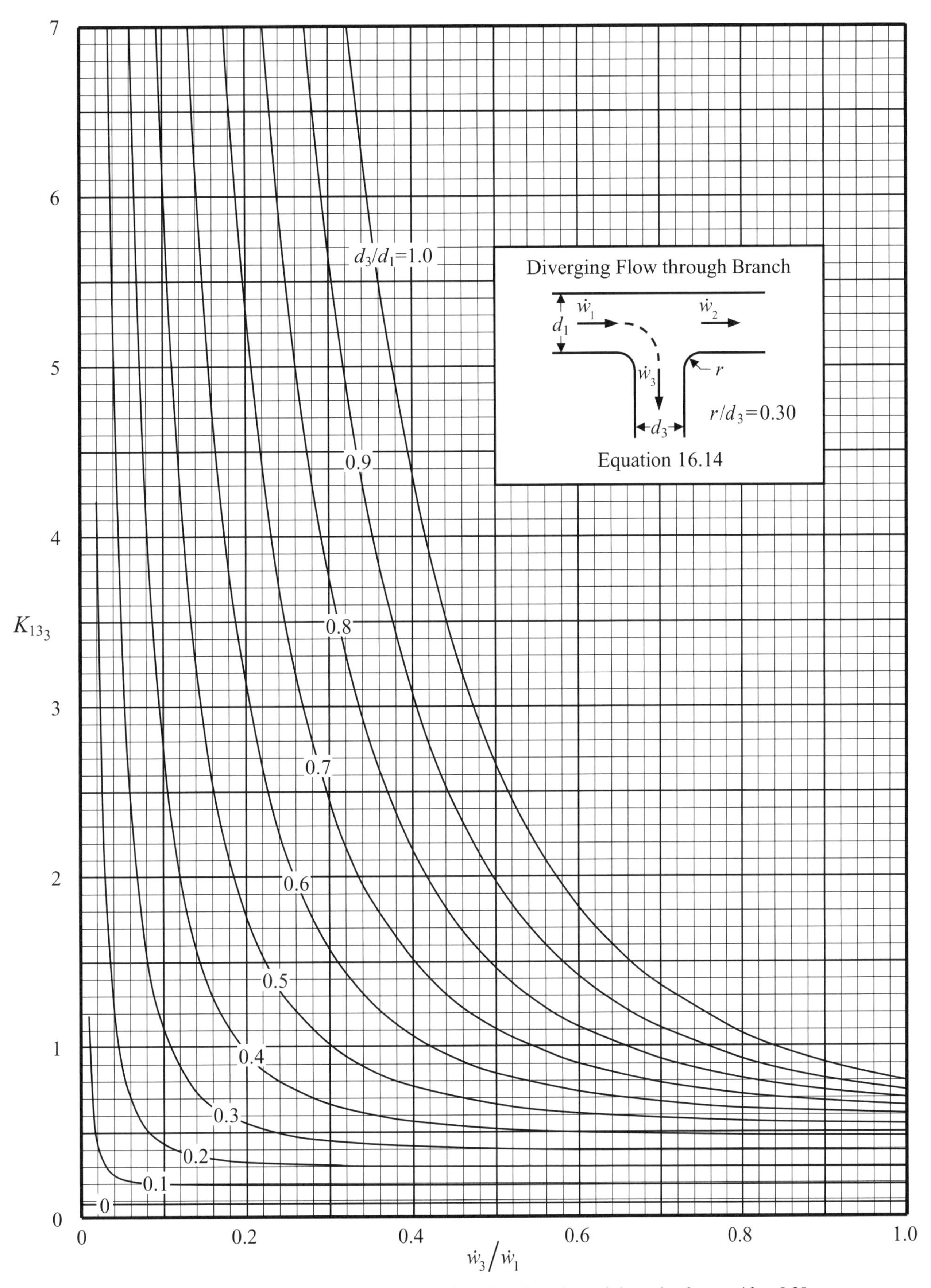

DIAGRAM 16.7. Loss coefficient K_{13_3} for diverging flow through branch of tee—$r/d_3 = 0.30$.

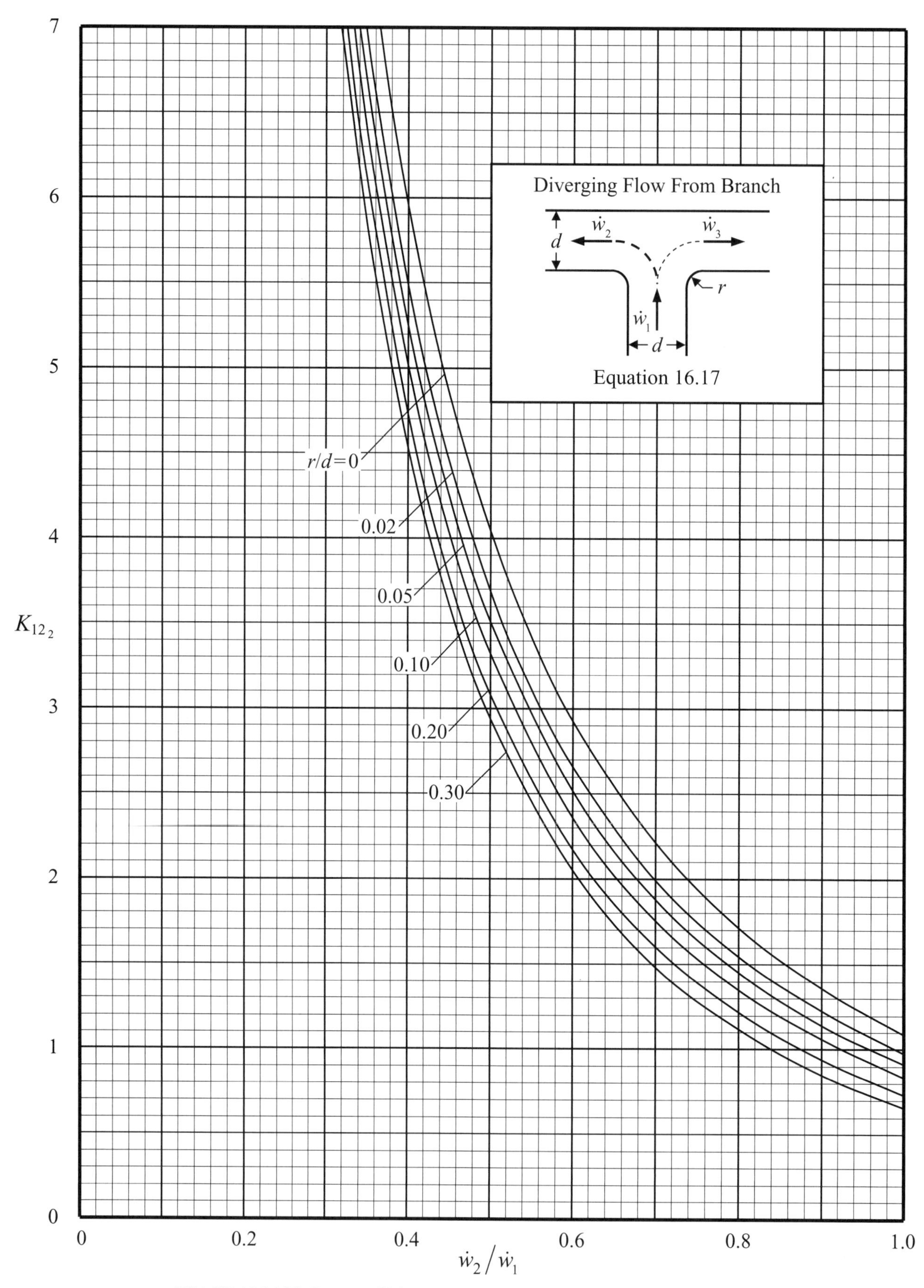

DIAGRAM 16.8. Loss coefficient K_{12_2} for diverging flow from branch of tee.

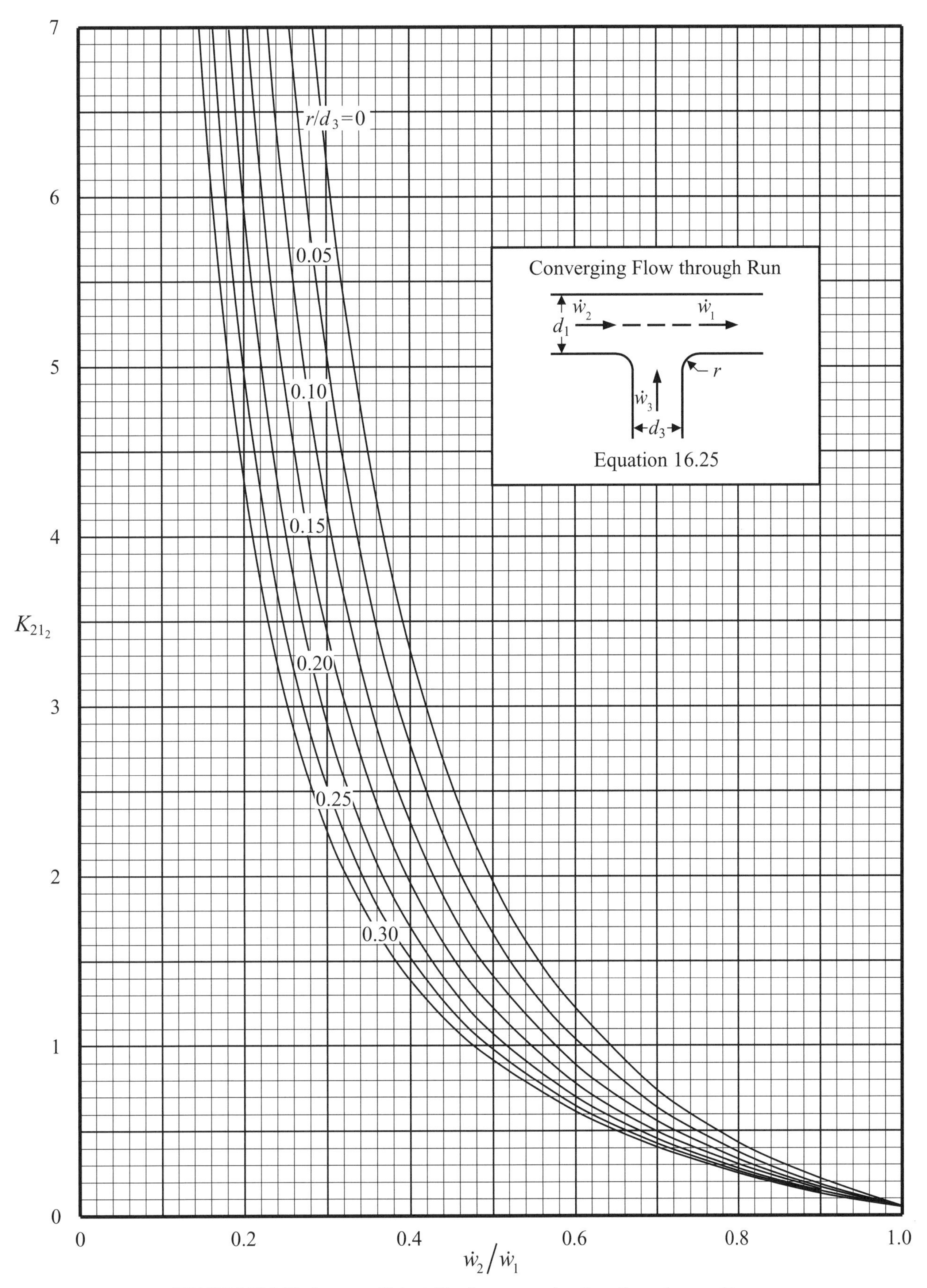

DIAGRAM 16.9. Loss coefficient K_{21_1} for converging flow through run of tee.

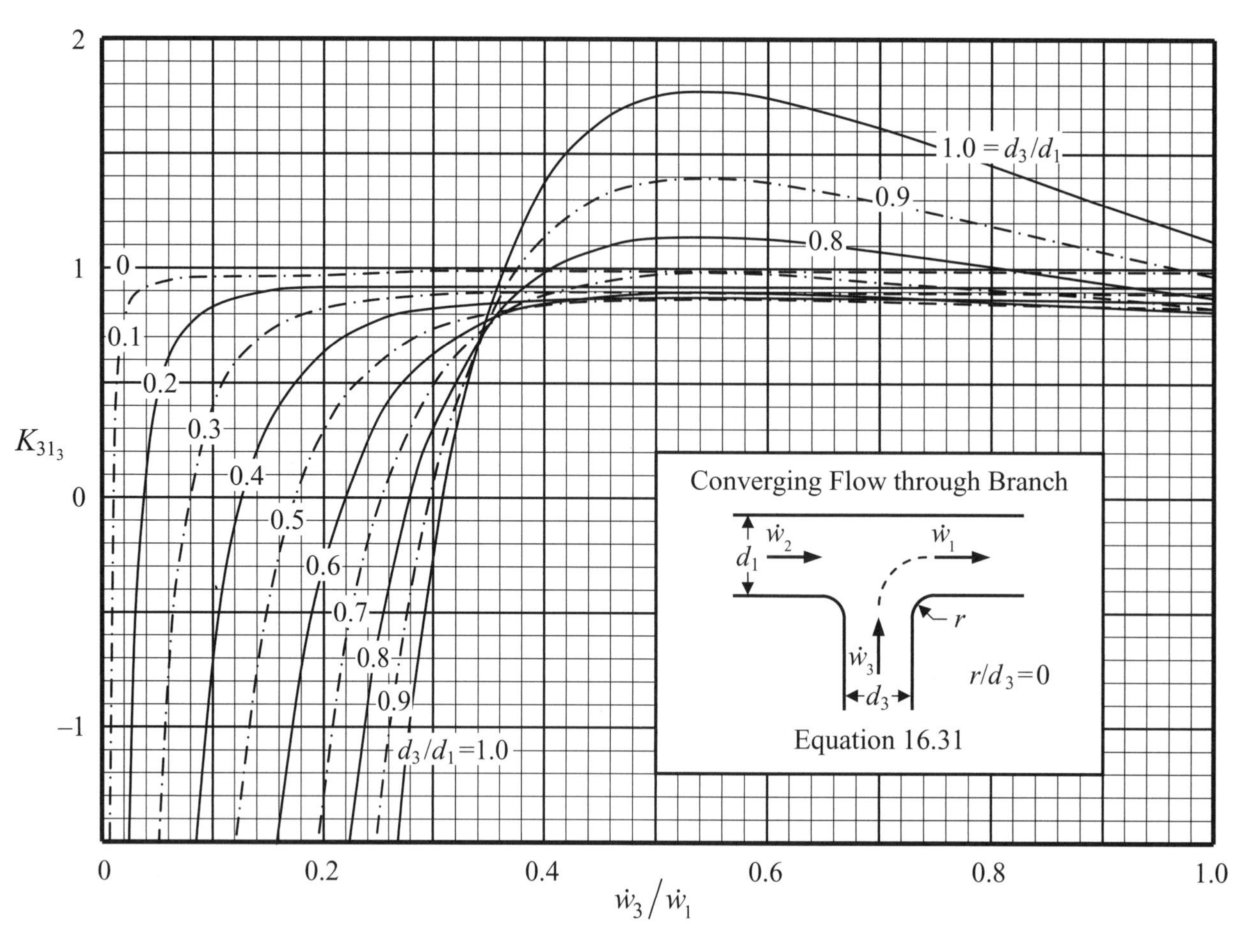

DIAGRAM 16.10. Loss coefficient K_{31_3} for converging flow through branch of tee—$r/d_3 = 0$.

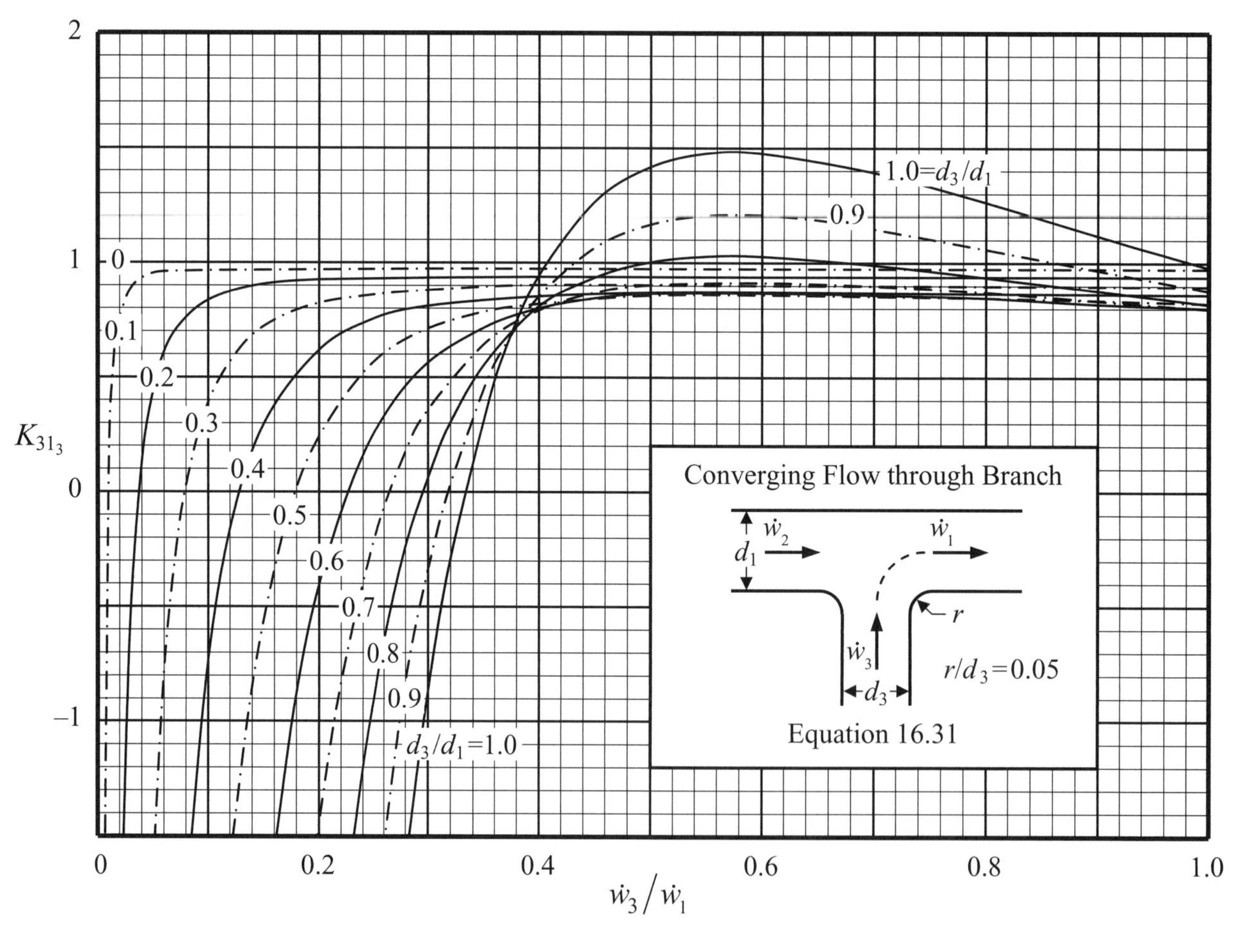

DIAGRAM 16.11. Loss coefficient K_{31_3} for converging flow through branch of tee—$r/d_3 = 0.05$.

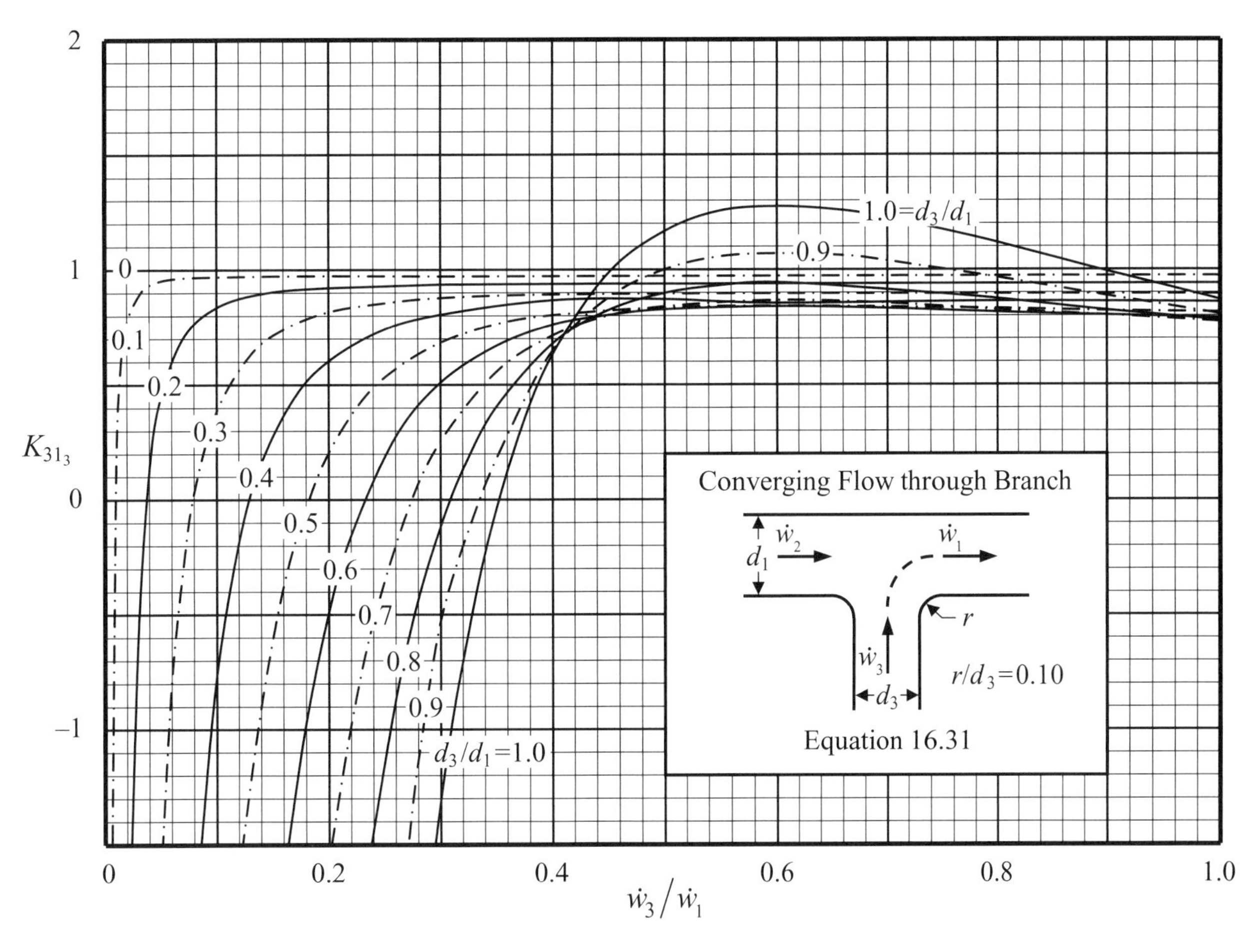

DIAGRAM 16.12. Loss coefficient K_{31_3} for converging flow through branch of tee—$r/d_3 = 0.10$.

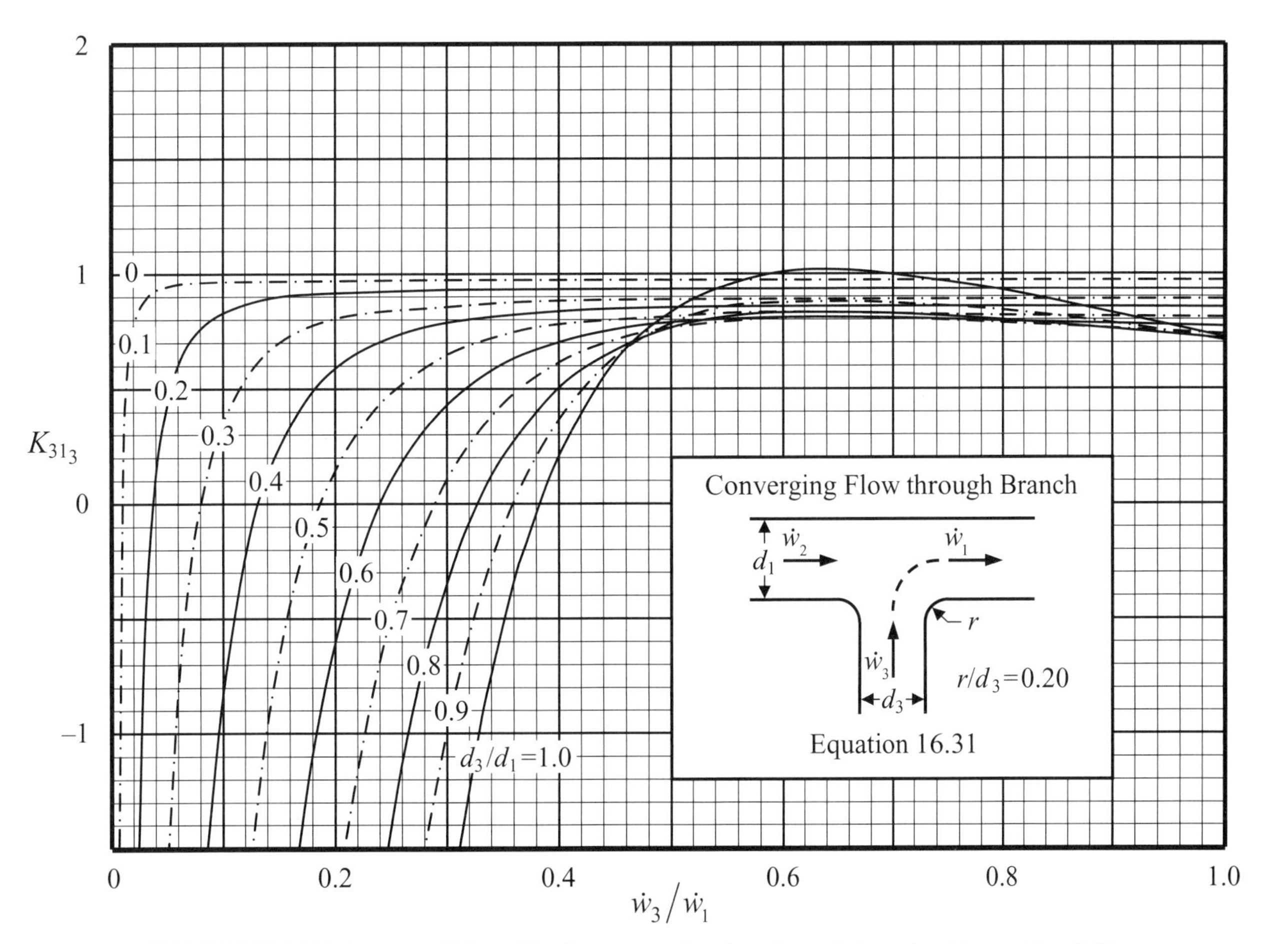

DIAGRAM 16.13. Loss coefficient K_{31_3} for converging flow through branch of tee—$r/d_3 = 0.20$.

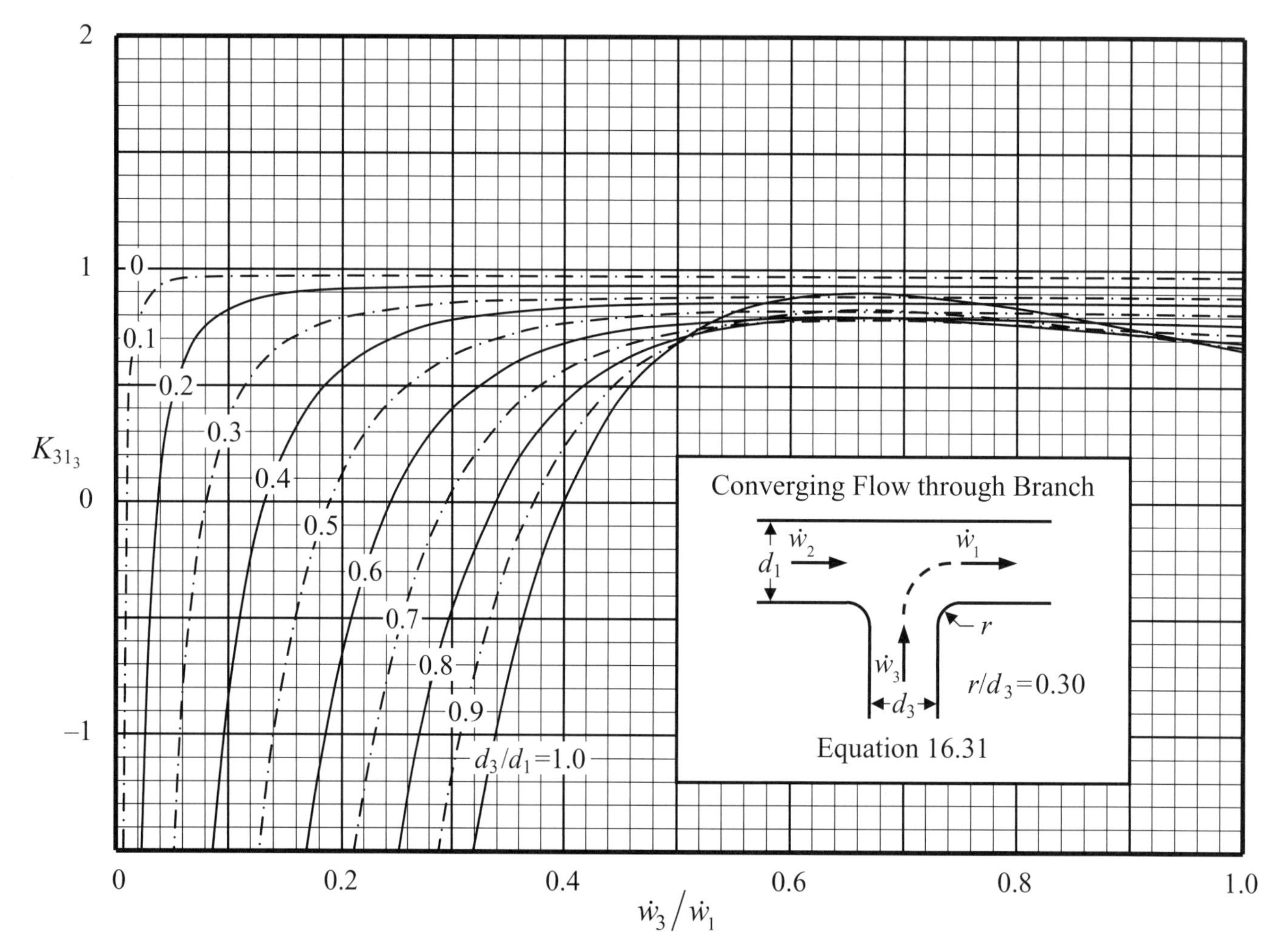

DIAGRAM 16.14. Loss coefficient K_{31_3} for converging flow through branch of tee—$r/d_3 = 0.30$.

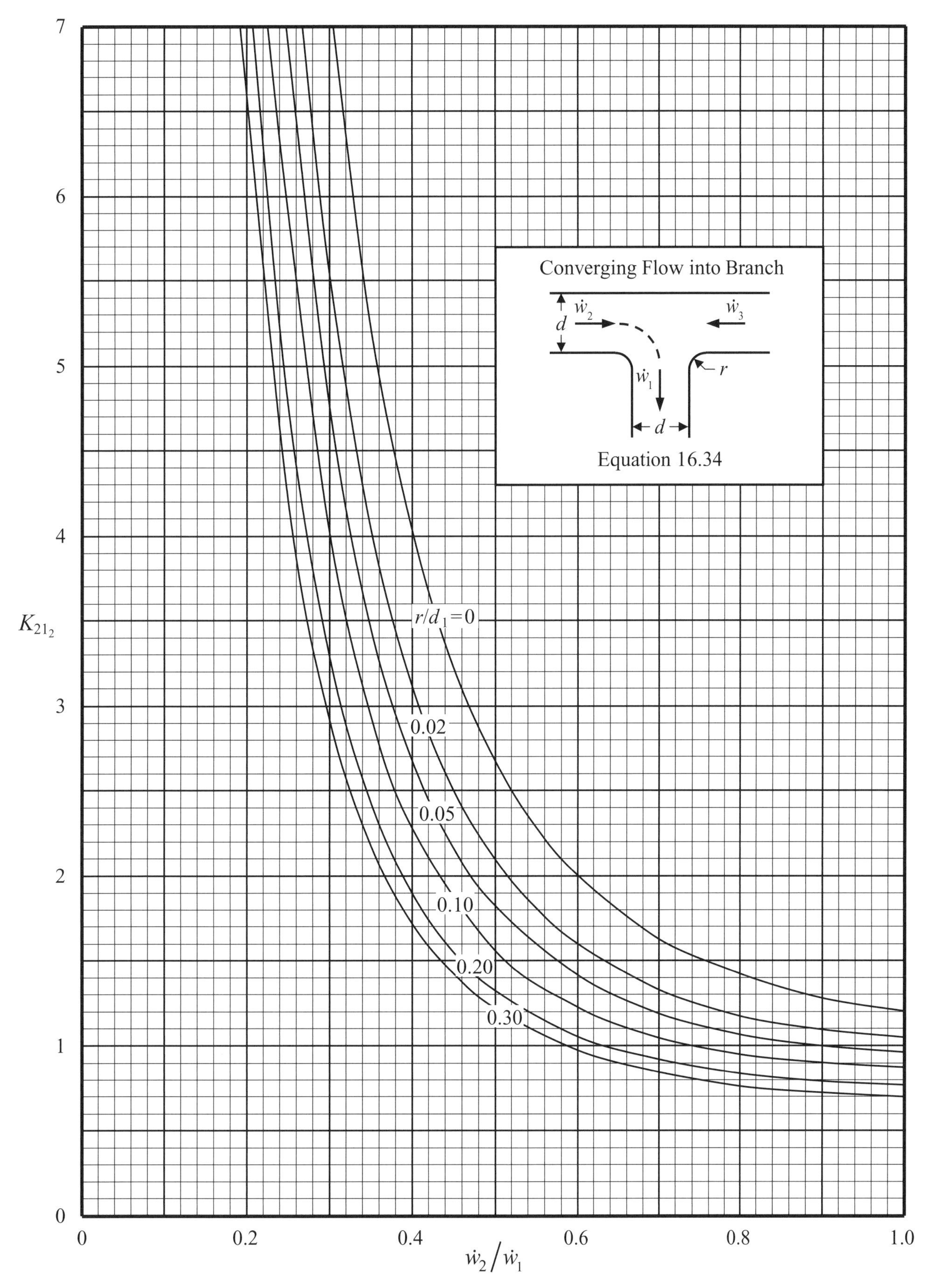

DIAGRAM 16.15. Loss coefficient K_{21_2} for converging flow into branch of tee.

REFERENCES

1. Gardel, A., Chambres d'équilibre, (doctorate thesis in French) École Polytechnique de l'Université de Lausanne, Librairie de l'Universite, 1956.
2. Gardel, A., Pressure drops in flows through T-shaped fittings, (in French), *Bulletin Technique de la Suisse Romande*, 83(9), April 1957, pp. 123–130, and (10), May 1957, pp. 143–148.
3. Ito, H. and K. Imai, Energy losses at 90° pipe junctions, *Journal of the Hydraulics Division, Proceedings of the American Society of Civil Engineers*, HY9, September 1973, 1353–1368.
4. Idel'chik, I. E., *Handbook of Hydraulic Resistance—Coefficients of Local Resistance and of Friction*, Moskva-Leningrad, 1960, (Translated from Russian. Published for the U.S. Atomic Energy Commission and the National Science Foundation, Washington, D.C. by the Israel Program for Scientific Translations, Jerusalem, 1966.)
5. Kinne, E., Contribution to the knowledge of hydraulic losses in branches, *Hydro Institute of Technology Hoschule-Munchen*, (4), 1931, 70–93. (Translation No. 323 by Bureau of Reclamation, U.S. Dept. of the Interior, Washington, 1955.)
6. Gardel, A. and G. F. Rechsteiner, *Les Pertes de Charge dans les Branchments en Te des Conduites de Section Circulaire*, (in French), Publication No, 118, Ecole Polytechnique Federale de Lausanne, Lausanne, Switzerland, 1971.

FURTHER READING

This list includes books and papers that may be helpful to those who wish to pursue further study.

Vogel, G., Investigation of the loss in right-angled pipe branches, *Hydro Institute of Technology Hoschule-Munchen*, (1), 1926, pp. 75–90, and (2), 1928, pp. 61–64.

Corp, C. I., Experiments on loss of head in U, S, and Twisted S pipe bends, *Bulletin of the University of Wisconsin, Engineering Experiment Station Series No. 66*, Madison, 1927.

Petermann, F., Loss in oblique-angled pipe branches, *Hydr. Inst. Tech. Hoschule-Munchen No. 3*, 1929, pp. 65–77. (Translated by N. H. Eaton and K. H. Beij, American Society of Mechanical Engineer Special Publication, 1935.)

Naramoto, I. and T. Kasai, On the loss of energy at impact of two confined streams of water, *Kyushu Imperial University College of Engineering Memory*, Fukuoka, Japan, 6(3), 1931, 189–261.

Farve, H., On the laws governing the movement of fluids in conduits under pressure with lateral flow, *Univ. des Mines Rev.*, Belgium, (ser 8, t. 13) 12: 1937, pp. 502–512. (Translation supplied by Agr. Res. Serv., St. Anthony Falls Hydr. Lab., Minneapolis.)

Vazsonyi, A., Pressure loss in elbows and duct branches, *Transactions of the American Society of Mechanical Engineers*, 66, April 1944, 177–183.

Keller, J. D., The manifold problem, *Journal of Applied Mechanics, Transactions of the American Society of Mechanical Engineers*, 71, March 1949, 77–85.

McKnown, J. S., Mechanics of manifold flow, *Transactions of the American Society of Civil Engineers*, 119, 1954, 1103–1142.

Starosolszky, O., Pressure conditions in pipe branches, *Vízügyi Közlemények*, Imprimerie University, Budapest, No. 1958/1, 1958, pp. 115–121 (translated by Language Service Bureau, Washington, D.C.).

Levine, S., Collision of incompressible-liquid streams in pipe, *Leningrad Tech. Inst. No. 8*, 1958. (Published by Dunod, Paris, 1968).

Zeisser, M. H., Summary report of single-tube branch and multi-tube branch water flow tests conducted by the University of Connecticut, Pratt and Whitney Aircraft Division, United Aircraft Corporation, Report No. PWAC-231 USAEC Contract AT (11-1)-229. May, 1963.

Blaisdell, F. W. and P. W. Manson, Loss of energy at sharp-edged pipe junctions in water conveyance systems, U.S. Dept. of Agriculture, Tech. Bulletin No, 1283, 163 pp., 106 figs, August, 1963.

Lakshmana Rao, N. S., B. C. Syamala Rao, and M. S. Shivaswamy, Distribution of energy losses at conduit tribufurcations, *Journal of the Hydraulics Division, Proceedings of the American Society of Civil Engineers*, HY6, November 1968, 1363–1374.

Ruus, E., Head losses in wyes and manifolds, *Journal of the Hydraulics Division, Proceedings of the American Society of Civil Engineers*, HY3, March 1970, 593–608.

Bajura, R. A., A model for flow distribution in manifolds, *Journal of Engineering for Power, Transactions of the American Society of Mechanical Engineers*, 93, January 1971, 7–12.

Miller, D. S., *Internal Flow, a Guide to Losses in Pipe and Duct Systems*, The British Hydromechanics Research Association, 1971.

Bajura, R. A., V. F. LeRose, and L. E. Williams, Fluid distribution in combining, dividing and reverse flow manifolds, *Journal of Engineering for Power, Transactions of the American Society of Mechanical Engineers*, Paper No. 73-Pwr-1, 1973, 1–11.

Williamson, J. V. and T. J. Rhone, Dividing flow in branches and wyes, *Journal of the Hydraulics Division, Proceedings of the American Society of Civil Engineers*, HY5, May 1973, 747–769.

Bajura, R. A. and E. H. Jones, Flow distribution manifolds, *Journal of Fluids Engineering, Transactions of the American Society of Mechanical Engineers*, 88, 1976, 654–666.

Ito, H., M. Sato, and K. Oka, Energy losses due to division and combination of flow at 90° Wyes (in Japanese), *Transactions of the Japanese Society of Mechanical Engineers*, 50(450), 1978, 342–350.

Shen, P. I., The effect of friction on flow distribution in dividing and combining flow manifolds, *Journal of Fluids Engineering, Transactions of the American Society of Mechanical Engineers*, 114, March 1992, 121–123.

17

PIPE JOINTS

Pressure loss due to pipe connections, or joints, is usually ignored or neglected. Yet, the various joints used to assemble pipe components can sometimes give rise to significant pressure loss.

Two aspects of butt weld connections, weld protrusion and backing rings, result in relatively small pressure loss. Even so, for long pipelines where there are few sources of pressure loss other than pipe friction, the pressure loss due to weld connections may be significant. Socket weld and flanged connections offer minimal pressure loss unless they are terribly misaligned. The internal geometry of threaded (screwed) pipe fittings is discontinuous, creating additional pressure loss, and they are covered separately in Chapter 19.

17.1 WELD PROTRUSION

In achieving full penetration butt welds, the root pass normally protrudes through the inside surface of the pipe (drop through) to form a slight and somewhat irregular orifice. The orifice effect may be further heightened by radial shrinkage of the pipe wall during the welding process (see Fig. 17.1). Well-planned weld procedures, as well as a skilled welder, can minimize the combined affect.

In the case of short, compact piping sections, welds connect pipe components that are closely spaced. In this case, pressure loss due to weld protrusion is small compared to pressure loss in the various pipe components. Moreover, interaction effects with the various pipe components are difficult to quantify and may be minus as well as plus. However, in the case of long pipelines containing mainly straight sections of pipe with few fittings and valves, pressure loss due to weld protrusion may be significant.

The following formula can be employed to account for weld protrusion loss in a long, straight section of pipe:

$$K \approx K_W C_L, \tag{17.1}$$

where K_W is the loss coefficient of a butt weld joint separated from other weld joints or pipe fittings by a relative distance l/d equal to or greater than 40, and C_L is a correction factor to adjust for welds separated by less than a relative distance of 40.

The protruding weld bead tends to be somewhat rounded so that the loss may be treated as a rounded contraction. Assuming that the effective rounding radius r is equal to 1/2 the depth Δ_W of the protrusion, the loss coefficient K_W was evaluated as a round-edged orifice in a straight pipe by employing Equation 13.6. The calculated results were multiplied by d^4/d_o^4 to relate K_W to the velocity in the pipe rather than to the velocity in the orifice restriction. As such, the loss coefficient K_W of a butt weld joint separated by a relative distance l/d equal to or greater than 40 from the closest butt weld joint upstream is shown in Diagram 17.1.

Weld specifications usually stipulate maximum allowable weld protrusion. For lack of actual data on particular welds, the allowable value may be used as a first-order estimate.

Pipe Flow: A Practical and Comprehensive Guide, First Edition. Donald C. Rennels and Hobart M. Hudson.

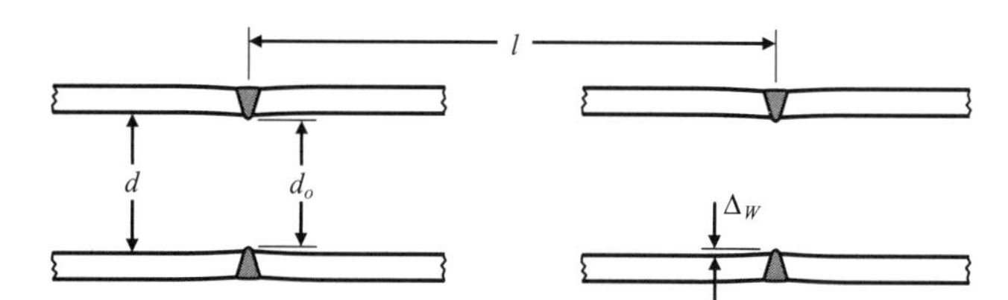

FIGURE 17.1. Weld protrusion.

FIGURE 17.2. Backing ring cross section.

The correction coefficient C_L in Equation 17.1 is another matter. Assuming that about 60% and 95% of full loss is attained at relative distances l/d of 10 and 30, respectively,* C_L can be tentatively determined from Equation 17.2 or from Diagram 17.2:

$$C_L \approx 0.1221\frac{l}{d} - 0.0237\left(\frac{l}{d}\right)^{1.5} + 0.00132\left(\frac{l}{d}\right)^2 \quad (l/d < 40). \tag{17.2}$$

17.2 BACKING RINGS

Backing rings are designed to provide quick, easy fit-up of pipe in order to simplify pipe welding and reduce costs. A number of cylindrical (sometimes spherical) spacer nubs, equally spaced around the backing ring, are often incorporated into the design to set the gap for the root pass of the weld. Standard backing rings are beveled about 15° to improve fluid flow as shown in Figure 17.2. On the downside, the rings are usually not fully consumed in the welding process so crevices between the ring and pipe surfaces can be a source for chemical corrosion or possible cracking from thermal or mechanical fatigue.

The following formula can be employed to account for the loss due to a backing ring in a long, straight section of pipe:

$$K \approx K_{BR} C_L.$$

K_{BR} represents backing rings separated by a relative distance l/d equal to or greater than 40. C_L is a correction factor to adjust for backing rings separated by less than a relative distance of 40.

* This estimation is based on examination of pressure distribution downstream of various pipe components such as those illustrated in Figure 15.2 ("Static pressure distribution in the neighborhood of a bend with long tangents").

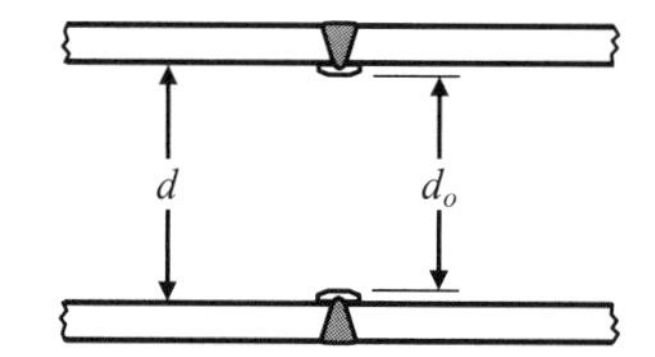

FIGURE 17.3. Pipe weld with backing ring.

TABLE 17.1. Approximate Loss Coefficient K_{BR} of Standard Backing Rings in Straight Pipe Separated by a Relative Distance Equal to or Greater Than 40

Nominal Pipe Size (inch)	Ring Thickness (inch)	Ring Width (inch)	K_{BR}
1	3/32	5/8	0.54
1-1/4	3/32	5/8	0.30
1-1/2	3/32	5/8	0.22
2	3/32	5/8	0.14
2-1/2	3/32	5/8	0.10
3	3/32	5/8	0.072
3-1/2	3/32	5/8	0.052
4	3/32	5/8	0.042
5	1/8	1	0.046
6	1/8	1	0.034
8	1/8	1	0.022
10	1/8	1	0.015
12	1/8	1	0.012
14	1/8	1	0.010
16	1/8	1	0.008
18	1/8	1	0.007
20	1/8	1	0.006
24	1/8	1	0.005
32	1/8	1	0.004
36	1/8	1	0.003

Calculated values of K_{BR} are based on schedule 40 pipe.

Backing rings, as illustrated in Figure 17.3, were evaluated as a bevel-edged orifice using Equation 13.8.† It was assumed that each 15° bevel takes up 25% of the width of the ring. The calculated results were multiplied by d^4/d_o^4 to relate K_W to the velocity in the pipe rather than to the velocity in the orifice restriction.

Loss coefficients for standard backing rings with a bevel are shown in Table 17.1. Note that pressure loss due to backing rings is chiefly significant for small pipes. The correction factor C_L to account for distance between backing rings may be determined from Equation 17.2 or from Figure 17.1.

† Radial shrinkage of the pipe wall during the welding process was ignored. The downstream bevel offers a small reduction of loss, but its effect was also ignored. The competing effects tend to cancel out.

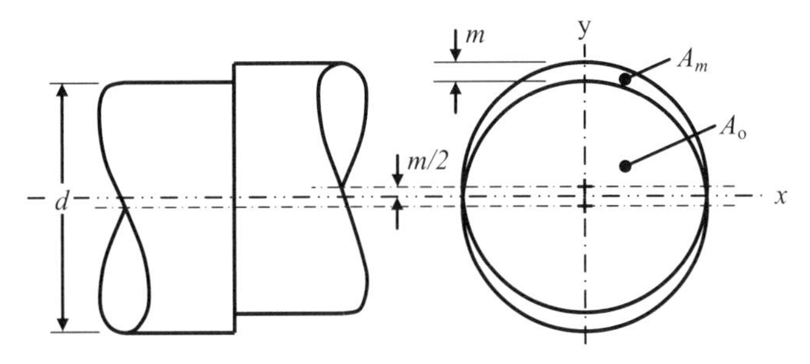

FIGURE 17.4. Misaligned pipe joint.

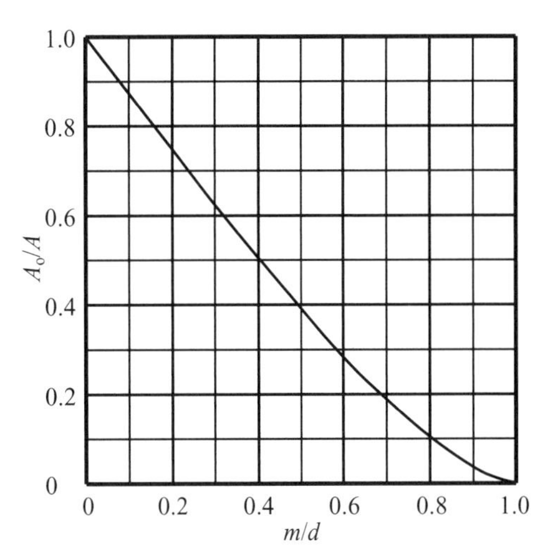

FIGURE 17.5. Open area ratio A_o/A as a function of m/d.

17.3 MISALIGNMENT

Appreciable pressure loss may result when piping components are misaligned during assembly, or when flange gaskets are not properly fitted.

17.3.1 Misaligned Pipe Joint

Misalignment of a pipe joint is illustrated in Figure 17.4. It is rational to treat the loss as a sharp-edged contraction in parallel with a sudden expansion. But first, we must determine the geometric relationship between misalignment m and open flow area A_o through the pipe joint.

Area A_m, and thereby open flow area A_o, was determined as a function of misalignment m by integration between the curves formed by the pipe walls. From this, the ratio of flow area A_o to flow area A of the pipe is presented in Figure 17.5 as a function of misalignment.

Next, the problem was treated as a sudden contraction and sudden expansion in parallel. First, the beta ratio was determined as $\beta = (A_o/A)^{1/2}$. The loss coefficient K_{contr} of the sudden contraction flow path was determined from Equation 10.5 and the loss coefficient K_{exp} of the sudden expansion flow path was determined from Equation 11.8. Assigning one-half the open flow area A_o to each flow path, Equation 5.9 was employed to determine the combined loss coefficient of the misaligned joint. The calculated results were multiplied by A^2/A_o^2 to relate K to the velocity in the pipe rather than to the velocity in the restriction. From this, the loss coefficient of a misaligned pipe joint in terms of the velocity in the pipe can be obtained from Diagram 17.3.

17.3.2 Misaligned Gasket

A misalignment gasket in a flanged connection is illustrated in Figure 17.6. It is reasonable to treat the loss as a sharp-edged orifice in a straight pipe.

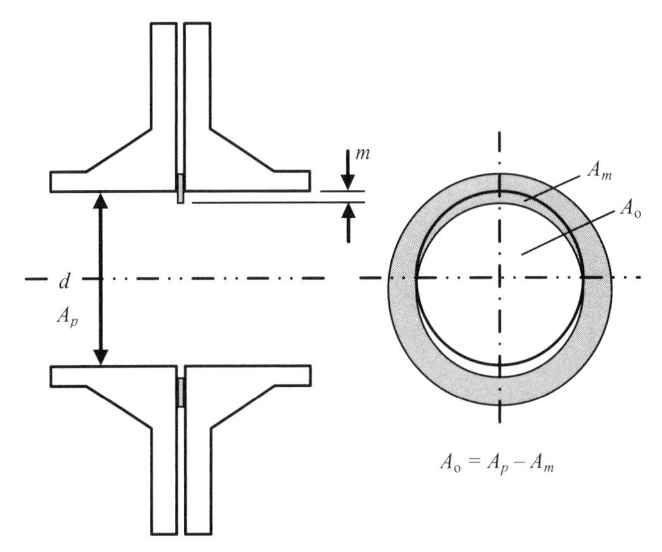

FIGURE 17.6. Misaligned flange gasket.

Open flow area A_o was determined as a function of misalignment m in Section 17.3.1 and was presented in Figure 17.5 as a function of m/d. Now we treat the misalignment gasket as a sharp-edged orifice. The beta ratio can be determined as $\beta = (A_o/A)^{1/2}$. The loss coefficient K of the misaligned gasket is determined from Equation 12.6 for a sharp-edged orifice. Once again, the calculated results were multiplied by A^2/A_o^2 to relate K to the velocity in the pipe rather than to the velocity in the orifice restriction. The approximate loss coefficient of a misaligned gasket can be obtained from Diagram 17.4.

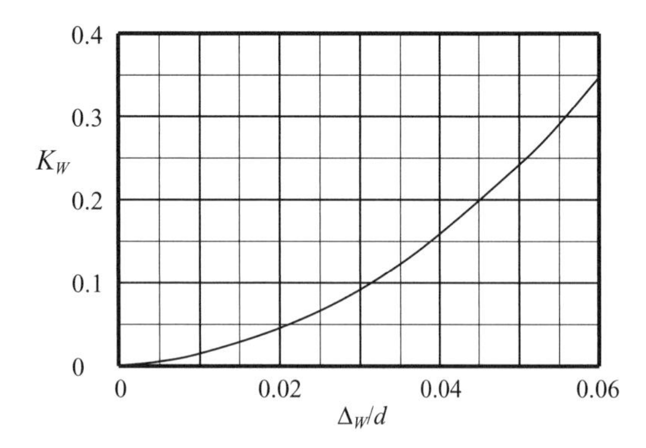

DIAGRAM 17.1. Approximate loss coefficient K_W of a butt weld joint separated by a relative distance l/d equal to or greater than 40.

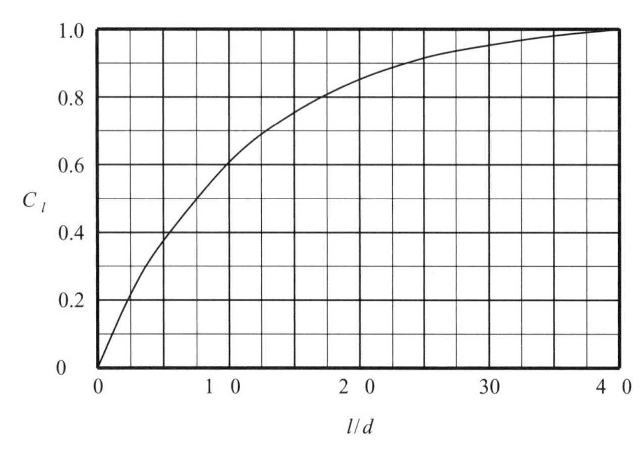

DIAGRAM 17.2. Correction coefficient for a weld joint separated by a relative distance l/d less than 40.

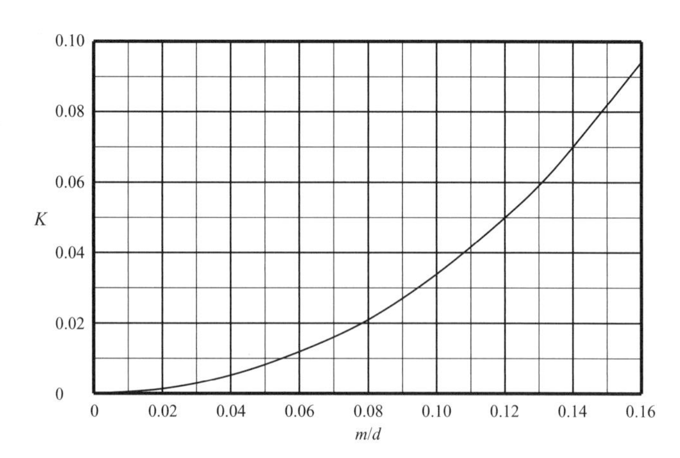

DIAGRAM 17.3. Approximate loss coefficient K of a misaligned pipe joint.

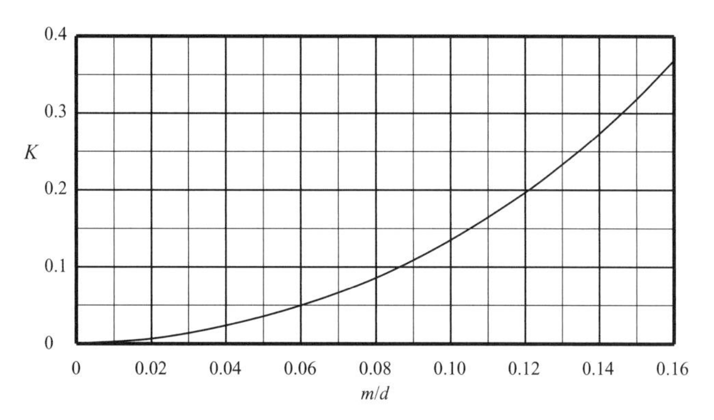

DIAGRAM 17.4. Approximate loss coefficient K of a misaligned gasket.

18

VALVES

A valve is a device that regulates the flow of a fluid by opening, closing, or partially obstructing the passageway in which it is installed. In general, valves used for on–off purposes have full or slightly reduced ports and are of symmetrical, straightway design such as ball valves and most gate valves. Valves used for flow control or throttling normally have reduced ports and are usually of asymmetrical design such as globe or needle valves. Their attendant high resistance to flow, necessary for control or throttling purposes, makes them undesirable for strictly on–off application.

Many valves are operated manually—usually by a hand wheel. Devices called actuators may also operate valves. They can be electromechanical actuators such as an electric motor or solenoid, or a pneumatic actuator motorized by air pressure; or hydraulic actuators that are powered by the pressure of a liquid such as oil or water. Actuators can be used for the purpose of automatic control driven by changes in pressure, temperature, or flow. They may be used when manual control is too difficult; for example, the valve is too large, or is generally inaccessible. Valves that are used to control the supply of air or other fluid going to the actuators are called pilot valves.

Large-size valves are normally supplied with flanged or butt-weld end connections. Smaller-size valves are often supplied with socket-weld or threaded (screwed) end connections. Valves are rated for maximum temperature and pressure as well as for flow capacity by the manufacturer. In addition to the standard valve products, many valve manufacturers produce custom-designed valves and actuators for specific applications. Valves are available in a broad spectrum of sizes and materials. Each design has its own advantages, and the selection of the proper valve for a particular application is critical. In this chapter, we discuss the characteristics and uses of the various types of valves and examine their flow characteristics.

18.1 MULTITURN VALVES

The closure members of some types of valves are traditionally called gates, disks, wedges, plungers, and so on. A linear motion equal to the pipe diameter is generally needed to fully open and fully close these valves. The linear motion is usually provided by a number of turns of a screw mechanism. Thus, these valves are referred to as multiturn or linear motion valves.

18.1.1 Diaphragm Valve

Diaphragm valves (see Fig. 18.1) have three simple elements: the valve body, the diaphragm, and the bonnet. The diaphragm serves as the closing member as well as a partition to seal the body fluid from the bonnet region. This eliminates the need for conventional valve stem packing material. A plunger is lowered by the valve stem onto the diaphragm to force it against a "weir" or wall in the valve body to seal and cut off flow.

Diaphragm valves are relatively low in cost, have low pressure drops, and can be tightly closed. The flexible

Pipe Flow: A Practical and Comprehensive Guide, First Edition. Donald C. Rennels and Hobart M. Hudson.

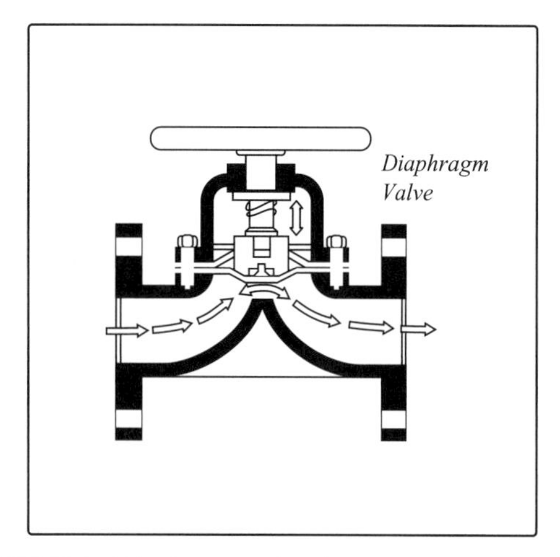

FIGURE 18.1. Diaphragm valve (courtesy of Valve Manufacturers Association of America, Washington, DC).

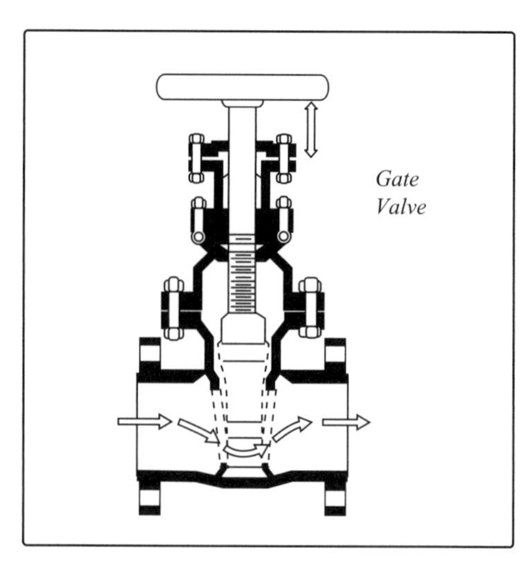

FIGURE 18.2. Gate valve (courtesy of Valve Manufacturers Association of America).

members are subject to wear and hence periodic replacement. They are limited to low pressure applications and may require high actuation forces to cut off flow. Diaphragm valves are limited to maximum temperatures ranging from 180°F to 300°F depending on the elastomeric diaphragm material that is used.

18.1.2 Gate Valve

In a gate valve (see Fig. 18.2), moving a gate—also known as a wedge or disk—directly into the fluid path stops the flow of fluid. Sealing is accomplished by metal-to-metal contact between the gate and the valve body in a plane perpendicular to the flow path. The linear motion of the gate will generally be equal to or greater than the pipe diameter.

There are two types of gate elements: the solid wedge and the flexible wedge. The solid wedge gates with matching tapered body are simpler in construction and stronger than the split or flexible wedge type. However, the flexible wedge-type gate, which consists of a pair of disks joined together by a center hub, has greater ability to accommodate housing distortion and is often used in high temperature applications where differential expansion could be a problem.

Gate valves are used primarily for on–off applications, that is, fully open or fully closed. In the open position, the gate valve presents very little restriction to the flowing fluid; thus, the primary advantage of the gate valve is a low pressure drop. As is the case with ball valves, manufacturers often offer as standard a port diameter one size smaller than the valve nominal pipe size so that the gate valve is more competitively priced with the other less efficient types of valves.

Gate valves are not normally considered for throttling purposes. They are prone to vibration when in a partially open position and are also subject to seat and disk wear. The relatively large gate travel results in a large envelope or overall size. They generally require large actuating forces.

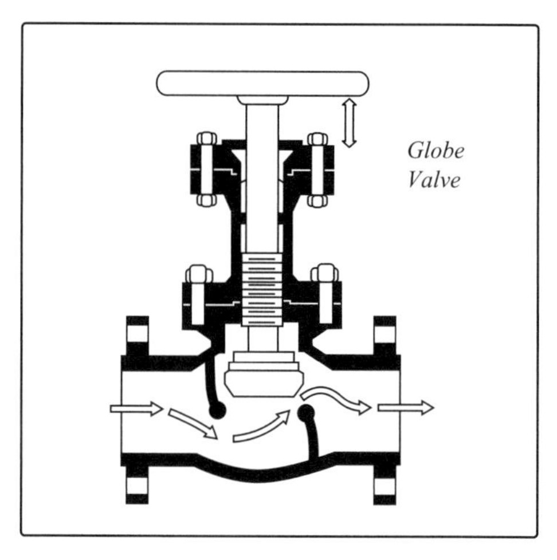

FIGURE 18.3. Globe valve (courtesy of Valve Manufacturers Association of America).

18.1.3 Globe Valve

The term "globe valve" (see Fig. 18.3) is applied to a large variety of valves whose internal body construction provides a fixed solid barrier between the inlet and outlet side of the valve. Fluid flows through a hole or port machined in the barrier. Shut-off sealing is accomplished by a closure member in the form of a plug or disk that is moved in a direction perpendicular to a ring-shaped seat. A threaded stem generally moves the disk.

In most applications, the valve is installed so that fluid enters and flows through the valve from the underside of the disk. This is done so that in the fully closed position fluid pressure is not continuously applied to the

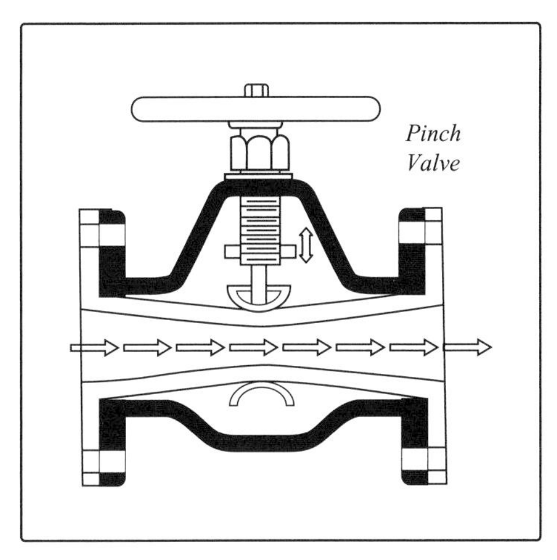

FIGURE 18.4. Pinch valve (courtesy of Valve Manufacturers Association of America).

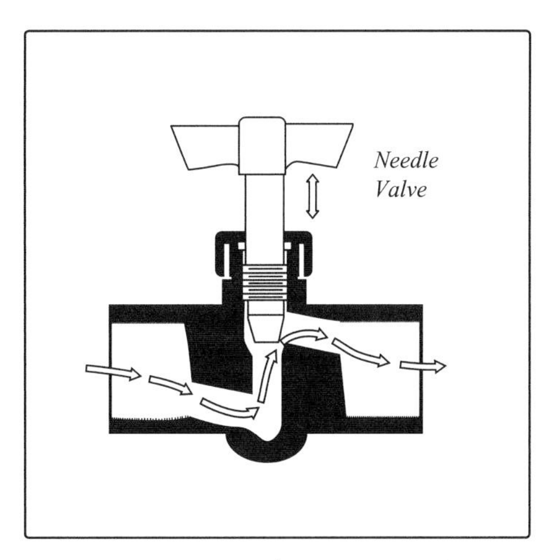

FIGURE 18.5. Needle valve (courtesy of Valve Manufacturers Association of America).

stem and stem packing. In some applications, the valve is installed so that fluid enters and flows through the valve from the top of the disk. This is done so that the fluid flow and upstream pressure aid valve closure and sealing. There is almost no sliding motion between the disk and the seat when the valve is opening or closing. This provides minimum wear in the case of frequent or continuous operation.

Globe valves are often used for flow control or throttling applications. The relatively tortuous flow path of the fluid through the valve due to the fixed barrier type of construction causes a high pressure drop that is acceptable for flow control but is a serious limitation in strictly on–off applications. High pressure drop in globe valves makes them undesirable in many piping applications. They may require considerable power to operate and are usually heavier than other valves of the same flow and pressure rating. Housing and stem distortion can be a problem providing sealing in high temperature applications.

In the standard globe valve, the fixed solid seat is at a 90° angle to the inlet and outlet ports. In the Y-pattern globe valve, the seat is at a 45° or 60° angle, which streamlines the flow path through the valve and significantly reduces pressure drop.

The angle valve is a special form of globe valve where the seat is perpendicular to the inlet of the valve. Angle valves are used where the pipeline changes 90°. It offers less resistance to flow. An angle valve reduces the number of joints in a line and saves on installation.

18.1.4 Pinch Valve

The pinch valve (see Fig. 18.4) seals by means of a flexible sleeve that can be mechanically pinched to shut off flow. The sleeve encloses the flow media and isolates it from the environment, hence reducing contamination of the environment. A suitable synthetic material can be selected for the sleeve to overcome the corrosiveness and abrasiveness of the flow media.

The pinch valve is especially suited for application of slurries of liquids with large amounts of suspended solids because the flexible sleeve allows the valve to close drop-tight around solids—solids that would otherwise be trapped in the seat or stuck in crevices in other types of valves. The advantages and disadvantages of the pinch valve are similar to those of the diaphragm valve.

18.1.5 Needle Valve

The needle valve (see Fig. 18.5) is a volume-control valve that restricts flow in small lines. It allows precise regulation of flow, although it is generally used for, and is capable of, only relatively small flow rates. It has a relatively small orifice, often with a long, conical seat. A needle-shaped plunger, on the end of a screw, exactly fits this seat. As the screw is turned and the plunger retracted, flow between the seat and the plunger is possible; however, until the plunger is fully retracted the flow is markedly impeded. Because it takes many turns of the fine-threaded screw to retract the plunger, precise regulation of flow rate is possible.

Needle valves are normally used in flow metering applications, especially when a constant, calibrated, low flow rate must be maintained. Needle valves are usually easy to shut off completely, although they are not used for simple shutoff purposes.

18.2 QUARTER-TURN VALVES

A quarter turn opens or closes these valves. The closure member is in the form of a sphere, cylinder, or tapered plug that can be rotated to direct flow.

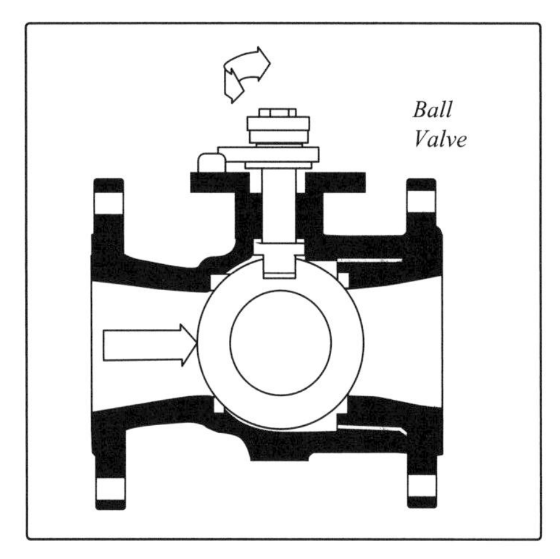

FIGURE 18.6. Ball valve (courtesy of Valve Manufacturers Association of America).

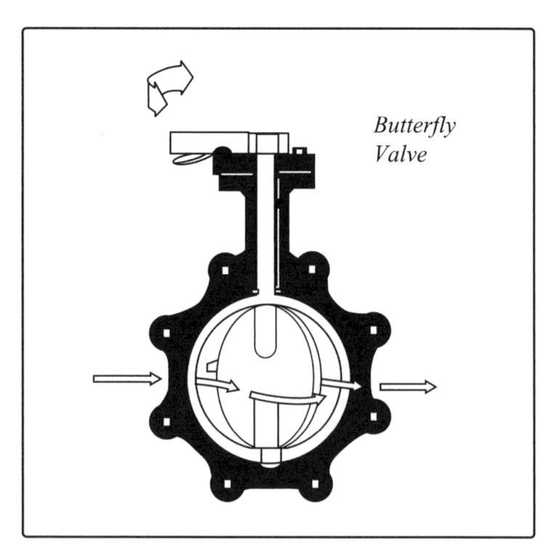

FIGURE 18.7. Butterfly valve (courtesy of Valve Manufacturers Association of America).

18.2.1 Ball Valve

The ball valve (see Fig. 18.6) consists of a spherical element with a cylindrical hole or port that allows straight through flow in the open position. They can be fully opened and closed by a quarter turn.

The distinctive feature of ball valves is that the diameter of the port can be the same diameter as the connecting pipe so that the full port ball valve offers virtually no more pressure drop than the equivalent length of straight pipe. In practice, many manufacturers offer as standard a port diameter one size smaller than the valve nominal pipe size so that the ball valve is more competitively priced with the other less efficient types of valves.

Ball valves are used in on–off applications where low pressure drop and quick opening and closing are required. They are sometimes used for flow control, or pressure control, purposes. Ball valves exhibit a nonlinear flow versus percent opening characteristic as the ball goes from fully closed to fully open. The opening can be modified to obtain a more linear flow control characteristic. Seat wear may be minimized by not requiring full shut-off capability.

18.2.2 Butterfly Valve

The butterfly valve (see Fig. 18.7) controls flow by using a circular disk within a housing with its pivot axis at right angle to the direction of flow in the pipe. The disk closes against a ring seal to shut off flow. The butterfly valve is used both for on–off and throttling services.

Butterfly valves are generally used in low pressure, large diameter lines where leakage is not important. They have a very low pressure drop and are relatively lightweight. Their length can be quite small. Yet, they usually require high actuation forces.

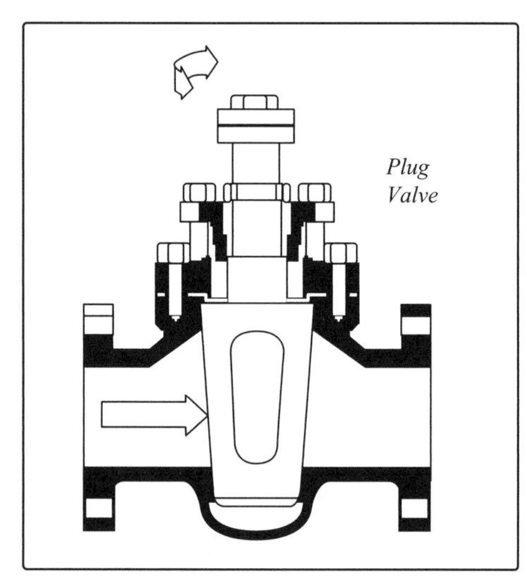

FIGURE 18.8. Plug valve (courtesy of Valve Manufacturers Association of America).

The metal-to-metal throttling type is primarily used for throttling control where positive shutoff is not required. The resilient-lined type provides positive shutoff and may be used for throttling service. The metal-seated, elastomer-sealed type provides positive shutoff but only limited throttling capability.

Swing valves are similar to butterfly valves except that they are hinged on one edge rather than along a diameter. They are primarily used as check valves to block flow in one direction. They have many of the advantages and disadvantages of butterfly valves.

18.2.3 Plug Valve

The term "plug valve" (see Fig. 18.8) applies to a category of valves in which a cylindrical or tapered plug with a hole is inserted directly into the basic fluid flow path.

The plug valve is very similar to the ball valve except that the closure member is a plug instead of a ball. Flow is smooth, straight, and uninterrupted, which means pressure drop is low. Plug valves are used primarily as shutoff valves where low valve profile and quick operation are required. Some are used for flow control applications. They are normally compact in size and require less headroom than other valves. They are fairly low in cost and provide a leak-proof seal.

Nonlubricated plug valves depend on wedging of a tapered plug against the valve body for seating. In order to reduce operating force requirements and eliminate high wear, nonlubricated plug valves usually employ a mechanism that lifts the plug from its seat before rotating it. In lubricated plug valves, the seating surfaces of the plug and its barrel are lubricated.

Their limitations are a tendency of the nonlubricated type to stick or gall. Lubricated plug valves require periodic lubrication and the lubrication may react with the fluid being carried.

18.3 SELF-ACTUATED VALVES

Valves may be automatically driven by changes in pressure, temperature, or flow.

18.3.1 Check Valve

The term "check valve" (see Fig. 18.9) is applied to a large variety of valves designed to close upon cessation or reversal of flow. The closure device may be a free-swinging disk, a free but guided plug or ball, or a spring-assisted disk, piston, or ball.

Conventional swing check valves employ a free-swinging disk. The disk is forced against an internal stop at full flow conditions to give the proper degree of opening, and at the same time to keep the edge of the disk within the stream flow so that a reversal of flow will cause closing. The loss coefficient of conventional swing check valves is low compared to other types of check valves and remains constant except at extremely low flows.

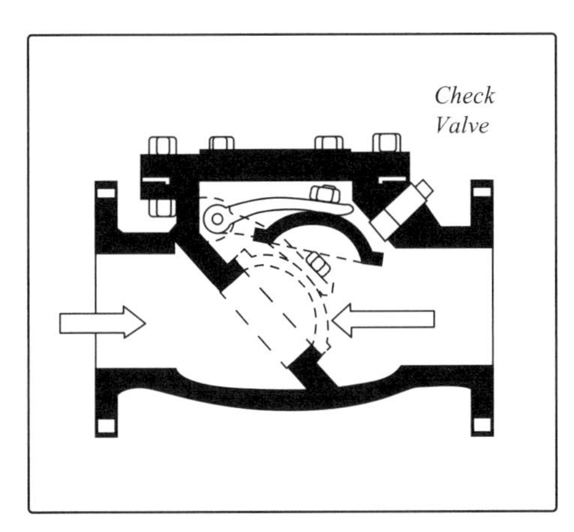

FIGURE 18.9. Check valve (courtesy of Valve Manufacturers Association of America).

The closure devices of spring-assisted check valves normally do not have sufficient force to close the valve against normal flow. At reduced flow, the spring does draw the closure device well down into the flow, however, and creates a back pressure so that the closure device will be seated just before cessation of flow and well ahead of a reversal of flow. The closure devices vary their position at reduced flow and the loss coefficient varies inversely with the amount of the opening. Just what the final position is for any given flow depends on jet pressure, spring gradient, back pressure, and velocity conversions. The loss coefficients of check valves shown in Table 18.1 are for full flow conditions. Higher values are to be expected at low flow conditions for spring-assisted valves.

Closure-assisted swing check valves are provided with an externally mounted closing cylinder to give a secondary source of power to assist the closure motion.

TABLE 18.1. Valve Loss Coefficient K

Type of Valve		Range of Values[a]	Representative Value[a]
	Multiturn Valves		
Diaphragm	–	1.0–3.0	2.0
Gate	Full port	0.1–0.4	0.2
	Reduced port	0.5–1.3	0.8
Globe	Standard	2–10	3.5
	60° Y-pattern	1.5–4	2.5
	45° Y-pattern	1.0–3.0	1.6
	Angle	2–5	4
Pinch	–	1–2	1.5
Needle	–	3–15	6
	Quarter-Turn Valves		
Ball	Full port	0.01–0.03	0.02
	Reduced port	0.1–0.3	0.2
Butterfly	–	0.04–0.6	0.2
Plug	Full port	0.05–0.2	0.1
	Reduced port	0.2–1.0	0.7
	Self-Actuated Valves		
Check	Swing	0.6–2	1.5
	Lift	1.5–3	2.4
	Globe	(Same as globe types)	
	Ball	1.5–3	2.4
Relief	–	0.2–5	1.5

[a] For valves in the fully open position. Or, in the case of check valves, in the full flow condition.

The cylinder consists of a spring-loaded piston actuated hydraulically or pneumatically upon demand. The cylinder acts to close the valve; it has no ability to open it. Swing check valves can also be provided with a top closing mechanism that combines check valve features with shutoff features. The closing mechanism can be either manually or motively operated upon demand.

Globe-type check valves are similar in housing construction to the standard globe valves and exhibit the same high pressure drop characteristics. The closure device is a free but guided plug. As in swing check valves, additional features may be added to assist closure or to provide positive shutoff.

Ball check valves exhibit approximately the same high pressure drop characteristic as the globe type. The closure device is a free but guided ball. Closure assist features may also be added to ball check valves.

18.3.2 Relief Valve

The relief valve (RV) (see Fig. 18.10) is a self-actuated, fast-opening valve used for quick relief of excessive pressure. The RV is designed or set to open at a predetermined set pressure to protect pressure vessels and other equipment from being subject to pressures that exceed their design limits. As the fluid is diverted, the pressure inside the vessel will drop. Once it reaches the valve's reseating pressure, the valve will close.

The RV is part of a bigger set that includes the safety valve (SV) and the safety relief valve (SRV). It should be noted that in practice people often do not stick to the technical distinction between the common names—they just use the term they are comfortable with. Most valves are spring operated. At lower pressures, some use a diaphragm in place of a spring. The oldest designs use a weight to seal the valve.

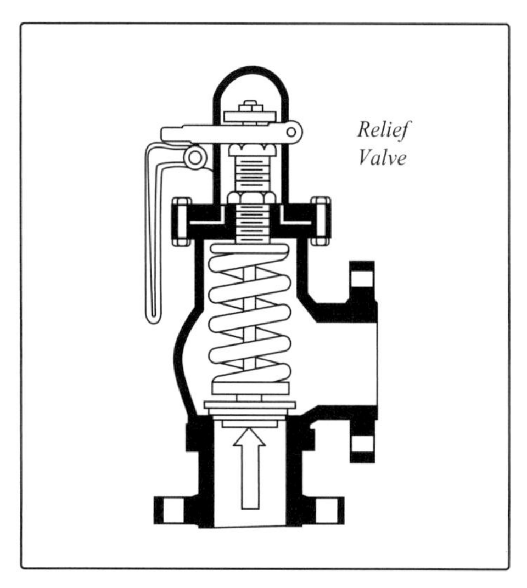

FIGURE 18.10. Relief valve (courtesy of Valve Manufacturers Association of America).

Technically, the RV is a valve used in liquid service; it opens proportionally as the increasing pressure overcomes the spring pressure. SVs are used for gas service. Most are full lift or snap acting; they pop open all the way. SRVs can be used for gas or liquid service, but set pressure will usually only be accurate for one type of fluid at a time (the type it was set with). The pilot-operated safety relief valve (POSRV) relieves by remote command from a pilot on which the static pressure (from the equipment to protect) is connected.

The poppet valve is one in which the closure member moves parallel to the fluid flow and perpendicular to the sealing surface. The closure element is usually flat, conical, or spherical on the sealing end. They may have many kinds of actuating elements, including springs, screws, and so on. Their main uses are for pressure control, check, safety, and relief functions.

Poppet valves generally provide large flow with very little actuator travel, excellent leakage control, and low pressure drop. For rapid opening, a spring-loaded poppet valve is nearly always used. They are subject to pressure imbalances that may cause chattering in some applications.

In some cases, equipment must be protected against being subjected to an internal vacuum (i.e., low pressure) that is lower than the equipment can withstand. In such cases, vacuum relief valves are used to open at a predetermined low pressure limit to admit air or an inert gas into the equipment so as to control the amount of vacuum.

18.4 CONTROL VALVES

The control valve (see Fig. 18.11) regulates the flow or pressure of a medium by fully or partially opening or closing in response to signals received from independent sensing devices in a continuous process. The opening or closing of control valves is done by means of control mechanisms powered electrically, pneumatically, electrohydraulically, and so on. They are used to control conditions such as flow, pressure, temperature, and liquid level.

Some valves are designed specifically as control valves. However, most types of valves can be used as control valves, both multiturn and quarter turn, by the addition of power actuators, positioners, and other accessories. The most common and versatile types of control valves are globe and angle valves. Their popularity derives from rugged construction and the many options available that make them suitable for a variety of process applications.

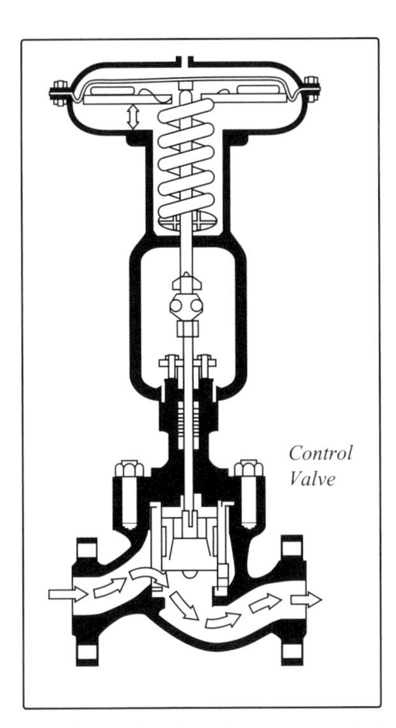

FIGURE 18.11. Control valve (courtesy of Valve Manufacturers Association of America).

18.5 VALVE LOSS COEFFICIENTS

The quality of surface finish of the inside of the body influences friction losses. Local losses depend largely on the relative size and the detailed geometry of the port and on the position of the shutoff member. In order to reduce the size of a valve, and the magnitude of forces and torques necessary to control it, the flow section in the valve body is often contracted. This contraction may be symmetrical, as in ball valves and most gate valves, or it may be asymmetrical with abrupt and complex variations of direction as in globe valves.

It has been found convenient in the valve industry, particularly in connection with flow control valves, to express the valve capacity in terms of the flow coefficient. In the English system, the flow coefficient C_V of a valve is defined as the flow of water at 60°F, in gallons per minute, at a pressure drop of 1 pound per square inch across the valve. The relationship between the loss coefficient K and the flow coefficient C_V in the English system is expressed as:

$$K = \frac{890.4d^4}{C_V^2},$$

or

$$C_V = \frac{29.84d^2}{\sqrt{K}},$$

where d is the inside diameter, in inches, of the connecting pipe.

In metric units, the flow coefficient K_V is defined as the flow rate in cubic meters per hour (m^3/h) of water at a temperature of 20°C with a pressure drop across the valve of 1 bar.* The relationship between the loss coefficient K and the flow coefficient K_V in metric units is expressed as:

$$K = \frac{16.0d^4}{K_V^2},$$

or

$$K_V = \frac{4.00d^2}{\sqrt{K}},$$

where d is the inside diameter, in millimeters, of the connecting pipe. K_V is related to C_V by the following expression:

$$K_V = 0.865C_V.$$

The flow coefficient varies considerably with size and with type of valve, and also varies between valves of the same type so that it is difficult to quantify the value of flow coefficient of the various types of valves. Valve manufacturer's handbooks and catalogs are the best source of pressure drop data for a particular valve. In lieu of manufacturer's data, the loss coefficient of various types and sizes of valves can be estimated using formulas provided by Idel'chik [1] and Crane [2]. Idel'chik's data include pressure loss as a function of valve opening for several types of valves.

In place of specific data, ranges of loss coefficient values for various types of valves are available in Table 18.1. The lower values are applicable to full port flanged or welded valves; the higher values are applicable to reduced port and threaded valves. The loss coefficient values were compiled from References [2–5]. Representative values are also shown. The data in Table 18.1 may be used for preliminary design purposes, where the actual value has not been specified or is otherwise unknown.

REFERENCES

1. Idel'chik, I. E., *Handbook of Hydraulic Resistance—Coefficients of Local Resistance and of Friction*, Moskva-Leningrad, 1960. (Translated from Russian; Published for

* The bar is a unit of pressure equal to 10^5 N/m^2 and is roughly equivalent to the atmospheric pressure on Earth at sea level.

the U.S. Atomic Energy Commission and the National Science Foundation, Washington, D.C. by the Israel Program for Scientific Translations, Jerusalem, 1966.)
2. Crane, Flow of Fluids Through Valves, Fittings and Pipe, Technical Paper No. 410, Cranc Company, 1985.
3. Streeter, V. L., Fluid flow friction factors for pipes, valves and fittings, Product Engineering, July 1947, pp. 89–91.
4. Tube Turns, Flow of fluids, Bulletin TT 725, 1952.
5. Fluid Controls, Ball Valve Handbook, 1971.

FURTHER READING

This list includes books and papers that may be helpful to those who wish to pursue further study.

Stone, J. A., Discharge coefficients and steady-state flow forces for hydraulic poppet valves, *Journal of Basic Engineering, Transactions of the American Society of Mechanical Engineers*, 82, March 1960, 144–154.

Tullis, J. P. and M. M. Skinner, Reducing cavitation in valves, *Journal of the Hydraulics Division, Transactions of the American Society of Civil Engineers*, 94, November 1968, 1475–1488.

Tullis, J. P., Cavitation scale effects for valves, *Journal of the Hydraulics Division, Proceedings of the American Society of Civil Engineers*, 99, July 1973, 1109–1128.

Kirik, M. J. and R. J. Gradie, A model for check valve/feedwater system waterhammer analysis, *Pressure Vessels & Piping Division, Transactions of the American Society of Mechanical Engineers*, May 1981, 1–9.

Goldberg, D. E. and C. L. Karr, Quick stroking: Design of time-optimal valve motions, *Journal of Hydraulic Engineering, Transactions of the American Society of Mechanical Engineers*, 109, June 1987, 780–795.

Fluid Controls, Piping and valves engineering reference guide, Plant Engineering, March 1987.

Kaisi, M. S., C. L. Horst, and J. K. Wang, Prediction of valve performance and degradation in nuclear power plant systems, Kaisi Engineering, Inc., KEI No. 1559, (Prepared for Division of Engineering, Office of Nuclear Regulatory Research, U. S. Nuclear Regulatory Commission, Washington, DC 20666, NRC FIN D2042), 1988.

Eom, K., Performance of butterfly valves as a flow controller, *Journal of Fluids Engineering, Transactions of the American Society of Mechanical Engineers*, 110, March 1988, 16–19.

Thorley, A. R. D., Check valve behavior under transient flow conditions: A state-of-the-art review, *Journal of Fluids Engineering, Transactions of the American Society of Mechanical Engineers*, 111, June 1989, 178–1893.

Morris, M. J. and J. C. Dutton, An experimental investigation of butterfly valve performance downstream of an elbow, *Journal of Fluids Engineering, Transactions of the American Society of Mechanical Engineers*, 113, March 1991, 81–85.

Morris, M. J. and J. C. Dutton, The performance of two butterfly valves mounted in series, *Journal of Fluids Engineering, Transactions of the American Society of Mechanical Engineers*, 113, September 1991, 419–423.

Liou, C. P., Maximum pressure head due to linear valve closure, *Journal of Fluids Engineering, Transactions of the American Society of Mechanical Engineers*, 113, December 1991, 643–647.

Kuehn, S. E., Valve reliability: Industry challenge for the '90s, Power Engineering, January 1993, pp. 20–26.

Lyons, J. L. and C. L. Askland, *Lyons' Encyclopedia of Valves*, Krieger Publishing Company, 1993.

Stojkov, B., *Valve Primer*, Industrial Press, 1997.

Zappe, R. W., *Valve Selection Handbook*, 4th ed., Gulf Publishing, 1998.

Nesbitt, B., *Handbook of Valves and Actuators*, Elsevier, 2007.

Skousen, P. L., *Valve Handbook*, 3rd ed., McGraw-Hill, 2011.

19

THREADED FITTINGS

Pressure loss through threaded (or screwed) pipe fittings is generally higher than through welded, or otherwise more smoothly connected pipe fittings. The internal geometry of threaded pipe fittings is discontinuous, creating additional pressure loss in the form of a partial expansion followed by a contraction. The actual loss is subject to fabrication and installation differences. The edge of the downstream pipe may have burrs or may be chamfered to some extent. The insertion length of the threaded pipe into the upstream and downstream sockets of the fitting can be widely variable.

19.1 REDUCERS: CONTRACTING

The contraction loss through concentric and eccentric threaded pipe reducers (see Fig. 19.1) is much higher than for a welded pipe reducer. The internal geometry creates additional pressure loss in the form of an incomplete sudden expansion preceding a gradual contraction, followed by a somewhat sudden expansion into the downstream pipe.

In practice, the initial sudden expansion is not pronounced because of the short length. All in all, it is reasonable, and most likely conservative, to model the loss simply as a sharp-edged contraction (see Section 10.1). Thus, the loss coefficient based on the downstream velocity head may be computed as:

$$K_2 \approx 0.0696\left(1-\beta^5\right)\lambda^2+(\lambda-1)^2, \qquad \text{(10.4, repeated)}$$

where λ is given by:

$$\lambda = 1+0.622(1-0.215\beta^2-0.785\beta^5), \qquad \text{(10.3, repeated)}$$

and where β is the ratio of the inside diameter of the outlet pipe to the inside diameter of the inlet pipe. Surface friction loss is small and may be ignored. There is some question as to whether eccentric reducers produce more head loss than do concentric reducers. When conservatism is desired, consider adding 20% to the sharp-edged contraction loss coefficient values for eccentric reducers.

19.2 REDUCERS: EXPANDING

Concentric and eccentric threaded pipe reducers in the expanding mode are shown in Figure 19.2. The pressure loss is much higher for threaded joints than for butt-welded joints. The internal flow path is discontinuous, creating additional pressure loss in the form of an initial sudden expansion into a more or less ineffective diffuser section, followed by a somewhat sudden contraction.

Simply treating the configuration as a sudden expansion ignores the added resistance to flow due to the somewhat abrupt contraction into the downstream pipe. Adding a multiplier to the sudden expansion equation should provide reasonable results. Thus, the loss coefficient K_1 based on the upstream velocity head may be computed as:

Pipe Flow: A Practical and Comprehensive Guide, First Edition. Donald C. Rennels and Hobart M. Hudson.

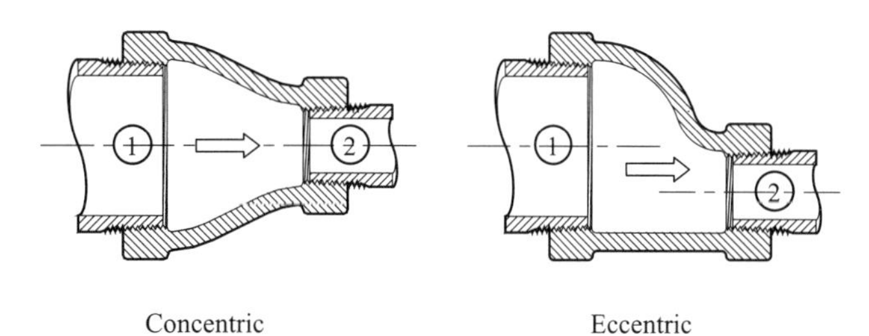

FIGURE 19.1. Threaded pipe reducer—contracting.

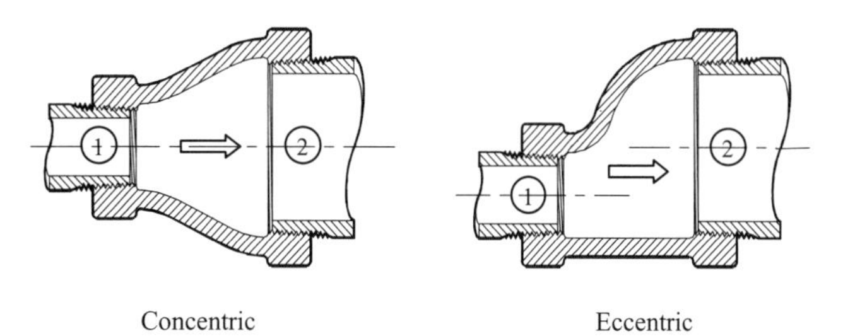

FIGURE 19.2. Threaded pipe reducer—expanding.

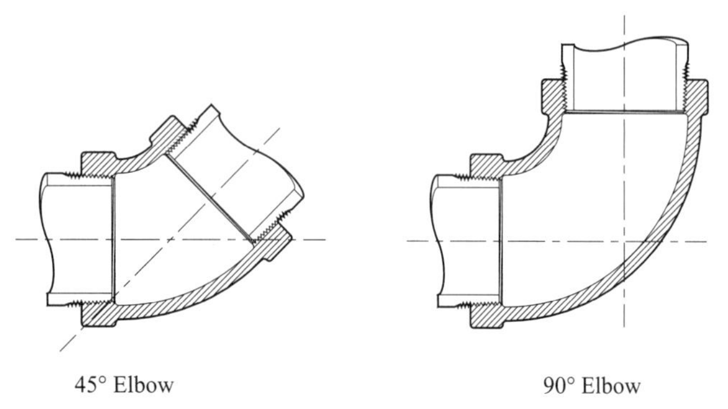

FIGURE 19.3. Threaded elbows.

$$K_1 \approx 1.25\left(1-\beta^2\right)^2,$$

where β is the ratio of the inside diameter of the inlet pipe to the inside diameter of the outlet pipe. Surface friction loss is small and may be ignored. It appears there is little difference in head loss between concentric and eccentric threaded pipe reducers in the expansion mode.

19.3 ELBOWS

Ninety and 45° standard threaded elbows are shown in Figure 19.3. The internal flow path is discontinuous, creating additional pressure loss in the form of a sudden expansion into the turning section, followed by a more or less sudden contraction. As noted in Chapter 15, the loss through elbows and bends is practically a direct function of friction factor. Therefore, the long-held practice of assuming $K = 16 f_T$ and $K = 30 f_T$ for 45° and 90° threaded elbows, respectively, appears to be reasonable [1].

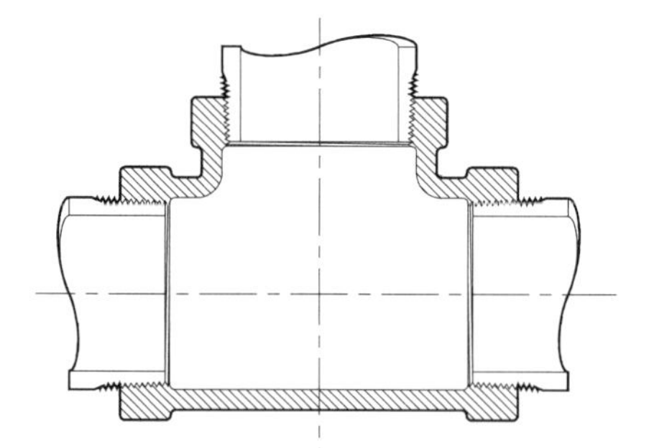

FIGURE 19.4. Threaded tee.

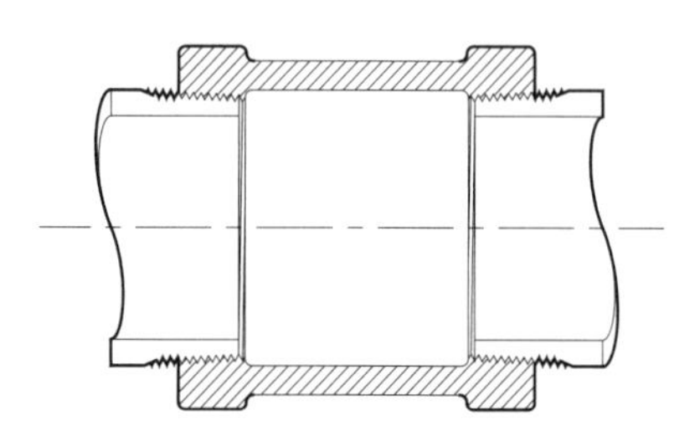

FIGURE 19.5. Threaded coupling.

19.4 TEES

A threaded tee is shown in Figure 19.4. The internal flow path is discontinuous compared to a tee with smooth connections, creating new pressure loss in the form of sudden expansion and sudden contraction losses. However, these losses may be more or less offset because the enlarged cross-sectional area at the branch connection effectively increases the radius ratio of the branch inlet (or outlet). With this trade-off in mind, it is reasonable to assume that the loss for threaded tees is approximately the same as for smooth tees and that the data in Chapter 16 for smooth tees may be applied to threaded tees as well. The question then arises as to what rounding radius to assume. The authors suggest assuming a rounding radius ratio r/d of 0.20 unless data are available to justify a different value.

Treating threaded tees the same as smooth tees should provide much greater accuracy than the long-held practice of assuming fixed values, $K = 20 f_T$ for flow through run and $K = 60\ f_T$ for flow through branch, regardless of flow configuration, flow rate ratio, and diameter ratio.

19.5 COUPLINGS

A threaded coupling is shown in Figure 19.5. A gap normally exists between the faces of the upstream and downstream pipes. This gap creates expansion and contraction losses. The losses are often neglected.

The loss coefficient is highly indeterminate. The insertion length of threaded pipe into the upstream and downstream sockets is variable, the expansion loss is relatively incomplete because of the short length of the fitting, and the degree of "sharpness" of the inlet edge of the downstream pipe is unpredictable. Nonetheless, it is reasonable to model the loss simply as a sudden expansion. Thus, the loss coefficient may be computed as:

$$K \approx \left(1 - \beta^2\right)^2, \qquad (19.5)$$

where β is the ratio of the inside to outside diameter of the pipe (overprediction of the sudden expansion loss may compensate for neglecting the contraction loss). Surface friction loss is small and may be ignored. This is a crude method, but it should provide better results than ignoring the loss entirely.

19.6 VALVES

The pressure loss through valves with threaded end connections is necessarily higher than through valves with smooth connections. In lieu of specific data from valve manufacturers, ranges of loss coefficient values for various types of valves are presented in Table 18.1. The higher values are applicable to reduced port and threaded valves; the lower values are applicable to full port, flanged, or welded valves.

REFERENCE

1. Flow of Fluids Through Valves, Fittings and Pipe, Technical Paper No. 410, Crane Company, Chicago, 1985, p. A-26.

PART III

FLOW PHENOMENA

PROLOGUE

There are a number of flow phenomena that can affect the performance of piping systems. In Part III, we investigate several interesting phenomena: cavitation, flow-induced vibration, temperature rise, and flow to run full. The phenomena are related to occurrences in the nuclear power industry. Of course, the information can be applied to flow conditions that may exist in any industry.

The phenomenon of cavitation is of great importance in the design and operation of hydraulic equipment. In Chapter 20, we study its nature and learn how to design and analyze piping systems to avoid its potentially damaging effects.

A brief categorization of flow-induced vibration in piping systems is presented in Chapter 21. Water hammer and column separation can create significant loads on pipe, its components, and its supports. Ways and means to prevent or mitigate such events are presented.

We learned in Chapter 2 that head loss is a loss of useful energy by conversion of mechanical energy to heat energy, and that in liquid (or incompressible) systems, the heat energy is usually of no consequence. In Chapter 22, we consider some situations where the heat energy may be of interest in liquid systems.

Whether or not horizontal flow passages or openings run full at low flow rates may be an important design consideration. This topic is treated in Chapter 23.

Pipe Flow: A Practical and Comprehensive Guide, First Edition. Donald C. Rennels and Hobart M. Hudson.

20

CAVITATION

Understanding the phenomenon of *cavitation* is of great importance in the design and operation of hydraulic equipment—turbines, pumps, valves, and other piping components. Cavitation may be expected in a flowing liquid whenever the absolute pressure at a point falls below the *vapor pressure* of the liquid. Local vaporization of the liquid will then result, causing a void or cavity in the flow field. The void eventually collapses—oftentimes accompanied by erosion (pitting) of nearby metal surfaces; loss of efficiency, excessive vibration, fluctuations of flow and pressure, and calculated flow rates are inaccurate.

20.1 THE NATURE OF CAVITATION

When the pressure falls below the vapor pressure of the liquid, a cavity of vapor is formed and moves along with the stream. The cavity contains a swirling mass of droplets and vapor and, although appearing steady to the naked eye, actually forms and reforms many times a second. The low-pressure cavity is swept downstream into a region of high pressure where it suddenly collapses—the surrounding liquid rushing in to fill the void. At the point of disappearance of the cavity, the onrushing liquid comes together, momentarily raising the local pressure within the liquid to a very high value. If the point of collapse of the cavity is in contact with or very near the boundary wall, the wall receives a blow as from a tiny hammer, and its surface may be stressed locally beyond its elastic limit, resulting eventually in fatigue and destruction of wall material. In the case of rotating machinery, the action predictably takes place in close proximity to the blades or sides of an impeller or draft tube, and particles of the metal may be gradually removed.

The nature of cavitation can be easily observed by study of the flow of a liquid through a constriction in a pipe as shown in Figure 20.1. Under low flow conditions, the variation of pressure through the pipe and constriction is given by the hydraulic grade line (HGL) A, the point of lowest pressure occurring at the minimum area where the velocity is highest. As the downstream pressure is reduced (by opening a valve, increasing the speed of a pump, etc.) the flow rate will increase to produce HGL B, for which the absolute pressure in the throat of the constriction falls to the vapor pressure of the liquid, causing the inception of cavitation. Further reduction in downstream pressure will then not result in further increase in flow rate, but will serve to extend the zone of vapor pressure downstream from the throat of the constriction to produce HGL C. Here the flow stream of liquid separates from the boundary walls, producing a cavity in which the mean pressure is the vapor pressure of the liquid. The cavity contains a swirling mass of droplets and vapor and, although appearing steady to the naked eye, actually forms and reforms many times a second.*

The sound emitted by the flow system changes at the onset of cavitation. At first, it sounds as if sand were

* The nature of cavitation is taken largely from Vennard [1].

Pipe Flow: A Practical and Comprehensive Guide, First Edition. Donald C. Rennels and Hobart M. Hudson.

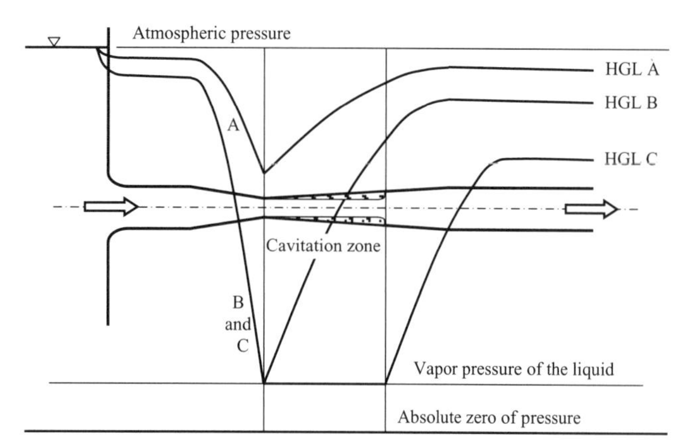

FIGURE 20.1. Cavitation (after Vennard [1]).

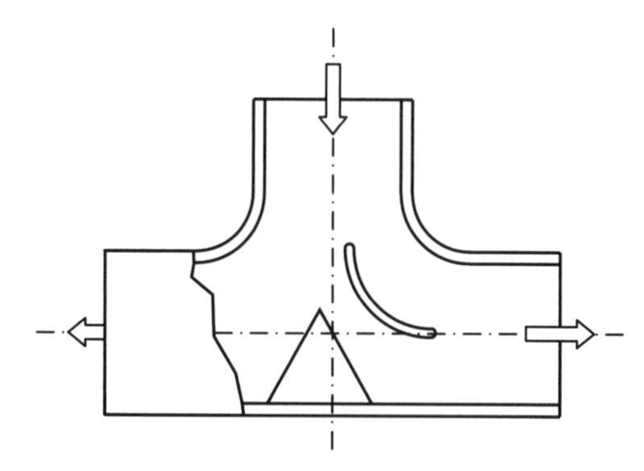

FIGURE 20.2. Inlet tee with flow splitter and turning vane.

passing through the system. As the flow is increased the sound or noise may increase, to give the impression of gravel or rocks passing through the system, or (at higher flow) of a machine gun barrage. In many cases, the collapse of the vapor cavity may take place away from the wall so that structural damage is not a problem. Nonetheless, normal flow patterns are disrupted which can result in decreased efficiency and, potentially, create excessive vibration at acoustic frequencies.

20.2 PIPELINE DESIGN

The piping designer should always be aware of the possibility of cavitation, particularly in high flow rate pipelines that connect to the atmosphere or otherwise operate at low pressure. What's more, because vapor pressure increases with temperature, the likelihood of cavitation increases with increase in temperature. Cavitation may take place wherever the flow stream is contracted, such as at a valve, bend, or tee. The resulting increase in flow stream velocity may reduce the local pressure below the vapor pressure of the liquid.

Cavitation may be a problem in bends, orifices, flow control valves, or partially open shutoff valves. Cavitation may be a problem in converging and diverging tees. As an example, cavitation was detected at the inlet tee of a sparger during flow testing at atmospheric conditions. A turning vane and flow splitter (see Fig. 20.2) were added to the inlet tee to reduce flow separation into the sparger arms. The flow splitter, a simple wedge, was located off-center because the sparger was designed to deliver more flow into one arm than the other. The wedge was located so that the projected inlet area to each arm was in the same ratio as the expected flow rate into each arm. Similarly, the turning vane was located so as to direct the flow more smoothly into the high flow arm of the tee. These added features eliminated cavitation and substantially decreased pressure loss into the run of the tee.

20.3 NET POSITIVE SUCTION HEAD

Operation of a centrifugal pump with a suction or inlet pressure that is close to the vapor pressure of the liquid may cause cavitation within the pump at the impeller. In practice, the term *NPSHR*, or *net positive suction head required*, has been established by the pump industry as an aid in evaluating the likelihood of cavitation. NPSHR, expressed in feet (or meters) of liquid, is the allowable difference between the absolute pressure of the liquid at the pump suction inlet and the vapor pressure of the liquid. It is impossible to design a centrifugal pump to exhibit absolutely no pressure drop between the suction inlet and its minimum pressure point that normally occurs at the entrance to the impeller vanes. Therefore, all pumping systems must maintain a positive suction pressure that is sufficient to overcome this pressure drop. If the pressure is not sufficient, cavitation is initiated.

The NPSHR value or rating increases with increased pump flow. NPSHR values are a function of centrifugal pump design and pump manufacturers publish pump performance curves that typically include a curve of NPSHR. The pump is run throughout its operating range at constant flow rate and constant speed with the suction condition varied to produce cavitation. The current industry standard for this test specifies that NPSHR values are determined as that value of NPSH that causes a reduction of total head of a pump by 3% due to blockage of flow through the impeller due to cavitation. The definition used to be that NPSHR was the suction pressure required to *prevent* cavitation. The current standard raises some concerns. Pump experts recommend an NPSHA (see below) to NPSHR margin of several feet to preclude pump damage. The margin may depend on the type of pump, the type of liquid and condition of the liquid, and other variables. Check with your pump manufacturer for its specific margin requirements.

Meanwhile, the term *NPSHA*, or *net positive suction head available*, also expressed in feet (or meters) of liquid, depends on knowledge of the liquid vapor pressure and the pressure of the liquid at the pump suction inlet during operation. NPSHA can be determined analytically. Let:

- A represent the inlet port area of the pump (usually the same as inlet pipe area).
- g represent the acceleration of gravity.
- H_S represent the elevation head of the surface of the pump suction supply above the pump datum elevation*. If it is below the pump datum elevation, it is negative.
- K_L represent the loss coefficient of the suction line from the suction vessel to the pump inlet, in terms of the velocity V_P in the pipe. This includes losses due to surface friction, fittings, strainers, and valves.
- P_P represent the absolute static pressure at the pump inlet.
- P_S represent the absolute static pressure at the *surface* of the pump suction supply. This will be atmospheric pressure if the suction vessel is open to the atmosphere. If the suction is taken from an enclosed vessel, P_S is the absolute pressure in the vessel.
- P_V represent the pressure that is required to keep the fluid in the liquid state at the prevailing liquid temperature. It is obtained from a vapor pressure table.
- V_P represent the velocity of the liquid at the pump inlet.
- $\dot{w}$ represent the weight flow rate.
- ρ_w represent the weight density of the liquid.

The energy equation from the surface of the pump suction supply to the suction inlet of the pump is:

$$\frac{P_S}{\rho_w}+H_S=\frac{P_P}{\rho_w}+\frac{V_P^2}{2g}+K_L\frac{V_P^2}{2g}.$$

* For horizontal pumps the datum elevation is the centerline of the pump shaft; for vertical single-suction pumps, both volute and diffusion vane type, it is the entrance eye to the first stage impeller; for vertical double-suction pumps it is the impeller discharge horizontal centerline. (Vertical and horizontal refer to the direction of the axis of the pump shaft.)

Rearranging Equation 20.1 gives:

$$\frac{P_P}{\rho_w}+\frac{V_P^2}{2g}=\frac{P_S}{\rho_w}+H_S-K_L\frac{V_P^2}{2g},$$

where the term on the left side of the equals sign represents the total pressure head available at the pump suction.† NPSHA represents the total suction head available minus the vapor pressure P_v of the liquid. Therefore, NPSHA expressed in feet of liquid is:

$$\text{NPSHA}=\frac{P_S}{\rho_w}+H_S-K_L\frac{V_P^2}{2g}-\frac{P_V}{\rho_w}$$

or

$$\text{NPSHA}=\frac{P_S}{\rho_w}+H_S-K_L\frac{\dot{w}^2}{2g\rho_w^2A^2}-\frac{P_V}{\rho_w}. \quad (20.1)$$

Clearly, the pumping system should be designed so NPSHA (available) is greater than NPSHR (required). Keep in mind that pumping system parameters (flow rate, fluid temperature, supply pressure, and supply elevation) can vary over the operating range of the pumping system. "Worst-case" values are usually used when calculating NPSHA.

NPSHA can be increased by:

- Raising the pump suction supply elevation or lowering the elevation of the pump.
- Increasing the pressure at the surface of the pump suction supply.
- Increasing suction pipe size or decreasing its length.
- Utilizing and maintaining low pressure drop valves, strainers, pipe bends, and so on in the pump suction line.

20.4 EXAMPLE PROBLEM: CORE SPRAY PUMP

The flow performance of a nuclear reactor core spray system during a postulated loss of coolant accident (LOCA) was evaluated in Section 5.6. The injection valve was opened and vessel pressure was progressively decreased from 120 to 14.7 psia to simulate core spray injection during vessel blowdown during the postulated event. Two valve lineups were considered: (1) the pump bypass valve remained open and (2) the pump bypass valve was closed. As an exercise to take into account the

† Total pressure head at the pump suction, *as determined by test*, is the reading of a pressure gauge at the suction inlet of the pump referred to the pump centerline, plus the velocity head at this point.

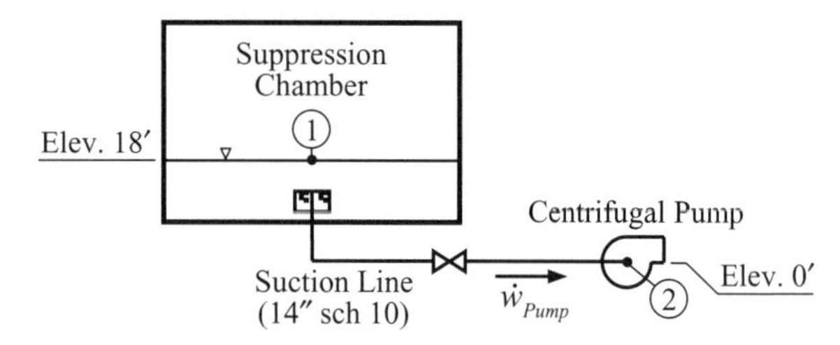

FIGURE 20.3. Pump suction line.

effects of age and usage, the evaluation assumed two dissimilar pipe surface roughnesses: (1) new, clean steel pipe and (2) moderately corroded steel pipe.

20.4.1 New, Clean Steel Pipe

Here, we will calculate NPSHA at the suction entrance to the core spray pump to assure that it is greater than NPSHR during the postulated LOCA. The pump suction line portion of the core spray system is shown in Figure 20.3. Input parameters from Section 5.6.1 applicable to this evaluation, as well as vapor pressure of water at 120°F,* are listed below.

20.4.1.1 Input Parameters All loss coefficients are in terms of velocity in 14″ schedule 10 pipe.

$\rho_w = 61.7$ lb/ft^3	Weight density of water at 120°F during postulated LOCA.
$A_{14} = 0.9940$ ft^2	Flow area of 14″ schedule 10 pipe.
$d_{14} = 13.500$ in	Inside diameter of 14″ schedule 10 pipe.
$e = 0.001800$ in	Absolute roughness of new, clean steel pipe.
$Elev_1 = 18.0$ ft	Elevation of minimum water level in suppression chamber.
$Elev_2 = 0$ ft	Elevation of core spray pump suction inlet.
$f_{14} = 0.0134$	Adjusted friction factor for flow in 14″ schedule 10 pipe.
$g = 32.174$ ft/s^2	Acceleration of gravity.
$K_{Valve14} = 0.20$	Loss coefficient of gate valve in pump suction line.
$K_{LREll14} = 0.177$	Adjusted loss coefficient of 14″ schedule 10, 90° long radius elbow.
$K_{Strainer} = 6.0$	Loss coefficient of "dirty" strainer.
$L_{1,2} = 40$ft	Pump suction line straight pipe length.
$p_1 = 14.7$ psia	Suppression chamber pressure.
$p_8 = 120–14.7$ psia	Decreasing reactor vessel pressure.
$p_V = 1.693$ psia	Vapor pressure of water at 120°F.
$\dot{w}_{Pump} =$ (varies) lb/s	Core spray pump flow rate varies with vessel pressure (from Table 5.3).

* Over time, the water in the suppression chamber may heat up to 120°F during the LOCA event.

20.4.1.2 Solution Use Equation 20.1 to calculate NPSHA as a function of core spray pump flow rate $\dot{w}_{Pump}$ (or vessel pressure) during the postulated LOCA:

$$p_S = p_1 = 14.7 \text{ psia},$$

$$H_S = Elev_1 - Elev_2 = 18.0 - 0 = 18.0 \text{ ft},$$

$$\begin{aligned} K_L &= K_{Strainer} + K_{Valve\,14} + 4K_{KREll14} + f_{14}\frac{L_{1,2}}{D_{14}} \\ &= 6.0 + 0.20 + 4 \times 0.177 + 0.0134\frac{40}{13.500/12} \\ &= 7.38, \end{aligned}$$

$$\begin{aligned} \text{NPSHA} = &\frac{144 \times 14.7}{61.71} + 18.0 - \\ &7.38\frac{\dot{w}_{Pump}^2}{2 \times 32.174 \times 61.71^2 \times 0.9940^2} - \\ &\frac{144 \times 1.693}{61.71} \\ &= 48.38 - 2.938 \times 10^{-5} \times \dot{w}_{Pump}^2. \end{aligned}$$

20.4.1.3 Results As shown in Table 20.1, core spray pump flow rate as a function of vessel pressure during the postulated LOCA event was calculated in Section 5.6.1. From this, NPSHA was calculated as a function of pump flow rate.

Calculated NPSHA is compared to NPSHR data from the pump manufacturer in Figure 20.4. The NPSHA curves for the open and closed bypass condition overlap each other as they should. There is significant NPSH margin over the operating range of the core spray pump during the postulated LOCA event. NPSH margin is gained by closing the bypass valve because maximum pump flow is reduced by about 100 gpm.

20.4.2 Moderately Corroded Steel Pipe

Input parameters are the same as in Section 20.4.1 except as listed below.

TABLE 20.1. NPSHA as a Function of Vessel Pressure (New, Clean Steel Pipe—e = 0.001,80 in)

Bypass valve position	Open					Closed				
Vessel pressure, psia	120	90	60	30	14.7	120	90	60	30	14.7
q_{Pump}, gpm (lb/s)	4622 (635)	4905 (674)	5163 (710)	5400 (742)	5514 (758)	4471 (615)	4768 (656)	5038 (693)	5287 (727)	5407 (744)
NPSHA, ft	35.9	34.3	32.8	31.4	30.6	36.7	35.1	33.6	32.1	31.3

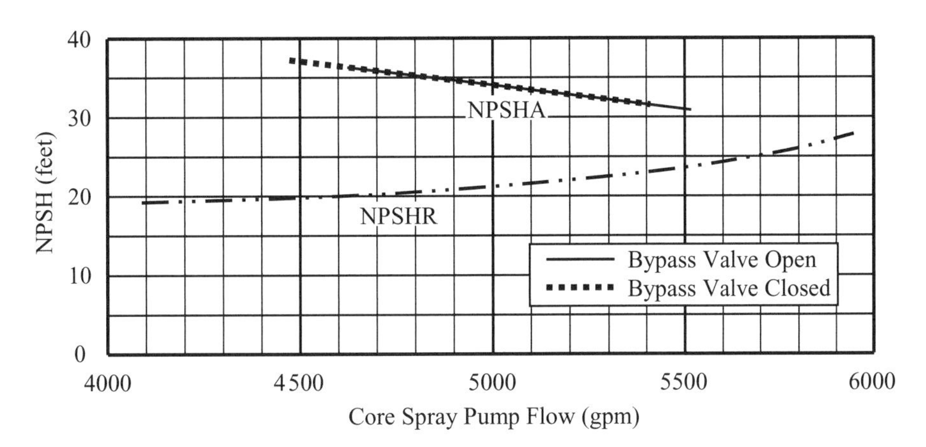

FIGURE 20.4. NPSHA compared to NPSHR during core spray system operation (new, clean steel pipe—e = 0.001,80 in).

TABLE 20.2. NPSHA as a Function of Vessel Pressure (Moderately Corroded Steel Pipe—e = 0.0130 in)

Bypass valve position	Open					Closed				
Vessel pressure, psia	120	90	60	30	14.7	120	90	60	30	14.7
q_{Pump}, gpm (lb/s)	4550 (626)	4831 (664)	5088 (699)	5323 (732)	5436 (747)	4393 (604)	4687 (645)	4955 (681)	5202 (715)	5322 (732)
NPSHA, ft	35.3	33.6	32.0	30.5	29.7	36.2	34.5	32.9	31.3	30.5

20.4.2.1 Input Parameters

$e = 0.01300$ in	Absolute roughness of moderately corroded steel pipe.
$f_{14} = 0.0207$	Adjusted friction factor for flow in 14″ schedule 10 pipe.
$K_{LREll14} = 0.289$	Adjusted loss coefficient of 14″ schedule 10, 90° long radius elbow.
$\dot{w}_{Pump}$ = (varies) lb/s	Core spray pump flow rate varies with vessel pressure (from Table 5.4).

20.4.2.2 Solution Equation 20.1 is used to calculate NPSHA as a function of core spray pump flow rate $\dot{w}_{Pump}$ (or vessel pressure) during the postulated LOCA:

$$p_S = p_1 = 14.7 \text{ psia},$$

$$H_S = Elev_1 - Elev_2 = 18.0 - 0 = 18.0 \text{ ft},$$

$$K_L = K_{Strainer} + K_{Valve\,14} + 4K_{KREll14} + f_{14}\frac{L_{1,2}}{D_{14}}$$
$$= 6.0 + 0.20 + 4 \times 0.289 + 0.0207\frac{40}{13.500/12}$$
$$= 8.09,$$

$$\text{NPSHA} = \frac{144 \times 14.7}{61.71} + 18.0 - 8/09\frac{\dot{w}_{Pump}^2}{2 \times 32.174 \times 61.71^2 \times 0.9940^2} - \frac{144 \times 1.693}{61.71}$$
$$= 48.38 - 3.336 \times 10^{-5} \times \dot{w}_{Pump}^2.$$

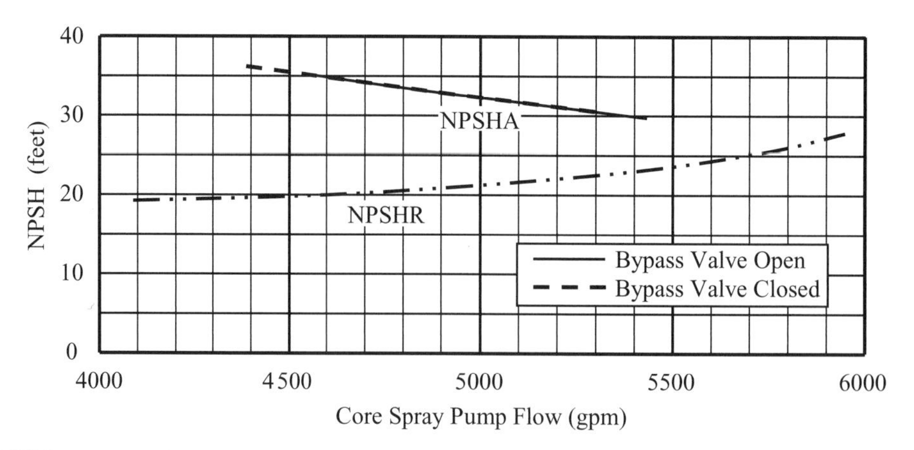

FIGURE 20.5. NPSHA compared to NPSHR during core spray system operation (moderately corroded steel pipe—$e = 0.0130$ in).

20.4.2.3 Results As shown in Table 20.2, core spray pump flow rate as a function of vessel pressure during the postulated LOCA event was calculated in Section 5.6.2. From this, NPSHA was calculated as a function of pump flow rate.

Calculated NPSHA is compared to NPSHR data from the pump manufacturer in Figure 20.5. Even with moderately corroded pipe, there is significant NPSH margin over the operating range of the core spray pump during the postulated LOCA event. NPSH margin is gained by closing the bypass valve because maximum pump flow is reduced by about 100 gpm.

REFERENCE

1. Vennard, J. K., *Elementary Fluid Mechanics*, 4th ed., John Wiley & Sons, New York, 1961.

FURTHER READING

This list includes books and articles that may be helpful to those who wish to pursue further study.

Stepanoff, A. J., *Centrifugal and Axial Flow Pumps, Theory, Design, and Application*, 2nd ed., John Wiley & Sons, 1957.

Keller, G. R., *Hydraulic System Analysis*, The Editors of Hydraulics and Pneumatics Magazine, 1978.

Hydraulic Institute Standards for Centrifugal, Rotary & Reciprocating Pumps, 14th ed., Hydraulic Institute, 1983.

21

FLOW-INDUCED VIBRATION

Flow-induced vibration (FIV) is the structural and mechanical vibration of structures immersed in or conveying fluid. Many engineering structures are susceptible to the interaction between the fluid's dynamic forces and the structures' inertial, damping, and elastic forces. Because designers are using materials to their limits, causing structures to become lighter and more flexible, FIV considerations have become increasingly important.

21.1 STEADY INTERNAL FLOW

Steady fluid flow through a pipe can impose pressures on the pipe walls that deflect the pipe and cause instabilities. The pipe may become susceptible to resonance or fatigue failure if its natural frequency falls below certain limits. If the fluid velocity becomes large enough, the pipe can become unstable. The most familiar form of this instability is the flailing about of an unrestricted garden hose.

If a pipe ruptures through its cross section, a flexible length of unsupported pipe is left spewing out fluid and is free to whip about and impact on other structures. This occurrence, called *pipe whip*, was a major consideration in the design of nuclear reactor main steam, recirculation, and feed water piping systems, and in the design of other auxiliary piping systems. During the 1970s the nuclear industry responded to the need to evaluate the dynamics of the highly nonlinear pipe whip event and to develop pipe whip restraints. Criteria were developed for the location of postulated pipe breaks (those points of high relative stress and high relative fatigue). Complying with the criteria resulted in the application of over 100 restraints on the aggregate piping systems inside the typical nuclear reactor containment.[*,†]

21.2 STEADY EXTERNAL FLOW

Any structure with a sufficiently bluff trailing edge sheds vortices in a subsonic flow; cylindrical structures are particularly susceptible. Periodic forces on the structure are generated as the vortices are alternately shed from side to side of the structure. The large amplitude vibrations that can be induced in elastic structures by vortex shedding are of great practical importance because of their destructive effect on suspension bridges, power transmission lines, television antennas, pipelines, heat exchanger tubes, and nuclear fuel assemblies.

If the frequency of vortex shedding coincides with the natural frequency of the structure, then the forces can induce large amplitude structural vibration normal

* A later industry-wide evaluation, called *leak before break*, resulted in a significant reduction of the number of pipe restraints. It was proven that fluid leaking from cracks in the ductile pipe material could be detected, and preventive action taken, long before complete pipe rupture would occur.

† See the section "Further Reading" at the end of this chapter for further information on this flow phenomenon.

Pipe Flow: A Practical and Comprehensive Guide, First Edition. Donald C. Rennels and Hobart M. Hudson.

to the free stream. If contained in a cavity, sound waves reflect off the cavity walls. Acoustic resonance has produced intense sound pressure levels in tubular heat exchangers that have damaged heat exchanger shells.

When a tube in a tube array in a cross flow is displaced from its equilibrium position, a fluid force, owing to the asymmetry of the flow field, may be exerted on the tube and the tube may vibrate with a large amplitude. In closely spaced tube arrays often used in heat exchangers, the distinct vortex shedding frequency degenerates into broadband turbulence, which buffets tubes. Such vibrations can be classified as axial flow induces of cross flow induces, depending on the incident angle of the incoming flow with respect to the cylinder's axes.

Damping of structures, avoidance of resonance, and the streamlining of structures are the primary mechanisms for limiting FIV.*

21.3 WATER HAMMER†

Pressure changes in a closed conduit produced by changes in fluid flow are called *fluid hammer* (or, more generally, *water hammer*). The fluid is usually a liquid, but sometimes can be a gas. The pressure change can create significant loads on pipe, its components, and its supports.

Velocity and pressure changes or disturbances can be expressed as $V_d - V_i$, and $P_d - P_i$, where the subscripts $_i$ and $_d$ designate initial and disturbed values, respectively. If the velocity of a fluid in a pipe is disturbed, it causes a corresponding disturbance of pressure, related by the classical water hammer relationship:

$$P_d - P_i = \frac{A_{pipe}\rho_w(V_d - V_i)}{g},$$

where A_{pipe} is the speed of sound of the fluid in the pipe, and ρ_w is the density of the fluid.

A flow *disturbance time* can be identified with occurrences such as full or partial valve opening or closure, pipe rupture, or the period of cyclic pulses caused by hydraulic machinery. If a flow disturbance at the system boundary occurs over a time interval of the same order as that required for an acoustic wave to pass through the system, propagation will be important. Most water hammer problems involve a pipe section of arbitrary length L. The pipe *acoustic response time* t_{ar}, allowing for acoustic propagation throughout its length, is given as

$$t_{ar} = \frac{L}{A_{pipe}}.$$

If the velocity of water or other fluid in a pipe is suddenly diminished or stopped, the energy given up by the fluid will be divided between compressing the liquid itself, stretching the pipe walls, and frictional resistance to wave propagation. Water hammer is manifest as a series of shocks, sounding like hammer blows, which may have sufficient magnitude to rupture the pipe or damage connected equipment. It may be caused by the nearly instantaneous or too rapid closing of a valve in the line, or by an equivalent stoppage of flow that would take place with the sudden failure of electricity supply to a motor-driven pump.

The shock pressure is not concentrated at the valve, and if rupture occurs it may take place near the valve simply because it acts there first. The pressure wave due to water hammer travels back upstream to the inlet of the pipe; there, it reverses and surges back and forth through the pipe. This cycle would continue indefinitely were it not for viscosity of the fluid and friction against the pipe walls.

The excess pressure due to water hammer is additive to the normal pressure in the pipe. Complete stoppage of flow is not necessary to produce water hammer as any sudden change in velocity will create it to a greater or lesser degree.

Under normal conditions the flow through the pipe is steady, having a velocity V past all sections. If the valve is made to close instantaneously, the particles of fluid in immediate proximity to it will have their velocity at once reduced to zero. If the whole mass of fluid in the pipe were inelastic (rigid) and contained in pipe walls that were also inelastic, then all the particles of fluid would likewise be instantaneously brought to rest and the pressure against the valve and all through the pipe would be infinite. That the pressure does not become infinite is due to the compressibility of the fluid and to the elasticity of the pipe wall.

Accounting for compressibility of the fluid and the elasticity of the pipe wall, it can be shown that the velocity of the pressure wave (or the speed of sound) of the fluid in the pipe is‡:

$$A_{pipe} = \sqrt{\frac{144gB}{\rho_w}}\sqrt{\frac{1}{1+\frac{B\ d}{E\ t_w}}}, \tag{21.1}$$

where B is the bulk modulus of the fluid (B is about 300,000 pounds per square inch for fresh water at

* See previous footnote.

† This section on water hammer is largely taken from Russell [1] and Crocker [2].

‡ In International System of Units (SI), delete 144 from the equation.

ordinary conditions), E is the elastic modulus of the pipe material (E is approximately 30,000,000 pounds per square inch for steel), and d and t_w are the diameter and wall thickness of the pipe, respectively.

The water hammer pressure ΔP_{max}, which is the intensity of the *excess* pressure produced by extinguishing the velocity V, can be determined as:

$$\Delta P_{\max} = \frac{\rho_w V}{g} A_{pipe}. \qquad (21.2)$$

If the expression for A_{pipe}, Equation 21.1, is substituted into Equation 21.2, the water hammer pressure ΔP_{max}, is*:

$$\Delta P_{max} = V\sqrt{\frac{144\rho_w B}{g}}\sqrt{\frac{1}{1+\frac{B}{E}\frac{d}{t_w}}}. \qquad (21.3)$$

As an example, the sudden valve closure of 60°F water flowing through a 16-in schedule 40 steel pipe at a velocity of 20 ft/s results in an excess pressure in the pipe:

V = 20 ft/s
ρ_w = 62.37 lb/ft^3
B = 300,000 lb/in^2
E = 30,000,000 lb/in^2
d = 15.000 in
t_w = 0.500 in

$$\Delta P_{max} = 20\sqrt{\frac{144\times 62.37\times 300{,}000}{32.174}}\sqrt{\frac{1}{1+\frac{300{,}000}{30{,}000{,}000}\times\frac{15.000}{0.500}}}$$
$$= 180{,}500\ \text{lb/ft}^2\ (=1115\ \text{lb/in}^2)$$

This example results in a substantial increase above the normal pressure in the pipe. However, it should be noted that the derivation of ΔP_{max} assumed instantaneous closure of the valve. This maximum pressure rise at the valve is maintained during the time t_{rt} required for the pressure wave to make a round trip of the pipe. For a pipe of length L, this time is:

$$t_{rt} = \frac{2L}{A_{pipe}}. \qquad (21.4)$$

If the valve is closed gradually, but within this time, the excess pressure at the gate will build up to the maximum value as before. This is because the first small pressure wave, generated as the valve starts to close, will not have had time to make a round trip and return to the valve as a wave lowering the pressure to normal. Equation 21.3 therefore applies for any time of closure up to time t_{rt} in Equation 21.4.

In 1898, N. Joukovsky [3] of Moscow was the first to demonstrate the validity of this equation. Joukovsky experimented with the effects of slow closure time. He concluded that for slow closure times (where closure time t_{cl} is greater than round-trip time t_{rt}) the excess pressure is reduced in intensity according to the proportion:

$$\frac{\Delta P}{\Delta P_{max}} = \frac{t_{rt}}{t_{cl}} = \frac{2L}{t_{cl} A_{pipe}}.$$

It has been proven since that this relation results in values greater than the actual pressure. Values computed from it therefore err on the side of safety.

The simplest method of protecting pipes from water hammer is found to be slowly closing the valve. In the case of long, cross-country pipelines, including gas lines, several minutes closing time may be necessary to alleviate water hammer.

In addition, the rise of pressure caused by water hammer may be minimized by the use of pressure relief valves of adequate size. They should be designed to open quickly and close slowly. In the case of liquid systems, air chambers or surge tanks of adequate size connected to the pipe near the valve may prevent pressure waves of significant magnitude from passing up the pipe. Such chambers should be kept filled with gas, perhaps with the aid of a diaphragm, since the liquid may readily absorb gas under pressure. To obtain greatest effectiveness, these devices should be located as close as possible to the source of the disturbance.

21.4 COLUMN SEPARATION

A water hammer-type event called *column separation* may occur in a pipeline filled with liquid when a vapor cavity forms and suddenly collapses. This results in a large and nearly instantaneous rise in pressure due to the collision of two liquid columns, or the collision of onc liquid column with a closed end. Another related phenomenon, known as *steam hammer*, might occur in vapor distribution systems. Some vapor may condense into liquid in a section of piping and form a slug. Subsequently, the vapor may hurl the slug at high velocity into hydraulic equipment or into pipe fittings and cause major problems.

Early on, licensees of operating nuclear reactors in the United States reported a number of column

* See previous footnote.

separation events during commercial operation. Many of these events resulted in damage to piping supports and restraints. A few cases involved small cracks or ruptures. None of the events affected the health and safety of the public.

In 1977 the U.S. Nuclear Regulatory Commission (NRC) staff initiated a review of reported water hammer events. The most serious and numerous column separation concern was line voiding. This generic cause included: (1) sudden water flow into a voided line, (2) steam bubble formation, and (3) steam bubble collapse.* Line voiding generally occurred in standby systems such as in emergency core cooling systems that are normally idle. The presence of these voids or steam bubbles was not readily detectable by plant operators. Other major causes of column separation events were steam-water entrainment in the high-pressure coolant injection turbine inlet and outlet lines and in isolation condenser lines.

The U.S. NRC reported the findings and recommendations of the industry-wide review in 1982 [4]. Design and operating recommendations for the prevention or mitigation of column separation included:

1. Provide keep-full provisions in standby systems.
2. Provide line void detection and alarm.
3. Train plant operators and maintenance personnel in the causes and prevention of column separation.
4. Reappraise plant operating and maintenance procedures.
5. Always account for column separation in the design of piping, its support system, and other components, such as valves.

These recommendations were incorporated into every operating reactor in the United States and the number of column separation events has significantly declined in the last three decades. The piping designer working in any industry should always consider the above recommendations, particularly when designing systems that are normally idle, or when working with a fluid system operating at or near saturation pressure and temperature.

REFERENCES

1. Russell, G. E., *Hydraulics*, Henry Holt and Company, New York, 1942.
2. Crocker, S., *Piping Handbook*, McGraw-Hill, New York, 1945.
3. Joukovsky, N., A translation of Joukovsky's paper on water-hammer experiments, *Journal of the American Waterworks Association*, 1904, 335.
4. NUREG/CR-2781, Quad-1-82-018, EGG-2203, Evaluation of Waterhammer Events in Light Water Reactor Plants, U.S. Nuclear Regulatory Commission, July 1982.

* Steam bubble *formation* occurs where a drop in pressure causes hot water to flash to steam. Steam bubble *collapse* occurs due to rapid condensation at steam–water interfaces.

FURTHER READING

This list includes works that may be helpful to those who wish to pursue further study.

Parmakian, J., *Waterhammer Analysis*, Dover Publications, 1963.

Thorley, A. R. D., Pressure transients in hydraulic pipelines, *Transactions of the American Society of Mechanical Engineers, Journal of Fluids Engineering*, 91, September 1969, 453–461.

Esswein, G. A., Development of a Plastic Strain Energy Absorbing Pipe Whip Restraint Design, Special Conference on Structural Design of Nuclear Plant Facilities, American Society of Civil Engineers, December 17–18, Volume II, 1973, pp. 171–200.

Blevins, R. D., *Flow-Induced Vibration*, Van Nostrand Reinhold Co., New York, 1977.

Rockwell, D. and E. Naudascher, Self-sustaining oscillations in flow past cavities, *Transactions of the American Society of Mechanical Engineers, Journal of Fluids Engineering*, 104, June 1982, 152–165.

GEAP-24158, COO/4175-7, Preliminary Design Handbook for Flow-Induced Vibration of Light Water Reactors, U.S. Department of Energy, November 1978.

NUREG-0582, Waterhammer in Nuclear Plants, U.S. Nuclear Regulatory Commission, July 1979.

Kirik, M. J. and R. J. Gradle, A Model for Check Valve/Feedwater System Waterhammer Analysis, Contributed by *the Pressure Vessels & Piping Division* of the *American Society of Mechanical Engineers* at the Century 2 Pressure Vessels & Piping Conference, San Francisco, Calif., August 12–15, 1980.

Chapman, R. L., D. D. Christensen, R. E. Dafoe, O. M. Hanner, and M. E. Wells, NUREG/CR-2059, CAAD-5629, Compilation of Data Concerning Known and Suspected Waterhammer Events in Nuclear Power Plants, EG&G Idaho Inc., April 1982.

Hatfield, F. J., D. C. Wiggert, and R. S. Otwell, Fluid structure interaction in piping by component synthesis, *Transactions of the American Society of Mechanical Engineers, Journal of Fluids Engineering*, 111, September 1982, 318–325.

Baldwin, R. M. and H. R. Simonns, Flow induced vibrations in safety valves, *Transactions of the American Society of Mechanical Engineers, Journal of Pressure Vessel Technology*, 108, August 1986, 267–272.

Fraas, A. P., *Heat Exchanger Design*, John Wiley & Sons, 1989.

Bruggeman, J. C., A. Hirschberg, M. E. H. van Dongen, and A. P. J. Wijnands, Flow induced pulsations in gas transport systems: Analysis of the influence of closed side branches, *Transactions of the American Society of Mechanical Engineers, Journal of Fluids Engineering*, 111, December 1989, 484–491.

Moody, F. J., *Introduction to Unsteady Thermofluid Mechanics*, John Wiley & Sons, New York, 1990.

Liou, C. P., Maximum pressure head due to linear valve closure, *Transactions of the American Society of Mechanical Engineers, Journal of Fluids Engineering*, 113, December 1991, 643–647.

Au-Wang, M. K., *Flow-Induced Vibration of Power and Process Plant Components: A Practical Workbook*, ASME Books, 2001.

Sekulic, D. P. and R. K. Shah, *Fundamentals of Heat Exchanger Design*, John Wiley & Sons, 2003.

Kaneko, S., R. Nakamura, F. Inada, M. Kato, and N. W. Mureithi, *Flow Induced Vibrations, Classifications and Lessons from Practical Experiences*, Elsevier Ltd., 2008.

22

TEMPERATURE RISE

We learned in Chapter 2 that head loss is a loss of useful energy by conversion of mechanical energy to heat energy, and that in liquid (or incompressible) systems, the heat energy is usually of no interest. Now we consider some situations where the heat energy may be of interest.

Head loss in the English system can be expressed as:

$$H_L = JU_2 - JU_1 + \frac{E_W}{\dot{w}},$$

where E_W is the heat energy passing *out of* the liquid through the walls of the pipe.*

This offers confirmation that head loss is not a loss of total energy but rather a conversion of mechanical energy into heat energy, part of which may leave the fluid, the remainder serving to increase its internal energy U. Assuming that all heat generated remains in the liquid (letting heat flow E_W through the walls of the flow system equal zero), the temperature rise ΔT due to head loss can be calculated as:

$$\Delta T = \frac{H_L}{778 c_P} = \frac{\Delta P}{778 \rho_w c_P}.$$

As an example, calculate the temperature rise of water initially at 200 psia and 120°F undergoing a head loss of 100 psid. From Table A.1 , the initial density of the water is 61.71 lb/ft^3. The heat capacity c_P of water can be taken as 1.0 Btu/lb-°F:

$$\Delta T = \frac{144 \times 100}{778 \times 61.71 \times 1.0} = 0.3°\text{F}.$$

A 0.3°F increase in water temperature produces little change in density, viscosity, and vapor pressure and can be neglected in most engineering applications.

In a pump, friction and work of compression increase the temperature of the liquid as it flows from suction to discharge. The temperature rise ΔT due to pump operation is†:

$$\Delta T = 2545 \frac{bhp}{c_P W} - \frac{E_L}{c_P} = \frac{H_P}{778 c_P \eta} - \frac{E_L}{c_P} = \frac{\Delta P_P}{778 c_P \rho_w \eta} - \frac{E_L}{c_P}, \tag{22.1}$$

where

bhp is brake horsepower of the pump;
E_L is heat loss from the pump through radiation, bearing, and external seal losses;
H_p is pump head;
ΔP_P is pump differential pressure;
c_p is heat capacity of the liquid;
W is pump flow rate;
ρ_w is density of the liquid; and
η is pump efficiency.

* If heat energy passes *into* the liquid, E_W will be negative or will appear on the opposite side of the equation.

† Equation 22.1 was derived from pump temperature rise equations given by Stepanoff [1] and by Karassik et al. [2].

Pipe Flow: A Practical and Comprehensive Guide, First Edition. Donald C. Rennels and Hobart M. Hudson.

The heat loss term E_L is usually small in comparison with the pump power and is most often ignored.

Three example problems follow.

22.1 REACTOR HEAT BALANCE

A reactor system heat balance is prepared during the early design stage of each nuclear power plant. This document provides reactor hydraulic and thermodynamic conditions at rated power for plant design and warranty purposes. The temperature rise, and ensuing enthalpy increase due to recirculation pump operation, is accounted for in the heat balance.

Determine the temperature rise in the recirculation pump loop of a nuclear plant considering that the pump head is 710 ft and the hydraulic efficiency of the pump is 0.87 at the rated power condition.

From Equation 22.1 (ignoring the heat loss term), the temperature increase is:

$$\Delta T = \frac{710}{778 \times 1.0 \times 0.87} = 1.0°\text{F}.$$

Accordingly, the rated power heat balance for the plant would indicate a 1°F increase in temperature across the recirculation pumps and a corresponding 1 Btu/lb increase in enthalpy.

22.2 VESSEL HEAT UP

Preoperational tests are performed at operating pressure and temperature conditions prior to loading fuel at nuclear reactors during plant startup. At that time, the reactor recirculation pumps are used to heat and, by way of isolation, pressurize the reactor vessel.

A reactor vessel and adjoining piping contains approximately 150,000 gallons of water. Assuming no heat loss, determine the time required to heat the reactor vessel from 70°F to 545°F using the recirculation pumps described in Section 22.1. The combined flow rate of the two recirculation pumps is 90,000 gpm, and the density of saturated water at 545°F is 46.3 lb/ft^3.

The amount of heat required to heat the vessel from 70°F to 545°F is:

$$\Delta Heat \approx \frac{150{,}000\ \text{gal} \times (545°\text{F} - 70°\text{F}) \times 46.3\ \text{lb/ft}^3 \times 1.0\ \text{Btu}/(\text{lb}°\text{F})}{7.48\ \text{gal/ft}^3}$$
$$\approx 441{,}000{,}000\ \text{Btu}.$$

The heat rate is:

$$HR_{Pump} = \frac{90{,}000\ \text{gal/min} \times 46.3\ \text{lb/ft}^3 \times 1.0\ \text{Btu/lb}}{0.87 \times 7.48\ \text{gal/ft}^3}$$
$$= 640{,}000\ \text{Btu/min}.$$

Finally, the time required to heat the reactor from 70°F to 545°F is:

$$\text{Time} \approx \frac{441{,}000{,}000\ \text{Btu}}{640{,}000\ \text{Btu/min}} \approx 689\ \text{minutes} \approx 11.5\ \text{hours}.$$

The actual time will be longer than 11.5 hours because heat loss from the reactor vessel and adjoining piping, as well as heat loss due to cooling water flow to maintain pump bearing and seal temperature, was ignored in the above calculation. In practice, it normally takes almost 2 days to raise the temperature of the reactor vessel to 545°F using heat generated by recirculation pump operation.*

22.3 PUMPING SYSTEM TEMPERATURE

As seen above, friction and the work of compression increase the temperature of the liquid as it flows from suction to discharge of a pump. A further temperature increase derives from liquid returned to the pump suction through a minimum-flow bypass line that may protect a pump when operating at or near its shutoff head. These temperature increases must be determined in order to specify the design temperature of the pumping system.

Figure 22.1 shows a portion of a control rod drive (CRD) system. The high pressure CRD pump continually supplies 80 gpm water to the hydraulic control units (HCUs) during normal plant operation. The minimum flow bypass line is designed to maintain a bypass flow rate of 61 gpm.

* At one nuclear power plant, reactor vessel heat up was attempted before the reactor vessel and adjoining piping were completely insulated. Because of excessive heat loss, heat up had to be postponed until insulation efforts were completed.

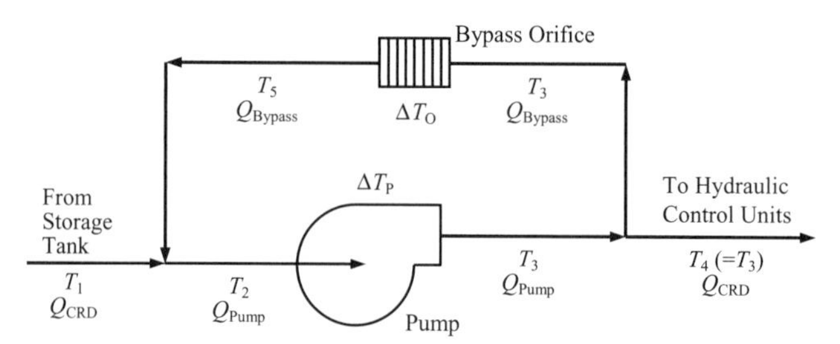

FIGURE 22.1. High pressure pump with minimum flow bypass line.

At full flow operation during the summer, the maximum temperature T_1 of water from the storage tank is 120°F. The pump head is 5100 ft. The pump efficiency is 0.94. The temperature increase across the pump from Equation 22.1 (ignoring the heat loss term) is:

$$\Delta T_P = \frac{5100}{778 \times 1.0 \times 0.94} = 7.0°\text{F}.$$

The temperature increase across the bypass line is:

$$\Delta T_O = \frac{5100}{778 \times 1.0} = 6.6°\text{F}.$$

Simple heat balances across the pump, across the bypass orifice, and at the cooling tower/bypass flow junction result in the following equations:

$$T_3 = T_2 + \Delta T_P, \quad \text{(a)}$$

$$T_5 = T_3 + \Delta T_O, \quad \text{(b)}$$

$$T_1 Q_{CRD} + T_5 Q_{Bypass} = T_2 (Q_{CRD} + Q_{Bypass}). \quad \text{(c)}$$

Substituting Equations (a) and (b) into Equation (c) and rearranging gives:

$$T_5 = T_1 + \frac{(\Delta T_P + \Delta T_O)(Q_{CRD} + Q_{Bypass})}{Q_{CRD}},$$

and, by substitution of values,

$$T_5 = 120 + \frac{(7.0 + 6.6)(80 + 61)}{80} = 144.0°\text{F}.$$

Rearrangement and substitution of values into Equation (b) gives:

$$T_3 = T_5 - \Delta T_O = 144.0 - 6.6 = 137.4°\text{F},$$

and likewise for Equation (a) gives:

$$T_2 = T_3 - \Delta T_P = 137.4 - 7.0 = 130.4°\text{F}.$$

Thus, the design temperature of the bypass portion of the CRD system should be at least 144°F. The design temperature of the lines downstream of the pump (and to the HCUs) should be at least 137.4°F.

REFERENCES

1. Stepanoff, A. J., *Centrifugal and Axial Flow Pumps*, 2nd ed., John Wiley & Sons, Inc., New York, 1957.
2. Karassik, I. J., W. C. Krutzsch, W. H. Fraser, and J. P. Messina, *Pump Handbook*, 2nd ed., McGraw-Hill, Inc., 1976.

23

FLOW TO RUN FULL

Whether or not horizontal flow passages run full at low flow rates may be an important design consideration. For example, relatively cold fluid may discharge into a vessel filled with a hot fluid through a horizontal opening. At low flow rates, the space above the cold fluid at the opening may be occupied by hot fluid because of the difference in density of the two fluids.* If the flow tends to be unsteady, temperature cycling may occur in the discharge openings and cause thermal fatigue of the metal surfaces. A related but different example is stratification of cold fluid at the bottom of a horizontal pipe: hot fluid, or vapor, at the top of the pipe may produce excessive thermal loads on the pipe and its supports.

If the flow rate of the fluid is not sufficient to fill the opening, the condition is called *open flow*. If the flow rate is just enough to fill the opening, the condition is called *full flow* or *flow to run full*. *Submerged flow* is the condition in which the surface level of the heavy fluid is above the top of the opening. The governing relationships are developed using energy and continuity equations. Inertial and gravitational forces dominate. Shear stresses are considered to exert only a negligible effect on flow and thus viscous forces are ignored. Surface tension forces are also ignored. The solutions assume hydrostatic pressure distribution at the opening.

The Froude number, the ratio of inertial and gravitational forces (see Section 1.3.5), can be used to characterize open flow or, in this case, whether or not horizontal openings run full. The Froude number N_{Fr} is related to volumetric flow rate in cubic feet per second as follows:

$$Q = \pi\sqrt{2g}R^{5/2}N_{Fr}, \tag{23.1}$$

or in gallons per minute as

$$q = 448.83\pi\sqrt{2g}R^{5/2}N_{Fr}. \tag{23.2}$$

23.1 OPEN FLOW

Figure 23.1 depicts a heavy fluid discharging into a large space or chamber filled with a light fluid. In this case, the flow rate of the heavy fluid is not sufficient to fill the opening and the condition is called *open flow*.

Using the heavy to light fluid interface in the opening as the reference elevation and ignoring the approach velocity, the energy equation along a streamline crossing the exit of the opening is:

$$\rho_{heavy}h = \frac{\rho_{heavy}u^2}{2g} + \rho_{light}h. \tag{23.3}$$

Rearranging and letting $h = H - y$ gives:

$$u = \sqrt{2g\left(\frac{\rho_{heavy} - \rho_{light}}{\rho_{heavy}}\right)(H - y)}.$$

* Of course, the situation would be reversed if hot fluid were injected into a cold fluid through horizontal openings. At low flow rates, the cold fluid may occupy the space below the hot fluid.

Pipe Flow: A Practical and Comprehensive Guide, First Edition. Donald C. Rennels and Hobart M. Hudson.

The continuity equation can be written as:

$$dQ = u dA, \tag{23.4}$$

where $dA = 2x dy$ and $x = \sqrt{R^2 - y^2}$. Substitution gives:

$$dQ = 2\sqrt{2g\left(\frac{\rho_{heavy} - \rho_{light}}{\rho_{heavy}}\right)(H-y)}\sqrt{R^2 - y^2}\,dy.$$

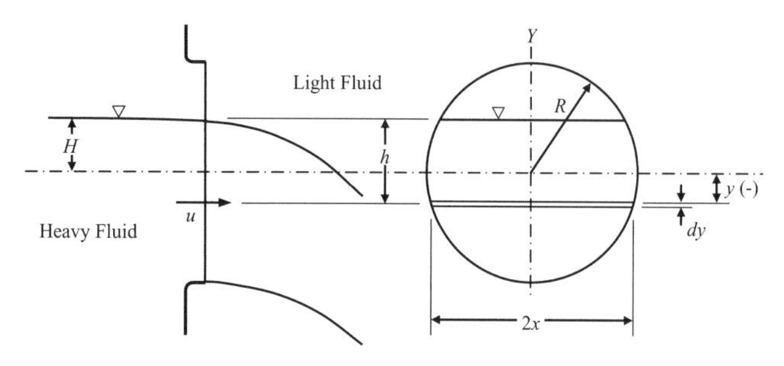

FIGURE 23.1. Open flow.

The integral form in the case of open flow ($H < R$) is:

$$Q = 2\sqrt{2g\left(\frac{\rho_{heavy} - \rho_{light}}{\rho_{heavy}}\right)}\int_{-R}^{H}\sqrt{(H-y)(R^2-y^2)}\,dy. \tag{23.5}$$

The Froude number can be expressed as:

$$N_{\mathrm{Fr}} = \frac{Q}{\pi\sqrt{2g}R^{5/2}}. \tag{23.6}$$

Substitution of Equation 23.5 into Equation 23.6 gives:

$$N_{\mathrm{Fr}} = \frac{2\sqrt{\dfrac{\rho_{heavy} - \rho_{light}}{\rho_{heavy}}}}{\pi R^{5/2}}\int_{-R}^{H}\sqrt{(H-y)(R^2-y^2)}\,dy. \tag{23.7}$$

The integral form of Equation 23.7 is indeterminate. A computer program was developed to perform numerical integration of Equation 23.7. The results are plotted in Figure 23.2 as a function of depth ratio $(R + H)/D$. The upper curve, at $(\rho_{heavy} - \rho_{light})/\rho_{heavy} = 1.0$, represents the

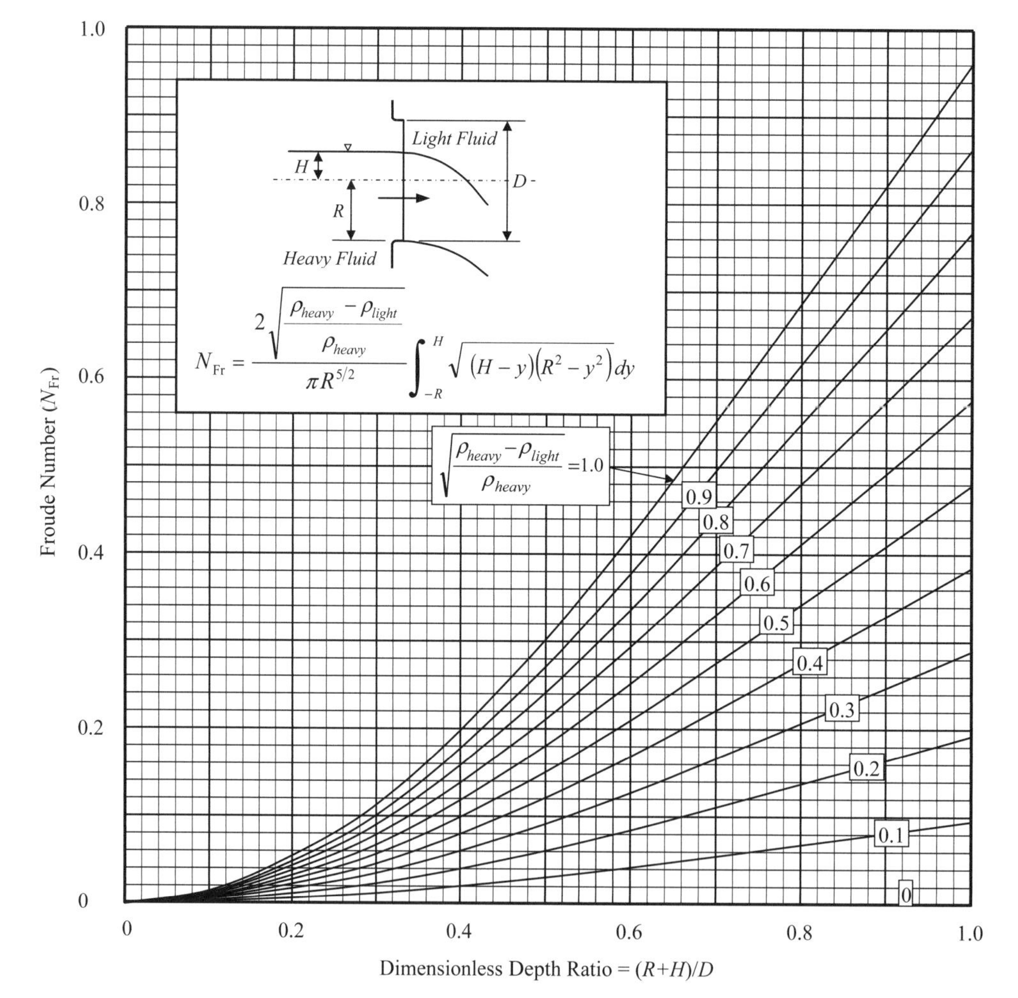

FIGURE 23.2. Open flow graph.

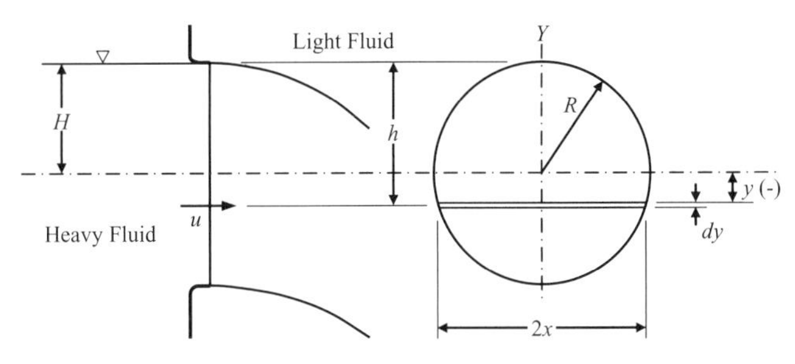

FIGURE 23.3. Full flow.

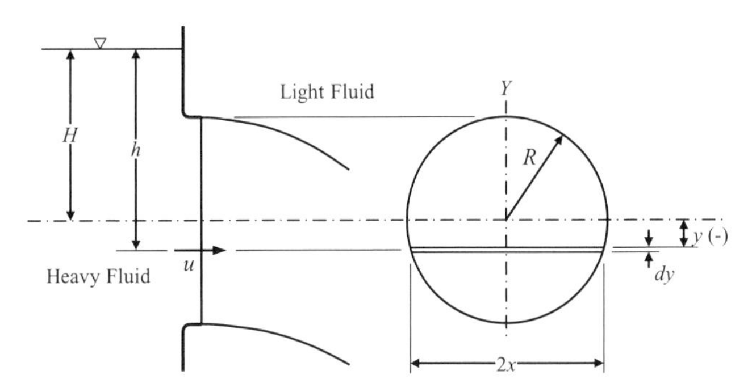

FIGURE 23.4. Submerged flow.

case of a liquid discharging into a gas or vapor where the density of the light fluid is negligible. Note that the Froude number (or flow rate) required to support any given depth approaches zero as the density of the lighter fluid approaches the density of the heavier fluid—as should be the case.

23.2 FULL FLOW

Figure 23.3 represents full flow (or flow to run full) for a heavy fluid discharging into a large space or chamber filled with a light fluid. The flow rate is just sufficient to fill the opening with heavy fluid; any decrease in flow rate would allow the formation of a pocket of light fluid at the top of the exit plane of the opening.

The governing relationships for full flow are the same as the open flow case except that $h = R - y$ (rather than $h = H - y$) and the integration is carried out from R to $-R$ (rather than from H to $-R$).

The energy equation along a streamline at the exit using the fluid interface as the reference elevation is:

$$\rho_{heavy} h = \frac{\rho_{heavy} u^2}{2g} + \rho_{light} h. \qquad (23.3, \text{repeated})$$

Rearranging and letting $h = R - y$ gives:

$$u = \sqrt{2g\left(\frac{\rho_{heavy} - \rho_{light}}{\rho_{heavy}}\right)(R - y)}.$$

The continuity equation can be written as:

$$dQ = u dA, \qquad (23.4, \text{repeated})$$

where $dA = 2x dy$ and $x = \sqrt{R^2 - y^2}$. Substitution gives:

$$dQ = 2 \cdot \sqrt{2g\left(\frac{\rho_{heavy} - \rho_{light}}{\rho_{heavy}}\right)(R - y)}\sqrt{R^2 - y^2}\, dy.$$

The integral form in the case of full flow is:

$$Q = 2 \cdot \sqrt{2g\left(\frac{\rho_{heavy} - \rho_{light}}{\rho_{heavy}}\right)} \int_{-R}^{R} \sqrt{(R - y)(R^2 - y^2)}\, dy,$$

and rearranging gives:

$$Q = 2 \cdot \sqrt{2g\left(\frac{\rho_{heavy} - \rho_{light}}{\rho_{heavy}}\right)} \int_{-R}^{R} \left(R\sqrt{R + y} - y\sqrt{R + y}\right) dy. \qquad (23.8)$$

Closed form integration of Equation 23.8 yields flow to run full in a horizontal opening:

$$Q = \frac{64\sqrt{g}}{15} \sqrt{\frac{\rho_{heavy} - \rho_{light}}{\rho_{heavy}}} R^{5/2}. \qquad (23.9)$$

Substitution of Equation 23.9 into Equation 23.6 gives flow to run full in terms of the Froude number:

$$N_{\text{Fr}} = \frac{64}{15\sqrt{2}\pi} \sqrt{\frac{\rho_{heavy} - \rho_{light}}{\rho_{heavy}}} = 0.9603 \sqrt{\frac{\rho_{heavy} - \rho_{light}}{\rho_{heavy}}}. \qquad (23.10)$$

Flow to run full in a horizontal opening is represented in Figure 23.2 at the intercepts of the various curves with $(R + H)/D = 1$ (at the right-hand side of the figure). The actual flow may then be obtained from Equation 23.1 (or 23.2), or may be directly obtained from Equation 23.8.

23.3 SUBMERGED FLOW

Figure 23.4 describes a submerged flow condition in which the surface level of the heavy fluid is above the top of the flow nozzle or horizontal opening. The development of the governing relationships is the same as for open flow (see Section 23.1) except that the integration is carried out from R to $-R$, rather than from H to $-R$. Thus, Equation 23.5 becomes:

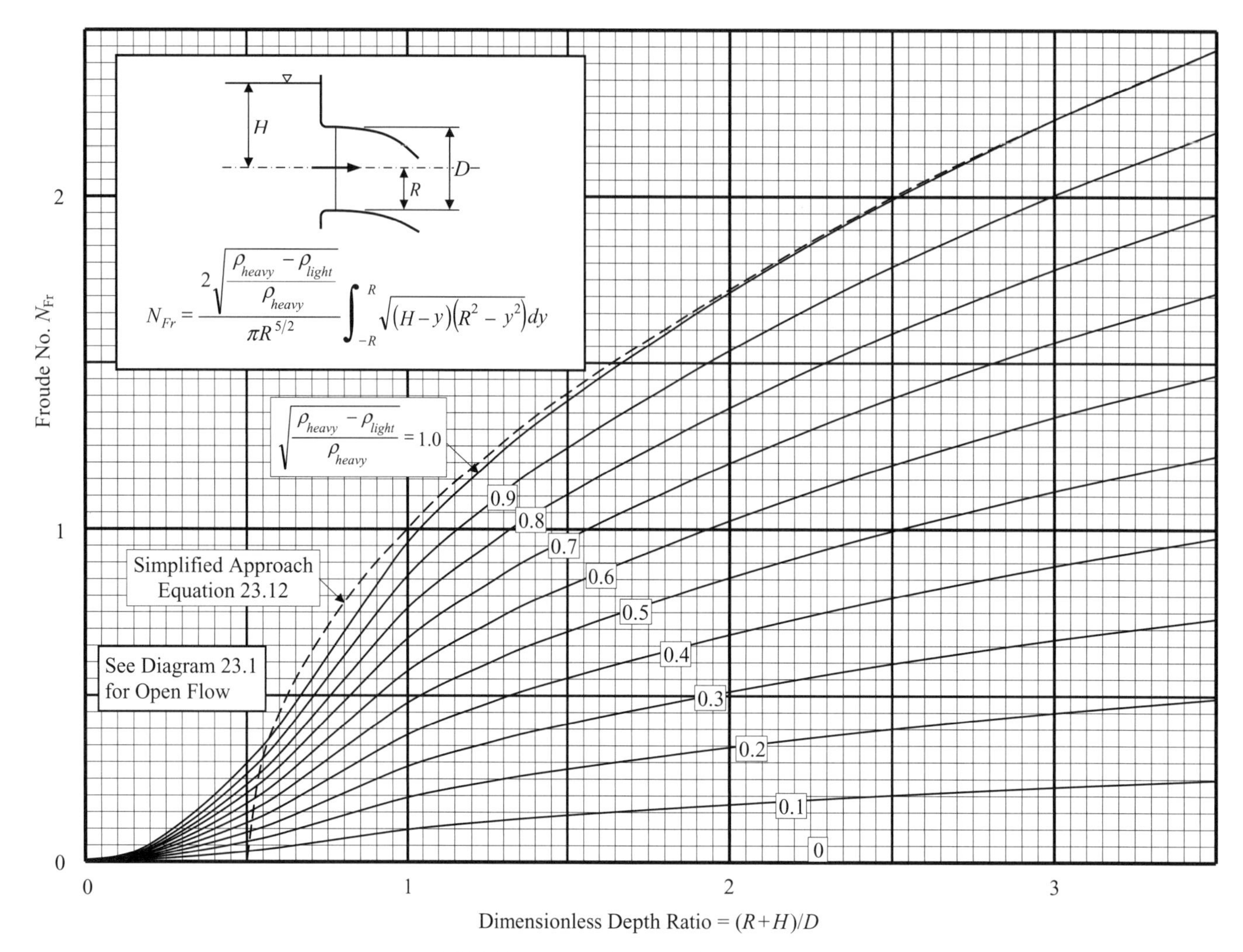

FIGURE 23.5. Submerged flow graph.

$$N_{Fr} = \frac{2\sqrt{\frac{\rho_{heavy} - \rho_{light}}{\rho_{heavy}}}}{\pi R^{5/2}} \int_{-R}^{R} \sqrt{(H-Y)(R^2 - y^2)}dy. \tag{23.11}$$

A computer program was developed to perform numerical integration of Equation 23.11. The Froude number for submerged flow is plotted in Figure 23.5 as a function of depth ratio $(R + H)/D$ and as a function of ρ_{heavy} and ρ_{light}.

As a means of validating the above results, a simplified approach to submerged flow was taken (see Fig. 23.6). This is the classic vessel drain problem, where $V = \sqrt{2gH}$. The vessel is draining to atmosphere so that the density of the lighter fluid (air in this case) is neglected.

Substituting velocity $V = \sqrt{2gH}$ into Equation 1.3 yields the Froude number for a vessel draining to atmosphere (or for submerged flow in this case) as:

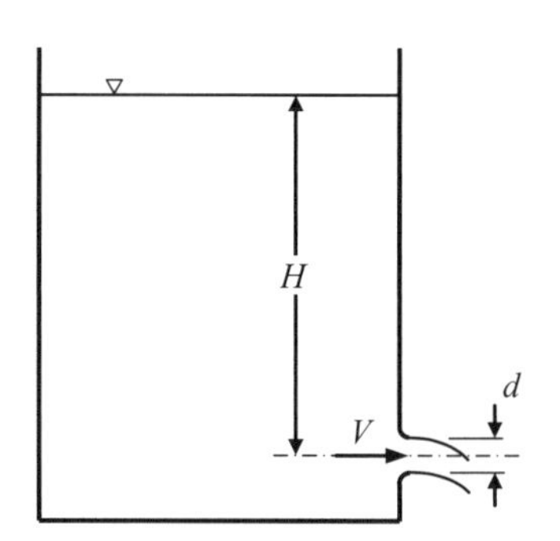

FIGURE 23.6. Simplified approach.

$$N_{Fr} = \sqrt{\frac{H}{R}}. \tag{23.11}$$

This function is shown as a dashed line in Figure 23.5. The function is accurate for large values of H/R, but quickly loses accuracy below $(R + H)/D$ less than 2.

23.4 REACTOR APPLICATION

Cracks were discovered in the outlet "nozzles" of spargers* that deliver feedwater flow into a boiling water reactor vessel. It was understood that thermal (hot and cold) cycling at low feedwater flow rates caused the cracks. The relationships in Figures 23.2 and 23.5 were used to estimate the flow rate required to ensure full flow from the feedwater sparger nozzles and thus avoid thermal cycling.

As a simple example of the analysis method, the left arm of a simple feedwater sparger discharging water into air is depicted in Figure 23.7. Applying the relationships of Figure 23.2 and Equation 23.2, the volumetric flow rate for full flow through the nozzles in the left arm is calculated as 21.7 gpm. Because of symmetry, the total sparger flow rate is 43.4 gpm, and given that there are four feedwater spargers in the reactor vessel, the total flow to run full is 173.6 gpm. In this example, it is assumed that the sparger inlet centerline elevations are uniform and that the sparger arms are truly horizontal.

* The design of early feedwater spargers featured simple side holes located along the length of the sparger arms to deliver water horizontally into the reactor vessel.

A more rigorous analysis than depicted above was performed for the nuclear plant. Photographs of the feedwater spargers, taken at plant startup during a water flow test discharging to air, were reviewed. The combination of low flow rate, sparger centerline elevation differences, three different nozzle sizes, and canting of the sparger arms resulted in unique sparger flow patterns, including no flow through several nozzles. As-built drawings were reviewed to reconstruct the relative elevations of the sparger inlet centerlines as well as departure of the individual sparger arms from horizontal. Using the open flow relationships of Figure 23.2, the assumed water level elevation was adjusted until total flow rate from the four spargers (138 gpm in this case) closely approximated the test flow rate (140 gpm). At this condition, the calculated nozzle flow rates closely resembled the nozzle flows patterns observed in the photographs. This comparison validated the methodology.

Next, in order to simulate reactor startup conditions, water level was adjusted to the top of the nozzle located at the highest elevation. The submerged flow relationships of Figure 23.2 were then used to calculate the flow to run full in the reconstructed installed condition. The calculated value of 1230 gpm was about 9% of rated

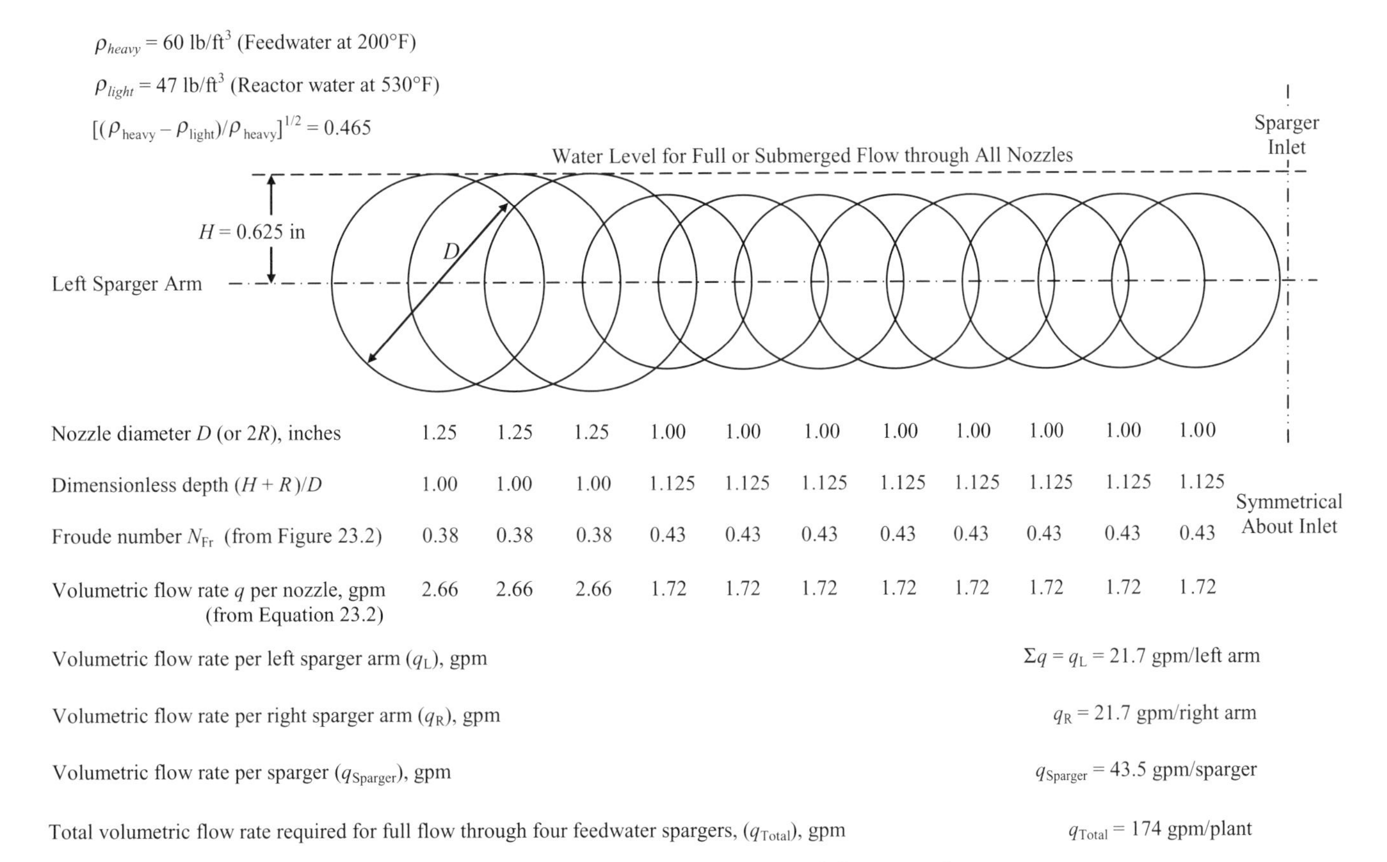

Nozzle diameter D (or 2R), inches	1.25	1.25	1.25	1.00	1.00	1.00	1.00	1.00	1.00	1.00	1.00
Dimensionless depth (H + R)/D	1.00	1.00	1.00	1.125	1.125	1.125	1.125	1.125	1.125	1.125	1.125
Froude number N_{Fr} (from Figure 23.2)	0.38	0.38	0.38	0.43	0.43	0.43	0.43	0.43	0.43	0.43	0.43
Volumetric flow rate q per nozzle, gpm (from Equation 23.2)	2.66	2.66	2.66	1.72	1.72	1.72	1.72	1.72	1.72	1.72	1.72

FIGURE 23.7. Flow rate required for full sparger flow.

feedwater flow assuming water discharging into air. However, during reactor startup the reactor water temperature is about 530°F (ρ_{light} = 47 lb/ft^3) and the temperature of the feedwater is about 200°F (ρ_{heavy} = 60 lb/ft^3). Thus, the above results were adjusted by a factor of:

$$\sqrt{(\rho_{heavy} - \rho_{light})/\rho_{heavy}} = \sqrt{(60-47)/60} = 0.465.$$

This adjustment resulted in a flow to run full of 572 gpm (0.465 × 1230 gpm). This turned out to be about 4% of rated feedwater flow. A review of plant operating data revealed that feedwater flow was maintained at less than 4% rated flow for significant periods during plant startup and shutdown. Procedures were established to minimize operating at or below 4% feedwater flow.

FURTHER READING

This list includes books and an article that may be helpful to those who wish to pursue further study.

Chow, V. T., *Open Channel Hydraulics*, McGraw-Hill Book Company, 1959, (the classic text).

Wallis, G. B., S. J. Crowley, and Y. Hagi, Conditions for a pipe to run full when discharging liquid into a space filled with gas, *Journal of Fluids Engineering, American Society of Mechanical Engineers*, 99, June 1977, 405.

Chandhry, M. H., *Open-Channel Flow*, Prentice-Hall, 1993.

Munson, B. R., D. F. Young, and T. H. Okiishi, *Fundamentals of Fluid Mechanics*, 3rd ed., John Wiley & Sons, 1998.

Jain, S. C., *Open-Channel Flow*, John Wiley & Sons, 2000.

APPENDIX A

PHYSICAL PROPERTIES OF WATER AT 1 ATMOSPHERE

These tables present the values of five physical properties of fresh water at various temperatures using units familiar to engineers. In Table A.1, Fahrenheit temperatures are displayed in whole numbers, and properties given in English units are shown unitalicized. International System of Units (SI) quantities are given in italics. In Table A.2, Celsius temperatures are displayed in whole numbers, and properties given in SI units are shown unitalicized. English quantities are given in italics.

Because the number of Celsius degrees between freezing and boiling is 100, and the number of Fahrenheit degrees in the same span is 180, the ratio between the two is 5/9, which gives a repeating decimal when converting from °F to °C.

The SI unit for absolute viscosity, N-s/m^2, is exactly 1000 times the derived unit centipoise, which is often used in calculations using the English system. Vapor pressure in SI is given in N/m^2, a unit sometimes called the pascal.

The specific heat is also known as heat capacity. As English and SI units for heat are based on the heat required to raise the temperature of a unit mass of water by a unit degree, the numerical values of the specific heat are the same in both systems.

Pipe Flow: A Practical and Comprehensive Guide, First Edition. Donald C. Rennels and Hobart M. Hudson.

TABLE A.1 Physical Properties of Water for Temperatures from 32°F to 212°F

Temperature		Density		Speed of Sound		Absolute Viscosity		Vapor Pressure		Specific Heat	
						lb-sec/ft² × 10⁵	*N-sec/m² × 10³*				
°F	*°C*	lb/ft³	*kg/m³*	ft/sec	*m/sec*			lb/in²	*N/m²*	Btu/lb°F	*Kcal/kg°C*
32	*0.000*	62.418	*999.84*	4603	*1403*	3.7445	*1.793*	0.089	*613.5*	1.0073	*1.0073*
35	*1.667*	62.424	*999.93*	4630	*1411*	3.5367	*1.693*	0.100	*691.3*	1.0060	*1.0060*
40	*4.444*	62.426	*999.97*	4673	*1424*	3.2290	*1.546*	0.122	*840.8*	1.0041	*1.0041*
45	*7.222*	62.421	*999.89*	4712	*1436*	2.9621	*1.418*	0.148	*1,018*	1.0025	*1.0025*
50	*10.000*	62.410	*999.70*	4748	*1447*	2.7290	*1.307*	0.178	*1,228*	1.0013	*1.0013*
55	*12.778*	62.391	*999.41*	4782	*1457*	2.5240	*1.208*	0.214	*1,476*	1.0004	*1.0004*
60	*15.556*	62.367	*999.02*	4813	*1467*	2.3426	*1.122*	0.256	*1,766*	0.9996	*0.9996*
65	*18.333*	62.337	*998.54*	4842	*1476*	2.1814	*1.044*	0.305	*2,106*	0.9990	*0.9990*
70	*21.111*	62.302	*997.98*	4869	*1484*	2.0373	*0.9754*	0.363	*2,502*	0.9986	*0.9986*
75	*23.889*	62.261	*997.33*	4895	*1492*	1.9079	*0.9135*	0.430	*2,961*	0.9983	*0.9983*
80	*26.667*	62.216	*996.61*	4918	*1499*	1.7913	*0.8577*	0.507	*3,493*	0.9981	*0.9981*
85	*29.444*	62.167	*995.82*	4940	*1506*	1.6858	*0.8072*	0.596	*4,107*	0.9980	*0.9980*
90	*32.222*	62.113	*994.96*	4961	*1512*	1.5900	*0.7613*	0.698	*4,812*	0.9979	*0.9979*
95	*35.000*	62.055	*994.03*	4980	*1518*	1.5028	*0.7195*	0.815	*5,620*	0.9979	*0.9979*
100	*37.778*	61.994	*993.05*	4997	*1523*	1.4230	*0.6814*	0.949	*6,543*	0.9980	*0.9980*
105	*40.556*	61.929	*992.00*	5013	*1528*	1.3500	*0.6464*	1.102	*7,595*	0.9980	*0.9980*
110	*43.333*	61.860	*990.90*	5027	*1532*	1.2829	*0.6142*	1.275	*8,790*	0.9982	*0.9982*
115	*46.111*	61.788	*989.74*	5041	*1536*	1.2210	*0.5846*	1.471	*10,143*	0.9983	*0.9983*
120	*48.889*	61.712	*988.53*	5052	*1540*	1.1639	*0.5573*	1.693	*11,671*	0.9985	*0.9985*
125	*51.667*	61.633	*987.27*	5063	*1543*	1.1111	*0.5320*	1.943	*13,393*	0.9987	*0.9987*
130	*54.444*	61.552	*985.96*	5072	*1546*	1.0621	*0.5085*	2.223	*15,328*	0.9989	*0.9989*
135	*57.222*	61.467	*984.60*	5080	*1548*	1.0166	*0.4868*	2.538	*17,497*	0.9991	*0.9991*
140	*60.000*	61.379	*983.20*	5086	*1550*	0.9743	*0.4665*	2.889	*19,922*	0.9994	*0.9994*
145	*62.778*	61.289	*981.75*	5092	*1552*	0.9348	*0.4476*	3.282	*22,628*	0.9997	*0.9997*
150	*65.556*	61.196	*980.26*	5096	*1553*	0.8979	*0.4299*	3.719	*25,639*	1.0000	*1.0000*
155	*68.333*	61.100	*978.72*	5099	*1554*	0.8634	*0.4134*	4.204	*28,985*	1.0004	*1.0004*
160	*71.111*	61.001	*977.14*	5101	*1555*	0.8310	*0.3979*	4.742	*32,692*	1.0008	*1.0008*
165	*73.889*	60.900	*975.52*	5102	*1555*	0.8007	*0.3834*	5.336	*36,792*	1.0012	*1.0012*
170	*76.667*	60.796	*973.86*	5102	*1555*	0.7722	*0.3697*	5.993	*41,317*	1.0017	*1.0017*
175	*79.444*	60.690	*972.16*	5100	*1555*	0.7454	*0.3569*	6.716	*46,303*	1.0022	*1.0022*
180	*82.222*	60.582	*970.42*	5098	*1554*	0.7201	*0.3448*	7.511	*51,784*	1.0027	*1.0027*
185	*85.000*	60.471	*968.65*	5095	*1553*	0.6963	*0.3334*	8.383	*57,799*	1.0032	*1.0032*
190	*87.778*	60.357	*966.83*	5090	*1552*	0.6739	*0.3226*	9.339	*64,389*	1.0038	*1.0038*
195	*90.556*	60.242	*964.98*	5085	*1550*	0.6526	*0.3125*	10.384	*71,596*	1.0045	*1.0045*
200	*93.333*	60.124	*963.09*	5079	*1548*	0.6325	*0.3029*	11.525	*79,463*	1.0052	*1.0052*
205	*96.111*	60.003	*961.16*	5073	*1546*	0.6135	*0.2938*	12.769	*88,038*	1.0059	*1.0059*
210	*98.889*	59.881	*959.20*	5065	*1544*	0.5955	*0.2851*	14.123	*97,369*	1.0066	*1.0066*
212	*100.000*	59.831	*958.40*	5062	*1543*	0.5885	*0.2818*	14.696	*101,325*	1.0070	*1.0070*

Listed values of lb-sec/ft² have been multiplied by 10^5, and those of N-sec/m² have been multiplied by 10^3. To obtain the actual values, the listed values must be divided by their respective multipliers..

TABLE A.2 Physical Properties of Water for Temperatures from 0°C to 100°C

Temperature		Density		Speed of Sound		Absolute Viscosity		Vapor Pressure		Specific Heat	
°C	*°F*	kg/m³	*lb/ft³*	m/sec	*ft/sec*	N-sec/m² × 10³	*lb-sec/ft² × 10⁵*	N/m²	*lb/in²*	Kcal/ kg°C	*Btu/ lb°F*
0	*32.0*	999.84	*62.418*	1403	*4603*	1.793	*3.7445*	613.51	*0.089*	1.0073	*1.0073*
2	*35.6*	999.94	*62.424*	1413	*4636*	1.675	*3.4973*	707.92	*0.103*	1.0057	*1.0057*
4	*39.2*	999.97	*62.426*	1422	*4666*	1.568	*3.2753*	815.08	*0.118*	1.0043	*1.0043*
6	*42.8*	999.94	*62.424*	1431	*4695*	1.472	*3.0750*	936.44	*0.136*	1.0032	*1.0032*
8	*46.4*	999.85	*62.419*	1439	*4722*	1.386	*2.8937*	1,073.6	*0.156*	1.0022	*1.0022*
10	*50.0*	999.70	*62.410*	1447	*4748*	1.307	*2.7290*	1,228.3	*0.178*	1.0013	*1.0013*
12	*53.6*	999.50	*62.397*	1455	*4773*	1.235	*2.5788*	1,402.4	*0.203*	1.0006	*1.0006*
14	*57.2*	999.25	*62.381*	1462	*4796*	1.169	*2.4415*	1,597.9	*0.232*	1.0000	*1.0000*
16	*60.8*	998.95	*62.362*	1468	*4818*	1.109	*2.3156*	1,817.1	*0.264*	0.9995	*0.9995*
18	*64.4*	998.60	*62.341*	1475	*4839*	1.053	*2.1998*	2,062.4	*0.299*	0.9991	*0.9991*
20	*68.0*	998.21	*62.316*	1481	*4859*	1.002	*2.0930*	2,336.2	*0.339*	0.9988	*0.9988*
22	*71.6*	997.78	*62.289*	1487	*4878*	0.9549	*1.9944*	2,641.6	*0.383*	0.9985	*0.9985*
24	*75.2*	997.30	*62.260*	1492	*4896*	0.9112	*1.9030*	2,981.3	*0.432*	0.9983	*0.9983*
26	*78.8*	996.79	*62.228*	1497	*4913*	0.8706	*1.8182*	3,358.6	*0.487*	0.9982	*0.9982*
28	*82.4*	996.24	*62.193*	1502	*4929*	0.8328	*1.7394*	3,777.0	*0.548*	0.9980	*0.9980*
30	*86.0*	995.65	*62.156*	1507	*4945*	0.7976	*1.6659*	4,240.1	*0.615*	0.9980	*0.9980*
32	*89.6*	995.03	*62.118*	1512	*4959*	0.7648	*1.5974*	4,751.9	*0.689*	0.9979	*0.9979*
34	*91.4*	994.37	*62.077*	1516	*4973*	0.7341	*1.5333*	5,316.6	*0.771*	0.9979	*0.9979*
36	*93.2*	993.68	*62.034*	1520	*4986*	0.7054	*1.4733*	5,938.6	*0.861*	0.9979	*0.9979*
38	*96.8*	992.96	*61.989*	1523	*4998*	0.6784	*1.4170*	6,622.6	*0.961*	0.9980	*0.9980*
40	*100.4*	992.21	*61.942*	1527	*5010*	0.6531	*1.3641*	7,373.8	*1.070*	0.9980	*0.9980*
42	*104.0*	991.43	*61.893*	1530	*5021*	0.6293	*1.3144*	8,197.5	*1.189*	0.9981	*0.9981*
44	*107.6*	990.62	*61.843*	1533	*5031*	0.6069	*1.2676*	9,099.3	*1.320*	0.9982	*0.9982*
46	*111.2*	989.79	*61.790*	1536	*5040*	0.5858	*1.2234*	10,085	*1.463*	0.9983	*0.9983*
48	*118.4*	988.92	*61.736*	1539	*5049*	0.5658	*1.1817*	11,162	*1.619*	0.9984	*0.9984*
50	*122.0*	988.03	*61.681*	1541	*5057*	0.5469	*1.1423*	12,335	*1.789*	0.9985	*0.9985*
52	*125.6*	987.12	*61.624*	1543	*5064*	0.5291	*1.1050*	13,613	*1.974*	0.9987	*0.9987*
54	*129.2*	986.17	*61.565*	1545	*5071*	0.5122	*1.0697*	15,003	*2.176*	0.9989	*0.9989*
56	*132.8*	985.21	*61.504*	1547	*5076*	0.4962	*1.0362*	16,512	*2.395*	0.9990	*0.9990*
58	*136.4*	984.22	*61.443*	1549	*5082*	0.4809	*1.0045*	18,149	*2.632*	0.9992	*0.9992*
60	*140.0*	983.20	*61.379*	1550	*5086*	0.4665	*0.9743*	19,922	*2.889*	0.9994	*0.9994*
62	*143.6*	982.16	*61.314*	1552	*5090*	0.4527	*0.9455*	21,840	*3.168*	0.9996	*0.9996*
64	*147.2*	981.10	*61.248*	1553	*5094*	0.4396	*0.9182*	23,914	*3.468*	0.9999	*0.9999*
66	*150.8*	980.02	*61.180*	1553	*5097*	0.4272	*0.8922*	26,152	*3.793*	1.0001	*1.0001*
68	*154.4*	978.91	*61.111*	1554	*5099*	0.4153	*0.8674*	28,565	*4.143*	1.0004	*1.0004*
70	*158.0*	977.78	*61.041*	1555	*5100*	0.4040	*0.8437*	31,163	*4.520*	1.0006	*1.0006*
72	*161.6*	976.63	*60.969*	1555	*5101*	0.3932	*0.8211*	33,960	*4.926*	1.0009	*1.0009*
74	*165.2*	975.46	*60.896*	1555	*5102*	0.3828	*0.7995*	36,965	*5.361*	1.0012	*1.0012*
76	*168.8*	974.27	*60.821*	1555	*5102*	0.3729	*0.7789*	40,191	*5.829*	1.0016	*1.0016*
78	*172.4*	973.05	*60.746*	1555	*5101*	0.3635	*0.7591*	43,651	*6.331*	1.0019	*1.0019*
80	*176.0*	971.82	*60.669*	1554	*5100*	0.3544	*0.7402*	47,358	*6.869*	1.0023	*1.0023*
82	*179.6*	970.57	*60.590*	1554	*5098*	0.3457	*0.7221*	51,326	*7.444*	1.0026	*1.0026*
84	*183.2*	969.29	*60.511*	1553	*5096*	0.3374	*0.7047*	55,570	*8.060*	1.0030	*1.0030*
86	*186.8*	968.00	*60.430*	1552	*5093*	0.3295	*0.6881*	60,103	*8.717*	1.0034	*1.0034*
88	*190.4*	966.68	*60.348*	1551	*5090*	0.3218	*0.6721*	64,942	*9.419*	1.0039	*1.0039*
90	*194.0*	965.35	*60.265*	1550	*5086*	0.3145	*0.6568*	70,103	*10.168*	1.0043	*1.0043*
92	*197.6*	964.00	*60.180*	1549	*5082*	0.3074	*0.6420*	75,601	*10.965*	1.0048	*1.0048*
94	*201.2*	962.63	*60.095*	1548	*5078*	0.3006	*0.6279*	81,455	*11.814*	1.0053	*1.0053*
96	*204.8*	961.24	*60.008*	1546	*5073*	0.2941	*0.6143*	87,681	*12.717*	1.0058	*1.0058*
98	*208.4*	959.83	*59.920*	1545	*5067*	0.2878	*0.6011*	94,298	*13.677*	1.0064	*1.0064*
100	*212.0*	958.40	*59.831*	1543	*5062*	0.2818	*0.5885*	101,325	*14.696*	1.0070	*1.0070*

Listed values of lb-sec/ft² have been multiplied by 10^5, and those of N-sec/m² have been multiplied by 10^3. To obtain the actual values, the listed values must be divided by their respective multipliers..

APPENDIX B

PIPE SIZE DATA

Pipe sizes can be confusing because the terminology may relate to historical dimensions that are loosely related to actual dimensions. Early on, pipe was sized by inside diameter. This practice was abandoned to improve compatibility with pipe fittings that must usually fit the outside diameter of the pipe, but it has had a lasting impact on modern standards around the world. Presently, the pipe size designation generally includes two numbers: one that indicates the outside diameter or nominal diameter and another that identifies the wall thickness.

Historically, only a small selection of pipe wall thickness was in use: standard weight (Std), extra strong (XS), and double extra strong (XXS), based on the iron pipe size system of the day. In 1927, the American Standards Association created a system of schedule numbers that designated steel pipe wall thickness based on smaller steps between sizes. In the mid-twentieth century, stainless steel pipe—which permitted the use of thinner walls with much less risk of failure due to corrosion—came into more common use. Consequently, schedules 5S and 10S were created in 1949, and other "S" sizes followed.

Pipe is specified by its nominal inside diameter through 12 in and smaller. In this range, nominal size refers to the approximate inside diameter of a schedule 40 (or standard weight) pipe. Pipe 14 in and larger is specified by the actual outside diameter. For a given pipe size, the outside diameter is the same for all weights and schedules. In North America, pipe size is specified by Nominal Pipe Size (NPS) and is based on inches. The European version is called Diametre Nominal (DN) and is based on millimeters. Japan has its own set of standard pipe sizes, called Japanese Industrial Standards (JIS) pipe, that are based on millimeters.

Many different national and international standards exist for pipe sizes, including American Petroleum Institute (API) 5 L, American National Standards Institute (ANSI)/American Society of Mechanical Engineers (ASME) B32.10M and B36.19M in the United States, and British Standard (BS) 1600 and BS European Standard (EN) 10255 in the United Kingdom and Europe. Manufacturing standards commonly require a test of chemical composition and a series of mechanical strength tests for each heat of pipe.

Pipe is made of a wide variety of materials including ceramic, fiberglass, metals, concrete, and plastic. Metallic pipes are commonly made of steel or iron, such as carbon steel, stainless steel, galvanized steel, cast iron, and ductile iron. Inconel, chrome moly, and titanium alloys are used in high temperature and pressure applications. Copper and aluminum pipe are frequently employed.

The following NPS and DN table gives selected data for commercial pipe.

Pipe Flow: A Practical and Comprehensive Guide, First Edition. Donald C. Rennels and Hobart M. Hudson.

B.1 COMMERCIAL PIPE DATA

Nominal Pipe Size (NPS)	Diametre Nominal (DN)	Outside Diameter		Wall Identification			Wall Thickness		Inside Diameter			Flow Area		
				Iron Pipe Size	Steel Schedule Number	Stainless Steel Schedule Number								
(in)	*(mm)*	(in)	*(mm)*				(in)	*(mm)*	(in)	(ft)	*(mm)*	(in^2)	(ft^2)	*(cm^2)*
1/8	*3*	0.405	*10.29*	–	–	10S	0.049	*1.24*	0.307	0.0256	*7.80*	0.0740	0.000514	*0.478*
				Std	40	40S	0.068	*1.73*	0.269	0.0216	*6.83*	0.0568	0.000395	*0.367*
				XS	80	80S	0.095	*2.41*	0.215	0.0179	*5.46*	0.0363	0.000252	*0.234*
1/4	*6*	0.540	*13.72*	–	–	10S	0.065	*1.65*	0.410	0.0342	*10.41*	0.1320	0.000917	*0.852*
				Std	40	40S	0.088	*2.24*	0.364	0.0303	*9.25*	0.1041	0.000723	*0.671*
				XS	80	80S	0.119	*3.02*	0.302	0.0252	*7.67*	0.0716	0.000497	*0.462*
3/8	*10*	0.675	*17.15*	–	–	10S	0.065	*1.65*	0.545	0.0454	*13.84*	0.2333	0.001620	*1.505*
				Std	40	40S	0.091	*2.31*	0.493	0.0411	*12.52*	0.1909	0.001326	*1.232*
				XS	80	80S	0.126	*3.20*	0.423	0.0353	*10.74*	0.1405	0.000976	*0.907*
1/2	*15*	0.840	*21.34*	–	–	5S	0.065	*1.65*	0.710	0.0592	*18.03*	0.3959	0.002749	*2.554*
				–	–	10S	0.083	*2.11*	0.674	0.0562	*17.12*	0.3568	0.002478	*2.302*
				Std	40	40S	0.109	*2.77*	0.622	0.0518	*15.80*	0.3039	0.002110	*1.960*
				XS	80	80S	0.147	*3.73*	0.546	0.0455	*13.87*	0.2341	0.001626	*1.511*
				–	160	–	0.187	*4.75*	0.466	0.0388	*11.84*	0.1706	0.001184	*1.100*
				XXS	–	–	0.294	*7.47*	0.252	0.0210	*6.40*	0.0499	0.000346	*0.322*
3/4	*20*	1.050	*26.67*	–	–	5S	0.065	*1.65*	0.920	0.0767	*23.37*	0.6648	0.004616	*4.289*
				–	–	10S	0.083	*2.11*	0.884	0.0614	*22.45*	0.6138	0.004262	*3.960*
				Std	40	40S	0.133	*2.87*	0.824	0.0687	*20.93*	0.5333	0.003703	*3.440*
				XS	80	80S	0.154	*3.91*	0.742	0.0618	*18.85*	0.4324	0.003003	*2.790*
				–	160	–	0.219	*5.56*	0.612	0.0510	*15.54*	0.2942	0.002043	*1.898*
				XXS	–	–	0.308	*7.82*	0.434	0.0362	*11.02*	0.1479	0.001027	*0.954*
1	*25*	1.315	*33.40*	–	–	5S	0.065	*1.65*	1.185	0.0988	*30.10*	1.1029	0.007659	*7.115*
				–	–	10S	0.109	*2.77*	1.097	0.0914	*27.86*	0.9452	0.006564	*6.098*
				Std	40	40S	0.133	*3.38*	1.049	0.0874	*26.64*	0.8643	0.006002	*5.576*
				XS	80	80S	0.179	*4.55*	0.957	0.0797	*24.31*	0.7193	0.004995	*4.641*
				–	160	–	0.250	*6.35*	0.815	0.0679	*20.70*	0.5217	0.003623	*3.366*
				XXS	–	–	0.358	*9.09*	0.599	0.0499	*15.21*	0.2818	0.001957	*1.818*
1-1/4	*32*	1.660	*42.16*	–	–	5S	0.065	*1.65*	1.530	0.1275	*38.86*	1.839	0.01277	*11.86*
				–	–	10S	0.109	*2.77*	1.442	0.1202	*36.63*	1.633	0.01134	*10.54*
				Std	40	40S	0.140	*3.56*	1.380	0.1150	*35.05*	1.496	0.01039	*9.65*
				XS	80	80S	0.191	*4.85*	1.278	0.1065	*32.46*	1.283	0.00891	*8.28*
				–	160	–	0.250	*6.35*	1.160	0.0967	*29.46*	1.057	0.00734	*6.82*
				XXS	–	–	0.382	*9.70*	0.896	0.0747	*22.76*	0.631	0.00438	*4.07*

1-1/2	40	1.900	48.26	–	–	5S	0.065	1.65	1.770	0.1475	44.96	2.461	0.01709	15.88
				–	–	10S	0.109	2.77	1.682	0.1402	42.72	2.222	0.01543	14.34
				Std	40	40S	0.145	3.68	1.610	0.1342	40.89	2.036	0.01414	13.13
				XS	80	80S	0.200	5.08	1.500	0.1250	38.10	1.767	0.01227	11.40
				–	160	–	0.281	7.14	1.338	0.1115	33.99	1.406	0.00976	9.07
				XXS	–	–	0.400	10.16	1.100	0.0917	27.94	0.950	0.00660	6.15
2	50	2.375	60.32	–	–	5S	0.065	1.65	2.245	0.1871	57.02	3.958	0.02749	25.54
				–	–	10S	0.109	2.77	2.157	0.1797	54.79	3.654	0.02538	23.58
				Std	40	40S	0.154	3.91	2.067	0.1723	52.50	3.356	0.02330	21.65
				XS	80	80S	0.218	5.54	1.939	0.1616	49.25	2.953	0.02051	19.05
				–	160	–	0.344	8.34	1.687	0.1406	42.85	2.235	0.01552	14.42
				XXS	–	–	0.436	11.07	1.503	0.1252	38.18	1.774	0.01232	11.45
2-1/2	65	2.875	73.02	–	–	5S	0.083	2.11	2.709	0.2258	68.81	5.764	0.04003	37.19
				–	–	10S	0.120	3.05	2.635	0.2196	66.93	5.453	0.03787	35.18
				Std	40	40S	0.203	5.16	2.469	0.2058	62.71	4.788	0.03325	30.89
				XS	80	80S	0.276	7.01	2.323	0.1836	59.00	4.238	0.02943	27.34
				–	160	–	0.375	9.52	2.125	0.1771	53.97	3.547	0.02463	22.88
				XXS	–	–	0.552	14.02	1.771	0.1476	44.98	2.463	0.01711	15.89
3	80	3.500	88.90	–	–	5S	0.083	2.11	3.334	0.2778	84.68	8.730	0.06063	56.32
				–	–	10S	0.120	3.05	3.260	0.2717	82.80	8.347	0.05796	53.85
				Std	40	40S	0.216	5.49	3.068	0.2557	77.93	7.393	0.05134	47.69
				XS	80	80S	0.300	7.62	2.900	0.2417	73.66	6.605	0.04587	42.61
				–	160	–	0.438	11.13	2.624	0.2187	66.65	5.408	0.03755	34.89
				XXS	–	–	0.600	15.24	2.300	0.1917	58.42	4.155	0.02885	26.80
3-1/2	90	4.000	101.60	–	–	5S	0.083	2.11	3.834	0.3195	97.38	11.54	0.0802	74.48
				–	–	10S	0.120	3.05	3.760	0.3133	95.50	11.10	0.0771	71.64
				Std	40	40S	0.226	5.74	3.548	0.2957	90.12	9.89	0.0687	63.79
				XS	80	80S	0.318	8.08	3.364	0.2803	85.45	8.89	0.0617	57.34
4	100	4.500	114.30	–	–	5S	0.083	2.11	4.334	0.3612	108.08	14.75	0.1024	95.18
				–	–	10S	0.120	3.05	4.260	0.3550	108.20	14.25	0.0990	91.96
				Std	40	40S	0.237	6.02	4.026	0.3355	102.26	12.73	0.0884	82.13
				XS	80	80S	0.337	8.56	3.826	0.3188	97.18	11.50	0.0798	74.17
				–	120	–	0.438	11.12	3.624	0.3020	92.05	10.31	0.0716	66.55
				–	160	–	0.531	13.49	3.438	0.2865	87.33	9.28	0.0645	59.89
				XXS	–	–	0.674	17.12	3.152	0.2627	80.06	7.80	0.0542	50.34
5	125	5.563	141.30	–	–	5S	0.109	2.77	5.345	0.4454	135.76	22.44	0.1558	144.76
				–	–	10S	0.134	3.40	5.295	0.4412	134.49	22.02	0.1529	142.07
				Std	40	40S	0.258	6.55	5.047	0.4206	128.19	20.01	0.1389	129.07
				XS	80	80S	0.375	9.52	4.813	0.4011	122.25	18.19	0.1263	117.39
				–	120	–	0.500	12.70	4.563	0.3802	115.90	16.35	0.1136	105.50
				–	160	–	0.625	15.88	4.313	0.3594	109.55	14.61	0.1015	94.26
				XXS	–	–	0.750	19.05	4.063	0.3386	103.20	12.96	0.0900	83.65

(Continued)

B.1 COMMERCIAL PIPE DATA (Continued)

Nominal Pipe Size (NPS)	Diametre Nominal (DN)	Outside Diameter		Wall Identification			Wall Thickness		Inside Diameter			Flow Area		
				Iron Pipe Size	Steel Schedule Number	Stainless Steel Schedule Number								
(in)	*(mm)*	(in)	*(mm)*				(in)	*(mm)*	(in)	(ft)	*(mm)*	(in^2)	(ft^2)	*(cm^2)*
6	*150*	6.625	*168.27*	–	–	5S	0.109	*2.77*	6.407	0.5339	*162.74*	32.24	0.2239	*208.0*
				–	–	10S	0.134	*3.40*	6.357	0.5298	*161.47*	31.74	0.2204	*204.8*
				Std	40	40S	0.280	*7.11*	6.065	0.5054	*154.05*	28.89	0.2006	*186.4*
				XS	80	80S	0.432	*10.97*	5.761	0.4801	*146.33*	26.07	0.1810	*168.2*
				–	120	–	0.562	*14.27*	5.501	0.4584	*139.73*	23.77	0.1650	*153.3*
				–	160	–	0.719	*18.26*	5.187	0.4322	*131.75*	21.13	0.1467	*136.3*
				XXS	–	–	0.864	*21.95*	4.897	0.4081	*124.38*	18.83	0.1308	*121.5*
8	*200*	8.625	*219.08*	–	–	5S	0.109	*2.77*	8.407	0.7006	*213.54*	55.51	0.3855	*358.1*
				–	–	10S	0.148	*3.76*	8.329	0.6941	*211.56*	54.48	0.3784	*351.5*
				–	20	–	0.250	*6.35*	8.125	0.6771	*206.38*	51.85	0.3601	*334.5*
				–	30	–	0.277	*7.04*	8.071	0.6726	*205.00*	51.16	0.3553	*330.1*
				Std	40	40S	0.322	*8.18*	7.891	0.6651	*202.72*	50.03	0.3474	*322.8*
				–	60	–	0.406	*10.31*	7.813	0.6511	*198.45*	47.94	0.3329	*309.3*
				XS	80	80S	0.500	*12.70*	7.625	0.6354	*193.67*	45.66	0.3171	*294.6*
				–	100	–	0.594	*15.09*	7.437	0.6197	*188,90*	43.44	0.3017	*208.3*
				–	120	–	0.719	*18.26*	7.187	0.5989	*182.55*	40.59	0.2817	*261.7*
				–	140	–	0.812	*20.62*	7.001	0.5834	*177.83*	38.50	0.2673	*248.4*
				XXS	–	–	0.975	*22.22*	6.875	0.5729	*174.63*	37.12	0.2578	*239.5*
				–	160	–	0.906	*23.01*	6.813	0.5678	*173.05*	36.46	0.2532	*235.2*
10	*250*	10.750	*273.05*	–	–	5S	0.134	*3.40*	10.482	0.8735	*266.24*	86.29	0.5993	*556.7*
				–	–	10S	0.165	*4.19*	10.420	0.8683	*264.67*	85.28	0.5922	*550.2*
				–	20	–	0.250	*6.35*	10.250	0.8542	*260.35*	82.52	0.5730	*532.4*
				–	30	–	0.307	*7.80*	10.136	0.8447	*257.45*	80.69	0.5604	*520.6*
				Std	40	40S	0.365	*9.27*	10.020	0.8350	*254.51*	78.85	0.5476	*508.7*
				XS	60	80S	0.500	*12.70*	9.750	0.8125	*247.65*	74.66	0.5185	*481.7*
				–	80	–	0.594	*15.09*	9.562	0.7968	*242.87*	71.81	0.4987	*463.3*
				–	100	–	0.719	*18.26*	9.312	0.7760	*236.52*	68.10	0.4729	*439.4*
				–	120	–	0.844	*21.44*	9.062	0.7552	*230.17*	64.50	0.4479	*416.1*
				XXS	140	–	1.000	*25.40*	8.750	0.7292	*222.25*	60.13	0.4176	*387.9*
				–	150	–	1.125	*28.57*	8.500	0.7083	*215.90*	56.75	0.3941	*366.1*

12	*300*	12.75	*323.9*	–	–	5S	0.156	*3.96*	12.438	1.0365	*315.93*	121.50	0.8438	*783.9*
				–	–	10S	0.180	*4.57*	12.390	1.0325	*314.71*	120.57	0.8373	*777.9*
				–	20	–	0.250	*6.35*	12.250	1.0208	*311.15*	117.86	0.8185	*760.4*
				–	30	–	0.330	*8.38*	12.090	1.0075	*307.09*	114.80	0.7972	*740.6*
				Std	–	40S	0.375	*9.52*	12.000	1.0000	*304.80*	113.10	0.7854	*729.7*
					40	–	0.406	*10.31*	11.938	0.9948	*303.23*	111.93	0.7773	*722.1*
				XS	–	80S	0.500	*12.70*	11.750	0.9792	*298.45*	108.43	0.7530	*699.6*
				–	60	–	0.562	*14.27*	11.626	0.9688	*295.30*	106.16	0.7372	*684.9*
				–	80	–	0.688	*17.47*	11.374	0.9578	*288.90*	101.61	0.7056	*665.6*
				–	100	–	0.844	*21.44*	11.062	0.9218	*280.97*	96.11	0.6674	*620.0*
				XXS	120	–	1.000	*25.40*	10.750	0.8958	*273.05*	90.76	0.6303	*585.6*
				–	140	–	0.125	*28.57*	10.500	0.8750	*266.70*	86.59	0.6013	*558.6*
				–	160	–	0.312	*33.32*	10.126	0.8438	*257.20*	80.53	0.5592	*519.6*
14	*350*	14.00	*355.6*	–	–	5S	0.156	*3.96*	13.688	1.1407	*347.68*	147.15	1.0219	*949.4*
				–	–	10S	0.188	*4.77*	13.624	1.1353	*346.05*	145.78	1.0124	*940.5*
				–	10	–	0.250	*6.35*	13.500	1.1250	*342.90*	143.14	0.9940	*923.5*
				–	20	–	0.312	*7.92*	13.376	1.1147	*339.75*	140.52	0.9758	*906.6*
				Std	30	–	0.375	*9.52*	13.250	1.1042	*336.55*	137.89	0.9575	*889.6*
				–	40	–	0.438	*11.13*	13.124	1.0937	*333.35*	135.28	0.9394	*872.7*
				XS	–	–	0.500	*12.70*	13.000	1.0833	*330.20*	132.73	0.9218	*856.3*
				–	60	–	0.594	*15.09*	12.812	1.0677	*325.42*	128.92	0.8953	*831.7*
				–	80	–	0.750	*19.05*	12.500	1.0417	*317.50*	122.72	0.8522	*791.7*
				–	100	–	0.938	*23.82*	12.124	1.0103	*307.95*	115.45	0.8017	*744.8*
				–	120	–	1.094	*27.79*	11.812	0.9843	*300.02*	109.58	0.7610	*707.0*
				–	140	–	1.250	*31.75*	11.500	0.9583	*292.10*	103.87	0.7213	*670.1*
				–	160	–	1.406	*35.71*	11.188	0.9323	*284.18*	98.31	0.6827	*634.3*
16	*400*	16.00	*406.4*	–	–	5S	0.165	*4.19*	15.670	1.3058	*398.02*	192.85	1.3393	*1244.2*
				–	–	10S	0.188	*4.77*	15.624	1.3020	*396.85*	191.72	1.3314	*1236.9*
				–	10	–	0.250	*6.35*	15.500	1.2917	*393.70*	188.69	1.3104	*1217.4*
				–	20	–	0.312	*7.92*	15.376	1.2813	*390.55*	185.68	1.2895	*1198.0*
				Std	30	–	0.375	*9.52*	15.250	1.2708	*387.35*	182.65	1.2684	*1178.4*
				XS	40	–	0.500	*12.70*	15.000	1.2500	*381.00*	176.71	1.2272	*1140.1*
				–	60	–	0.656	*16.66*	14.688	1.2240	*373.08*	169.44	1.1767	*1093.2*
				–	80	–	0.844	*22.45*	14.232	1.1860	*361.49*	159.08	1.1047	*1026.3*
				–	100	–	1.031	*26.19*	13.938	1.1615	*354.03*	152.58	1.0596	*984.4*
				–	120	–	1.219	*30.96*	13.562	1.1302	*344.47*	144.46	1.0032	*932.0*
				–	140	–	1.438	*36.52*	13.124	1.0937	*333.35*	135.28	0.9394	*827.7*
				–	160	–	1.594	*40.49*	12.812	1.0677	*325.42*	128.92	0.8953	*831.7*

(Continued)

B.1 COMMERCIAL PIPE DATA (Continued)

Nominal Pipe Size (NPS)	Diametre Nominal (DN)	Outside Diameter		Wall Identification			Wall Thickness		Inside Diameter			Flow Area		
				Iron Pipe Size	Steel Schedule Number	Stainless Steel Schedule Number								
(in)	*(mm)*	(in)	*(mm)*				(in)	*(mm)*	(in)	(ft)	*(mm)*	(in^2)	(ft^2)	*(cm^2)*
18	*450*	18.00	*457.2*	–	–	5S	0.165	*4.19*	17.670	1.4725	*448.82*	245.22	1.7029	*1582.1*
				–	–	10S	0.188	*4.77*	17.624	1.4687	*447.65*	243.95	1.6941	*1573.9*
				–	10	–	0.250	*6.35*	17.500	1.4583	*444.50*	240.53	1.6703	*1551.8*
				–	20	–	0.312	*7.92*	17.376	1.4480	*441.35*	237.13	1.6467	*1529.9*
				Std	–	–	0.375	*9.52*	17.250	1.4375	*438.15*	233.71	1.6230	*1507.8*
				–	30	–	0.438	*11.13*	17.124	1.4270	*434.95*	230.30	1.5993	*1485.8*
				XS	–	–	0.500	*12.70*	17.000	1.4167	*431.80*	226.98	1.5763	*1464.4*
				–	40	–	0.562	*14.27*	16.876	1.4063	*427.65*	223.68	1.5533	*1443.1*
				–	60	–	0.750	*19.05*	16.500	1.3750	*419.10*	213.82	1.5849	*1379.5*
				–	80	–	0.938	*23.82*	16.124	1.3437	*409.55*	204.19	1.4180	*1317.4*
				–	100	–	1.156	*29.36*	15.688	1.3073	*398.48*	183.30	1.3423	*1247.1*
				–	120	–	1.375	*34.92*	15.250	1.2708	*387.35*	182.65	1.2684	*1178.4*
				–	140	–	1.562	*39.67*	14.876	1.2397	*377.85*	173.80	1.2070	*1121.3*
				–	160	–	1.781	*45.24*	14.438	1.2032	*366.73*	163.72	1.1370	*1056.3*
20	*500*	20.00	*508.0*	–	–	5S	0.188	*4.77*	19.624	1.6353	*498.45*	302.46	2.1004	*1951.3*
				–	–	10S	0.218	*5.54*	19.564	1.6303	*496.93*	300.61	2.0876	*1939.4*
				–	10	–	0.250	*6.35*	19.500	1.6250	*495.30*	298.65	2.0739	*1926.8*
				Std	20	–	0.375	*9.52*	19.250	1.6042	*488.95*	291.04	2.0211	*1877.7*
				XS	30	–	0.500	*12.70*	19.000	1.5833	*482.60*	283.53	1.9689	*1829.2*
				–	40	–	0.594	*15.09*	18.812	1.5677	*477.82*	277.95	1.9302	*1793.2*
				–	60	–	0.812	*20.62*	18.376	1.5313	*466.75*	265.21	1.8417	*1711.0*
				–	80	–	1.031	*26.19*	17.938	1.4948	*455.63*	252.72	1.7550	*1630.4*
				–	100	–	1.281	*32.54*	17.438	1.4532	*442.93*	238.83	1.6585	*1540.8*
				–	120	–	1.500	*38.10*	17.000	1.4167	*431.80*	226.98	1.5763	*1464.4*
				–	140	–	1.750	*44.45*	16.500	1.3750	*419.10*	213.82	1.4849	*1379.5*
				–	160	–	1.969	*50.01*	16.062	1.3385	*407.97*	201.62	1.4071	*1307.2*
22	*550*	22.00	*558.8*	–	–	5S	0.188	*4.77*	21.624	1.8020	*549.25*	367.25	2.5503	*2369.4*
				–	–	10S	0.218	*5.54*	21.564	1.7970	*547.73*	365.21	2.5362	*2356.2*
				–	10	–	0.250	*6.35*	21.500	1.7917	*546.10*	363.05	2.5212	*2342.3*
				Std	20	–	0.375	*9.52*	21.250	1.7708	*539.75*	354.66	2.4629	*2288.1*
				XS	30	–	0.500	*12.70*	21.000	1.7500	*533.40*	346.36	2.4053	*2234.6*
				–	40	–	0.594	*15.09*	20.812	1.7343	*528.62*	340.19	2.3524	*2194.7*
				–	60	–	0.875	*22.22*	20.250	1.6875	*514.35*	322.06	2.2365	*2077.8*
				–	80	–	1.125	*28.57*	19.750	1.6458	*501.65*	306.35	2.1275	*1976.5*
				–	100	–	1.375	*34.92*	19.250	1.6042	*488.95*	291.04	2.0211	*1877.7*
				–	120	–	1.625	*41.27*	18.750	1.5625	*476.25*	276.12	1.9175	*1781.4*
				–	140	–	1.875	*47.62*	18.250	1.5208	*463.55*	261.59	1.8166	*1687.7*
				–	160	–	2.125	*53.97*	17.750	1.4792	*450.85*	247.45	1.7184	*1596.4*

24	*600*	24.00	*609.6*	–	–	5S	0.218	*5.54*	23.564	1.9637	*598.53*	436.10	3.0285	*2813.6*
				–	10	10S	0.250	*6.35*	23.500	1.9583	*596.90*	433.74	3.0121	*2798.3*
				Std	20	–	0.375	*9.52*	23.250	1.9375	*590.55*	424.56	2.9483	*2739.1*
				XS	–	–	0.599	*12.70*	23.000	1.9167	*584.20*	415.48	2.8852	*2680.5*
				–	30	–	0.562	*14.27*	22.876	1.9063	*581.05*	411.01	2.8542	*2651.7*
				–	40	–	0.688	*17.48*	22.624	1.8853	*574.65*	402.00	2.7917	*2593.6*
				–	60	–	0.969	*24.61*	22.062	1.8385	*560.37*	382.28	2.6547	*2466.3*
				–	80	–	1.219	*30.96*	21.562	1.7968	*547.67*	365.15	2.5357	*2355.8*
				–	100	–	1.531	*38.89*	20.938	1.7448	*531.83*	344.32	2.3911	*2221.4*
				–	120	–	1.812	*46.02*	20.376	1.6980	*517.55*	326.08	2.2645	*2103.8*
				–	140	–	2.062	*52.37*	19.876	1.6563	*504.85*	310.28	2.1547	*2001.8*
				–	160	–	2.344	*59.54*	19.312	1.6093	*490.52*	292.92	2.0341	*1889.8*
26	*650*	26.00	*660.4*	–	10	–	0.312	*7.92*	25.376	2.1147	*644.55*	505.75	3.5122	*3262.9*
				Std	–	–	0.375	*9.52*	25.250	2.1942	*641.35*	500.74	3.4774	*3230.6*
				XS	20	–	0.500	*12.70*	25.000	2.0833	*635.00*	490.87	3.4088	*3166.9*
28	*700*	28.00	*711.2*	–	10	–	0.312	*7.92*	27.376	2.2813	*697.35*	588.61	4.0876	*3797.5*
				Std	–	–	0.375	*9.52*	27.250	2.2708	*692.15*	583.21	4.0501	*3762.6*
				XS	20	–	0.500	*12.70*	27.000	2.2500	*685.80*	572.56	3.9761	*3693.9*
				–	30	–	0.625	*15.87*	26.750	2.2292	*679.45*	562.00	3.9028	*3625.8*
30	*750*	30.00	*762.0*			5S	0.250	*6.35*	29.500	2.4583	*749.30*	683.49	4.7465	*4409.6*
				–	10	10S	0.312	*7.92*	29.376	2.4480	*746.15*	677.76	4.7067	*4372.6*
				Std	–	–	0.375	*9.52*	29.250	2.4375	*742.95*	671.96	4.6664	*4335.2*
				XS	20	–	0.500	*12.70*	29.000	2.4167	*736.60*	660.52	4.5869	*4261.4*
				–	30	–	0.625	*15.87*	28.759	2.3958	*730.25*	649.18	4.5082	*4188.3*
32	*800*	32.00	*812.8*	–	10	–	0.312	*7.92*	31.376	2.6147	*796.95*	773.19	5.3694	*4988.3*
				Std	–	–	0.375	*9.52*	31.250	2.6042	*793.75*	766.99	5.3263	*4948.3*
				XS	20	–	0.500	*12.70*	31.000	2.5833	*787.40*	754.77	5.2414	*4869.5*
				–	30	–	0.650	*15.87*	30.750	2.5625	*781.05*	742.64	5.1572	*4791.2*
				–	40	–	0.688	*17.47*	30.624	2.5520	*777.85*	736.57	5.1151	*4752.1*
34	*850*	34.00	*863.6*	–	10	–	0.344	*8.74*	33.312	2.7760	*846.12*	871.55	6.0524	*5622.9*
				Std	–	–	0.375	*9.52*	33.250	2.7708	*844.55*	868.31	6.0299	*5602.0*
				XS	20	–	0.500	*12.70*	33.000	2.7500	*838.20*	855.30	5.9396	*5518.0*
				–	30	–	0.625	*15.87*	32.750	2.7292	*831.85*	842.39	5.8499	*5434.8*
				–	40	–	0.688	*17.47*	22.624	2.7187	*828.65*	835.92	5.8050	*5393.0*
36	*900*	36.00	*914.4*	–	10	–	0.312	*7.92*	35.376	2.9480	*898.55*	982.90	6.8257	*6341.2*
				Std	–	–	0.375	*9.52*	35.250	2.9375	*895.35*	975.91	6.7771	*6296.2*
				XS	20	–	0.500	*12.70*	35.000	2.9167	*889.00*	962.11	6.6813	*6207.2*
				–	30	–	0.625	*15.87*	34.750	2.8958	*882.65*	948.42	6.5862	*6118.8*
				–	40	–	0.750	*19.05*	34.500	2.8750	*876.30*	934.82	6.4910	*6031.1*
42	*1050*	42.00	*1066.8*	Std	–	–	0.375	*9.52*	41.250	3.4375	*1047.75*	1336.40	9.2806	*8621.9*
				XS	20	–	0.500	*12.70*	41.000	3.4167	*1041.40*	1320.25	9.1894	*8517.8*
				–	30	–	0.625	*15.87*	40.750	3.3958	*1035.05*	1304.20	9.0570	*8414.3*
				–	40	–	0.750	*19.05*	40.500	3.3750	*1028.70*	1288.25	8.9462	*8311.3*

The italicized numbers indicate metric measure.

APPENDIX C

PHYSICAL CONSTANTS AND UNIT CONVERSIONS

C.1 IMPORTANT PHYSICAL CONSTANTS

Acceleration of gravity, standard	32.1740 ft/s^2
	980.665* cm/s^2
	9.80665* m/s^2
Universal gas constant	1545.31 ft-lb/mol$_{lb}$°R
	49,718.8 ft-lb/mol$_{slug}$°R
	8314.34 J/mol$_{kg}$°K
Pi	3.141592653…
e	2.718281828…
Standard atmospheric pressure at sea level	14.69597 lb/in^2
	2116.217 lb/ft^2
	759.9998 mm Hg (0°C)
	29.92125 in Hg
	101,325* N/m^2 (pascals)
	1.01325* bar
Mechanical equivalent of heat	778.169 ft-lb/Btu (Int. Table, IT)
	777.649 ft-lb/Btu (thermochemical)
	1.000000* N-m/J
Length of year	365.24220 days (tropical)
	365.25635 days (sidereal)
Speed of light in vacuum	2.997925 m/s
	186,282 mi/s
Avogadro's number	6.022169 × 10^{23} items/mol
H_2O latent heat of vaporization (212°F)	970.3 Btu (IT)/lb (ΔH)
	897.5 Btu (IT)/lb (ΔIE)
Maximum density of H_2O (1 atm)	62.4266 lb/ft^3 (in vacuo)
Density of H_2O (32°F, 0°C)	62.4183 lb/ft^3 (in vacuo)
(62°F, 16.6667°C)	62.3554 lb/ft^3 (in vacuo)
Density of Hg (32°F, 0°C)	848.714 lb/ft^3 (in vacuo)
	13.5951 g/cm^3 (in vacuo)

The asterisk symbol (*) indicates "by definition."
Source: Mechtly, E. A., *The International System of Units*, 2nd rev., National Aeronautics and Space Administration, Washington, D.C., 1973, NASA SP-7012.

Pipe Flow: A Practical and Comprehensive Guide, First Edition. Donald C. Rennels and Hobart M. Hudson.

C.2 UNIT CONVERSIONS

Multiply	By	To Obtain
Acre (U.S. survey)	1.000004	acre
Acre	0.404686	hectare
	10*	chain^2 (Gunter)
	43,560*	ft^2
	0.00156250	mi^2
	4840*	yd^2
	160*	rod^2
Acre-feet	43,560*	ft^3
	325,851	gallons (U.S.)
	1233.48	m^3
	1.23348×10^6	liters
Acre-feet per hour (Acre-ft/h)	726*	ft^3/min
	5430.86	gal/min
Angstroms (Å)	10^{-10}*	m
Ares	0.01*	hectares
	1076.39	ft^2
	0.0247105	acres
Atmospheres	760.000	mm Hg (32°F)
	29.9213	in Hg (32°F)
	33.9380	ft H_2O (62°F)
	1.01325*	bars
	1013.25*	millibars
	2116.22	lb/ft^2
	14.6960	lb/in^2
	235.136	oz/in^2
Bars	0.986923	atm
Barrels of oil	42*	gallon of oil (U.S.)
Boiler horsepower (hp)	33,471.4	Btu/h
	9.80950	kW
	34.496 ± 0.001	lb/h H_2O (212°F)
British thermal unit (Btu [thermochemical])	0.99933084	Btu(IST)
Btu (IST)	251.996	cal (IST)
	0.251996	kcal
	778.169	ft-lb
	0.000393015	hp-h
	0.000293071	kW-h
Btu per hour-feet-degree Fahrenheit (Btu/h-ft-°F)	2.31481×10^{-5}	Btu/s-in-°F
	0.00413379	cal/s-cm-°C
	1.48816	kcal/h-m-°C
Btu per minute (Btu/min)	12.9695	ft-lb/s
	0.0235809	hp
	0.0175843	kW
	17.5843	watts
Btu per pound (Btu/lb)	0.555556	kcal/kg
Bushels (U.S.)	2150.42	in^3
	35.2391	liters
	4*	pecks
	32*	quarts (dry)
Calories (cal, thermochemical)	0.99933084	cal (IST)
Calories (IST)	0.00396832	Btu
	0.001*	kcal
	3.08803	ft-lb
	1.55961×10^{-6}	hp-h
	4.18680	joules
	1.16300×10^{-6}	kW-h
	0.00116300	watt-h
Calories per second-centimeter-degree Celsius (cal/s-cm-°C)	241.909	Btu/h-ft-°F
	0.00559974	Btu/s-in-°F
Carats	200*	mg
Centares	1*	m^2
Centigram	0.01*	g
Centiliters	0.01*	liters
Centimeters	0.393701	in
	0.0328084	ft
	0.01*	m
	10*	mm
Centimeters of mercury (cm Hg, 0°C)	0.0131579	atm
	0.446553	ft H_2O (62°F)
	27.8450	lb/ft^2
	0.193368	lb/in^2
Centimeters per second (cm/s)	1.96850	ft/min
	0.0328084	ft/s
	0.036*	km/h
	0.6*	m/min
	0.0223694	mi/h
	0.000372823	mi/min
Centipoise	0.000671969	lb_m/ft-s
	2.41909	lb_m/h-ft
	0.01*	poise
	0.001*	N-s/m^2
	0.001*	Pa-s
	2.08854×10^{-5}	lb-s/ft^2
Centistoke	1.07639×10^{-5}	ft^2/s
	0.000001*	m^2/s
Chains (Gunter's)	4*	rods
	66*	ft
	65.9999	ft (U.S. Survey)
	100*	links

Multiply	By	To Obtain
Circular inches	10^6*	circular mils
	0.785398	in^2
	785,398	$mils^2$
Circular mils	0.785398	$mils^2$
	10^{-6}*	circular inches
	7.85398×10^{-7}	in^2
Square centimeters (cm^2)	0.00107639	ft^2
	0.155000	in^2
	0.0001*	m^2
	100*	mm^2
Cubic centimeters (cm^3)	3.53147×10^{-5}	ft^3
	0.0610237	in^3
	10^{-6}*	m^3
	1.30795×10^{-6}	yd^3
	0.000264172	gal (U.S. liq.)
	0.001*	liters
	0.00211338	pints (U.S. liq.)
	0.00105669	qt (U.S. liq.)
	0.0338140	fl oz
Cubit (English)	18*	in
Day (mean solar)	1440*	minutes
	24*	hours
	86,400*	seconds
Day (sidereal)	86,164.1	seconds
Decigrams	0.1*	g
Deciliters	0.1*	liters
Decimeters	0.1*	m
Degrees (arc)	60*	minutes
	0.0174533	rad
	3600*	seconds
Degrees(arc) per second	0.0174533	rad/s
	0.1666667	rev/min
	0.00277778	rev/s
Degrees Fahrenheit	(°F + 459.67) =	°R
	(°F − 32) × 5/9 =	°C
	(°F − 32) × 5/9 + 273.15 =	°K
Degrees Celsius	(°C + 273.15) =	°K
	(°C × 9/5) + 32 =	°F
	(°C × 9/5) + 32 + 459.67 =	°R
Degrees Rankine	(°R − 459.67) =	°F
	(°R − 491.67) × 5/9 =	°C
	(°R × 5/9) =	°K
Degrees Kelvin	(°K) − 273.15° =	°C
	(°K × 9/5) − 459.67 =	°F
	(°K × 9/5) =	°R
Decagrams	10*	g
Decameters	10*	m
Diameter	3.14159265 . . .	Circumference
(approximately)	3.14	within 0.05%
(better)	22/7	within 0.04%
(better)	355/113	within 8×10^{-6}%
	0.886227	side eq area sq
	0.707107	side inscrib sq
Cubic diameter (dia^3) (sphere)	0.523599	vol (sphere)
Square diameter (dia^2) (circle)	0.785398	area (circle)
Diameter (dia) (semimajor) × dia (semiminor)	0.785398	area (ellipse)
Square diameter (dia^2) (sphere)	3.14159	area (sphere)
Drams (avoirdupois, avd.)	27.34375	grains
	0.0625*	oz (avd.)
	1.771845	g_f
Dynes	10^{-5}*	newtons
	2.24809×10^{-6}	lb_f
	0.0157366	$grains_f$
	3.59694×10^{-5}	oz_f (avoir.)
	3.27846×10^{-5}	oz_f (troy)
Dynes per centimeter	0.001*	N/m
	5.71015×10^{-6}	lb/in
Fathoms	6*	ft
Feet (U.S. survey)	1.000002	ft
Feet	30.48*	cm
	12*	in
	0.3048*	m
	1/3*	yd
	0.0606061	rods
Foot water (ft H_2O) (62°F)	0.0294655	atm
	0.881644	in Hg (32°F)
	2.23938	cm Hg (0°C)
	62.3554	lb/ft^2
	0.433024	lb/in^2
Foot-pound (ft-lb)	0.00128507	Btu
	0.323831	cal
	3.23831×10^{-4}	kcal
	5.05051×10^{-7}	hp-h
	1.35582	joules
	3.76616×10^{-7}	kW-h
	0.000376616	W-h

(*Continued*)

Multiply	By	To Obtain
Foot pound per minute (ft-lb/min)	0.00128507	Btu/min
	0.0166667	ft-lb/s
	3.03030×10^{-5}	hp
	3.23831×10^{-4}	kcal/min
	2.25970×10^{-5}	kW
Foot pound per second (ft-lb/s)	0.0771042	Btu/min
	0.00181818	hp
	0.0194299	kcal/min
	0.00135582	kW
Foot per minute (ft/min)	0.508*	cm/s
	0.0166667	ft/s
	0.018288*	km/h
	0.3048*	m/min
	0.0113636	mi/h
Foot per second (ft/s)	30.48*	cm/s
	1.09728*	km/h
	0.592484	knots
	18.288*	m/min
	0.3048*	m/s
	0.681818	mi/h
	0.0113636	mi/min
Foot per square second (ft/s^2)	30.48*	cm/s^2
	0.3048*	m/s^2
Square feet (ft^2)	2.29568×10^{-5}	acres
	929.030	cm^2
	144*	in^2
	0.0929030	m^2
	3.58701×10^{-8}	mi^2
	0.111111	yd^2
Cubic feet (ft^3)	28,316.8	cm^3
	1728*	in^3
	0.0283168	m^3
	0.0370370	yd^3
	7.48052	gal (U.S. liq.)
	28.3168	liters
	59.8442	pints (U.S. liq.)
	29.9221	qt (U.S. liq.)
	2.29568×10^{-5}	acre-ft
	0.803564	bushels
Cubic feet water (ft^3 H_2O)	62.4266	lb (39.2°F)
	62.3554	lb (62°F)
Cubic feet per minute (ft^3/min)	471.947	cm^3/s
	0.124675	gal (U.S.)/s
	0.471947	L/s
	62.3554	lb H_2O/min (62°F)
	7.48052	gal(U.S.)/min
	10,771.9	gal(U.S.)/day
	0.0330579	acre-ft/day
Cubic feet per second (ft^3/s)	646,317	gal(U.S.)/day
	448.831	gal(U.S.)/min
	1.98347	acre-ft/day
Cubic feet per pound (ft^3/lb$_m$)	0.0624280	m^3/kg
	1728*	in^3/lb$_m$
Furlongs	40*	rods
	39.9999	rods (U.S. Survey)
	220*	yards
	660*	ft
	0.125*	mi
	201.168*	m
	0.201168*	km
Gallons (imperial)	277.42*	in^3
	4.54609	liters
	1.20095	gal (U.S.)
Gallons (U.S.)	3.06888×10^{-6}	acre-ft
	3,785.41	cm^3
	0.133681	ft^3
	231*	in^3
	0.00378541	m^3
	0.00495113	yd^3
	3.78541	liters
	8*	pints (liq. U.S.)
	4*	qt (liq. U.S.)
	0.832672	gal (imperial)
Gallon (U.S.) water	8.33570	lb H_2O (62°F)
Gallon (U.S.) per minute	6.00171	ton H_2O/da (62°F)
	0.00222801	ft^3/s
	0.133681	ft^3/min
	8.02083	ft^3/h
	0.0630902	L/s
	3.78541	L/min
	0.00441919	acre-ft/day
Grains	1*	grains (avoir.)
	1*	grains (apoth.)
	1*	grains (troy)
	0.00208333	oz (troy)
	0.00228571	oz (avoir.)
Grains per gallon (U.S.)	17.1380	ppm (62°F)
	142.857	lb/10^6 gal (U.S.)
Grams	0.00220462	lb$_m$
	0.001*	kg
	1000*	mg
	6.85218×10^{-5}	slug
Grams per cubic centimeter	62.4280	lb$_m$/ft^3
	1.94032	slugs/ft^3
	0.00112287	slugs/in^3
Grams per liter	0.0259383	slugs/100 gal
	0.00194032	slugs/ft^3
	1000	parts per million

Multiply	By	To Obtain
Gravity, std.	32.17405	ft/s^2
	980.665*	cm/s^2
	9.80665*	m/s^2
Hectares	2.47105	acres
	107,639	ft^2
	100*	acres
Hectograms	100*	g
Hectoliters	100*	Liters
Hectometers	100*	m
Hectowatts	100*	watts
Hex across flats	1.154701	Across corners
Hogsheads	63*	gal (U.S.)
	238.481	Liters
Horsepower	42.4072	Btu/min
	33,000*	ft-lb/min
	550*	ft-lb/s
	1.01387	metric hp (cheval-vapeur)
	10.6864	kcal/min
	0.745700	kW
	745.700	watts
Horsepower (boiler)	33,471.4	Btu/h
	9.80950	kW
	34.496 ± 0.001	lb/h H_2O (212°F)
Horsepower-hour	2544.43	Btu
	6.41186×10^5	cal
	641.186	kcal
	1,980,000*	ft-lb
	2.68452×10^6	joules
	0.745700	kW-h
	745.700	watt-h
Inches	2.54*	cm
	0.0833333	ft
	1000*	mils
Inch of mercury (in Hg, 32°F)	0.0334211	atm
	70.7262	lb/ft^2
	0.491154	lb/in^2
	1.13424	ft H_2O (62°F)
	13.6109	in H_2O (62°F)
	7.85847	oz/in^2
Inch of water (in H_2O, 62°F)	0.00245546	atm
	0.577365	oz/in^2
	5.19628	lb/ft^2
	0.0360853	lb/in^2
	0.0734703	in Hg (32°F)
Square inch (in^2)	6.4516*	cm^2
	0.00694444	ft^2
	645.16*	mm^2
	1.27324	circular in
	1,273,240	circular mils
	1,000,000*	$mils^2$

Multiply	By	To Obtain
Cubic inch (in^3)	16.3871	cm^3
	0.000578704	ft^3
	1.63871×10^{-5}	m^3
	2.14335×10^{-5}	yd^3
	0.00432900	gal (U.S.)
	0.0163871	liters
	0.0346320	pints (liq. U.S.)
	0.0173160	qt (liq. U.S.)
Cubic inch per pound (in^3/lb_m)	3.61273×10^{-5}	m^3/kg
	0.000578704	ft^3/lb_m
Joules	0.000947817	Btu
	0.238846	cal
	0.000238846	kcal
	0.737562	ft-lb
	3.72506×10^{-7}	hp-h
	2.77778×10^{-7}	kW-h
	0.000277778	watt-h
	1*	watt-s
Kilocalories (kcal)	3.96832	Btu
	1000*	cal
	3088.03	ft-lb
	0.00155961	hp-h
	4186.80	joules
	0.00116300	kW-h
	1.16300	watt-h
Kilocalories per cubic meter ($kcal/m^3$)	0.112370	Btu/ft^3
Kilocalories per hour-meter-degree Celsius (kcal/h-m-°C)	0.671969	Btu/h-ft-°F
Kilocalories per kilogram (kcal/kg)	1.8*	Btu/lb
Kilocalories per minute (kcal/min)	51.4671	ft-lb/s
	0.0935766	hp
	0.0697800	kW
Kilograms	1000*	g
	2.20462	lb_m
	0.0685218	slugs
	0.001*	metric tons
Kilograms per square meter (kg_f/m^2)	0.00142233	lb/in^2
	0.204816	lb/ft^2
Kilograms per square centimeter (kg_f/cm^2)	14.2233	lb/in^2
	2048.16	lb/ft^2
Kilograms per hour (kg/h)	0.000612395	lb_m/s
	0.0367437	lb_m/min
	2.20462	lb_m/h

(*Continued*)

Multiply	By	To Obtain
Kilograms per minute (kg/min)	0.0367437	lb_m/s
	2.20462	lb_m/min
	132.277	lb_m/h
Kilograms per second (kg/s)	2.20462	lb_m/s
	132.277	lb_m/min
	7936.64	lb_m/h
Kilograms per cubic meter (kg/m^3)	0.0624280	lb_m/ft^3
	0.00194032	slugs/ft^3
Kilolitres	1000*	liters
Kilometers (km)	100,000*	cm
	1000*	m
	3280.84	ft
	0.621371	mi
	1093.61	yd
Kilometers per hour (km/h)	27.7778	cm/s
	54.6807	ft/min
	0.911344	ft/s
	16.6667	m/min
	0.277778	m/s
	0.539957	knots
Kilometers per minute (km/min)	1666.67	cm/s
	3280.84	ft/min
	54.6807	ft/s
	1000*	m/min
	16.6667	m/s
	32.3974	knots
Kilometers per second (km/s)	100,000*	cm/s
	196,850	ft/min
	3280.84	ft/s
	60,000*	m/min
	1000*	m/s
	1943.84	knots
Kilometers per hour-second (km/h-s)	27.7778	cm/s^2
	0.911344	ft/s^2
	0.277778	m/s^2
Kilowatt (kW)	56.8690	Btu/min
	44,253.7	ft-lb/min
	737.562	ft-lb/s
	1.34102	hp
	14.3307	kcal/min
	1000*	watts
Kilowatt-hour (kW-h)	3412.14	Btu
	8.59845×10^5	cal
	859.845	kcal
	2.65522×10^6	ft-lb
	1000*	watt-h
Kilowatt per square meter (kW/m^2)	5.28330	Btu/ft^2-min
	4111.31	ft-lb/ft^2-min
	0.124585	hp/ft^2
Knots	1*	nautical mi/h
	1.15078	mi/h
	1.85200	km/h

Multiply	By	To Obtain
Leagues	3*	nautical mi
Liters	1000*	cm^3
	0.0353147	ft^3
	61.0237	in^3
	0.001*	m^3
	0.00130795	ud^3
	0.264172	gal (U.S. liq.)
	0.219969	gal (Imp.)
	2.11338	pints (U.S. liq.)
	1.05669	qt (U.S. liq.)
	8.10713×10^{-7}	acre-ft
	2.20206	lb H_2O (62°F)
Liters per minute (L/min)	0.000588578	ft^3/s
	0.00440287	gal (U.S.)/s
	0.264172	gal (U.S.)/min
Meters	10^6	microns
	100*	cm
	3.28084	ft
	39.3701	in
	1.09361	yd
	0.001*	km
	1000*	mm
Meters per minute (m/min)	1.66667	cm/s
	3.28084	ft/min
	0.0546807	ft/s
	0.06*	km/h
	0.0372823	mi/h
Meters per second (m/s)	100*	cm/s
	3.6*	km/h
	0.06*	km/min
	0.001*	km/s
	196.850	ft/min
	3.28084	ft/s
	2.23694	mi/h
	0.0372823	mi/min
	6.21371×10^{-4}	mi/s
Square meter (m^2)	2.47105×10^{-4}	acres
	10.7639	ft^2
	1.19599	yd^2
	1*	centares
Square meter per second (m^2/s)	10^6*	centistokes
	10^4*	stokes
Cubic meter (m^3)	10^6*	cm^3
	35.3147	ft^3
	61,023.7	in^3
	1.30795	yd^3
	264.172	gal (U.S.)
	1000*	liters
	2113.38	pints (liq. U.S.)
	1056.69	qt (liq. U.S.)

Multiply	By	To Obtain
Cubic meter per hour (m^3/h)	10^6*	cm^3/h
	16,666.7	cm^3/min
	277.778	cm^3/s
	35.3147	ft^3/h
	0.588578	ft^3/min
	0.00980963	ft^3/s
	61,023.7	in^3/h
	1,017.06	in^3/min
	16.9510	in^3/s
	1.30795	yd^3/h
	0.0217992	yd^3/min
	0.000363320	yd^3/s
	264.172	gal (U.S.)/h
	4.40287	gal (U.S.)/min
	0.0733811	gal (U.S.)/s
	1000*	L/h
	16.6667	L/min
	0.277778	L/s
	2113.38	pints (liq. U.S.)/h
	35.2229	pints (liq. U.S.)/min
	0.587049	pints (liq. U.S.)/s
	1056.69	qt (liq. U.S.)/h
	17.6115	qt (liq. U.S.)/min
	0.293525	qt (liq. U.S.)/s
Cubic meter per minute (m^3/min)	6×10^7*	cm^3/h
	10^6*	cm^3/min
	16,666.7	cm^3/s
	2,118.88	ft^3/h
	35.3147	ft^3/min
	0.588578	ft^3/s
	3.66142×10^6	in^3/h
	61,023.7	in^3/min
	1,017.06	in^3/s
	78.4770	yd^3/h
	1.30795	yd^3/min
	0.0217992	yd^3/s
	15,850.3	gal (U.S.)/h
	264.172	gal (U.S.)/min
	4.40287	gal (U.S.)/s
	60,000*	L/h
	1000*	L/min
	16.6667	L/s
	126,803	pints (liq. U.S.)/h
	2,113.38	pints (liq. U.S.)/min
	35.2229	pints (liq. U.S.)/s
	63,401.3	qt (liq. U.S.)/h
	1,056.69	qt (liq. U.S.)/min
	17.6115	qt (liq. U.S.)/s
Cubic meter per second (m^3/s)	3.6×10^9*	cm^3/h
	6×10^7*	cm^3/min
	10^6*	cm^3/s
	127,133	ft^3/h
	2118.88	ft^3/min
	35.3147	ft^3/s
	2.19685×10^8	in^3/h
	3.66142×10^6	in^3/min
	61,023.7	in^3/s
	4708.62	yd^3/h
	78.4770	yd^3/min
	1.30795	yd^3/s
	951,019	gal (U.S.)/h
	15,850.3	gal (U.S.)/min
	264.172	gal (U.S.)/s
	3.6×10^6*	L/h
	60,000*	L/min
	1000*	L/s
	7.60816×10^6	pints (liq. U.S.)/h
	126,803	pints (liq. U.S.)/min
	2113.38	pints (liq. U.S.)/s
	3.80408×10^6	qt (liq. U.S.)/h
	63,401.3	qt (liq. U.S.)/min
	1056.69	qt (liq. U.S.)/s
Microns	1*	micrometers
	10^{-6}*	m
	0.001*	mm
	0.0393701	mils
Mils	0.001*	in
	0.0254*	mm
	25.4*	microns
Square mils (mils^2)	1.27324	circular mils
	0.000645160	mm^2
	10^{-6}*	in^2
Miles	160,934	cm
	1609.34	m
	1.60934	km
	5280*	ft
	63,360*	in
	1760*	yd
	80*	chains
	320*	rods
	0.868976	nautical mi
Miles per hour (mi/h)	44.704*	cm/s
	88*	ft/min
	1.46667	ft/s
	1.60934	km/h
	0.868976	knots
	26.8224*	m/min

(*Continued*)

Multiply	By	To Obtain
Miles per minute (mi/min)	2682.24	cm/s
	88*	ft/s
	1.60934	km/min
	60*	mi/h
Square mile (mi^2)	640*	acres
	27,878,400*	ft^2
	2.58999	km^2
	258.999	hectares
	3,097,600*	yd^2
	102,400*	rod^2
	1*	sections
Millibars	9.86923×10^{-4}	atm
Milligrams	0.001*	g
Milliliters	0.001*	liters
Million gallon per day	1.54723	ft^3/s
Millimeters	0.1*	cm
	0.0393701	in
	39.3701	mils
	1000*	microns
Square millimeter (mm^2)	0.01*	cm^2
	0.00155000	in^2
	1550.00	$mils^2$
	1973.53	circular mils
Minutes (arc)	0.000290888	radians
Nautical miles	6076.12	ft
	1.15078	miles
	1852*	m
	1.852*	km
Newtons	100,000*	dynes
	0.224809	lb (avoir.)
	0.273205	lb (troy)
	3.59694	oz (avoir.)
	3.27846	oz (troy)
Newtons per square meter (N/m^2)	1.45038×10^{-4}	lb/in^2
	0.0208854	lb/ft^2
	1*	pascals
Newton-second per square meter ($N\text{-}s/m^2$)	1000*	centipoise
	10*	poise
	0.0208854	$lb\text{-}s/ft^2$
	1*	Pa-s
Ounce (avoirdupois)	16*	drams
	437.5*	grains
	0.0625*	lb
	0.911458	oz (troy)
Ounce (fluid)	1.80469	in^3
	0.0295735	Liters
	29.5735	cm^3
	0.25*	gills

Multiply	By	To Obtain
Ounce (troy)	480*	grains
	20*	pennyweight (troy)
	0.083333	lb (troy)
	1.09714	oz (avoir.)
Ounce per square inch (oz/in^2)	0.0625*	lb/in^2
	1.73201	in H_2O (62°F)
	4.39930	cm H_2O (62°F)
	0.127251	in Hg (32°F)
	0.00425287	atm
Pascals	0.000145038	lb/in^2
	0.0208854	lb/ft^2
	1*	N/m^2
Pascal-second	1*	$N\text{-}s/m^2$
	0.0208854	$lb\text{-}s/ft^2$
Pennyweights (troy)	24*	grains
	0.05*	oz (troy)
	0.00416667	lb (troy)
Pints (liq. U.S.)	4*	gills
	16*	oz (fluid)
	0.5*	qt (liq. U.S.)
	28.875*	in^3
	473.176	cm^3
Poise	0.0671969	$lb_m/ft\text{-}s$
	241.909	$lb_m/ft\text{-}h$
	100*	centipoise
Pounds (avoir.)	16*	oz (avoir.)
	256*	drams (avoir.)
	444,822	dynes
	7000*	grains
	0.0005*	tons (short)
	4.46429×10^{-4}	tons (long)
	1.21528	lb (troy)
	4.44822	newtons
	14.5833	oz (troy)
Pounds (lb_m)	0.453592	kg
Pounds (troy)	5760*	grains
	366,025	dynes
	240*	pennyweight. (troy)
	12*	oz (troy)
	3.66025	newtons
	0.822857	lb (avoirdupois)
	13.1657	oz (avoirdupois)
	3.67347×10^{-4}	tons (long)
	4.11429×10^{-4}	tons (short)
Pound water (lb H_2O, 62°F)	0.0160371	ft^3
	27.7121	in^3
	0.119966	gal (U.S.)

Multiply	By	To Obtain
Pound water per minute (lb H_2O/min, 62°F)	2.67285×10^{-4}	ft^3/s
Pound per foot-second (lb_m/ft-s)	1.488164	N-s/m^2
	14.88164	poise
	1488.164	centipoise
	1.488164	Pa-s
Pound per square feet (lb/ft^2)	0.0160371	ft H_2O (62°F)
	0.00694444	lb/in^2
	0.0141390	in Hg (32°F)
	0.000472541	atm
Pound per square inch (lb/in^2)	0.0680460	atm
	2.30934	ft H_2O (62°F)
	2.03602	in Hg (32°F)
	27.7121	in H_2O (62°F)
	0.0703070	kg_f/cm^2
	6,894.76	N/m^2
	6,894.76	pascals
Pound per cubic feet (lb/ft^3)	5.78704×10^{-4}	lb/in^3
Pound per cubic feet (lb_m/ft^3)	16.0185	kg/m^3
	1.60185×10^{-5}	kg/cm^3
Pound per cubic inch (lb/in^3)	1728*	lb/ft^3
Pound per cubic inch (lb_m/in^3)	27,679.9	kg/m^3
	0.0276799	kg/cm^3
Pound per hour (lb_m/h)	0.453592	kg/h
	0.00755987	kg/min
	1.25998×10^{-4}	kg/s
Pound per minute (lb_m/min)	27.2155	kg/h
	0.453592	kg/min
	0.00755987	kg/s
Pound per second (lb_m/s)	1,632.93	kg/h
	27.2155	kg/min
	0.453592	kg/s
Pounds-second per square feet (lb_f-s/ft^2)	47.8803	N-s/m^2
	47,880.3	centipoise
Quadrants (arc)	90*	degrees
	5400*	minutes
	324,000*	seconds
	1.57080	radians
Quarts (dry)	67.2006	in^3
Quarts (liq. U.S.)	2*	pints (liq. U.S.)
	0.946353	liters
	32*	oz (fluid)
	57.75*	in^3
	946.353	cm^3
Radians	57.2958	degrees
	3437.75	minutes
	206,265	seconds
	0.636620	quadrants
Radians per second (radians/s)	57.2958	degrees/s
	0.159155	revolutions/s
	9.54930	revolutions/min
Radians per square second (radians/s^2)	572.958	revolutions/min^2
	0.159155	revolutions/s^2
Revolutions	360*	degrees
	4*	quadrants
	6.28319	radians
Revolutions per minute	6*	degrees/s
	0.104720	radians/s
	0.0166667	revolutions/s
Revolutions per square minute (revolutions/min^2)	0.00174533	radians/s^2
	0.000277778	revolutions/s^2
Revolutions per second (revolutions/s)	360*	degrees/s
	6.28319	radians/s
	60*	revolutions/min
Revolutions per square second (revolutions/s^2)	6.28319	radians/s^2
	3600*	revolutions/min^2
Rods	16.5*	ft
	5.5*	yd
Seconds (arc)	4.84814×10^{-6}	radians
Sections	1*	mi^2
Side of square	1.41421	dia circumscribed circle
	1.12838	dia equal area circle
Square across flats	1.414214	across corners
Stere	1*	m^3
Stone	14*	lb
	62.2751	newtons
Tons (long)	2240*	lb
	9964.02	newtons
	1.12*	tons (short)
Tons (metric)	1000*	kg
Tons (short)	2000*	lb
	32,000*	oz (avoir.)
	0.892857	tons (long)
Tons (refrig)	12,000*	Btu/h
	288,000*	Btu/day
	200*	Btu/min

(*Continued*)

Multiply	By	To Obtain
Tons water per day (tons H_2O/day)	83.3333	lb H_2O/h
	0.166619	gal (U.S.)/min (62°F)
	1.33643	ft^3/h (62°F)
Watts	0.0568690	Btu/min
	44.2537	ft-lb/min
	0.737562	ft-lb/s
	0.00134102	hp
	0.0143307	kcal/min
	0.001*	kW
	1*	J/s
Watt-hour (watt-h)	3.41214	Btu
	859.845	cal
	0.859845	kcal
	2655.22	ft-lb
	0.00134102	hp-h
	3600*	joules
	0.001*	kW-h
Watt per square inch (watt/in^2)	8.18914	Btu/ft^2-min
	6372.54	ft-lb/ft^2-min
	0.193107	hp/ft^2
Watt per square centimeter (watt/cm^2)	52.8330	Btu/ft^2-min
	41,113.1	ft-lb/ft^2-min
	1.24585	hp/ft^2

Multiply	By	To Obtain
Yards	91.44*	cm
	3*	ft
	36*	in
	0.9144*	m
	0.181818	rods
Square yard (yd^2)	0.000206612	acres
	9*	ft^2
	0.836127	m^2
	3.22831×10^{-7}	mi^2
Cubic yard (yd^3)	764.555	liters
	1615.79	pints (liq. U.S.)
	807.896	qt (liq. U.S.)
	201.947	gal (U.S.)
	0.764555	m^3
	764,555	cm^3
	27*	ft^3
	46,656*	in^3
Cubic yard per minute (yd^3/min)	0.45*	ft^3/s
	3.36623	gal (U.S.)/s
	12.7426	L/s
Year (365 days)	8760*	hours
Year (sidereal)	8766.1528	hours
	365.2564 (365 days, 6 hours, 9 minutes, 9 seconds)	day (mean sol)

All calories and Btus are International Steam Table (IST) values to six significant figures unless noted (see footnote a). The asterisk symbol (*) indicates conversion is exact.

Each Btu listed in this table is the International Steam Table Btu, and every calorie is the International Steam Table calorie, unless otherwise noted. The International Steam Table values are 1.0006696 times the thermochemical values, and the thermochemical values are 0.99933084 times the International Steam Table values.

Source: Mechtly, E. A., *The International System of Units*, 2nd rev., National Aeronautics and Space Administration, Washington, D.C., 1973, NASA SP-7012.

APPENDIX D

COMPRESSIBILITY FACTOR EQUATIONS

D.1 THE REDLICH–KWONG EQUATION

The Redlich-Kwong equation is actually an equation of state. It was formulated by Otto Redlich and Joseph N. S. Kwong in 1949 [*Chemical Review*, 44, 1949, 233-244]. Their equation is

$$P = \frac{n\bar{R}T}{V - b} - \frac{a}{V(V + b)\sqrt{T}},$$

$$a = \frac{\Omega_a (n\bar{R})^2 T_c^{2.5}}{P_c},$$

$$b = \frac{\Omega_b n\bar{R}T_c}{P_c},$$

where T_c and P_c are the critical temperature and pressure of the gas being considered. The constants in the equation are derived as

$$\Omega_a = [(9)(2)^{1/3} - 1]^{-1} = 0.42748,$$

$$\Omega_b = [(2)^{1/3} - 1]/3 = 0.08664.$$

This equation is quite accurate at the critical temperature for $P/P_c = 4$ to 40 (error < 2.5%). At higher temperatures and at P/P_c above about 5 the equation becomes increasingly inaccurate.

The compressibility factor may be found explicitly from the Redlich-Kwong equation. It is the principal root of a cubic. Robert C. Reid, John M. Prausnitz, and Thomas K. Sherwood [*The Properties of Gases and Liquids*, 3rd Ed., McGraw-Hill Book Company, 1977] give the cubic as

$$z^3 - z^2 + (A^* - B^{*2} - B^*)z - A^*B^* = 0,$$

where

$$A^* = \frac{\Omega_a P_r}{T_r^{5/2}} = 0.42748 \frac{P_r}{T_r^{5/2}},$$

$$B^* = \frac{\Omega_b P_r}{T_r} = 0.086640 \frac{P_r}{T_r}.$$

In the equations below, P_r is the reduced pressure (that is, the ratio of actual pressure to critical pressure), and T_r is the reduced temperature (that is, the ratio of actual temperature to critical temperature). Let A, B, C, D, and E be constants defined by

$$A = 0.08664 \frac{P_r}{T_r},$$

$$B = \frac{4.9340}{T_r^{3/2}},$$

$$C = A^2 + A - AB,$$

Pipe Flow: A Practical and Comprehensive Guide, First Edition. Donald C. Rennels and Hobart M. Hudson.

TABLE D.1. Lee–Kesler Constants

Constant	Simple	Reference	Constant	Simple	Reference
B_1	0.1181193	0.2026579	c_3	0.0	0.016901
B_2	0.265728	0.331511	c_4	0.042724	0.041577
B_3	0.154790	0.027655	$d_1 \times 10^4$	0.155488	0.48736
B_4	0.030323	0.203488	$d_2 \times 10^4$	0.623689	0.0740366
c_1	0.0236744	0.0313385	β	0.65392	1.226
c_2	0.0186984	0.0503618	γ	0.060167	0.03754

$$D = \frac{-C}{3} - \frac{1}{9},$$

$$E = \frac{C}{6} + \frac{A^2B}{2} + \frac{1}{27}.$$

If $D^3 + E^2 \geq 0$, then

$$z = \sqrt[3]{E + \sqrt{D^3 + E^2}} + \sqrt[3]{E - \sqrt{D^3 + E^2}} + \frac{1}{3}.$$

If $D^3 + E^2 < 0$, then

$$z = 2\sqrt{-D}\cos\left[\frac{-1}{3}\cos^{-1}\left(\frac{E}{\sqrt{-D^3}}\right)\right] + \frac{1}{3}.$$

Critical constants for selected gases are given in Table D.3 in Section D.3. *Chemical Engineers' Handbook* (5th Ed., McGraw Hill, 1973) indicates that the Redlich-Kwong equation of state fits the data for helium and hydrogen only for reduced temperatures of 2.5 and higher when their critical temperatures are increased by 8°C, and their critical pressures are increased by 8 atmospheres.

D.2 THE LEE–KESLER EQUATION

Another equation of state that is much more accurate than the Redlich–Kwong equation is the Lee–Kesler equation. This is a generalized Benedict–Webb–Rubin equation developed by B.I. Lee and M.G. Kesler in 1975 from which the compressibility factor may be found. The solution is formidable, but with a computer it can be obtained without much difficulty using the Newton–Raphson trial-and-error solution technique. Their equation is:

$$\frac{P_r V_r^{(0)}}{T_r} = 1 + \frac{B}{V_r^{(0)}} + \frac{C}{(V_r^{(0)})^2} + \frac{D}{(V_r^{(0)})^5} + \frac{c_4}{T_r^3 (V_r^{(0)})^2}\left[\beta + \frac{\gamma}{(V_r^{(0)})^2}\right]\exp\left[\frac{-\gamma}{(V_r^{(0)})^2}\right], \quad \text{(D.1)}$$

where

$$B = b_1 - \frac{b_2}{T_r} - \frac{b_3}{T_r^2} - \frac{b_4}{T_r^3},$$

$$C = c_1 - \frac{c_2}{T_r} + \frac{c_3}{T_r^3},$$

$$D = d_1 + \frac{d_2}{T_r},$$

$$V_r^{(0)} = \frac{P_c V^{(0)}}{RT_c},$$

and

$$\exp\left[\frac{-\gamma}{(V_r^{(0)})^2}\right] = e^{-\gamma/(V_r^{(0)})^2}.$$

The constants for these equations for a simple fluid are given in Table D.1. (The β and γ shown are for the Lee–Kesler equation and should not be confused with those in the Nomenclature.) The equation is solved for $V_r^{(0)}$, the ideal reduced volume for a simple fluid, and then the simple fluid compressibility factor is calculated:

$$z^{(0)} = \frac{P_r V_r^{(0)}}{T_r}. \quad \text{(D.2)}$$

Next, using the same reduced pressure and temperature, the equation is solved again for $V_r^{(0)}$, but using the reference fluid constants from the table; therefore, call this value $V_r^{(R)}$. Then:

$$z^{(R)} = \frac{P_r V_r^{(R)}}{T_r}.$$

The compressibility factor z for the fluid of interest is then calculated from the following formula:

$$z = z^{(0)} + \left(\frac{\omega}{\omega^R}\right)\left(z^{(R)} - z^{(0)}\right),$$

where ω is Pitzer's acentric factor, and for the reference gas, $\omega^R = 0.3978$. The definition of the acentric factor is:

$$\omega = -\log_{10}(P_{vap_r})_{T_r=0.7} - 1.000,$$

where the pressure term is the *reduced* vapor pressure at $T_r = 0.7$. (Values of ω are given for selected gases in Table D.2 in Section D.3.)

In order to solve the Lee–Kesler equation by the Newton–Raphson method, we must devise a function from it whose value is zero. This may be done by moving the $P_r V^{(0)}/T_r$ term to the right side of Equation D.1:

$$0 = 1 - \frac{P_r V_r^{(0)}}{T_r} + \frac{B}{V_r^{(0)}} + \frac{C}{(V_r^{(0)})^2} + \frac{D}{(V_r^{(0)})^5} + \frac{c_4\beta}{T_r^3 (V_r^{(0)})^2}\exp\left[\frac{-\gamma}{(V_r^{(0)})^2}\right] + \frac{c_4\gamma}{T_r^3 (V_r^{(0)})^4}\exp\left[\frac{-\gamma}{(V_r^{(0)})^2}\right].$$

Call this function $f(V_r^{(0)})$ by substituting $f(V_r^{(0)})$ for the zero:

$$f(V_r^{(0)}) = 1 - \frac{P_r V_r^{(0)}}{T_r} + \frac{B}{V_r^{(0)}} + \frac{C}{(V_r^{(0)})^2} + \frac{D}{(V_r^{(0)})^5} + \frac{c_4\beta}{T_r^3 (V_r^{(0)})^2}\exp\left[\frac{-\gamma}{(V_r^{(0)})^2}\right] + \frac{c_4\gamma}{T_r^3 (V_r^{(0)})^4}\exp\left[\frac{-\gamma}{(V_r^{(0)})^2}\right]. \quad \text{(D.3)}$$

$f(V_r^{(0)})$ is *supposed* to equal zero. Of course, it is not likely to equal zero if we don't know the correct value for $(V_r^{(0)})$ but have to guess it instead. Any nonzero value for $f(V_r^{(0)})$ is the error incurred by using an incorrect value for $(V_r^{(0)})$ in it. It may be considered to be the required correction for the function. Using the Newton–Raphson method, we can refine our guesses very easily. In order to do this we need the derivative of the function $f(V_r^{(0)})$. The derivatives of the seven terms of the function are given below:

$$f_1' = 0, \qquad f_2' = -\frac{P_r}{T_r}, \quad \text{(D.4)}$$

$$f_3' = -\frac{B}{(V_r^{(0)})^2}, \quad \text{(D.5)}$$

$$f_4' = -2\frac{C}{(V_r^{(0)})^3}, \quad \text{(D.6)}$$

$$f_5' = -5\frac{D}{(V_r^{(0)})^6}, \quad \text{(D.7)}$$

$$f_6' = \left[2\frac{c_4\beta}{T_r^3 (V_r^{(0)})^3}\right]\left[\frac{\gamma}{(V_r^{(0)})^2} - 1\right]\exp\left[\frac{-\gamma}{(V_r^{(0)})^2}\right], \quad \text{(D.8)}$$

$$f_7' = \left[2\frac{c_4\gamma}{T_r^3 (V_r^{(0)})^5}\right]\left[\frac{\gamma}{(V_r^{(0)})^2} - 2\right]\exp\left[\frac{-\gamma}{(V_r^{(0)})^2}\right]. \quad \text{(D.9)}$$

The derivative of the function $f(V_r^{(0)})$ is then the sum of the derivatives of its terms, or

$$f'(V_r^{(0)}) = f_2' + f_3' + f_4' + f_5' + f_6' + f_7'. \quad \text{(D.10)}$$

The solution technique for the $f(V_r^{(0)})$ equation is to guess an initial $V_r^{(0)}$—call it $(V_r^{(0)})_{i=1}$. An initial guess for $V_r^{(0)}$ may be obtained by finding the Redlich–Kwong compressibility factor z_{RK} and assuming that it is approximately equal to the simple fluid compressibility factor. Then the $z^{(0)}$ equation (Eq. D.2) may be solved for $(V_r^{(0)})_{i=1}$:

$$(V_r^{(0)})_{i=1} \approx \frac{z_{\text{RK}} T_r}{P_r}. \quad \text{(D.11)}$$

This guess for $(V_r^{(0)})_{i=1}$ is then inserted into the function (Eq. D.3) and into the equations for the terms of its derivative (Eqs. D.4–D.11).

These three values ($[V_r^{(0)}]_{i=1}$, $f[V_r^{(0)}]_{i=1}$, and $f'[V_r^{(0)}]_{i=1}$) are then used to find $(V_r^{(0)})_{i=2}$. By dividing the value of the function (which is the required correction to the function, that is, to the *dependent* variable), we transform it into an estimate of the required correction in the *independent* variable, $V_r^{(0)}$. Equation D.12 applies the correction. The result is a much closer value of the independent variable, as shown in Figure D.1:

$$(V_r^{(0)})_{i+1} = (V_r^{(0)})_i - \frac{f(V_r^{(0)})_i}{f'(V_r^{(0)})_i}. \quad \text{(D.12)}$$

The procedure is then repeated with this better estimate of $(V_r^{(0)})$. After each repetition the correction term $f(V_r^{(0)})_n / f'(V_r^{(0)})_n$ will become smaller and smaller,

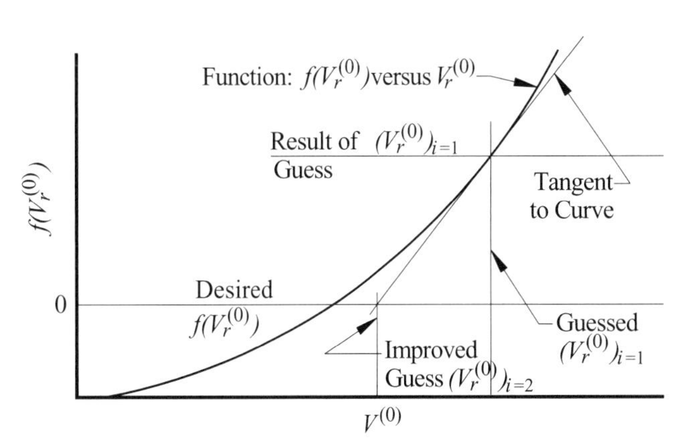

FIGURE D.1. Solution technique for $V_r^{(0)}$.

until it becomes small enough—as small a value as desired or allowed by the computational precision of the computer—that the solution may be considered to have been found.

The solution technique described above has one caveat—it works well except in the region of the critical point. There the derivative approaches zero and the procedure usually gets caught in a loop. To circumvent this a different technique must be substituted on the first occurrence of a change in sign of the correction term $f(V_r^{(0)})/f'(V_r^{(0)})$. One method is to interpolate between the $V_r^{(0)}$ that caused the sign of $f(V_r^{(0)})$ to change and the one just previous to it to estimate the $V_r^{(0)}$ where the curve crosses the $f(V_r^{(0)})=0$ line.

The reader will note that Pitzer's acentric factor ω is used in the solution for the compressibility factor z. Reid et al. (*The Properties of Gases and Liquids*, 3rd ed., McGraw-Hill Book Company, New York, 1977) state that application of correlations employing the acentric factor should be limited to normal fluids; in no case should such correlations be used for H_2, He, Ne, or for strongly polar and/or hydrogen-bonded fluids. Therefore, for these nonnormal fluids it is suggested that the compressibility factor yielded by the Lee–Kesler equation be compared with actual fluid data compressibility factors on a plot such as Figure 1.3 in Chapter 1. From this plot it may be seen what shift in critical constants will bring the Lee–Kesler compressibility factor into congruence with the real compressibility factor for the largest region on the chart. Several modifications of the acentric factor may be necessary to achieve the best agreement.

D.3 IMPORTANT CONSTANTS FOR SELECTED GASES

Following are two tables of important constants necessary to implement the Redlich–Kwong and Lee–Kesler compressibility factor equations. Table D.2 gives the acentric factor ω for selected gases.

Table D.3 gives critical constants for the same gases. Data are included for ammonia, hydrogen, and helium for use in these equations; the data for these gases should be amended as described in Section D.1 for best results in the Redlich–Kwong equation, and as described above in Section D.2 for best results in the Lee–Kesler equation.

Table D.3 includes critical constants from various authorities. It is suggested that consensus values or averages of all the values be used for each constant.

TABLE D.2. Acentric Factor ω for Selected Gases

Acetylene	0.184	CO_2	0.225	Methane	0.008
Air	0.036	Helium	–0.387	Nitrogen	0.040
Ammonia	0.250	n-Hydrogen	–0.22	Oxygen	0.021
Argon	–0.004	p-Hydrogen	–0.219	Propane	0.152

See Table D.3 footnote d for bibliography. Parahydrogen ω is calculated from footnote g data.

TABLE D.3. Critical Constants for Selected Gases According to Various Authorities

Gas	Parameter	*Marks' Handbook*[a]	Perry and Chilton[b]	Ražnjević[c]	Poling et al.[d]	*Handbook of Chemistry and Physics*[e]	ASME Fluid Meters[f]	National Bureau of Standards[g]
Acetylene	T_c °R	556.0	556.5	555.93	554.94	554.9	557.1	–
	P_c psia	911	911	920	886.8	890.2	905	–
Air	T_c °R	239.4	238.4	238.41	–	–	238.4	–
	P_c psia	547	547	546	–	–	547	–
Ammonia	T_c °R	730	730.0	729.99	729.72	729.9	731.1	–
	P_c psia	1639	1639	1639	1646.6	1646	1657	–
Argon	T_c °R	272.0	272	271.35	271.55	271.56	272.08	272.0
	P_c psia	705	705	705	710.4	710.4	705.4	711.5
CO_2	T_c °R	547.7	547.7	547.47	547.42	547.43	547.7	–
	P_c psia	1073	1073	1067	1070	1070.00	1073	–
Helium	T_c °R	9.5	9.5	9.45	9.34	9.34	9.4	9.363
	P_c psia	33.2	33.2	33.1	32.9	32.9	33.0	32.99
n-Hydrogen	T_c °R	59.9	59.9	59.85	59.85	59.35	59.9	–
	P_c psia	188	188	187	188.1	187.5	188	–
p-Hydrogen	T_c °R	–	–	–	–	–	–	59.29
	P_c psia	–	–	–	–	–	–	186.2
Methane	T_c °R	343.2	343.2	343.17	343.01	343.01	343.2	343.00
	P_c psia	673	673	671	667.0	667.0	673.1	666.9
Nitrogen	T_c °R	226.9	226.9	226.89	227.16	227.18	226.9	227.27
	P_c psia	492	492	492	492.8	492	492	493.0
Oxygen	T_c °R	277.9	277.8	277.83	278.24	278.26	277.9	278.25
	P_c psia	730	730	731	731.4	731.4	730	731.4
Propane	T_c °R	665.93	665.9	665.91	665.69	665.69	666	665.73
	P_c psia	617.4	617	616	616.1	616.1	617.4	616.1

[a] Baumeister, T., E. A. Avallone, and T. Baumeister III, eds., *Marks' Standard Handbook for Mechanical Engineers*, 8th ed., McGraw-Hill, 1978.
[b] Perry, R. H. and C. H. Chilton, *Chemical Engineers' Handbook*, 5th ed., McGraw Hill, 1973.
[c] Ražnjević, K., *Handbook of Thermodynamic Tables and Charts,* Hemisphere Publishing Corporation, 1976.
[d] Poling, B. E., J. M. Prausnitz, and J. P. O'Connel, *The Properties of Gases and Liquids*, 5th ed., McGraw-Hill, 2001.
[e] Lide, D. R., ed., *CRC Handbook of Chemistry and Physics*, 85th ed., CRC Press Inc., 2004.
[f] *Interim Supplement No. 19.5 on Instruments and Apparatus (Application, Part II of Fluid Meters,* 6th ed.), American Society of Mechanical Engineers, 1971.
[g] McCarty, R. D., *NBS Standard Database 12* (MIPROPS), National Bureau of Standards, 1986.

APPENDIX E

ADIABATIC COMPRESSIBLE FLOW WITH FRICTION, USING MACH NUMBER AS A PARAMETER

This appendix gives derivations for application equations presented in Chapter 4.

Street et al. [1] and Shapiro [2] give the following relation for a constant-area duct flowing a gas *with sonic velocity at the exit*:

$$f_{ave}\frac{L_{max}}{D}=\frac{1-M^2}{\gamma M^2}+\frac{\gamma+1}{2\gamma}\ln\left[\frac{(\gamma+1)M^2}{2\left(1+\frac{\gamma-1}{2}M^2\right)}\right], \quad (4.16, \text{repeated})$$

where

f_{ave} = average Darcy friction factor along the duct,

L_{max} = maximum attainable duct length with M at the inlet, ft (or m),

D = duct diameter, ft (or m),

γ = ratio of specific heats of flowing gas, and

M = Mach number of the gas flow at the duct inlet.

In the development of this equation, f is assumed to be a constant, and f_{ave} is taken as a reasonable value for f. In actuality, of course, since fluid temperature changes continuously along the duct, the fluid viscosity also changes, and then so does Reynolds Number—resulting in a varying friction factor. But it turns out that the variation is modest enough to be handled by using the average friction factor.

E.1 SOLUTION WHEN *STATIC* PRESSURE AND *STATIC* TEMPERATURE ARE KNOWN

Equation 4.16 may be used to find the L_{max} of the duct if the essential duct data are available: flow rate, inlet static pressure, inlet static temperature, duct diameter, friction factor, and gas ratio of specific heats, molecular weight, and compressibility factor. The Mach number of a gas flowing in a duct (assuming a flat velocity profile) is:

$$M=\frac{u}{A}\approx\frac{V}{A}. \quad (1.4, \text{repeated})$$

The equation for the acoustic velocity A is:

$$A=\sqrt{\frac{\gamma P}{\rho_m}}(\text{mass units}) \quad \text{or} \quad A=\sqrt{\frac{\gamma Pg}{\rho_w}}(\text{weight units}).$$

Utilizing the equation of state, Equations 1.6 in Chapter 1, the acoustic velocity may be expressed as:

$$A=\sqrt{\frac{\gamma P}{\rho_m}}=\sqrt{\frac{\gamma P}{P/zRT}}=\sqrt{\gamma zRT}=\sqrt{\gamma z\frac{\bar{R}}{m}T} \quad (\text{E.1a})$$

$$\left(\bar{R}=8314.34\ \text{J}/\text{mol}_{\text{Kg}}{}^\circ\text{K}\right),$$

or

$$A=\sqrt{\frac{\gamma P}{\rho_w}}=\sqrt{\frac{\gamma Pg}{P/zRT}}=\sqrt{\gamma gzRT}=\sqrt{\gamma gz\frac{\bar{R}}{m}T} \quad (\text{E.1b})$$

$$\left(\bar{R}=1545.31\ \text{ft-lb/mol}_{\text{lb}}{}^\circ\text{R}\right).$$

Pipe Flow: A Practical and Comprehensive Guide, First Edition. Donald C. Rennels and Hobart M. Hudson.

(In these two equations, m represents molecular weight, not mass.) The compressibility factor z may be evaluated using one of the formulas found in Appendix D. Utilizing Equations E.1a and E.1b, and $\dot{m} = AV\rho_m$ and $\dot{w} = AV\rho_w$ from Chapter 2, we may write:

$$M = \frac{\dot{m}/A\rho_m}{\sqrt{\gamma z \frac{\bar{R}}{m} T}} = \frac{\dot{m}}{A\frac{Pm}{z\bar{R}T}\sqrt{\gamma z \frac{\bar{R}}{m} T}} = \frac{\dot{m}}{AP\sqrt{\gamma \frac{m}{z\bar{R}T}}} = \frac{\dot{m}}{AP}\sqrt{\frac{z\bar{R}T}{\gamma m}} \quad \text{(E.2a)}$$

or

$$M = \frac{\dot{w}/A\rho_w}{\sqrt{\gamma g z \frac{\bar{R}}{m} T}} = \frac{\dot{w}}{A\frac{Pm}{z\bar{R}T}\sqrt{\gamma g z \frac{\bar{R}}{m} T}} = \frac{\dot{w}}{AP\sqrt{\gamma g \frac{m}{z\bar{R}T}}} = \frac{\dot{w}}{AP}\sqrt{\frac{z\bar{R}T}{\gamma g m}}. \quad \text{(E.2b)}$$

Using this Mach number, evaluated at the duct inlet, L_{max} becomes immediately available from Equation 4.16.

Equation 4.16 may not be violated.* The length of the duct may not exceed L_{max} with sonic velocity ($M = 1$) occurring at the exit. However, if the length of the duct is less than L_{max} as given by Equation 4.16, then the exit Mach number will be less than unity. This is the most frequently encountered case.

Consider a gas receiver discharging through a round duct of known length L_{line} to a lower pressure region and suppose that the pressure conditions are such that the discharging gas exits from the duct at subsonic velocity (see Fig. E.1). Assume that friction factor f and diameter D are constant. If we know the flowing conditions at one end—either end—of the duct (flow rate, duct diameter, pressure, and temperature), we may find the Mach number M there and then use Equation 4.16 to find the $(fL/D)_{limit}$ at that end of the duct. By Equation 8.1, this can be called K_{limit} at that end. (Remember that because f and D are constant, K in this context is simply length with a constant coefficient.) Note that since the flow exits from the duct subsonically, this K_{limit} includes a virtual length of duct at which the flow would attain sonic velocity (provided that the pressure at the virtual outlet was low enough). Now, because f/D is constant, K is proportional to L so that we can write:

$$(K_1)_{limit} = K_{line} + (K_2)_{limit}. \quad \text{(E.3)}$$

Knowing the line resistance coefficient K_{line} and limit resistance coefficient $(K)_{limit}$ at one end of the duct enables us to find the limit resistance coefficient at the other end of the duct. Then, since $(K)_{limit}$ is associated with M at that end by Equation 4.16, we may find M at that end by solving the equation.

Because Equation 4.16 cannot be solved for M explicitly, it must be solved by trial and error. The Newton–Raphson method is a convenient method for the solution. In order to implement it, we need to rearrange the equation so that we have an *expression that equals zero*. We can do this by subtracting $(K_1)_{limit}$ from both sides of Equation E.3:

$$0 = (K_2)_{limit} - [(K_1)_{limit} - K_{line}]. \quad \text{(E.3, rearranged)}$$

Now $(K_1)_{limit} - K_{line} = (K_2)_{limit}$, and while we know the values of $(K_1)_{limit}$, K_{line}, and $(K_2)_{limit}$, we do not know the value of the Mach number yielding $(K_2)_{limit}$, and we are interested in knowing this value so that we may find the flowing conditions at the actual duct outlet. Let us call K_i the guessed value of $(K_2)_{limit}$ and write:

$$f(M) = K_i - (K_2)_{limit} = 0. \quad \text{(E.4)}$$

This expression is supposed to equal zero, and it will be if we evaluate K_i using the right Mach number. If we guess a Mach number and evaluate K_i by Equation 4.16, the result is not likely to equal K_{limit} and $f(M)$ is not likely to equal zero. This is shown graphically in Figure E.2.

If we extrapolate down the function's tangent, it is clear that at the intersection with $f(M) = 0$, we will find a much better guess for M. To do this requires the derivative of K with respect to M:

$$\frac{dK_{limit}}{dM} = -\frac{2}{\gamma M^3} + \frac{\gamma+1}{\gamma M}\left[\frac{1}{1+\frac{\gamma-1}{2}M^2}\right]. \quad \text{(E.5)}$$

* This is not to say that supersonic flow cannot occur in a constant area duct; it can, but the flow must be introduced to the duct in a supersonic condition.

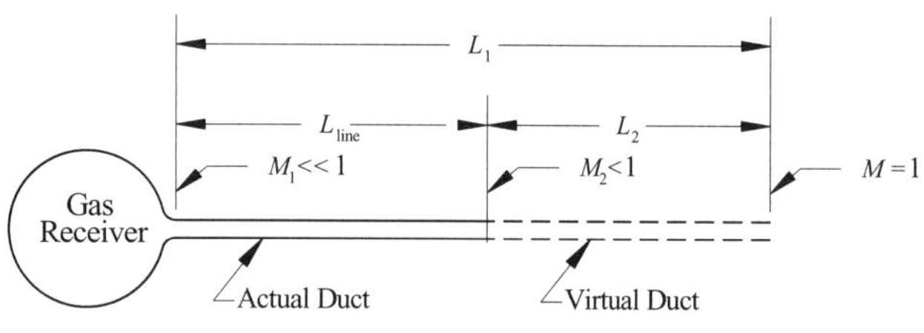

FIGURE E.1. Subsonic constant-area gas flow duct (Fig. 4.5, repeated).

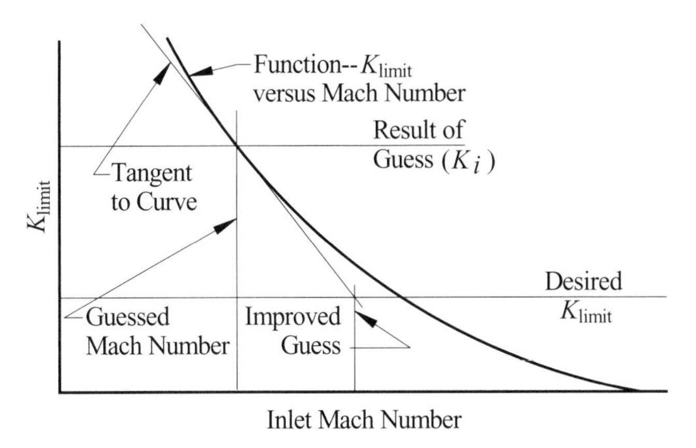

FIGURE E.2. Mach number solution by the Newton–Raphson method.

Now a better approximation of M may be found with the extrapolation formula:

$$M_{i+1} = M_i - \frac{K_i - K_{\text{limit}}}{(dK_{\text{limit}}/dM)}, \tag{E.6}$$

where M_{i+1} is the improved approximation and M_i is the earlier or guessed value. As the natural logarithm term in Equation 4.16 is much smaller than the preceding term, use the approximation:

$$K_{\text{limit}} \approx \frac{1-M^2}{\gamma M^2}$$

or

$$M \approx \sqrt{\frac{1}{1+\gamma K_{\text{limit}}}} \tag{E.7}$$

for the first guess of M. This guess for M may be entered in Equation 4.16 to find K_i, the estimated limit on K based on M. Enter it also in Equation E.5 to get dK_{limit}/dM. Then enter all three variables in Equation E.6 to obtain an improved estimate of M. Repeat the process to get K_i and dK_{limit}/dM at the new, better estimate of M, and then a much improved estimate of M.

After several iterations, the second term in the iteration formula will become quite small and the successive approximations of M will become more nearly alike. When the corrections become as small as desired (say, one part in a million), the iterations may be halted and the Mach number considered solved.

Once the unknown Mach number is found, the accompanying pressure and temperature may be found. The static pressure, in terms of the local Mach number and the static pressure P_* at the location where Mach number is unity (that is, where velocity is sonic) is given by:

$$\frac{P}{P_*} = \frac{1}{M}\sqrt{\frac{\gamma+1}{2\left(1+\frac{\gamma-1}{2}M^2\right)}}. \tag{E.8}$$

Taking the ratio of the expression evaluated for $M = M_1$ to that for $M = M_2$ yields:

$$\frac{P_1}{P_2} = \frac{M_2}{M_1}\sqrt{\frac{1+M_2^2(\gamma-1)/2}{1+M_1^2(\gamma-1)/2}}, \tag{E.9}$$

from which the desired pressure is easily found. The static temperature is available similarly from:

$$\frac{T}{T_*} = \frac{\gamma+1}{2\left(1+\frac{\gamma-1}{2}M^2\right)}. \tag{E.10}$$

The ratio of the inlet and outlet static temperatures is thus:

$$\frac{T_1}{T_2} = \frac{1+M_2^2(\gamma-1)/2}{1+M_1^2(\gamma-1)/2}, \tag{E.11}$$

from which the desired temperature is easily found.

The foregoing relationships are useful if the static pressure and static temperature at one end of the duct are known. If one or the other of the static values is not known, but the corresponding total value is known (and this is often, if not usually, the case) these equations may still be solved, but account must be made for the divergence between total and static values. For instance, if a gas in a pressurized vessel is allowed to escape to atmosphere through a duct and it attains sonic velocity at the end of the conduit, the static pressure at the outlet end of the duct may be as low as half its total pressure and static temperature may be as low as 80% of its total temperature.

There are three cases in which the required static values are not all known: (1) static pressure and total temperature are known; (2) total pressure and total temperature are known; and (3) total pressure and static temperature are known. These will be considered in order. We must make use of the following relationships:

$$T = \frac{T_t}{1+M^2(\gamma-1)/2}, \tag{E.12}$$

$$P = \frac{P_t}{\left[1+M^2(\gamma-1)/2\right]^{\gamma/(\gamma-1)}}, \tag{E.13}$$

where T, P, T_t, P_t, and M are local values (i.e., all at the same location).

In order to simplify the equations, let us recast the equation for Mach number (Eq. E.2a or E.2b) in the following form:

$$M = B\sqrt{T}/P, \tag{E.14}$$

where

$$B = \frac{\dot{m}}{A}\sqrt{\frac{z\bar{R}}{\gamma m}} \quad \left(\bar{R} = 8314.34 \text{ J/mol}_{\text{Kg}}\,^\circ\text{K}\right) \tag{E.15a}$$

or

$$B = \frac{\dot{w}}{A}\sqrt{\frac{z\bar{R}}{\gamma g m}} \quad \left(\bar{R} = 1545.31 \text{ ft-lb/mol}_{\text{lb}}\,^\circ\text{R}\right). \tag{E.15b}$$

E.2 SOLUTION WHEN *STATIC* PRESSURE AND *TOTAL* TEMPERATURE ARE KNOWN

Now, if *static pressure* and *total temperature* are known, substitute the expression for static temperature T (Eq. E.12), in terms of total temperature T_t, in place of T; then:

$$M = \frac{B}{P}\sqrt{\frac{T_t}{1+M^2(\gamma-1)/2}}. \tag{E.16}$$

This equation is a quadratic in M^2 whose solution is:

$$M^2 = \frac{\sqrt{1+2(\gamma-1)(B\sqrt{T_t}/P)^2}-1}{\gamma-1}. \tag{E.17}$$

Note the similarity of the expression $B\sqrt{T_t}/P$ in Equation E.17 to that for Mach number M in Equation E.14. They are identical except that the one above contains T_t while Equation E.14 contains simply T. Let us therefore call the expression (and similar expressions utilizing the available temperature and pressure, whether they be static or total) "core Mach number," M_{core}, because of its similarity to the simple expression for Mach number based on static values, and because it is the "core" of the expression for Mach number when other than static values are utilized. Then, for the *static pressure* and *total temperature* case, we may write:

$$M^2 = \frac{\sqrt{1+2(\gamma-1)M_{\text{core}}^2}-1}{\gamma-1}. \tag{E.18}$$

This M^2 may now be substituted into Equation 4.16 to find the $f_{\text{ave}}L_{\text{max}}/D$ or K_{limit}, and from thence to find the Mach number at the other end of the duct and the accompanying pressure and temperature.

E.3 SOLUTION WHEN *TOTAL* PRESSURE AND *TOTAL* TEMPERATURE ARE KNOWN

If *total pressure* and *total temperature* are known at one end of the duct, the expressions for static pressure in terms of total pressure and static temperature in terms of total temperature may be substituted into Equation E.14 to obtain the equation for M. But in order to simplify the algebra, let us simplify the equations for T_t and P_t (Eqs. E.12 and E.13) by substituting the parameter X for the expression $1 + M^2(\gamma - 1)/2$:

$$T = \frac{T_t}{1+M^2(\gamma-1)/2} = \frac{T_t}{X}, \tag{E.19}$$

$$P = \frac{P_t}{\left[1+M^2(\gamma-1)/2\right]^{\gamma/(\gamma-1)}} = \frac{P_t}{X^{\gamma/(\gamma-1)}}. \tag{E.20}$$

Now Equation E.14 may be written as:

$$M = B\frac{\sqrt{T}}{P} = B\sqrt{\frac{T_t}{X}}\frac{X^{\gamma/(\gamma-1)}}{P_t} = B\frac{\sqrt{T_t}}{P_t}X^{\gamma/(\gamma-1)}X^{-1/2} = M_{core}X^{(\gamma+1)/2(\gamma-1)}. \tag{E.21}$$

Squaring and substituting $1 + M^2(\gamma - 1)/2$ for X yields:

$$M^2 = M_{\text{core}}^2\left[1+M^2(\gamma-1)/2\right]^{(\gamma+1)/(\gamma-1)}. \tag{E.22}$$

Equation E.22 cannot be solved explicitly. Using the Newton–Raphson iterative method, however, it is easily solved. The solution is simpler if we use our parameter X as the variable. In the equation $X = 1 + M^2(\gamma - 1)/2$, solve for M^2:

$$M^2 = \frac{2(X-1)}{\gamma-1}. \tag{E.23}$$

Now substitute these expressions into Equation E.22 and solve for zero:

$$\frac{2(X-1)}{\gamma-1} = M_{\text{core}}^2 X^{(\gamma+1)/(\gamma-1)}, \tag{E.24}$$

$$0 = \frac{\gamma-1}{2}M_{\text{core}}^2 X^{(\gamma+1)/(\gamma-1)} - X + 1. \tag{E.25}$$

In the Newton–Raphson method, we need to set this function equal to $f(X)$ and differentiate in order to find the value of X when the function is equal to zero. The derivative of $f(X)$ is:

$$f'(X) = \frac{\gamma+1}{2}M_{\text{core}}^2 X^{2/(\gamma-1)} - 1. \tag{E.26}$$

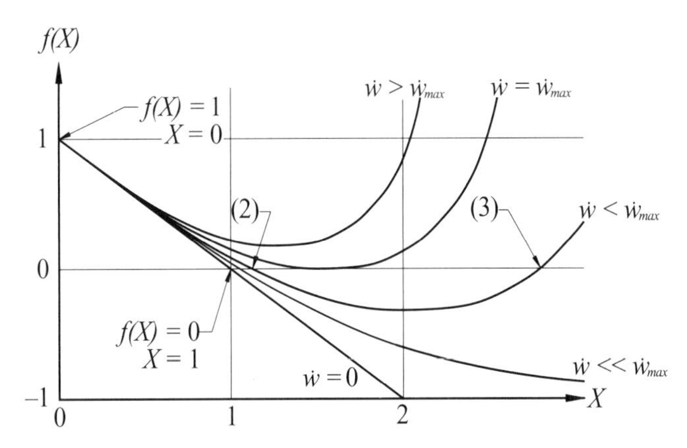

FIGURE E.3. Graph of $f(X)$ versus X.

Using the functions for $f(X)$ and $f'(X)$ defined above, any degree of precision may be obtained by repeated application of:

$$X_{i+1} = X_i - \frac{f(X_i)}{f'(X_i)}, \qquad \text{(E.27)}$$

where X_i is an estimate and X_{i+1} is a much closer estimate. After several successive iterations when the value of $f(X)$ is sufficiently close to zero, the value of X will be established. Then M may be found from Equation E.23 and Equation 4.16 evaluated for K.

A pitfall in employing this technique lies in assuming the equation has a solution. The graph of $f(X)$ versus X is illustrated in Figure E.3. If the flow rate is 0 then $M_{\text{core}} = 0$ and $f(X)$ crosses the zero axis at $X = 1$. As $\dot{w}$ is increased, the curve moves up and crosses the zero axis in two places, points (2) and (3) in the illustration, so there are actually two solutions—one is subsonic and one is supersonic. Depending on the value of your initial guess for $X_{i=0}$ your solution for M might be either the subsonic one or the supersonic one.

As $\dot{w}$ is increased more, the $f(X)$ curve intersections of the $f(X) = 0$ line become closer together; then, when the crossings coincide the $f(X)$ curve becomes tangent to the zero axis, and $M = 1$, the flow is sonic at the point of interest. At this point $\dot{w}$ is maximized and becomes $\dot{w}_{\max}$. If $\dot{w}$ is increased further, $f(X)$ does not intersect the zero axis and there is no solution. This indicates that for any given total pressure and total temperature condition, flow in a constant area duct cannot exceed a discrete value where Mach number at the outlet becomes unity.

The difficulty described above may be easily avoided by making the following test. At the minimum value of $f(X)$, $f'(X) = 0$:

$$f'(X) = \frac{\gamma+1}{2} M_{\text{core}}^2 X^{2/(\gamma-1)} - 1 = 0. \qquad \text{(E.28)}$$

Therefore, at $f'(X) = 0$, where $f(X) = f(X)_{\min}$, X is:

$$X = \left(\frac{\gamma+1}{2} M_{\text{core}}^2\right)^{-(\gamma-1)/2}. \qquad \text{(E.29)}$$

Substituting this value for X into the expression for $f(X)$ (see Eq. E.25) we find that:

$$f(X)_{\min} = \frac{\gamma-1}{2} M_{\text{core}}^2 \left(\frac{\gamma+1}{2} M_{\text{core}}^2\right)^{-(\gamma+1)/2} \left(\frac{\gamma+1}{2} M_{\text{core}}^2\right)^{-(\gamma-1)/2}. \qquad \text{(E.30)}$$

- If $f(X)_{\min} < 0$, two solutions exist as at (2) and (3), and since in duct flow we are interested in the subsonic solution, our initial guess for X, that is, $X_{i=0}$, must be less than X at $f(X)_{\min}$ (that is, X from Eq. E.29).
- If $f(X)_{\min} = 0$, this is the limiting condition, and may be treated accordingly.
- If $f(X)_{\min} > 0$, there is no solution, the input conditions are impossible, and the calculation may be halted or redirected, as, for instance, making the pipe diameter larger or reducing the flow rate, depending on what part of your design you are pursuing. If your design has a fixed flow rate, you can increase the pipe size. If your design has a fixed pipe size, you can reduce the flow rate to determine what flow it can handle and from this, you can determine the accompanying pressures and temperatures.

E.4 SOLUTION WHEN *TOTAL* PRESSURE AND *STATIC* TEMPERATURE ARE KNOWN

The equations for solving for M if *total pressure* and *static temperature* are given are similar to those derived above for total pressure and total temperature, and are derived similarly. Mach number is given by Equation E.14,

$$M = B\sqrt{T}/P, \qquad \text{(E.14, repeated)}$$

where B is defined by Equation E.15a or E.15b. In this case, static temperature is already known, but the known pressure is total pressure, from which static pressure must be determined using Equation E.13, which is

$$P = \frac{P_t}{\left[1 + M^2(\gamma-1)/2\right]^{\gamma/(\gamma-1)}}. \qquad \text{(E.13, repeated)}$$

Substituting this expression for P in Equation E.14 yields:

$$M = B\sqrt{T}/P = \frac{B\sqrt{T}}{P_t}\left[1+M^2(\gamma-1)/2\right]^{\gamma/(\gamma-1)}.$$

We have previously defined $B\sqrt{T}/P$ as M_{core} without regard as to whether T or P is total or static, so we can write the equation as:

$$M = M_{core}\left[1+M^2(\gamma-1)/2\right]^{\gamma/(\gamma-1)}.$$

Upon squaring and substituting X for $1 + M^2(\gamma - 1)/2$, the equation becomes:

$$M^2 = M_{\text{core}}^2\left[1+M^2(\gamma-1)/2\right]^{2\gamma/(\gamma-1)} = M_{\text{core}}^2 X^{2\gamma/(\gamma-1)}. \quad \text{(E.31)}$$

Solving the equation $X = 1 + M^2(\gamma - 1)/2$ for M^2 yielded Equation E.23:

$$M^2 = \frac{2(X-1)}{\gamma-1}, \quad \text{(E.23, repeated)}$$

which, when substituted in Equation E.31, gives:

$$\frac{2(X-1)}{\gamma-1} = M_{\text{core}}^2 X^{2\gamma/(\gamma-1)}.$$

If we rearrange this and make the rearrangement equal zero we obtain:

$$0 = \frac{\gamma-1}{2}M_{core}^2 X^{2\gamma/(\gamma-1)} - X + 1.$$

If we call the right side of this equation $f(X)$ we get a function that is supposed to equal zero (but it won't equal zero unless we discover the right value for X):

$$f(X) = \frac{\gamma-1}{2}M_{\text{core}}^2 X^{2\gamma/(\gamma-1)} - X + 1. \quad \text{(E.32)}$$

In order to find X we need the derivative of Equation E.32, which is

$$f'(X) = \gamma M_{\text{core}}^2 X^{(\gamma+1)/(\gamma-1)} - 1. \quad \text{(E.33)}$$

Equations E.32 through E.35 should be applied in the same fashion as Equations E.19 through E.30. Using the functions for $f(X)$ and $f'(X)$ defined above, any degree of precision may be obtained by repeated application of:

$$X_{i+1} = X_i - \frac{f(X_i)}{f'(X_i)}. \quad \text{(E.27, repeated)}$$

The value of X at $f'(X) = 0$ is:

$$X = \left(\gamma M_{\text{core}}^2\right)^{-(\gamma-1)/(\gamma+1)}. \quad \text{(E.34)}$$

The value of $f(X)_{\min}$ is:

$$f(X)_{\min} = \frac{\gamma-1}{2}M_{\text{core}}^2\left(\gamma M_{\text{core}}^2\right)^{-2/(\gamma+1)} - \left(\gamma M_{\text{core}}^2\right)^{-(\gamma-1)/(\gamma+1)}. \quad \text{(E.35)}$$

The caveats following those equations are also the same for this case:

- If $f(X)_{\min} < 0$, two solutions exist, and since in duct flow what we are interested in is the subsonic solution, our initial guess for X, that is, $X_{i=0}$, must be less than X at $f(X)_{\min}$ (that is, X from Eq. E.34 for this case). By making the first guess for X (i.e., $X_{i=0}$) less than X at $f(X)_{\min}$, Equation E.27 searches for the solution on the part of the curve where $f'(X)$ is negative, the descending part of the curve. The subsonic solution lies somewhere on the descending part of the curve and the supersonic solution lies on the ascending part of the curve.
- If $f(X)_{\min} = 0$, this is the limiting condition, and may be treated accordingly.
- If $f(X)_{\min} > 0$, there is no solution, the input conditions are impossible, and the calculation may be halted or redirected, as, for instance, making the pipe diameter larger or reducing the flow rate, depending on what part of your design you are pursuing. If your design has a fixed flow rate, you can increase the pipe size. If your design has a fixed pipe size, you can reduce the flow rate to determine what flow it can handle and from this you can determine the accompanying pressures and temperatures.

REFERENCES

1. Street, R. L., G. Z. Watters, and J. K. Vennard, *Elementary Fluid Mechanics*, John Wiley & Sons.
2. Shapiro, A. H., *The Dynamics and Thermodynamics of Compressible Flow*, Vol. 1, John Wiley & Sons, 1953.

APPENDIX F

VELOCITY PROFILE EQUATIONS

In this appendix, the derivations of the velocity profile equations presented in Chapter 2 are shown.

F.1 BENEDICT VELOCITY PROFILE DERIVATION

(Equation numbers in this section are from Benedict [1].)

In his chapter 5, section 5.4, "Turbulent Flow in Smooth Pipes," Robert P. Benedict gives equation 5.74 (on p. 221 of Reference [1]) that relates the kinetic energy correction factor to the friction factor in any flow situation. It is based in part on equations 5.33 and 5.36 (pp. 201, 203 of Reference [1]), namely:

$$u = V_c - 2.5V^* \ln \frac{R}{y}, \tag{5.33}$$

where
u = local fluid velocity,
V_c = fluid velocity at the center of the pipe,
V^* = friction velocity,
y = distance of u from the wall, and
R = radius of pipe;

and

$$V = V_c - 3.75V^*, \tag{5.36}$$

where
V = average velocity of fluid in the pipe and
V_c = velocity of fluid at the center of the pipe.

The friction velocity is defined (on p. 192) as:

$$V^* = \sqrt{\frac{\tau_0}{\rho_m}}, \tag{5.18}$$

where
τ_0 = fluid shear stress at the wall and
ρ_m = fluid mass density.

Then:

$$V_c = V + 3.75V^*. \qquad \text{(5.36 alternate)}$$

Then Benedict writes:

> But (5.36) is bound to define an average velocity that is greater than actual, because the log law of (5.33) does not yield a zero velocity gradient at the pipe center (see Figure 5.18).

One of our questions is, "Does equation 5.36 yield a profile in which $du/dy = 0$ at the center?" The answer is "no," as Hunter Rouse [2] notes, but this is of little practical effect. Benedict expresses equation 5.33 in slightly modified form as:

Pipe Flow: A Practical and Comprehensive Guide, First Edition. Donald C. Rennels and Hobart M. Hudson.

$$\frac{u}{V_c} = 1 + 2.5\left(\frac{V^*}{V_c}\right)\ln\frac{y}{R}. \quad (5.38)$$

Using the Darcy–Weisbach equation,

$$\Delta p = \left(f\frac{L}{D}\right)\frac{\rho V^2}{2g_c}, \quad (5.39)$$

with equation 5.2 (p. 182),

$$\tau = \frac{\Delta pr}{2L}, \quad (5.2)$$

and with equation 5.18 (equation for friction velocity, V^*),

$$V^* = \sqrt{\frac{\tau_0 g_c}{\rho}}, \quad (5.18)$$

Benedict obtains:

$$\frac{V^*}{V_c} = \sqrt{\frac{f}{8}}\left(\frac{V}{V_c}\right), \quad (5.40)$$

or

$$V^* = V\sqrt{\frac{f}{8}}. \quad (5.40 \text{ alternate})$$

We can write equation 5.38 (and Eq. 5.33) as:

$$u = V_c + 2.5V^*\ln\frac{y}{R}. \quad (5.38 \text{ or } 5.33 \text{ alternate})$$

Now, if we substitute equations 5.40 alternate and 5.36 alternate into equation 5.38 alternate, we obtain the explicit equation for velocity profile:

$$u = V_c + 2.5V^*\ln\frac{y}{R}. \quad (5.38 \text{ alternate})$$

But

$$V_c = V + 3.75V^*, \quad (5.36 \text{ alternate})$$

and

$$V^* = V\sqrt{\frac{f}{8}}. \quad (5.40 \text{ alternate})$$

Therefore,

$$\begin{aligned} u &= (V + 3.75V^*) + 2.5\,V^*\ln\frac{y}{R} \\ &= \left(V + 3.75\,V\sqrt{\frac{f}{8}}\right) + 2.5\,V\sqrt{\frac{f}{8}}\ln\frac{y}{R} \\ &= V\left(1 + 3.75\sqrt{\frac{f}{8}}\right) + 2.5V\sqrt{\frac{f}{8}}\ln\frac{y}{R} \\ &= V\left(1 + 3.75\sqrt{\frac{f}{8}} + 2.5\sqrt{\frac{f}{8}}\ln\frac{y}{R}\right). \end{aligned}$$

Then:

$$\frac{u}{V} = \left(1 + 3.75\sqrt{\frac{f}{8}} + 2.5\sqrt{\frac{f}{8}}\ln\frac{y}{R}\right). \quad (5.36 + 5.38 + 5.40)$$

Is this for smooth pipes only? Benedict gives for rough pipes:

$$u^+ = 2.5\ln\frac{y}{R} + 3.75 + \frac{V}{V^*}. \quad (5.52)$$

Benedict defines u^+ as follows:

$$u^+ = \frac{u}{V^*}. \quad (5.19)$$

Inserting this into Equation 5.52 yields:

$$\frac{u}{V^*} = 2.5\ln\frac{y}{R} + 3.75 + \frac{V}{V^*}$$

or

$$u = V^* 2.5\ln\frac{y}{R} + 3.75V^* + V.$$

By dividing by V we get:

$$\frac{u}{V} = \frac{V^*}{V}2.5\ln\frac{y}{R} + 3.75\frac{V^*}{V} + 1.$$

Then, remembering that $V^* = V\sqrt{f/8}$ (Eq. 5.40, alternate) or $V^*/V = \sqrt{f/8}$, we may insert it into the equation above to obtain:

$$\frac{u}{V} = \sqrt{\frac{f}{8}}2.5\ln\frac{y}{R} + 3.75\sqrt{\frac{f}{8}} + 1.$$

Rearranged, this is:

$$\frac{u}{V} = 1 + 3.75\sqrt{\frac{f}{8}} + 2.5\sqrt{\frac{f}{8}}\ln\frac{y}{R}.$$

This is exactly the same as given above as the smooth pipe velocity profile:

$$\frac{u}{V} = \left(1 + 3.75\sqrt{\frac{f}{8}} + 2.5\sqrt{\frac{f}{8}}\ln\frac{y}{R}\right).$$

So the velocity profile equation given above is valid (according to Benedict) for both smooth and rough pipes.

The plot of this equation is shown in Figure F.1.

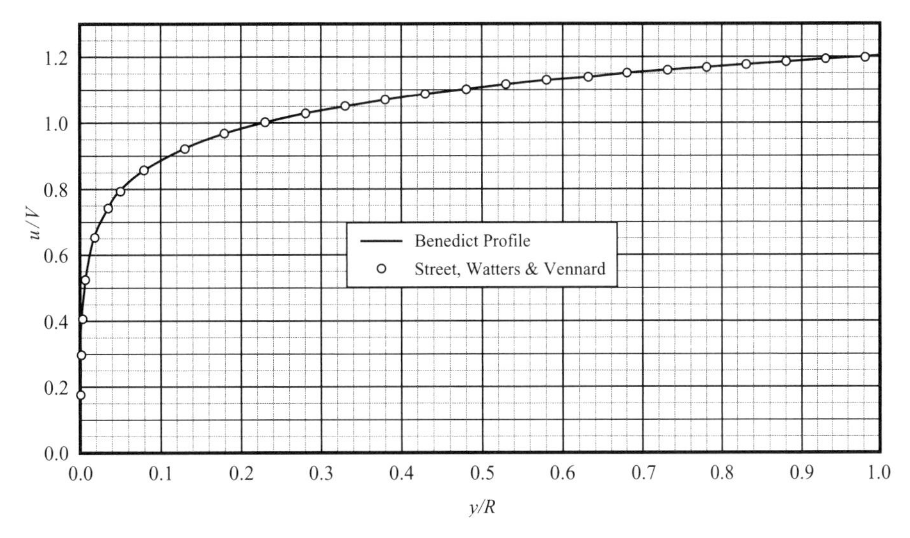

FIGURE F.1. Plot of fully turbulent velocity profile for $f = 0.024$ (Figure 2.3, repeated).

F.2 STREET, WATTERS, AND VENNARD VELOCITY PROFILE DERIVATION

Street et al. [3] give the following relations for rough pipes (their eqs. 9.29, 9.30, and 9.31, respectively):

$$\frac{u}{V*} = 5.75\log_{10}\frac{y}{\varepsilon} + 8.5, \tag{F.1}$$

$$\frac{V}{V*} = 5.75\log_{10}\frac{R}{\varepsilon} + 4.75, \tag{F.2}$$

$$\frac{1}{\sqrt{f}} = 2.0\log_{10}\frac{d}{\varepsilon} + 1.14. \tag{F.3}$$

By solving Equation F.3 for $1/\varepsilon$ we obtain:

$$\frac{1}{\varepsilon} = \frac{10^{\frac{1}{2\sqrt{f}}-0.57}}{d} = \frac{10^{\frac{1}{2\sqrt{f}}-0.57}}{2R} = \frac{10^{1/(2\sqrt{f})-0.57}}{2R}. \tag{F.4}$$

For simplicity let us represent $10^{1/(2\sqrt{f})-0.57}$ by the symbol ξ temporarily; then, we have:

$$\frac{1}{\varepsilon} = \frac{\xi}{d} = \frac{\xi}{2R}. \tag{F.5}$$

Substituting Equation F.5 into Equations F.1 and F.2, we have:

$$\frac{u}{V*} = 5.75\log_{10}\frac{y}{\varepsilon} + 8.5 = 5.75\log_{10}\left[y\frac{\xi}{2R}\right] + 8.5, \tag{F.1 Alt}$$

$$\frac{V}{V*} = 5.75\log_{10}\frac{R}{\varepsilon} + 4.75 = 5.75\log_{10}\left[R\frac{\xi}{2R}\right] + 4.75 = 5.75\log_{10}\left[\frac{\xi}{2}\right] + 4.75, \tag{F.2 Alt}$$

then dividing Equation F.1 Alt by Equation F.2 Alt we obtain:

$$\frac{u}{V} = \frac{u/V*}{V/V*} = \frac{5.75\log_{10}\left[\frac{y}{R}\frac{\xi}{2}\right] + 8.5}{5.75\log_{10}\left[\frac{\xi}{2}\right] + 4.75},$$

$$= \frac{\log_{10}\left[\frac{y}{R}\frac{\xi}{2}\right] + \frac{8.5}{5.75}}{\log_{10}\left[\frac{\xi}{2}\right] + \frac{4.75}{5.75}} = \frac{\log_{10}\frac{y}{R} + \log_{10}\left[\frac{\xi}{2}\right] + 1.478261}{\log_{10}\left[\frac{\xi}{2}\right] + 0.826087}. \tag{F.6}$$

But:

$$\log_{10}\left[\frac{\xi}{2}\right] = \log_{10}[\xi] - \log_{10} 2. \tag{F.7}$$

Remembering that $10^{1/(2\sqrt{f})-0.57} \equiv \xi$, we can rewrite Equation F.7 as:

$$\log_{10}\left[\frac{10^{1/(2\sqrt{f})-0.57}}{2}\right] = \log_{10}\left(10^{1/(2\sqrt{f})-0.57}\right) - \log_{10} 2.$$

Now the logarithm of $10^{1/(2\sqrt{f})-0.57}$ is simply $1/(2\sqrt{f}) - 0.57$, so that:

$$\log_{10}\left(10^{1/(2\sqrt{f}-0.57)}\right)-\log_{10}2=\frac{1}{2\sqrt{f}}-0.57-0.3010300$$
$$=\frac{1}{2\sqrt{f}}-0.8710300.$$

Now substitute $1/2\sqrt{f}-0.8710300$ into equation 2.26 wherever we find $\log_{10}(\xi/2)$:

$$\frac{u}{V}=\frac{\log_{10}\frac{y}{R}+\frac{1}{2\sqrt{f}}-0.8710300+1.478261}{\frac{1}{2\sqrt{f}}-0.8710300+0.826087},$$

$$\frac{u}{V}=\frac{\log_{10}\frac{y}{R}+\frac{1}{2\sqrt{f}}+0.60723}{\frac{1}{2\sqrt{f}}-0.044943}. \quad \text{(F.8)}$$

For convenience in plotting, change the $\log_{10}$ to ln (i.e., $\log_e$); The $\log_{10}$ term becomes:

$$\log_{10}\frac{y}{R}=0.43429448\ln\frac{y}{R}.$$

Then Equation F.8 may be written as:

$$\frac{u}{V}=\frac{0.43429\ln\frac{y}{R}+\frac{1}{2\sqrt{f}}+0.60723}{\frac{1}{2\sqrt{f}}-0.044943}.$$

From Figure F.1, it can be seen that the profile defined by the Street et al. equations falls on the profile defined by Benedict's equations.

REFERENCES

1. Benedict, R. P., *Fundamentals of Pipe Flow*, John Wiley & Sons, 1980.
2. Rouse, H., ed., *Engineering Hydraulics, Proceedings of the Fourth Hydraulics Conference, Iowa Institute of Hydraulic Research, June 12–15, 1949*, John Wiley & Sons, 1949.
3. Street, R. L., G. Z. Watters, and J. K. Vennard, *Elementary Fluid Mechanics*, 7th ed., John Wiley & Sons, 1996.

INDEX

Page references followed by d, f, or t denote diagrams, figures, or tables.

Pipe Flow: A Practical and Comprehensive Guide, First Edition. Donald C. Rennels and Hobart M. Hudson.